Paradigms in Theory Construction

Luciano L'Abate
Editor

Paradigms in Theory Construction

Springer

Editor
Luciano L'Abate
Department of Psychology
Georgia State University
University Plaza, Atlanta,
GA 30303, USA
llabate3@bellsouth.net

ISBN 978-1-4614-0913-7 e-ISBN 978-1-4614-0914-4
DOI 10.1007/978-1-4614-0914-4
Springer New York Dordrecht Heidelberg London

Library of Congress Control Number: 2011940429

Printed on acid-free paper

Springer is part of Springer Science+Business Media (www.springer.com

Preface

The purpose of this edited volume is to expand on an early classification of paradigms, theories, and models from a variety of contributors dedicated to each paradigm (L'Abate, 2009). Each contributor, in expanding on his/her favorite paradigm, implies or emphasizes her or his theoretical allegiance to the hegemonic importance of that paradigm over others. However, the wide range of paradigm available raises serious questions of whether one single paradigm can achieve such a superior position over others. Consequently, the major issue faced by such an embarrassment of paradigms raises the question on how one can choose one paradigm over others.

In a previous publication (L'Abate, 2009) I discussed the role of paradigms, theories, and models trying to make sense of a confusing, uncritical matching of paradigms with theories, paradigms with models, and theories with models. From this publication came forth a classification of paradigms that to my knowledge has not been attempted before, at least in psychology but, as we shall see, attempted frequently in other social science disciplines.

Paradigms include theories. Theories include models. Models include dimensions, according to a hierarchical framework. However, what is a paradigm? Many definitions equate a paradigm with a model, making it very difficult to differentiate among the different components of the proposed classification. Eventually, the best way I could define a paradigm was as "one way to look at reality." Or "a systematic system of values." This way of looking consists of different components chosen to perceive reality according to individual criteria. Different value systems for different individuals. Whether there is a unique, supra-ordinate paradigm, besides evolution, is a question that remains to be answered by most contributors. Is there a supra-ordinate paradigm, and if there is one, how would that supra-ordinate paradigm be chosen?

This classification, therefore, begs the question: Which rules, if any, control, govern, or link one paradigm with a theory? Thus far, as far as I know, the only rule relating to the link between a paradigm and a theory, has been by proclamation: a theorist would swear allegiance to a particular paradigm. However, how a particular paradigm was chosen seems by personal preferences of the theorist. Could any other paradigm be chosen instead? Why one particular paradigm instead of another? One cannot help wondering whether any other paradigm could have been chosen or proclaimed, reducing links among paradigms, theories, models, and dimensions. These choices seem arbitrary, questionable, and certainly tenuous.

Confusion about what is a paradigm, theory, or model continues to proliferate unabated in the psychological literature. For instance, in a recent chapter on legitimacy and social identity etc., the authors of that chapter (Spears, Greenwood, de Lemus, & Sweetman, 2010), introduced as "theories" what amounts to models. According to L'Abate's (2009) original definitions and differentiations among paradigms, theories, and models, these models, even though called theories, are: social identity, system justification, and social dominance. The reason for calling these models lies in their being defines by dimensions rather than by other models.

Therefore, the purpose of this edited work aims at exploring such relationships, whether these relationships follow any rational connection, whether there are hidden rules behind personal preferences, and whether paradigms are chosen at random or chosen according to general or specific criteria that have not been spelled out as yet. Can we find such criteria and, given such criteria, will they be qualified and/or perhaps quantified?

As far as I know, except for other social science disciplines, there is no competition in any equivalent publication I know of in psychology. Collaborators who signed up to write a chapter in this volume expressed their enthusiasm in doing something that was apparently never attempted heretofore, including a completely novel classification of paradigms, theories, models, and dimensions. If it were not for the enthusiasm of these collaborators, this volume would not have seen the light of publication.

To level the field for all contributors, including myself, I limited the absolute length of each chapter to 20 pages typed-single spaced. Some contributors tried to widen the lines horizontally to 6 and one-half inches. Sometimes I caught them, but sometimes I did not. I am not a good detective. Furthermore, as much as I tried to be inclusive, some paradigms were left out, such as ecology because at the last minute its contributor was not able to after complete it, due to family illness. Even evolution was left out, in spite of its being an already well established theory in psychology and in many other branches of biology. It will be interesting to see what claims will be made for the importance of paradigms not included here.

Sequence of Chapters

Concentration on the role of paradigms only in psychology would produce and reproduce a very parochial outlook limiting the whole enterprise. Consequently, in Part I introductory background, after Hillix and I in Chap. 1 discussed the role of paradigms in science and theory construction, I asked chemists/philosophers of science Corinna Guerra, Mario Capitelli, and Salvino Longo in Chap. 2 to give an historical perspective on the overall function of paradigms in science. Bill Willis volunteered to review evolving paradigms in Chap. 3 and I am grateful for his contribution.

Part II is dedicated to the arts and social sciences. For the sake of completeness, unfortunately I had to write Chap. 4 on Anthropology, while Laura G. Sweeny unexpectedly volunteered to write Chap. 5 on the role paradigms in artistic, verbal, and visual paradigms, an offer that delighted me and that I could not refuse. I had to write Chap. 6 on Economics because I could not find anyone in both Anthropology and Economic to collaborate on those chapters. Laura G. Sweeney in Chap. 7 reviewed the role of paradigms in educational theories and practices, while the same function in sociology was fulfilled by my brother, Alberto L'Abate in Chap. 8.

Part III is dedicated to General-Integrative Paradigms that include the biopsychosocial paradigm by Beatrice L. Wood in Chap. 9. In Chap. 10, Mitch Fryling and Linda J. Hayes cover Kantor's Inter-behavioral paradigm while in Chap. 11, Jeffrey Magnavita covers Systems from the viewpoint of personality theory and in Chap. 12, Vittorio Cigoli and Eugenia Scabini cover the intersubjective-narrative approach versus the relational-generational one. In Chap. 13, Alexander Riegler reviews the importance of Constructivism while in Chap. 14, Luciano L'Abate covers Materialism.

Part IV contains Particular-Specific Paradigms moving from the initial framework in the degree of abstraction and generality that permeated this whole volume. For instance, in Chap. 15, Sandra M. Loughlin and Patricia A. Alexander cover two coexisting cognitive and empirical paradigms in human learning. By the same token in Chap. 16, David Ryback covers Humanism and Behaviorism, while in Chap. 17, Mario Cusinato reviews Existentialism.

Going down the ladder of abstraction and generality, Part V includes even more specific and concrete Operational Paradigms such as Chap. 18 on Information Processing by Piero De Giacomo, Luisa Mich, Carlos Santamaria, Laura Sweeney, and Andrea De Giacomo, where I was forced to expand on the limits of only three authors per chapter, due to the complexity of the subject matter. Both specificity and concreteness are found in Chap. 19 about Reductionism as covered by Gary Berntson and John Cacioppo. The downward influence of reductionism is coupled by Walter Colesso in Chap. 20, Produced and Spontaneous Emergent Interactionism, while in Chap. 21, William A. Hillix, Duane M. Rumbaugh, and Sue Savage-Rumbaugh argue for the emergence of reason, intelligence, and language in humans and animals. Even though no longer a paradigm, in Chap. 22 on Essentialism, Luciano L'Abate had to show why this is a bona-fide theory with models and dimensions rather than a philosophical paradigm.

In Part VI, Chap. 23 about the Role of criteria and processes in paradigm selection to conclude the volume, Luciano L'Abate and Laura G. Sweeney declare that the only paradigm worthy of the link between paradigm, theory, models, and dimensions is the information processing one.

Readership

This topic would or should be of interest to graduate students and their teachers in at least six different disciplines where there is still an interest in theory-construction, as shown by courses and seminars on this topic. In addition to academic disciplines in the social sciences and philosophy of science, this volume might be of interest to the following academic disciplines: anthropology, communication science, economics, education, relationship science, sociology, family studies, and applied mental health professions, such as clinical psychology, psychiatry, and social work.

Atlanta, GA, USA Luciano L'Abate

References

L'Abate, L. (2009). Paradigms, theories, and models: Two hierarchical frameworks In L. L'Abate, P. De Giacomo, M. Capitelly, & S. Longo (Eds.), *Science, mind and creativity: The Bari symposium* (pp. 107–122). New York: Nova Science Publishers.

Spears, R., Greenwood, R., de Lemus, S., & Sweetman, J. (2010). Legitimacy, social identity and power. In A. Guinote & T. K. Vescio (Eds.), *The social psychology of power* (pp. 251–283). New York: Guilford.

Contents

Contributors

Patricia A. Alexander Disciplined Reading and Learning Research Laboratory, Department of Human Development, College of Education, University of Maryland, College Park, MD 20742–1131, USA

Gary G. Berntson Department of Psychology, Ohio State University, 1835 Neil Avenue, Columbus, OH 43210, USA

John T. Cacioppo Center for Cognitive and Social Neuroscience, The University of Chicago, 5848 S. University Avenue, Chicago, IL 60637, USA

Mario Capitelli Dipartimento di Chimica; Facoltà di Scienze MM. FF. NN, Università di Bari, via Orabona, 4, 70125 Bari, Italy

Vittorio Cigoli Centro Ricerche Famiglia, Universita' Cattolica, Via Nirone 15, 20123 Milan, Italy

Walter Colesso Pyschidiagnostic Laboratory, Family Center Foundation, Via San Nicolò, 60, 31100 Treviso, Italy

Mario Cusinato Centro della Famiglia, Via San Nicolo' 60 31100, Treviso, Italy

Interdepartmental Center of Family Research, University of Padua, Via Venezia 8, 35131, Padua, Italy

Andrea De Giacomo Dipartimento di Psichiatria Infantile, School of Medicine, University of Bari, Bari, Italy

Piero De Giacomo Department of Clinical Psychiatry, University of Bari, Via Fanelli, 239, Bari, Italy

Corinna Guerra Viale G. Di Vittorio, 115, 71121, Foggia, Italy

Mitch J. Fryling Department of Applied Behavior Analysis, The Chicago School of Professional Psychology, 617 W. 7th St., 8th Floor, Los Angeles, CA 90017, USA

Linda J. Hayes Psychology Department/296, University of Nevada Reno, Reno, NV 89557, USA

William A. Hillix 11028 Explorer Rd La Mesa, CA 91941, USA

Alberto L'Abate Via A. Mordini, 3, 50136, Florence, Italy

Luciano L'Abate Department of Psychology, Georgia State University, University Plaza, Atlanta, GA 30303, USA

Savino Longo Dipartimento di Chimica, Facoltà di Scienze MM. FF. NN, Università di Bari, via Orabona, 4, 70125 Bari, Italy

Sandra M. Loughlin Disciplined Reading and Learning Research Laboratory, Department of Human Development, College of Education, University of Maryland, College Park, MD 20742–1131, USA

Jeffrey J. Magnavita Glastonbury Medical Arts Center, 300 Hebron Ave. Suite, 215 Glastonbury, CT 06033, USA

Luisa Mich Department of Computer and Management Sciences, University of Trento, Via Inama 5, 38122, Trento, Italy

Willis F. Overton Department of Psychology, Temple University, Philadelphia, PA 19122, USA

Alexander Riegler Centre for Logic and Philosophy of Science and Leo Apostel Center for Interdisciplinary Studies, Vrije Universiteit Brussel, Krijgskundestraatt 33, 1160 Brussels, Belgium

Duane M. Rumbaugh 402 Lincoln Ave, Highland Park, NJ 08904, USA

E. Sue Savage-Rumbaugh Great Ape Trust of Iowa, 3131 Park Fleur, Unit 804, Des Moines, IA 50321, USA

David Ryback EQ Associates International, Suite 201, 1534 N. Decatur Rd., Atlanta, GA 30307–1022, USA

Carlos Santamaria Department of Cognitive Psychology, University of La Laguna, Campus de Guajara s/n, 38071, La, Laguna, Tenerife, Spain

Eugenia Scabini School of Psychology, Catholic University of Milan, Largo Agostino Gemelli 1, Milan, Italy

Laura G. Sweeney Avant Garde Books, 301 Briarvista Way N.E., Atlanta, GA 30329, USA

Beatrice L. Wood Women and Children's Hospital of Buffalo, 219 Bryant Street, Buffalo, NY 14222, USA

Part I
Introductory Background

Chapter 1
The Role of Paradigms in Science and Theory Construction

William A. Hillix and Luciano L'Abate

Values must also be recognized as operative within theory and practice

(Reisch 2005, p. 365).

The purpose of this introduction is to lay out the background for the chapters in this volume. The major issue discussed in this chapter, within the whole context of this volume, concerns the role of paradigms in the construction of any scientific or even artistic theory. What does it take to make a theory artistically and scientifically acceptable? Consequently, the first part of this chapter will cover how we should view the whole scientific enterprise. The second part will cover how paradigms need to be included within a hierarchical framework that clarifies how paradigms are related to theories, models, and dimensions (L'Abate, 2009a, 2009b).

How Should We View the Practice of Science?

Only 5 years after Beveridge (1957) urged scientists to study their history – and suggested that they could safely ignore the philosophy of science – Thomas Kuhn (1922–1996) created a revolution in the way people thought about the relationships between the two disciplines. Philosophers of science reacted strongly to Kuhn's *The structure of scientific revolutions* (1970; originally 1962) – about the way hornets do

W.A. Hillix (✉)
Emeritus Professor, San Diego State University, 11028 Explorer Road, La Mesa, CA 91941, USA
e-mail: ahillix@cox.net

L. L'Abate
Department of Psychology, Georgia State University, University Plaza, Atlanta, GA 30303, USA
e-mail: llabate3@bellsouth.net

L. L'Abate (ed.), *Paradigms in Theory Construction*, DOI 10.1007/978-1-4614-0914-4_1,

when their nests are whacked with a stick. There was a great busy, angry, defensive buzzing throughout the hallowed halls of philosophy. Kuhn, then a graduate student in physics at Harvard, claimed to be reporting what science was really like, the course that discovery really followed. He claimed that he had discovered, not invented, the new view by studying the history of science. What was this new picture of science that was so remarkable, so upsetting, and so long concealed under everyone's nose?

Kuhn said that science developed by alternately constructing and destroying what he called *paradigms*. He seems to have attached multiple meanings to the word. One critical analyst, Masterman (1970), thought that she had identified 21 different meanings in his book! Although Kuhn later responded to such criticism by tightening his definition, we will follow his earlier usage because it is richer in connotations.

A paradigm in Kuhn's inclusive sense encompassed everything necessary for practicing science, all the way from a set of metaphysical assumptions "at the top" through commitments to apparatus and experimental procedures "at the bottom." Between the top and the bottom are theories about the subject matter and specific examples of solved problems, called paradigmatic examples, which are very important components of a paradigm. The Copernican description of the revolution of the planets around the sun is one paradigmatic example.

It is clear that the inclusive view of a paradigm consists of many parts. Kuhn suggested that, in order to be regarded as a paradigm, the parts must be of high quality. Thus an approach to science, no matter how inclusive, is not a paradigm unless its accomplishments are impressive enough to attract the allegiance of essentially all of the practitioners of the branch of science involved. Usually the accomplishments must be practical, not just theoretical. Kuhn said that calendars and bombs are the kinds of results that have attracted practitioners. Thus, a system of clinical psychology would doubtless attract enough adherents to be classified as a paradigm if advocates of the position showed how clinical psychologists could cure all mental illness!

If a paradigm is to be attractive enough, the procedures it advocates must be precise and clearly communicated, and its predictions must be clear and unequivocal. Such a paradigm becomes very useful indeed. It tells the scientist what problems are worth studying and how they should be studied. A host of research projects are at the fingertips of every proponent of the paradigm. As Kuhn says, the paradigm presents puzzles, and, like the creator of a crossword puzzle, guarantees that every puzzle has a solution. And the amount of ingenuity required of the researcher need not be much greater than the amount of ingenuity required of a person solving a crossword puzzle.

Not surprisingly, considering the above properties, a paradigm may be regarded with great fondness by its beneficiaries. Specialized textbooks and history books in a discipline will rationalize and glorify the paradigm. Writers will feel justified in concentrating on the developments that led to the current, "correct," paradigmatic view. Little attention will be devoted to false starts, so the path of progress will appear to be more rational than it really was.

Given this background, it is not surprising that those nurtured within the paradigm come to accept it without bothering much about the preconceptions involved – to accept it, so to speak, unconsciously. Neither is it surprising that the paradigm is not given up easily; when occasional findings do not accord with what is expected within the paradigm, they are easily shrugged off as experimental errors, mistakes in observation, or, as a last resort, as unimportant exceptions not affecting the essential correctness of the paradigm as a whole.

As long as scientists can work in this comfortable "puzzle-solving" way within the paradigm, their science is within what Kuhn called a "normal" period. Unfortunately for those who prefer the easy life, the paradigm contains the seeds of its own destruction. Its vulnerability is a product of the necessary precision of the paradigm. On the one hand, the precision is an absolutely necessary feature, because imprecise and nonpredictive statements are not attractive to scientists. On the other hand, no body of scientific theory and practice has ever been so perfect that it contained no errors and so complete that it could predict all phenomena within an area of science. Errors and/or incompleteness lead to unexpected outcomes, which Kuhn called "anomalies." They eventually become so frequent and so obvious that they cannot be denied. Then, despite powerful resistance, they intrude upon the consciousness of the practitioners of the paradigm.

At this point, the paradigm goes into a "crisis" mode. The practitioners try desperately to patch up the old paradigm and perhaps still deny the reality of anomalous results. The higher order assumptions of the paradigm, unquestioned and unconsciously accepted during the normal period, suddenly are made explicit, challenged, and defended.

Scientists who are new to the science, perhaps younger than most of the paradigmatic scientists, or crossovers from neighboring disciplines, start to propose alternative paradigms. Eventually one of the new points of view triumphs by solving some of the problems posed by the anomalies. It will probably not solve all of the problems, nor is it likely to be as well developed as the paradigm it promises to displace. Nevertheless, the new paradigm appears to be more attractive because the old paradigm is no longer tenable. The new paradigm wins. It probably does not convert all of the proponents of the now "classical" paradigm. However, new people in the field tend to be attracted to it, and the stubborn devotees of the old paradigm will eventually die off and become a part of history.

Examples of successive paradigms are Ptolemaic astronomy, the old earth-centered astronomy, and Copernican astronomy, which placed the sun at the center of what we now call the solar system. In physics, Newtonian physics was essentially replaced by Einsteinian physics. Both replacements presented great difficulties, and only obvious and frequent anomalies, in Kuhn's view, are sufficient to force a paradigm shift.

We can see why Kuhn's book brought about such a buzzing and humming among historians of science. The traditional view of science immediately before Kuhn was that science was perfectly objective. Scientists were supposed to hold opinions lightly and abandon them at the first demonstration of error. However, anyone familiar with the controversies between scientists would recognize this description as a gross misrepresentation of the actual behavior of scientists.

Kuhn's analysis rested on the assumption that science is a socioculturally determined enterprise. A paradigm exists because the methods, problems, and "language" comprising the paradigm are accepted by the majority of the scientists and practitioners within a branch of science. Popper (1970) was critical of Kuhn's conclusions. Popper

(1957) believed that it is theories themselves, and the criticism of them, that provide the impetus for change. Popper attempted to create a "criterion for demarcation" for distinguishing scientific from nonscientific forms of explanation, regardless of whether or not those theories were accepted by a majority of scientists.

This criterion is the principle of "falsifiability": all sciences must be able to produce testable hypotheses that, if refuted, lead to the rejection of the theory. The best of these tests are "risky predictions," in the sense that they offer powerful evidence of the inadequacy of the theory if the results are not as predicted. Thus, scientists should seek anomalies rather than trying to rationalize them.

Nevertheless, Popper said that Kuhn had forced him to realize that something like normal science played a much larger role than he had previously realized. Popper and Kuhn have some fundamental agreements. One is that legitimate science has to make clear predictions. If it does not, then it offers nothing to disprove. Popper would describe unclear hypotheses as "not falsifiable," and for Popper falsifiability is a hallmark of science. And for Kuhn, nonpredictive theories would never attract all the practitioners a branch of science could never support paradigmatic, normal, science. Also, no anomalies would be possible because there would be no consensual expectations for new findings to contradict. Thus, paradigmatic science, like Popper's ideal science, depends on falsifications if it is to advance from one paradigm to another.

In addition to clarifying the roles of anomaly and revolution, Kuhn's picture leads to other fundamental conclusions. One is that he saw great difficulty in defining scientific progress. It is not always clear that a new paradigm is superior to the one it replaces. The problems attacked and the solutions attained are likely to be so different that it is not easy to compare paradigms on the dimension of "goodness." The world is seen differently, but it is not necessarily seen more clearly. Taken on its own terms, in its own time, the old paradigm was not less "scientific" than the new one is. Kuhn would have us show more respect for our past!

Kuhn's contribution appears to have directed attention to previously overlooked features of science, but no view of science is ever complete or final. Lakatos (1970) thought that Kuhn made two basic errors. First, it was not typical for a field of science to have only a single paradigm. Lakatos claimed that there were usually at least two of what he called "research programs" at a given time within a given field. Second, he believed, contrary to Kuhn, that practitioners in different research programs could talk meaningfully to each other, and data gathered by each set of practitioners could be used to test predictions by the other. Third, Kuhn's two errors led him to a faulty conclusion that progress in science was more subjective than objective. Lakatos believed that science's apparent progress was real progress, and that surviving research programs (Lakatos obviously did not like the word "paradigm") handled data better and were thus more progressive than their competitors.

Another critic, Laudan (1977), thought that Lakatos was more correct than Kuhn, and he added his own belief that research traditions gradually change their core assumptions and are therefore significantly more flexible than Kuhn would have us believe. Thus, Lakatos and Laudan would have us believe that evolution, not revolution, is the typical course of change within science.

Disagreements about Kuhn's picture of science show few signs of abating. Dar (1987) agreed with Popper that refutations of a theory count for more than confirmations. This follows from a comparison of the consequences of two claims of formal logic. The first is that if A implies B, and B is true, we still cannot conclude anything about whether A is true, because other conditions, C, D, etc., could also have implied (caused) B. If A is a theory, and B is an observation, then we might say that B "confirmed" A, but that would be extremely loose usage of the word "confirmed." It would mean only that A had correctly predicted B. This does not mean that a surprising confirmation will never be accepted as dramatic support for a theory, for example, when Einstein's theory correctly predicted that the path of light would be curved in the presence of the sun's gravitational pull.

The other claim of formal logic is that if A implies B, and B is not true, then A certainly is not true. This is the falsification that Popper believed lay at the foundation of scientific method. Thus, if theory A predicts observation B, and B does not occur, then certainly the theory is incorrect (falsified) and should be rejected. Only theories that have never been falsified survive.

Although the logic of falsification is unchallenged, it is not as simple to apply as one might imagine. The link between a theory and its predictions may be challenged, as may the validity of the observations. The looseness of these links makes it possible for a paradigm to survive an "anomaly" (i.e., a failure of prediction), or even a series of anomalies, unless a superior paradigm comes on the scene to replace it.

Leahy (1992) argued that no "behavioral revolution" and no "cognitive revolution" have occurred in psychology, because there was never a genuine paradigm in the first place. Kimble (2005) agreed with Leahy. Agassi (2002) reviewed the controversies from the point of view of a direct participant. Tsou (2006) believed that Piaget's view of gradual scientific progress is superior to Kuhn's discontinuous description. And the beat goes on.

Paradigms, Theories, and Models

The present volume is based on the Table of Contents (TOC) resulting from a previous publication (L'Abate, 2009a, 2009b) where the relative roles of paradigms, theories, models, and dimensions were discussed. Its author was trying to make sense of a frequent, chaotic, and confusing equating of paradigms with theories, paradigms with models, and theories with models. Therefore, he defined a paradigm as a worldview, how reality is perceived and validated by more than one individual or method, a value or an aggregate system of values about how to perceive reality. A theory is a conceptual framework constituted by models that allow direct empirical validation through empirical observations.

According to the APA Dictionary of Psychology (VandenBos, 2007), a theory is a principle or a body of interrelated principles that purports to explain or predict a number of interrelated phenomena. In philosophy of science, a theory is a set of logically related explanatory hypotheses that are consistent with a body of empirical facts and that may suggest more empirical relationships. In its best use, theory signifies a systematic account of some field of study derived from a set of general propositions. These propositions may be taken as postulates, or they may be principles

Table 1.1 A hierarchical framework for paradigms, theories, and models in the psychological literature

Paradigms		
Meta-theoretical		Operational
General-integrative	Particular-specific	
Biopsychosocial	Behaviorism	Essentialism
Ecological	Cognitivism	Information processing
Interbehaviorism	Contextualism	Reductionism proper
Levels of observation	Empiricism	Organic reductionism
Life-cycle perspective	Existentialism	Produced emergent interactions
Materialism	Rationalism	Salience: Spontaneous emergent interactions
Systems		Synthetic integration
Theories		
Informal-linear		Formal-hierarchical
Models		
Dimensionality	Directionality	
	Direct	Indirect
One-dimensional	Observations	Paper-and-pencil self-report tests
Two-dimensional	Mechanical simulation	
Three-dimensional	Measurement	
Four-dimensional		
Five or more dimensions		

Adapted from L'Abate (2009b)

more or less strongly confirmed by experience. As a science develops, the part played by deductive theory in it tends to become more and more important.

According to the APA Dictionary of Psychology (VandenBos, 2007), a model is a graphic, visual or verbal, theoretical, representation of a concept or construct that can be used for investigative or demonstrative purposes, such as an enhanced understanding of the concept, proposing hypotheses showing relationships, or identifying epidemiological patterns. Of course such abstract models represent only one kind of model; other models include model airplanes that can be used to test aerodynamic properties that apply to larger planes, rats as models for human responses to drugs, or bonobos as models for human language acquisition. In the latter cases, the model is not expected to enable exact predictions of how the phenomena modeled – the airplane or the human – will behave. According to L'Abate, as given in Table 1.1, models are composed of dimensions that vary in number from one to five or six at the most, and by directionality, how close they are to the behavior evaluated.

Explicit in these definitions is the notion of a pyramidal hierarchy, implicit in the nature of the table of organization in any conceptual or human enterprise, be it a framework like evolution, or a commercial, educational, industrial, military, or religious organization (Harkness, 2007). Hierarchies are inescapable in theory construction, as they are in real life, as Max Weber argued more than one century ago.

Confusion about what constitutes a paradigm, theory, or model continues to proliferate unabated in the psychological literature. For instance, in a recent chapter on legitimacy, social identity, and power, in a volume on the psychology of power, the authors of one chapter (Spears, Greenwood, de Lemus, & Sideman, 2010),

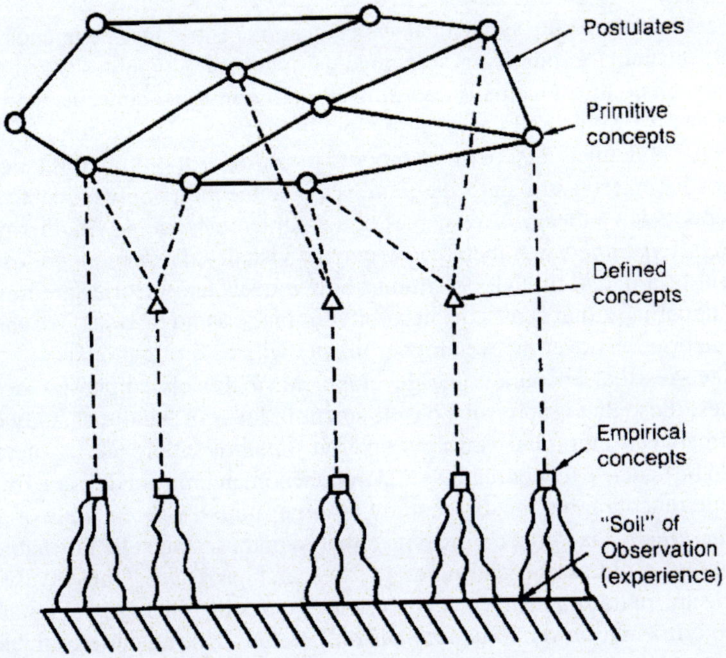

Fig. 1.1 Feigl's diagram of a theory

introduced as "theories" what may amount to models, according to the definitions of what theories or models are (Fig. 1.1).

The diagram above taken from Feigl (1970) is a useful centerpiece to clarify the multiple layers of a paradigm and to provide a distinction between a model and a theory. A postulate set, indicated by interconnected circles, "hovers" high above the "soil" of observation, indicated at the bottom. Postulates are verbal or mathematical statements that state relationships between concepts. Some concepts are called "primitive" because their meanings do not depend on definitions relating them to other concepts. The postulates can be combined to produce theorems, which are indicated by connections between the "primitive concept" circles and "defined concept" triangles, or by connections between the defined concepts. The circles and triangles representing the elements of the syntactic system are like the words of natural languages or the symbols in mathematical languages. Feigl derived this "picturesque" view of an "uninterpreted calculus" floating freely above an observation plane from Campbell (1920), who had a very similar view of theory.

The elements of Feigl's calculus may be defined via rules that relate the elements to each other. However, elements need not have a definition relating them to the outside world in order to have meaning; that is, some elements may be given meaning via strictly *internal* definitions. Their core meaning is syntactical. However, in a scientific theory some of the abstract elements must be given *external* meaning by relating them to empirical concepts. Feigl indicates such coordinations in his diagram by lines linking the hovering syntactical system with empirical concepts.

Feigl calls the hovering syntactical system a *calculus* because he intends for it to enable us to calculate something, to enable us to generate hypotheses and to make predictions. Once the calculus is coordinated with empirical concepts, and thus to observations, we would call the resulting totality a *theory.* The calculus is a *model.* Diesing (1971) defined a theory as "a model plus one or more interpretations"; an interpretation, in terms of Feigl's diagram, is the connection of the model to empirical concepts. Such a theory is a key part of a complete *paradigm.* Another part of a paradigm is a *metatheory.* A metatheory may be visualized as a more philosophical component of a paradigm that determines how models and theories are to be constructed, or perhaps, that denies the necessity for models and theories. We can imagine metatheories as hovering even higher than Feigl's hovering models.

Metatheories that encourage the development of models emphasize *deductive* approaches; those that ignore such development in favor of the direct study of relationships between empirical concepts take an *inductive* approach. A metatheory might also be analytic (encouraging breaking phenomena into small parts for study) or synthetic (encouraging the direct study of larger units). One example of a *metatheoretical directive* is Popper's insistence that hypotheses must be falsifiable.

Paradigms or, in Lakatos's terminology, "research programs," thus involve models, which are parts of theories, metatheories, and still other parts like attitudes toward experimentation versus naturalistic observation. During periods of "normal" science, researchers may not be conscious of many of their metatheoretical assumptions, proceeding with their daily work with little or no thought devoted to their global preconceptions. The authors of this chapter hope that this description of models, theories, metatheories, and paradigms will reduce confusion and clarify what we believe is the best way to distinguish between them.

Characteristics of Hierarchies

From the top down in any pyramidal hierarchy, be it a conceptual or human one, there are two overall characteristics that need explication. These characteristics are implicit in Kuhn's as well as in Feigl's model (Fig. 1.1). The first characteristic is *level of abstraction*, going from the very abstract, as in a paradigm, to the very concrete, as in the dimensions of a model (Table 1.1). The second characteristic is *generality*, going from the very general in a paradigm to the very specific in the dimensions of a model. In case one was wondering whether these two characteristics are not the same, the answer would consist of their being two sides of the same coin. However, what is the nature of this coin? This coin represents an implicit hierarchy, distinguished by function, rank, status, and *decision making* at various levels. Who makes the decisions, that is, who has the *authority* to make decisions and who has the *responsibility* to carry those decisions out? In any organization, including theories, such functions propagate from the top down, level by level. Supposedly, paradigms would direct and dictate what types of theories are more related to them than other theories. Theories, by the same token, would direct and dictate what models are relevant to the theory. Models would be defined by what dimensions are more relevant to a model than others.

As in any organization, conceptual as well as human, components in the same framework perform different functions according to who makes general decisions in the abstract and who carries out those decisions specifically in the concrete, according to printed blueprints, job descriptions, or written instructions. Given different functions implies also that each component part needs to get along with all the other parts for the whole framework to function effectively and efficiently. That is why most organizations have rules (job descriptions) to specify what particular function each component performs in a table that indicates clearly what is the particular relationship of each component to the others, i.e., how they are linked together. In the case of paradigms, theories, and models, how do they correlate with each other? How can one correlate a paradigm with a theory if paradigms are so abstract and general and there are no rules or regulations to link a given theory to a given paradigm?

Given such characteristics, the foregoing classification (Table 1.1), therefore, begs the question: Which rules (blueprints, correlations, job descriptions, instructions, etc.), if any, control, govern, or link one particular paradigm with a particular theory? Thus far, as far as we know, the only rule relating to the link between a paradigm and a theory has been by assertion or proclamation: a theorist would swear allegiance to a particular paradigm. However, how and why a particular paradigm was chosen seems determined by the personal preferences of a theorist. Could any other paradigm be chosen instead? Why one particular paradigm instead of another? One cannot help wonder whether any other paradigm could have been chosen or proclaimed, making links between paradigms and theories apparently arbitrary, usually questionable, and certainly tenuous. Paradigms in and of themselves cannot be evaluated directly. They are as valid as the theory or theories they are supposed to overlook. However, if there are no links or rules specifying how paradigms connect with theories, then the whole enterprise is strictly subjective.

Therefore, the purpose of this edited work aims at exploring such relationships, whether these relationships between paradigms and theories follow any rational or empirical connection, whether there are hidden rules behind personal preferences, and whether paradigms are chosen at random or chosen according to abstract/general or concrete/specific criteria that have not yet been spelled out. Can we find such criteria and, given such criteria, will they need to be qualified and/or, even better, quantified?

A Unifying Paradigm or Multiple Ones?

Another important issue relating to paradigms deals with the relative importance of one type of paradigm over other paradigms. For instance, are operational paradigms (Section III "General-Integrative Paradigms in Psychology," this volume) more important than general/integrative (Section I "Introductory Background") or particular specific (Section II "Paradigms in the Arts and Social Sciences," this volume) paradigms? Could the latter be more important than the former? One could argue, for instance, that operational paradigms could assume greater importance because they involve more observable and concrete operations than other paradigms.

The foregoing question implies also that a notion of a pyramidal hierarchy may be operational also within paradigms themselves. Can one paradigm claim hegemony over other paradigms? For instance, could evolution claim to be a more encompassing paradigm than others, positioning itself as a "super-duper paradigm"? To wit, Perilloux et al. (2010) claimed that:

> Evolutionary psychology currently provides the most powerful metatheory for psychological science. It is not a static metatheory, but one that continues to mature and deepen with new theoretical and empirical advances (p. 932).

Furthermore, could two different paradigms unite and coexist with each other? For instance, could an operational paradigm coexist with either a general/integrative or particular specific paradigm? Could essentialism (Chap. 22) coexist with materialism? (Chap. 14) Hopefully, all the questions asked in this chapter will be answered in one way or another by contributors to this volume.

As noted in the previous section, Kuhn (1970) considered only one existing paradigm as reigning up to a point until it would be displaced or replaced by another more encompassing or useful one, as would be evolution in biology. In the social sciences and psychology, however, as we have seen thus far, that is not the case. We have a plethora of existing points of view, and arguably no paradigms. As Liardi, Rand, and Kawasaki (2007, pp. 293–294) commented on this lack of a unifying paradigm:

> This unfortunate absence of a unifying theoretical framework, or *paradigm*, has been identified by science historians as one of the telltale signs that a discipline has not yet progressed beyond the practice of *immature science* … . Clinical psychology, of course, has at present a dizzying array of such theoretical functions … . Therefore, it is perhaps not surprising that the field has witnessed relative stagnation in recent years regarding the introduction of effective new interventions for psychological disorders.

Here we agree with Liardi et al. (2007) about the proliferation, not only of possible paradigms, as documented by the TOC and contributions to this volume, but also by a proliferation of accompanying models and theories (L'Abate, 2005, 2009a, 2009b). However, we beg to differ with Liardi et al.'s conclusion about the "relative stagnation" of effective "new treatments." Indeed, the last generation has seen the introduction of a variety of verbal and nonverbal *replicable* methods of treatment, including distance writing (L'Abate, 2001, 2002, 2011a, 2011b, 2011c, 2012; L'Abate & De Giacomo, 2003; L'Abate & Sweeney, 2011), homework assignments (Kazantzis & L'Abate, 2007), low cost approaches to promote physical and mental health (L'Abate, 2007), self-help (Harwood & L'Abate, 2010), not to speak about the introduction of innovative diagnostic and therapeutic technologies in the neuroscience disciplines (L'Abate & Kaiser, 2012).

The major innovation in mental health service delivery is, by all possible criteria, the Internet. This medium, in this century, will become one of the major ways to help and heal troubled people. Indeed, most promotion and prevention, if not face-to-face, talk-based psychotherapy, will be conducted online at a distance between professionals and participants, without one person ever talking with the other (L'Abate, 2012).

Furthermore, one could argue that Bowlby's Scottish school of object relations, overarching attachment theory, may have reached the level of a reigning paradigm in theory, if not in practice (Mikulincer & Shaver, 2007), if we were to judge it by the thousands of contributions produced all over the world in the last generation. Starting to compete with attachment theory now might be a relative newcomer, relational competence theory (RCT; Cusinato & L'Abate, 2011; L'Abate, Cusinato, Maino, Colesso, & Scilletta, 2010). The empirical connection between "internal working models" of attachment and four relational propensities of a selfhood model was argued by L'Abate and De Giacomo (2003) and demonstrated by L'Abate, De Giacomo, McCarty, De Giacomo, and Verrastro (2000).

Whether related with each other or not, both attachment and RCT do signify a major shift from an intra-psychic, intra-individual, person-in-vacuum paradigm to an individual-in-context paradigm (Bergman, Magnussen, & El-Koori, 2003; Capaldi & Proctor, 1999; Mesquita, Barrett, & Smith, 2010; Strathman & Joireman, 2005). Consequently, if there is a shift in psychology, and especially in personality theory and in clinical psychology, it would be in the direction of seeing individuals as relating continuously, in one way or another, with other individuals, if not in reality, certainly in their feelings and emotions (L'Abate, 2011a, 2011b, 2011c).

Conclusion

In this introductory chapter, we have attempted to lay the background for our contributors by clarifying the role that paradigms have played in the philosophy of science and more recently in psychology. Whether a unifying paradigm is necessary or desirable for psychology or the other social science disciplines included in this volume, like education and sociology, remains to be determined, possibly by what contributors to this volume have to say on this topic. As we have concluded, if there is a major shift in psychology, it may be well directed toward an individual in context rather than an individual-in-vacuum paradigm. If there is a major shift in mental health we will be working at a distance through the Internet through writing, rather than talking face to face.

Appendix

These comments came from the discussion that ensued between the authors during the writing of this chapter.

H: The preface indicates an admirable level of ambition, but also makes me question what my contribution could be. I am afraid that many of the key terms are not nailed down as solidly as they might be, and I think I would be highly skeptical about the details if they were more nailed down. My paradigmatic choices make me deeply suspicious of the validity of rigid hierarchical categories.

Part of my problem is a difficulty interpreting your figure. Of the two paradigms I consider myself most comfortable with, I am not sure what would make the Ecological approach "General-Integrative" nor what would make Behaviorism "Particular Specific". (This is especially true as the Ecological approach can be classified as an odd type of Behaviorism, and such nesting might even work in reverse.) Further, the emphasis on your model, and the novelty of your models, makes the goal of the individual chapters unclear to me. Am I to evaluate your model using our designated approach as a case study or Evaluate our designated approach in the context of your model or Am I to focus on the more specific questions listed in the preface, only some of which necessitate your model? Etc.

Another part of the problem is that I am uncertain how to discuss Ecological Psychology (or any of the other topics I might discuss) in a manner that would help answer the worthy questions you propose. The majority of your questions seem to demand idiographic case studies, and that the best I could do along those lines is a confessional. Such a confessional would address ecological psychology, behaviorism, radical empiricism, and more generally the unique virtues of "American Philosophy", which I believe counts as a paradigm within your scheme. That is, questions like "Why do some people choose one paradigm over another?" are best answered based on the idiosyncratic past history of the individual in question (I assert, again, from my paradigmatic standpoint).

On a separate note, the preface contains several odd over-generalizations. For example: You say, "From the top down in any pyramidal hierarchy, there are two overall characteristics that need explication … " but, the things that follow are certainly not true of ANY pyramidal hierarchy. You say, I can see "one existing paradigm as reigning up to a point until it would be replaced by another more encompassing one, as would be evolution in biology." But in evolutionary biology complete replacement is hardly ubiquitous (i.e., why there exists more than 1 species) … of course, this is only a draft of the preface … but I thought I should point it out.

Please let me know your thoughts,

L. These comments were made in answer to points made by another collaborator. Thank you so much for your detailed and thoughtful comments to what was my Preface which, with the help of Al Hillix, was transformed into Introductory Chapter 1, which I hope you received in PDF format. I don't know whether in that revision I answered some of your concerns. Nonetheless, I have appended them at the end of the chapter for future reference with my replies.

You question "rigid hierarchical categories" However, let me remind you that, as Max Weber pointed out more than a century ago, I do have hierarchies everywhere, including disciplinary levels of complexity in biology and psychology, going from the molecular to the societal, levels of observations, analysis, and interpretation, levels of development, etc. Whether the classification is rigid remains to be seen. If not rigid, how flexible should it be? and if flexible what would its limits be?

Yes, I agree with you about the relativity of the three categories I started with, and wonder myself whether the Particular Specific category, especially as far as Empiricism and Behaviorism are concerned, would not fit into the Operational Category and if Paradigms specific categories were done away with. I look forward to seeing more comments on the classification and questions about its validity and usefulness.

Could I think of the outline as tentative and as propositional? As one another contributor commented, my Table 1.1, in and of itself, represents a model in its own rights. If that is the case, how useful is it? How rigid is it or should it be? Clearly, the reason d'être for this work is to raise the kind of questions you raise, as in politics (ha), everything is on the table. Some contributors who have already contributed one chapter to the volume do not mention my previous chapter as well as the TOC or outline. You, like all other contributors, are free to use it, not use it, and criticize it. You are more than welcome to make your comments part of your chapter, at least as an introduction or conclusion to your chapter.

Perhaps my analogies about the characteristics of a hierarchy may be too over-generalized. Perhaps you may want to look at the latest work I published with my buddies from Padova U about Relational Competence Theory (RCT), where in one way or another I apply the two characteristics of a hierarchy. Can other characteristics be considered? Can other TOC's be considered? Of course, you, like all the other collaborators, are open to suggest alternative views and opinions. I would welcome them as part of the dialogue I have started with this work. OK?

I hope I have answered some of your comments, and, if I did not, please tell me. I appreciate and welcome your comments, and I hope we shall continue this dialogue in your chapter or through this correspondence.

Cordially,
Lu.

References

Agassi, J. (2002). Kuhn's way. *Philosophy of the Social Sciences, 32*(3), 394–430.

Bergman, L. R., Magnussen, D., & El-Koori, B. M. (2003). *Studying individual development in an inter-individual context: A person-oriented approach*. Mahwah, NJ: Erlbaum.

Beveridge, W. I. B. (1957). *The art of scientific investigation*. New York: New Vintage Books.

Capaldi, E. J., & Proctor, R. W. (1999). *Contextualize in psychological research? A critical review*. Thousand Oaks, CA: Sage.

Cusinato, M., & L'Abate, L. (Eds.). (2011). *Advances in relational competence theory: With special attention to alexithymia*. New York: Nova Science.

Dar, R. (1987). Another look at Meehl, Lakatos, and the scientific practices of psychologists. *The American Psychologist, 42*(2), 145–151.

Diesing, P. (1971). *Patterns of Discovery in the Social Sciences*. Chicago: Aldine Atherton.

Feigl, H. (1970). The "orthodox" view of theories. Remarks in defense as well as critique. In M. Radner & S. Winker (Eds.), *Analyses of theories and methods of physics and psychology*. Minneapolis: University of Minnesota Press.

Harkness, A. R. (2007). Personality traits are essential for a complete clinical science. In S. O. Scott & W. T. O'Donohue (Eds.), *The great ideas of clinical science: 17 principles that every mental health professional should understand* (pp. 263–290). New York: Rutledge.
Harwood, T. M., & L'Abate, L. (2010). *Self-help: A critical evaluation*. New York: Springer-Science.
KazantzIs, N., & L'Abate, L. (Eds.). (2007). *Handbook of homework assignments in psychotherapy: Research, practice, and prevention*. New York: Springer Science.
Kimble, G. A. (2005). Unity in psychology: Possibility or pipedream? In R. I. Sternberg (Ed.), *Paradigm lost, paradigm regained: Toward unity In psychology*. Washington, DC: American Psychological Association.
Kuhn, T. S. (1970). *The structure of scientific revolutions*. Chicago, IL: University of Chicago Press.
L'Abate, L. (Ed.). (2001). *Distance writing and computer-assisted interventions in psychiatry and mental health*. Westport, CT: Ablex.
L'Abate, L. (2002). *Beyond psychotherapy: Programmed writing and structured computer-assisted interventions*. Westport, CT: Ablex.
L'Abate, L. (2005). *Personality in intimate relationships: Socialization and psycho-pathology*. New York: Springer Science.
L'Abate, L. (Ed.). (2007). *Low-cost approaches to promote physical and mental health: Theory, research, and practice*. New York: Springer Science.
L'Abate, L. (2009a). In search of a relational theory. *The American Psychologist, 64*, 776–788.
L'Abate, L. (2009b). Paradigms, theories, and models: Two hierarchical frameworks. In L. L'Abate, P. De Giacomo, M. Capitelly, & S. Longo (Eds.), *Science, mind and creativity: The Bari symposium* (pp. 107–122). New York: Nova Science.
L'Abate, L. (2011a). *Hurt feelings: Theory, research, and applications in intimate relationships*. New York: Cambridge University Press.
L'Abate, L. (2011b). Psychotherapy consists of written homework assignments: A radical iconoclastic conviction. In H. Rosenthal (Ed.), *Favorite counseling and therapy homework techniques* (Classic Anniversary Edition, pp. 219–229). New York: Routledge.
L'Abate, L. (2011c). *Sourcebook of interactive exercises in mental health*. New York: Springer Science.
L'Abate, L. (2012). *Clinical psychology and psychotherapy as a science*. New York: Springer-Science.
L'Abate, L., Cusinato, M., Maino, E., Colesso, W., & Scilletta, C. (2010). *Relational competence theory: Research and mental health applications*. New York: Springer- Science.
L'Abate, L., & De Giacomo, P. (2003). *Intimate relationships and how to improve them: Integrating theoretical models with preventive and psychotherapeutic interventions*. Westport, CT: Praeger.
L'Abate, L., De Giacomo, P., McCarty, F., De Giacomo, A., & Verrastro, G. (2000). Testing three models of Intimate relationships. *Contemporary Family Therapy: An International Journal, 22*, 103–122.
L'Abate, L., & Kaiser, D. A. (Eds.). (2012). *Handbook of technology in psychology, psychiatry, and neurology: Theory, research, and practice*. New York: Nova Science.
L'Abate, L., & Sweeney, L. G. (2011). *Research on writing approaches in mental health*. Bingley: Emerald.
Lakatos, I. (1970). Falsification and the methodology of scientific research programmes. In A. Musgrave & I. Lakatos (Eds.), *Criticism and the growth of knowledge* (pp. 91–195). New York: Cambridge University Press.
Laudan, L. (1977). *Progress and Its problems*. Berkeley, CA: University of California Press.
Leahy, T. H. (1992). The mythical revolutions of American psychology. *The American Psychologist, 47*, 191–216.
Liardi, S. S., Rand, K., & Kawasaki, L. (2007). The cognitive neuroscience perspective allows us to understand abnormal behavior at multiple levels of complexity. In S. O. Scott & W. T. O'Donohue (Eds.), *The great ideas of clinical science: 17 principles that every mental health professional should understand* (pp. 291–309). New York: Rutledge.

Masterman, M. (1970). The nature of a paradigm. In I. Lakatos & A. Musgrave (Eds.), *Criticism and the growth of knowledge* (pp. 59–90). New York: Cambridge University Press.

Mesquita, B., Barrett, L. F., & Smith, E. R. (Eds.). (2010). *The mind in context*. New York: Guilford.

Mikulincer, M., & Shaver, P. R. (2007). *Adult attachment: Structure, dynamics, and change*. New York: Guilford.

Perilloux, C., Lewis, D. M. G., Goetz, C. D., Fleischman, D. S., Easton, J. A., Confer, J. C., et al. (2010). Trade-offs, individual differences, and misunderstan-dings about evolutionary psychology. *The American Psychologist, 65*, 930–932.

Popper, K. (1957; original German version, 1934). *Logic of Scientific Discovery*. New York: Harper & Row.

Popper, K. (1970). Normal science and Its dangers. In I. Lakatos & A. Musgrave (Eds.), *Criticism and the growth of knowledge* (pp. 51–58). New York: Cambridge University Press.

Reisch, G. A. (2005). *How the cold war transformed philosophy of science: To the icy slopes of logic*. New York: Cambridge University Press.

Spears, R., Greenwood, R., de Lemus, S., & Sideman, J. (2010). Legitimacy, social identity, and power. In A. Guinote & T. K. Vescio (Eds.), *The psychology of power* (pp. 251–283). New York: Guilford.

Strathman, A., & Joireman, J. (Eds.). (2005). *Understanding behavior in the context of time: Theory, research, and application*. Mahwah, NJ: Erlbaum.

Tsou, J. Y. (2006). Genetic epistemology and Piaget's philosophy of science: Piaget vs Kuhn on scientific progress. *Theory and Psychology, 16*(2), 203–224.

VandenBos, G. (2007). *Dictionary of Psychology*. Washington, D. C. American Psychological Association.

Chapter 2
The Role of Paradigms in Science: A Historical Perspective

Corinna Guerra, Mario Capitelli, and Savino Longo

Philosophy of science without history of science is empty; history of science without philosophy of science is blind

(Lakatos, 1971:91).

The studies of Thomas Kuhn gave a new interpretation of the philosophy of science (1977, 1978, 1983, 1995). This new kind of approach consists of a total reorganization even of the linguistic ways, to approach the whole scientific enterprise as well as that of individual scientists, so that even today the notion of "paradigm" is an essential one. To make use of the category "paradigm" implies, of course, the risk inherent in its polysemy, but this is also an advantage because it allows a more free and effective application of the concept in different historical periods and even more in different fields of knowledge.

When Kuhn elaborated his analysis of the particular form of human knowledge, i.e., science, he had in mind the developments in specific disciplines such as chemistry, physics, and astronomy, which were the sources of the vast majority of examples presenting it as evidence of its allegations in *The Structure of Scientific Revolutions*. Of these sciences he chose to treat also the moments of their historical development in which they had offered the paradigms that are predominantly presented as such to nonspecialist audiences. The reference here is of course the "new chemistry," Newtonian physics or Copernican revolution, for in the text even recurrence of the theme of relativity of Einstein could not have the same significance of the work of Newton, who entered the stock of more shared knowledge of humankind.

C. Guerra (✉)
Viale G. Di Vittorio, 115, 71121, Foggia, Italy
e-mail: guerra.corinna@gmail.com

M. Capitelli • S. Longo
Dipartimento di Chimica; Facoltà di Scienze MM. FF. NN, Università di Bari,
via Orabona, 4, 70125, Bari, Italy

L. L'Abate (ed.), *Paradigms in Theory Construction*, DOI 10.1007/978-1-4614-0914-4_2,

That it is just to observe the new epistemological thought of Kuhn was based on a deep knowledge of the history of science and of the hard sciences above all; but the system that belonged to it is efficient thanks to its surprising ductility, by which it was possible to apply it to the history of economics or social sciences, as many papers had showed in the past decades. So, for many scholars the term "paradigm" seems to be unused in the original sense in historiography, while one can meet it in a very different contexts (Cerretta, 2007) thanks to its elusive polysemic nature (Barbano, 1999:254).

Before exploring the role that paradigms have historically played in the evolution of the paradigms of science, it is appropriate, to avoid any opportunities for confusion, to distinguish some elements that are characteristic of the scientific paradigm, and therefore must remain clear for the entire chapter.

We stick to the idea that a paradigm is a sort of methodological and conceptual universe in which the scientist can operate. As a universe it includes a self-coherent repertoire of scientific theories, no exact models, tools and even styles of thought of a historical period, which delimit the set of concepts and methods which can provide the scientist to do his job. Often the paradigm stems from a successful system, or a discovery that has shown its fertility either theoretical or experimental, that is shared as examples of answers to his research, but even more questions. This is because the structure of the paradigm is already providing the codified part of the knowledge system which is shared by all scientists who adhere to the implied methodology to look for the result that, according to coordinates of the paradigm, is expected to further confirm the explanatory power of the paradigm itself.

As an example we can use the mechanistic paradigm which was much current especially in the seventeenth and eighteenth centuries, when most of the scientific community acknowledged as the sole explanation of natural phenomena the movement or combination of movements of bodies in space. This paradigm, when applied to physics, implied that all what occurs in nature is the revelation of the laws of mechanics, and everything was interpreted according to the terms of quantities like force, mass, and energy. Referring also to what is described in the Preface, the framework given by the mechanistic paradigm can accommodate different styles of interpretation or theories, starting from the same conceptual tools, diverging interpretation of a field of phenomena, but without leaving the boundaries of the system, which in the chosen example, explains the phenomena with the combination of matter and movement. The symbolic value of the mechanistic paradigm is that it was widely extended to all natural sciences and to physiology as psychology where it provided successful clues to investigate important phenomena like the transmission of sensations, in a purely mechanical way (Abbagnano, 2001). On the other hand, the "mechanistic paradigm" slowed down the acceptance of the revolutionary ideas of Arrhenius who at the end of the nineteenth century introduced the concept of electrolytic dissociation, strongly opposed by the classical thermodynamics advocate. As a paradox, Ostwald, who was developing another paradigm, that of energetics contributed to such acceptance (Guerra & Capitelli, 2009).

After these premises is now our intention to show schematically, rather than argue, the changing paradigms in the history of natural science going beyond the

Fig. 2.1 Niccolò Tornioli (Siena, 1598 – Roma, 1651), *Gli astronomi-The Astronomers, oil on canvas*, cm 148×218,5, Inv. 269, Galleria Spada, Rome, 1640 ca

Poincarè conventionalism (Poincarè, 1905). We will not discuss these systems completely for different reasons, especially to avoid two major risks: to stick to a narration of rather remote times, and stray too far from more familiar scientific ideas that best convey the general sense of the speech, and re-proposing the classical case studies such as Newton, Galileo, and Copernicus, who immediately illustrate the point, but for which reappears the evidence that Kuhn hooked to his theories. Finally, a historical narrative of the history of scientific paradigms will never be complete in the sense of creating a harmonious succession of stages that lead naturally into one another, because the gap between paradigms is generally accomplished with fractures, the emergence of new forms and new ways to view these forms, in short, the communication of knowledge is comparable to the translation of a rigid body in Euclidean space (Fleck, 1983:192) (Fig. 2.1).

We therefore opted for the selection of several topics of particular interest in the history of natural sciences, they are explained by the worldviews, which gave a specific direction to the research in those historical moments. In this way scientists specifically interpreted certain phenomena with reference to the general view, which would therefore also oriented research and future results. Indeed, it is possible to recognize the establishment of a paradigm in science when the members of the scientific community for a period no longer feel obliged to justify the foundation of their education or seek alternatives (Buzzoni, 2008).

We will reference to the paradigm with the terms "worldview" and "style of thinking," not only because they seem more intuitive without impoverishing the sense, but also because these terms are used by other philosophers of science who also read the way science as a succession of paradigms and the time of the alternation is condensed into what we call scientific revolutions. We would thus only briefly refer to the relationship between Kuhn's thought and reflections of the bacteriologist Ludwik Fleck (1896–1961), an issue that has occupied many epistemologists, given the formidable connections (Campelli, 1999), but we are here not so much interested in this debate, but rather to highlight how the idea of this volume deals with the paradigms in various fields which is then also eligible in the perspective of a partial (opposed to full) adhesion to the Kuhnian tradition.

Fleck already before Kuhn talked about a *Denkstil*, a collective style of thinking, that for every age is the criteria that discriminate concepts eligible for a certain field of knowledge and explains how researchers can work within it, which just as the paradigm must be a shared and prescriptive structure. So a scientific theory is not in itself a logical system, even if it aspires to be so, it is nothing but a unit provided with a style, and this restricts the creative freedom of the scientist, but at the same time it makes the job easier because certain limits, however, allow the researcher to take for granted the basic assumptions (Campelli, 1999:13). The outcome of this approach is that the scientist's own style of thinking is obvious, while the alternative ways of thinking appear to him like fancy constructions and it is here that scientific disputes on the interpretation of the same phenomenon start, because in this optics it is not possible to decide on the basis of a single experiment, considered as the typical concept of "experimentum crucis." Contrairly, in order to give a decisive opinion, we need a whole system of trial and error for compliance to style.

A somewhat extreme illustration of the effect of the above on the history of science lies in the consideration that an illness like syphilis could not exist by itself but a concept developed and historically determined in a collective thought.

So the language confusion (Baldini, 1986:78) which went around with time by creating the concept of paradigm is due to the ambiguity of the great communicators as T. Kuhn, that is, their willingness to invent subtle and elusive, although effective slogans, which actually end to be sources of separations in the community (Campelli, 1999:6).

For example, in the studies of black body radiation of the second half of the nineteenth century (Kirchhoff, 1860), one can see that the choice of the scientist to observe it from the viewpoint of thermodynamics or electromagnetism or the probabilistic theory of the kinetics of gas has changed the scientific interpretation of the phenomena observed. In 1873, Maxwell showed that radiation incident normally on a surface would exert pressure on it while Bartoli in 1875 gave a proof of the independence of black body radiation on thermodynamics, without reference to any particular theory of light, obtaining a significant methodological result. Boltzmann, in the wake of these two works, derived in 1884 the famous law that bears his name. It has been said that black body radiation was not the most pressing problems in those decades: first was the source of irreversibility, the problem of specific heat of polyatomic gases, and the problem of the luminiferous ether. And then there was the need to explain the nature of the electron, of X-ray (Roentgen, 1896), and radioactivity (Bequerel, 1896). However, in the work of Planck on the radiation from a cavity most clearly emerge those traits of this new way of doing science, the revolution in the scientific paradigm in which quantum mechanics has its full completion.

The new way of doing science asks to abandon the reductionist methodology, according to which the study of any reality problem is due to the fundamental laws of dynamics or electromagnetism. The new methodology will be to identify new physical principles, experimentally detectable (not too difficult given the vastness of the data in those years), which should not necessarily refer to specific models and, extrapolating from their original context, to elevate the general laws of physics. In this optics, the black body problem is resolved by Planck. While this originally was strictly a thermodynamic problem, and the initial studies concerned the distribution of energy and entropy between the oscillators of the wall without any use of statistical methods, in 1901, when it became clear that we need to give a foundation to the law of the theoretical black body spectrum, Planck proposed the quantization of energy and founded his remarks on the statistical significance of entropy introduced by Boltzmann. The derivation by Planck of radiation law is nowadays regarded as the beginning of modern physics, Planck's law assuming a revolutionary character against the Kuhn's ideas (Kuhn, 1978). The real revolution was made by Boltzmann 20 years before in the development of statistical mechanics when his energy packets can be considered as an anticipation of quantum packets. On the other hand, the mixing between Boltzmann and Planck ideas was at the basis of their conflict (Lindley, 2001).

We add a short reflection on the Newtonian paradigm. Of course it has been for several centuries the reference for all the sciences, reinforcing the idea that gravitational and electric forces were able to explain other types of attractions, such as the interparticle attraction in the event of chemical reactions. Even today the laws codified in 1687 in *Philosophiae Naturalis Principia Mathematica* represent for us the basis of classical physics. Anyway, after the work of Newton, his theories were not easily accepted due to the gravity, mainly interpretated as an occult force, like those successfully used in science during the 16th century. It means that theories of Newtonians were perceived less "modern" than the mechanistic ones, which needed only the concepts of matter and motion to explain phenomena. But with the wave theory and the resumption of the ether, was solved in a way beyond the embarrassment of the surreptitious return of the occult force.

The success of Newtonian physics had as one of the many effects that to establish for a long and fruitful period the idea that mechanics should be the culmination of all the others, in order to match its explanatory scope. The application of the paradigm of a Newtonian world also means that the revolution of Lavoisier was born, in a certain sense, with the ambition of being able to vindicate an idea of chemistry intended primarily as a record of quantitative aspects, that is why we always remember in texts on the history of chemistry the efforts of Antoine Laurent Lavoisier (1743–1794) to use the balance of the experiments (analysis + synthesis) and attempts to prepare the calorimeter. Yet earlier, under the current system based on phlogiston, a theory invalidated by Lavoisier (Guerra & Capitelli, 2009), the chemistry seemed to be more explanatory, since it had the possibility of referring to "principles" of quality carriers. The transition to the new paradigm of the French chemistry was more the result of attention given to measured quantities, inspired from the successes of physics at the time, that a renewed interest in the opportunities of chemistry.

Indeed, after the work of Lavoisier on the path taken by the chemical discipline would lead to a re-evaluation of qualitative aspects, in most studies on the foundations will then be aimed at the explanation of the qualities revealed by the substances. Moving shortly to contemporary science, the mechanistic paradigm is now current in chemistry. The possibility of using fast computers to calculate directly molecular properties from first principles convinced most scientists, especially in sectors like chemical physics and metal-organic chemistry, that accurate calculations of numerical data are essential to chemical understanding. A look to contemporary journal papers will show that the use of theoretical calculations based on quantum-mechanical methods like density functional theory (DFT) or similar is often considered an essential part of chemical understanding. While the range of possibilities offered by such a quantitative chemistry is incredibly vast, one should not forget that the mechanistic paradigm has its limits, and the history of science registered them, while it appears that contemporary chemists are again unaware of such limits in their enthusiasm.

To state a case, the use of fully quantitative methods reduces the chemical understanding to the recognition of results produced by an algorithm. Interpretive ideas such as the interaction of atomic orbitals, the Lewis theory, or more advanced logical-linguistic concepts used in inorganic and organic chemistry cannot be completely exchanged, in terms of semantic value of the chemical structure, with an approach that actually transforms the theory into a kind of numerical experiment, semantically poor, as it is often recognized by the specialists themselves. Although the use of such methods asks the chemist a considerable understanding of mathematics and physics as well as numerical techniques and computer programming, these competences are not, as such, enriching in terms of chemical understanding of nature. Therefore, some prudence is in order, and one should not forget Mach's advice on the risk of considering all phenomena as manifestations of a mechanics. Fortunately the Lewis' bonding concept, based on the two electron exchange, still resists towards "ab initio" quantum mechanical description of chemical bonding in chemical practice and interpretation (Guerra & Capitelli, 2009).

In a volume of 1978, Robotti significantly traced the dynamic development of the conception of the first atomic models, in order to connect them to the world-views prevailing in those years. The proof of this point of departure is given by studying the changes that these models have undergone, the intermediate stages, corresponding to changes in the broader vision of science and methodological rules which meant that individual scientists were geared toward a certain reading of the experimental data, the result of theoretical choices (Robotti, 1978:10).

At the end of the eighteenth century gained great prominence in the scientific world the research on electricity, especially after the discovery of Luigi Galvani (1737–1798) on animal electricity and the work of Alessandro Volta (1745–1827). Without going into too much detains, one can easily imagine how the scientific community was attracted by the potential of the electrical phenomenon in general. For many decades, the focus was on applications of the "electric fluid," but with epochal success, as when Humphry Davy (1778–1829) by observing the action of electricity on chemical compounds established the invalidity of the role of universal acidifying character that Lavoisier had attributed to oxygen (from which the very term "oxygen," the Greek for "acid principle," Russian "Kislorod" and German "Sauerstoff" bearing the same meaning). In fact, following the invention of the battery by Volta, Davy succeeded since 1807 in the most demanding decompositions, up to the most discussed of hydrochloric acid, which has always considered the most acidic substance at all, which is devoid of the element oxygen (Testi, 1940:106). For Lavosier in fact, the 33 elementary substances he identified were combined together to obtain compounds with increasing degrees of acidity on the amount of oxygen contained in them (Lavoisier, 1789).

However, despite the remarkable results of applications scientists during the first half of the nineteenth century, the first approaches to unveil the behavior of electricity do not illuminate its nature, theory still lacking a clear-cut formulation based on experimental data. A change came in 1873 when James Clerk Maxwell (1831–1879) was able to give a unitary explanation of the whole corpus of causes and effects of electrical phenomena in the *Treaty on electricity and electromagnetism;* according to Robotti this persuaded many scientists to focus on a particular type of investigation: the study of the behavior of a rarefied gas in the presence of an electric discharge. The scenario was then represented by a renewed interest for a number of phenomena that had managed to gather around the idea of electricity, on which he worked for some time, but that still does not, however, suggest a decisive solution to its nature.

Consequently, the search had to move towards the investigation of the interactions between electricity and matter, was regarded as a simpler, namely gas, and was part of experiments in this sense that Joseph John Thomson (1856–1949) had signs of the existence of charged particles smaller than the atom (electrons). This led to the real possibility of a complex structure of the atom, since until that time the term "atomism" still meant that matter was made up of tiny particles, but not further divisible, according to the classical etymology of the word. This does not mean that the opposite hypothesis of composite particles in them had never been contemplated for some time, e.g., in the kinetic theory of thermal phenomena, where the origin of the observed properties could not be understood without referring to an internal structure of the atom, as was suggested by a more general study of electricity.

The next step was the creation of two models of the atom at the beginning of the twentieth century due to Thomson and Ernest Rutherford (1871–1937), but although there were all conditions to make reference to Planck's theory on the quantization of energy this was not taken into account, according to Robotti, because those ideas were held in serious consideration only by a small contingent of scientists who dealt with the physics of black body (Robotti, 1978:24).

The "construction" of the atom was also influenced by certain conflicting positions on how to look at a hierarchical level lower than that of the paradigm, scientific theories. This directed the choices of the majority of scientists convinced that physics had to deal with the phenomena perceived by the senses and that the atom was a pure mathematical model. In this context, one can mention the famous debate between Ernst Mach (1838–1916) and Ludwig Boltzmann (1844–1906) on the nature of atoms (Guerra & Capitelli, 2009).

Let us return to the production of an electric discharge in a gaseous medium, which provides an illuminating case study. The structure of an electric discharge was found to have a quite complex structure with several luminous and dark regions which were explained only by a long interdisciplinary research involving atomic physics, electrostatics, and kinetic theory. In this interpretative effort a main step was to consider the productive value of the new mechanics, which allowed to determine the probability of atom ionization by electron impact, a crucial factor for the self-sustainment of the ionized condition of the gas (Loeb, 1934). An interesting open question is the very low recognition of the substantial relation of these studies with radioactivity: in chemistry two class phenomena: the chemical reactions produced by electric discharges (plasma chemistry) and those produced by ionizing radiations (radiation chemistry) have been traditionally and still are studied by two very weakly connected communities. Also laser chemistry, i.e., the chemistry activated by lasers, which could provide a link between plasma, radiation, and photochemistry (the science of chemical reaction produced by light absorption) is still a discipline basically independent of plasma and radiation chemistry.

It is interesting, in this context, to mention the attempt of the Soviet scientists L. T. Bugaenko, M. G. Kuzmin, and L. S. Polak to unify conceptually the three fields of photochemistry, radiation chemistry, and plasma chemistry in a single scientific discipline to be named "High Energy Chemistry" (Bugaenko, Kuzmin, & Polak, 1988). Not surprisingly, despite the foundation of a relatively important chemical journal with this title, such a bold effort was left basically unnoticed by the scientific community.

Experiments were also made in an indirect way on hot metal filaments, which were used to determine the charge/mass ratio of particles to lead to a direct proof of the existence of smaller constituents of the atom (Loeb, 1934). Dropping the assumption that the atom was the ultimate constituent of matter, it was necessary to establish the nature of such small particles (Jeans, 1901). This led back to the question of the interpretation of atomic spectra, which until then were considered, which much overlooking of the subject, to be possibly explained by the oscillation of the atom as an indivisible particle. This allowed the gradual introduction of the magnitude of these ions and other types of forces. The finding of charged material particles present equally in all atoms led to the hypothesis that the different chemical properties of elements were originated from the different electronic configurations of atoms (Robotti, 1978:58).

As mentioned earlier, another research parallel to the intrinsic nature of the atom was based on the theme of the radioactivity, as the only suitable means to obtain information about processes occurring inside the atom (Bellone, 1998:325; Soddy, 1912).

The decision to investigate a particular area of course suggested the desirability to promote a model or unconsciously represented adherence to a paradigm that favored a certain field of inquiry rather than another, e.g., the interest of Kelvin (1824–1907) to resolve questions related to electricity led him to devise a static model of the atom in which the "positive power" could be equally distributed uniformly within the atom or a smaller sphere which was part of it. Instead, Thomson's mathematical model based on dynamics orbits could better explain (still much less satisfactory than successive models) issues such as the frequency of the emitted radiation of the chemistry of the elements. In short, much depended on the studies in which the scientist had been formed, that they did see an issue as more decisive than any other.

A similarly complex situation arose in the effort to the interpretation of experimental data and the scattering of spectral lines: Bohr moved away from the strong paradigm of mechanics and classical electromagnetism and started to build the structure of Planck's law, intended as a postulate.

Probably, Niels Bohr (1885–1962) is left reasonably groped by new lines of research because previously he had devoted himself to study slightly different from his colleagues, had in fact studied the atomic theory in relation to metals, where he had frequently observed phenomena that made him doubt about the interpretive power of classical mechanics and electromagnetism, as demonstrated in several stages of his dissertation (Robotti, 1978:228). On this same light one could draw a difference between the chemical conception of atoms, strongly connected to observation, and that of physicists, which was mostly a dynamic object. This chemical attitude towards atoms and molecules is largely lost nowadays, as discussed earlier. It is also important to observe that the "physical atom," especially in the golden age of nuclear physics between 1930 and 1950, was much studied for its nucleus and for the nuclear transmutations which actualized the alchemical "Phylosopher's Stone" but whose negation was an essential interpretive principle in the classical chemistry. Of course there are exceptions to this view; chemical transmutation were and are essential in the important field of nuclear chemistry, while the specialist of what is now called "physics of matter" studied the structure of atoms and molecules by methods which are essentially equivalent to those used in chemical physics and theoretical inorganic chemistry. In this context we want to underline that until 1930s atomic and nuclear physics were treated at the same level. Majorana nuclear forces were derived by the corresponding quantum mechanical exchange forces used to explain chemical bonding. The statement "chemistry is completely understood on the basis of quantum mechanics" held by well famous scientists including Dirac and Fermi shifted atomic physics research toward nuclear physics, creating a sort of monadism between chemistry and physics. In a certain sense, the dualism between chemistry and physics comes back to the ideas of Berthelot who in the nineteenth century defined chemistry as a positive science while physics was too much based on speculative and metaphysical ideas.

To complete this chapter, we will introduce two modern paradigms arising in the spirit of energetics. One example is Haken's synergetic, which comprises, in turn, more or less explicit adherence to the Darwinian paradigm (Swenson, 1992), let us see how.

The set of phenomena that present themselves to our experience can be regarded as so many systems composed of parts that act together under the limitation imposed by boundary conditions that determine the changes or limitations to which the system can respond or not to qualitative changes of their structure or its behavior (Haken, 1990:39). Synergetics deals with these changes, which can be the occurrence of a new motion characteristic of a fluid such as water, the emergence of a new

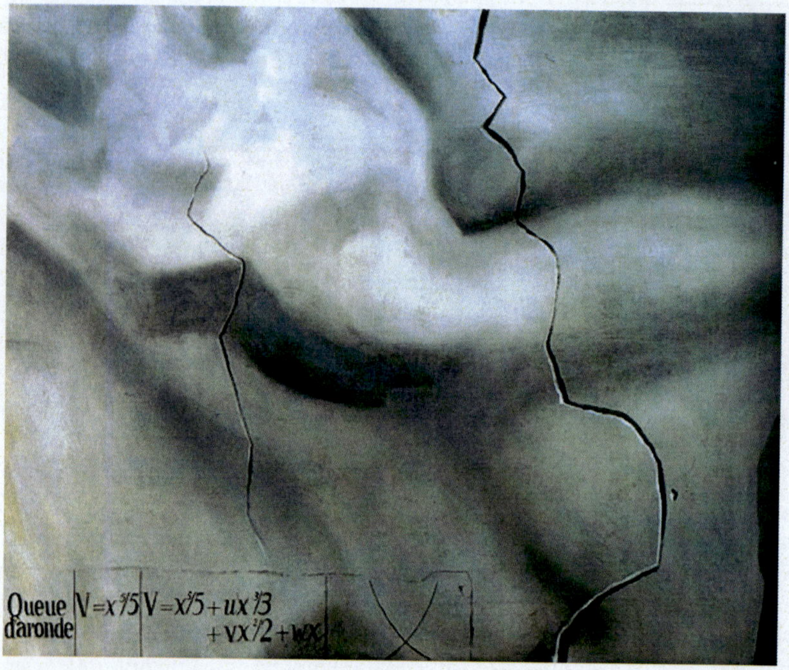

Fig. 2.2 Salvador Dalì, *Ratto topologico d'Europa – Omaggio a René Thom*, Oil on canvas, Fundación Gala-Salvador Dalí, Figueres, 1983

theory in a scientific discipline, or approval of a form or an image from our brains. The reaction of the system can be decomposed into modes, similar to the oscillation or Fourier modes in acoustics, only some of which continue to perpetuate their qualitative effects. Haken embraced the option of a generalized Darwinism because in his view only a few modes win the competition to determine the emergent properties of the observed system. In these terms, it is possible to use images and idea from the dynamic of continua to understand the development of structures in the world and even formulate a basic theory of creativity (Longo, 2009).

A second related attempt is the historically well known "catastrophe theory" (Thom, 1972) where the shapes of biological structures were interpreted as due to "regime conflicts" associated to a dynamics derived from a potential, in mechanical terms. While this attempt to explain by a single theory (although based on a non-trivial topology) phenomena like tissue specialization, earthquakes, social changes, language, mental illness, the action of man on matter fascinated scientists, laymen and even artists (Fig. 2.2) this theory is now mostly discredited from the technical point of view since its mathematical assumptions have been shown to be inadequate to describe most systems. In his most popular book, Haken discusses catastrophe theory as an example to illustrate the rise and decay of a scientific paradigm in terms of its synergetics.

The two theories share the ambition to provide a dynamic theory of mind. For example in the process of storing images in the brain, they store a lot for any perceived object, but then an image is identified with the observed object: "Once formed, the visual thinking behaves like a particle described by a state 'quantum' nonstationary in the attention potential (...) In simple terms, in the course of his wanderings, it may happen that the visual thinking to fall into a minimum of the attention potential" (Haken, 1990:45). Thom's book ends with reference to the Zeeman's model for memory, which is one of the inspiration of the theory (Zeeman will become later one of the public advocates of catastrophe theory of which offered a more defendable version closer in style to hard sciences).

Conclusion

We have shown a few cases of changing paradigms in sciences, focusing mainly on the problem of the nature of atoms and molecules. The reason for this choice is based on the easy evidence of the role of other structures, not experimental but paradigmatic, which drive scientist to devote attention to towards certain scientific topics and to consider those peculiar methodologies and solutions they actually adopt in specific works. We mostly ignore whether there is a logical or computable reason accounts for this state of matters, but for many events in the history of natural sciences this seems to be the right interpretation of the behavior of scholars' community. What we have written here may contribute to provide the reader with some arguments towards such demonstrations and with conceptual tools to profit of this important interpretive concept in different contexts.

Acknowledgment One of the authors (SL) thanks his former student Leonardo Malgieri for providing him material on the history of studies on black body radiation.

References

Abbagnano N. (2001). *Dizionario di filosofia*, Torino: UTET Libreria, 3rd edition.

Alferj, P., & Pilati, A. (1990). *Conoscenza e complessità: Strategie e prospettive della scienza contemporanea*. Roma: Theoria.

Baldini, M. (1986). *Congetture sull'epistemologia e sulla storia della scienza*. Roma: Armando Editore.

Barbano F. (1999). Thomas S. Kuhn e i paradigmi: Tra sistematica e storia della scienza. In E. Campelli (Ed.), *T.S. Kuhn: Come mutano le Idee sulla scienza*. Milano: FrancoAngeli.

Bellone E. (1988). In *Storia della scienza moderna e contemporanea*, Rossi Paolo (ed.), Torino: UTET, 5 voll.

Bugaenko, L. T., Kuzmin, M. G., & Polak, L. (1988). *S.* Khimiya Vysokikh Energii, Moskow: Khimiya.

Buzzoni M. (2008). *Filosofia della scienza*, Brescia: La scuola.

Campelli, E. (Ed.). (1999). *T.S. Kuhn: Come mutano le Idee sulla scienza*. Milano: FrancoAngeli.

Cerreta P. (2007). *Il passaggio di Kuhn dal paradigma a due concetti sostitutivi: la matrice disciplinare e l'esempio condiviso da un gruppo*, in M. Leone, N. Robotti, B. Preziosi (Eds.). *L'eredità di Fermi Majorana e altri temi: atti del 24. Congresso nazionale della Società italiana degli storici della fisica e dell'astronomia (SISFA)*, Napoli-Avellino, 3-6 giugno 2004, Napoli: Bibliopolis, pp. 223–227.

Fleck L. (1935). *Enstehung und Entwicklung einer wissenschaftlichen Tatsache*, Basel: Benno Schwabe & Co., ed. Consultata (1983) *Genesi e sviluppo di un fatto scientifico. Per una teoria dello stile e del collettivo di pensiero.* Bologna: Il Mulino.

Guerra, C., & Capitelli, M. (2009). Success and limits of reductionism: Case studies in chemistry. In L. L'Abate, P. De Giacomo, M. Capitelli, & S. Longo (Eds.), *Science, mind, and creativity: The Bari symposium*. New York: Nova Science.

Haken, H. (1981). *Erfolgsgeheimnisse der Natur.* Stoccarda: Deutsche Verlags.

Haken, H. (1983). *Sinergetica. Il segreto del successo della natura*. Torino: Boringhieri.

Kirchhoff, G. (1860). Über das Verhältnis zwischen dem Emissionsvermöogen und dem Absorptionsvermögen. der Körper für Wärme und Licht. *Poggendorfs Annalen der Physik und Chemie*, *109*, 275–301.

Kuhn, T. S. (1977). *The essential tension: Selected studies in tradition and change*. Chicago: Chicago University Press.

Kuhn, T. S. (1978). *Black body theory and the quantum discontinuity 1894–1912*. New York: Oxford Clarendon.

Kuhn, T. S. (1983). *Paradigmi e rivoluzioni nella scienza*. Roma: Armando.

Kuhn T. S. (1995). Un entretien avec Thomas S. Kuhn. *Le Monde*, 6 février.

Lakatos, I. (1971). History of science and its rational reconstructions. In R. Buck & R. Cohen (Eds.), *Boston studies in the philosophy of science* (Vol. 8, p. 91). Dordrecht: Riedel.

Lavoisier A. L. (1789). *Traité élémentaire de chimie, présenté dans un ordre nouveau et d'après les découvertes modernes*, Paris.

Leone, M., Robotti, N., & Preziosi, B. (Eds.). (2007). *L'eredità di Fermi Majorana e altri temi: Atti del XXIV Congresso nazionale della Società italiana degli storici della fisica e dell'astronomia (SISFA). Napoli-Avellino, 3–6 giugno 2004*. Napoli: Bibliopolis.

Lindley, D. (2001). *Boltzmann's atom. The great debate that debate launched a revolution in physics*. New York: The Free Press.

Loeb, L. B. (1934). *The kinetic theory of gases*. New York: Mc-Graw Hill.

Longo, S. (2009). Complex systems and creativity. In L. L'Abate, P. De Giacomo, M. Capitelli, & S. Longo (Eds.), *Science, mind, and creativity: The Bari symposium*. New York: Nova Science.

Poincarè, H. (1905). *La Valeur de la Science*. Paris: Ernest Flammarion Editeur.

Robotti, N. (1978). *I primi modelli dell'atomo. Dall'elettrone all'atomo di Bohr*. Torino: Loescher Editore.

Rossi, P. (Ed.). (1998). *Storia della scienza moderna e contemporanea*. Torino: UTET.

Soddy, F. (1912). *The interpretation of radium and the structure of the atom*. New York: G.P. Putnam.

Swenson, R. (1992). Order, evolution and natural law: Fundamental relations in complex system theory. In C. V. Negoita (Ed.), *Cybernetics and applied systems* (pp. 125–148). New York: Marcel Dekker.

Testi, G. (1940). *Storia della chimica con particolare riguardo all'opera degli italiani*. Roma: Casa editrice mediterranea.

Thom, R. (1972). *Stabilité structurelle et morphogenèse*. Paris: InterEditions.

Chapter 3
Evolving Scientific Paradigms: Retrospective and Prospective

Willis F. Overton

It is well known that the contemporary concept of *scientific paradigm* arrived on the scientific and philosophical scene in 1962 with the publication of Thomas Kuhn's *Structure of Scientific Revolutions*. The book became enormously popular and soon it was common to find it on the syllabi of university undergraduate and graduate courses. In scientific and philosophical circles, the publication reignited a long simmering debate and turned it into a raging firestorm of criticism and countercriticism (see, e.g., Lakatos & Musgrave, 1970). Partially, the debate was about the introduction of sociological matters into the body of scientific theory and method, and partially, it was about the nature of scientific change itself (e.g., normal vs. revolutionary science, scientific crises, anomalies, gestalt switches). The present essay will concern the former issue, rather than the latter.[1] Kuhn's use of the term *paradigm* was initially vague – a sympathetic critic (Masterman, 1970) noted that Kuhn had used the term in 22 different ways – but at the time of the second edition (1970) of *SSR* Kuhn wrote a *Postscript – 1969* defining two distinct meanings; one narrow, one broad (see also, Kuhn 1977a). The narrow meaning of scientific paradigm he called *exemplars*; examples shared by a community of scientists involving "the concrete problems-solutions that students encounter from the start of their scientific education, whether in laboratories, on examinations, or at the ends of chapters in science texts (p. 187)." For example, the student of physics learns "problems such as the inclined plane, the conical pendulum, and Keplerian orbits, instruments such as the vernier, the calorimeter, and the Wheatstone bridge" (Kuhn, 1970, p. 187). The broader meaning of the term he called *disciplinary matrix* and two of its key features were shared *metaphysical beliefs* – "from heuristic to ontological models,

[1] For a history of the impact of Kuhn's SST on the understanding of how the rules of science change, see Overton (1984).

W.F. Overton (✉)
Department of Psychology, Temple University, Philadelphia, PA 19122, USA
e-mail: overton@temple.edu

L. L'Abate (ed.), *Paradigms in Theory Construction*, DOI 10.1007/978-1-4614-0914-4_3,

… [which] supply the group with preferred or permissible analogies and metaphors (p. 184)" and *values*, which are especially important in choosing between incompatible ways of practicing a particular discipline (e.g., a good theory would be coherent, self-consistent, plausible). The present essay will focus primarily on scientific paradigm as interdisciplinary matrix. While Kuhn, who worked within the natural sciences, was ambivalent about the status of the social sciences, sometimes calling them *preparadigmatic*, he argued that even preparadigmatic sciences were guided by paradigms in the interdisciplinary matrix sense of the term (Kuhn, 1970, p. 179).

Background to Paradigms

Thus, a central thesis of Kuhn's work was the idea that paradigms, including metaphysical beliefs and values, *enter science as necessary feature* of the process. To appreciate the implication of this thesis, especially as it was to impact on psychology, it needs to be placed into an historical context. As one who, at the time, was an undergraduate and then graduate student observer of the changing state of psychology, I saw the introduction of the concept *paradigm* partially through the lens of what came to be called the *Cognitive Revolution*. From 1956 to 1960 as an undergraduate psychology major at Boston University my world of scientific (i.e., experimental) psychology was encapsulated by the logo *S–R* (stimulus–response) or sometimes *S–rg–sg–R* (stimulus–mediating response–mediating stimulus–response). To be scientific in this world, it was required that all the interesting actions of people be reduced to observable movements (responses) and the environmental or biological events (stimuli) that were believed to produce them. At times, *intervening variables* were introduced (e.g., the mediators), but these were necessarily ultimately defined in terms of, and only in terms of, the overt stimuli and responses that were believed to produce them. The prevailing behaviorism of the day was identified with the grand learning theories, especially the work of Clark Hull (1943) and Kenneth Spence (1956) – known as the *Hull*–Spence theory (Spiker, 1970), with B. F. Skinner (1953) as a relatively minor, although increasingly important, contributor. In this setting, cognition or *higher mental processes* (e.g., thinking, reasoning) as it was called in the day consisted of a set of problems to be ignored as nonscientific; problems that would find a naturalized solution as strictly empirical generalizations induced from the laws to be discovered in the field of stimuli and responses.

Although it was *de rigueur* at the time to deny the scientific value of any philosophical concepts, this hegemony of behaviorism was, in fact, supported by an epistemology of a *radical empiricism* (i.e., that knowledge comes through pristine [interpretation free] observations and only pristine observations), an *objectivism* or *scientific realism* (i.e., the assertion that objects of scientific knowledge exist independently of the minds or acts of scientists, and that scientific theories are true of that objective [mind-independent] world; also called a *God's eye view* by Hilary Putnam [1990]), and an ontological *atomism* and *foundationalism* (i.e., that there is an ultimate unchanging bedrock reality) and *physicalism* (i.e., the physical constitutes the ultimate nonreducible really Real).[2] These features of radical empiricism, in turn, formed the base for the philosophical/scientific methodology called *neopositivism*. According to the neopositivist creed two rules defined whether a proposition (e.g., a concept, an hypothesis, a theoretical proposition, a law) was scientifically meaningful: (1) The proposition was accepted as meaningful *if, and only if*, it could be reduced to words whose meaning could be directly observed and pointed to. "The meaning of the word must ultimately be *shown*, it has to be *given*. This takes place through an act of pointing or showing" (Schlick, 1991, p. 40). The words "whose meaning could be directly observed" constituted a *neutral observation language* – completely objective, and free from subjective or mind-dependent interpretation. Thus, all theoretical language required reduction to pristine observations and a neutral observational language (e.g., for *aggression* to be scientifically meaningful it would have to be defined as the response of hitting another person. Or for *intelligence* to be scientifically meaningful it would have to be defined as the score on a test, and nothing more than the score on a test). Because not each and every proposition in the major personality theories of the day was so reducible, these became prime targets as exemplars of the scientifically meaningless. (2) The proposition was acceptable as scientifically meaningful *if, and only if*, it could be shown to be a strictly inductive generalization, drawn directly from the pristine observations. Thus, to be scientifically meaningful, any universal propositions (e.g., hypotheses, theories, laws) had to be demonstrably nothing more than summary statements of the pristine observations. In today's vernacular, it would be said that a theory must be *data based* and only *data based* to be acceptable as scientifically meaningful. When the notion of science as a *Hypothetico-Deductive Method* was introduced by Hull (1943) it was understood that the *hypothetico* was not speculative interpretative hypotheses, but earlier inductively derived empirical generalizations, and the deduction was merely a formal heuristic for moving from empirical generalizations back down to pristine observations to repeat the observation – inductive generalization process.

Obviously, with the dual hegemony of behaviorism and neopositivism there was no room for paradigms in this world of scientific psychology. By 1960 when I entered the Ph.D. program at Clark University, I entered a new world and my introduction to this world was primarily guided by Heinz Werner and Bernie Kaplan (See Werner, 1948, 1957; Werner & Kaplan, 1963). This was a world whose past had represented an opposition to both behaviorism and neopositivism, whose present represented one of a number of newly respectable scientific developmental and cognitive developmental perspectives, and whose future would point the way to how this perspective might be further advanced. If was in this world that I first read Kuhn's *SSR* in 1963 and it was in the context of this world that my colleague Hayne Reese and I published in 1970 what was likely the first major work in psychology, and, quite

[2] Hilary Putnam (1987) makes a useful distinction between the term *real* as used in commonsense discussions, such as this is a real table, chair, book etc., and the *Real* with a capital R in referring to an ontological ultimate reality. This distinction will be used throughout this chapter.

certainly the first in developmental psychology, on the central and necessary role of paradigms and models in psychological and developmental psychological science.[3]

The Cognitive Revolution and Paradigms in Psychology

What had happened in science between 1956 and 1960? In psychology it was the cognitive revolution. Few agree upon a specific date for the revolutionary shift. Some date it to 1956 (Miller, 2003). This was the year that Bruner, Goodenough, and Austin (1956) published their influential *A Study of Thinking,* and the year after Jean founded the International Center for Genetic Epistemology (ICGE) in Geneva, Switzerland with the support of the Rockefeller Foundation (Vauclair & Perret, 2003). Certainly the 1957 publication of *Syntactic Structures* by Chomsky, and his 1959 review of Skinner's (1957) book *Verbal Behavior* were important dates (see Searle, 1972); so too was 1960 when, at Harvard, Bruner created the Center for Cognitive Studies.

Regardless of any specific date, and regardless of any later diversions into *information processing*, and *cognitive science*, we at Clark lived the revolution the way Bruner (1990) later described it

> That revolution was intended to bring "mind" back into the human sciences after a long cold winter of objectivism (p. 1). It was … an all-out effort to establish meaning as the central concept of psychology. – not stimuli and responses, not overtly observable behavior, not biological drives and their transformation, but meaning … It focused upon the symbolic activities that human beings employed in constructing and in making sense not only of the world, but of themselves … Its aim was to prompt psychology to join forces with its sister interpretive disciplines in the humanities and in the social sciences (p. 2). We were not out to 'reform' behaviorism, but to replace it (p. 3).

And the mind that was brought back into psychology was that of the living developing organism understood and described in terms of what today would be called a *relational developmental system's perspective* (Lerner, 2006, 2011; Lerner & Overton, 2008; Overton, 2006, 2010). This organism is a spontaneously active (not moved about by biological or environmental forces), complex, self-creating (auto-poetic), self-organizing, (i.e., operating according to its own principles, intertwined with its biology and environment, which it uses as resources), self-regulating, embodied, adaptive system that functions and develops epigenetically through co-acting with a world of sociocultural objects. At birth the psychological human organism is a relatively undifferentiated *relational–sensorimotor–emotional–motivational action system*. The acts of this system – *expressive, communicative, instrumental* – and the resistances they meet in the physical and sociocultural world, constitute – through complex negative and positive feedback loops – the mechanism of the organism's

[3] This paper Reese and Overton (1970) was based on a presentation to the first developmental life span conference at West Virginia University in 1969. This conference itself represented the beginning of the developmental life span movement in the United States.

development. Development, in turn, proceeds in a nonlinear fashion with increasing differentiation and reintegration of systems. In this way, cognitively, the organism moves recursively and epigenetically from the level of global action system to a level of symbolic acts, and then to a level of symbolically reflective acts, a level of logical symbolic acts, and further later levels of integration. Similar differentiations occur in the relational, emotional, and motivational subsystems as all subsystems function holistically as the relational developmental system.

Does this relational developmental system itself constitute a Kuhnian paradigm? As the outcome of a scientific crisis it meets one criterion. It also meets the criteria of having long functioned in various forms as an alternative to behaviorism. To mention only a few names, there were commitments to this disciplinary matrix going back to William James (1975), and John Dewey (1925), James Mark Baldwin (1895), William Stern (1938), Gordon Allport (1955), the early "new look in perception" by Bruner and Postman (1949), and continuing through the broad theories of Jean Piaget (1952, 1954, 1967), Heinz Werner (1948, 1957), Erik Erikson (1968) and the attachment and object relations theories of John Bowlby (1958), W. D. F. Fairbairn (1952), Harry Stack Sullivan (1953), and Donald Winnicott (1965, 1971).

The Scientific Revolution: From Neopositivism to a Paradigm-Based Relational Methodology

But this psychological vision was not the sum of what ruptured the hegemony of behaviorism and neopositivism. There was also a revolutionary shift in the epistemological and ontological roots of the nature of science, and this shift recursively provided the grounding for the psychological vision just described. Thus, if there was a scientific revolution in psychology, the alternative paradigm consisted of both a broad metatheory (relational developmental system) and new scientific methodology/epistemology (i.e., *relationism*, Latour, 1993, 2004; Overton, 2006, 2010). The main outline and central features of this revolution, beginning from neopositivism, are shown in Fig. 3.1. There has been virtually no question that neopositivism, which reached its zenith in the 1940s and early 1950s, failed. It failed because (1) It became clear, as demonstrated in the work of Quine (1953) and others (e.g., Lakatos, 1978a; Popper, 1959; Putnam, 1983), that rich scientific theories are not reducible to a neutral observational language involving pristine observations. (2) There was a demonstrated inadequacy of induction as the method for arriving at theoretical propositions (Hanson 1958, 1970; Lakatos, 1978a; Popper, 1959). (3) It was recognized that there are theories that warrant the attribution *scientific* despite the fact that they lead to no testable predictions (Putnam, 1983; Toulmin, 1961).

With the failure of neopositivism there was an attempt to maintain the epistemology of empiricism and the ontology of a fixed, mind-independent, (objectivism) bedrock Real. This attempt was termed *conventionalism* or *instrumentalism* and Karl Popper (1959, 1963) was among its most well-known advocates (see Fig. 3.1). Conventionalism maintained with neopositivism the empiricist stance that the way

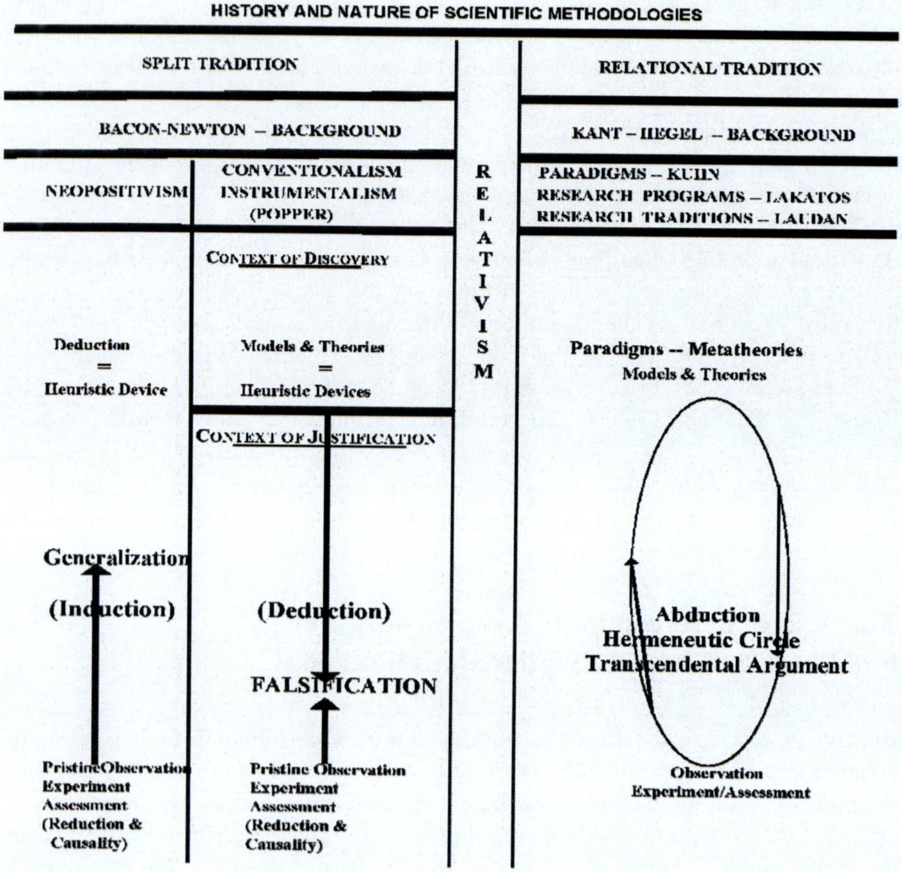

Fig. 3.1 The history and nature of scientific paradigms

to distinguish between science and nonscience is first and foremost for science to continue making the assertion that all scientific knowledge must be based on pristine data and only pristine data. However, conventionalism treated this base of *truth* of scientific realism as an ideal and ultimate goal. Conventionalism also recognized that, contra-neopositivism, not all scientific terms can be reduced to observational statements. As a consequence, conventionalism admitted into the scientific lexicon nonobservable propositions (e.g., models like the computer model in cognitive science and information processing, and other theoretical terms). However, and this is the crux of the matter, these propositions necessarily had to be incased in a separate split-off conceptual realm called the *context of discovery* (see Fig. 3.1), and most significantly they were to function *as and only as* convenient and conventional heuristic devices for ordering and organizing hard data (i.e., pristine observations), and making predictions. Non-observable propositions could not be allowed to influence the database itself and hence, they had no epistemological or cognitive value. Rather, they operated like pigeonholes or coat racks to classify, arrange, and organize hard data into coherent units. Following this formula it continued to be possible to retain empiricism and absolute objectivist truth as norms of science. And in the context of this formula, Popper would argue that science was conjecture and refutation (1963). That is, Popper argued it was scientifically legitimate for the scientist to offer speculative hypotheses drawn from any source, as long as those hypotheses were falsifiable by pristine observations [hard data] (1959, 1963) in the arena called the *context of justification* (see Fig. 3.1). Popper (1959) added a unique dimension to instrumentalism through the claim that theories and models should become acceptable in the body of science, *if and only if*, they specify observational results that, if found, would disprove or falsify a theory. Ultimately, it was claimed, *falsification* would lead to the objective truth of scientific realism.

It was into this world of a dying neopositivism along with its closely related conventionalism that Kuhn's *SSR* entered and participated in the scientific revolution characterized by a new methodology (i.e., the epistemological and ontological rules of how science is done) that is relational in character (see Fig. 3.1), and which forms the grounding context for the theoretical perspective of relational developmental systems in psychology. Recall that a central thesis of Kuhn's *SSR* was the idea that paradigms, including metaphysical beliefs and values, *enter science as necessary feature* of the process. Kuhn's assertion here is that nonobservable propositions do, indeed, influence the *data*. Paradigms are conceptual systems that are a part of the warp and woof of science, and conceptual systems are the constructions of active knowledge constructing organisms engaged in the world. As Kuhn stated, "we must account for … [scientific progress] by examining the nature of the scientific group, discovering what it values, what it tolerates, and what it disclaims. That position is intrinsically sociological and, as such, a major retreat from the canons of explanation licensed by [other] traditions (1970, p. 238)." Understood as conceptual constructions of actively engaged members of the scientific community that are intrinsically entwined in the scientific process, paradigms were vehicles that participated in moving the epistemology of science from that of *empiricism*[4] to *constructivism* and an *ontology* of a *fixed objectivist foundationalism* to one of *activity and engagement*.

Background to a New Scientific Methodology

As with the psychological component, the scientific methodological component of the broad revolution was centered in the 1950s. This component, like the psychological, had a history and this dated back at least to Kant (1781) and his insistence

[4] It should be noted that *empiricism* is a philosophical doctrine and *empiricists* are those committed to this doctrine. One can be committed to *empirical* science (i.e., science involving observational testing of hypotheses whenever possible) without being an *empiricist*. Thus, the new psychology to emerge from the cognitive revolution was an empirical science that did not follow the doctrine of empiricism.

that the activity of mind is a constitutive feature all knowing, followed by Hegel's (1807, 1830) dialectical argument that knower and known constitute a single indissociable complementarity that develops through history. Also central to this history were the contributions of Heidegger (1962) and Merleau-Ponty (1962, 1963) in their phenomenological analysis leading to the conclusion that all knowledge is the product of engaged human agents coping with the world, and Ernst Cassirer (1951) through his neo-Kantian analysis of the cognitive prerequisites of knowing.

Looking to the 1950s and early 1960s the list is long of the characters who participated in this scientific revolution that resulted in a relational methodology in which paradigms (i.e., interdisciplinary matrices) became a necessary feature of science. Some of the most central figures included, the later Wittgenstein who represented analytic philosophy and, whose seminal *Philosophical Investigations* was originally published in 1953, Wittgenstein's was later followed by his student Georg Henrik von Wright (e.g., 1971) as well as Hilary Putnam (e.g., 1983). Representing hermeneutics was Hans-Georg Gadamer, whose *Truth and Method* was first published in 1960, and later Jurgen Habermas (e.g., 1984), Richard Bernstein (e.g., 1983), and Paul Ricoeur (e.g., 1984, 1991). Representing the social sciences, Elizabeth Anscombe's *Intention* was published in 1957, as were William Dray's *Laws and Explanation in History* (1957), and Charles Frankel's *Explanation and Interpretation in History* (1957). These were followed by Peter Winch (1958), whose *The Idea of* a *Social Science and Its Relation to Philosophy* was published in 1958, followed later by Charles Taylor (e.g., 1964). Representing the natural sciences were Steven Toulmin, whose *Philosophy of Science* was published in 1953, and N. R. Hanson, whose *Patterns of Discovery* was published in 1958. It was at this point that Kuhn's *SST* (1962, 1970) entered the picture with his specific concept of *paradigms*, which along with the nature of scientific change itself were later followed and further developed by Imre Lakatos (e.g., 1978a, 1978b), Larry Laudan (e.g., 1977, 1984, 1996), and, most recently, Bruno Latour (e.g., 1993, 2004).

Relational Scientific Methodology

The narrative of the nature and development of the relational methodology is long and complex (see Overton, 2006, 2010). Here I will present an outline by focusing on a few central figures and their contributions. These include Wittgenstein's *Philosophical Investigations* (1958), Gadamer's *Truth and Method* (1960/1989), Hanson's *Patterns of Discovery* (1958), von Wright's *Explanation and Understanding* (1971), and Ricoeur's *Time and Narrative* (1984).

Wittgenstein (1958) and Gadamer (1989) provided the basic scaffolding for the construction of this relational methodology. Wittgenstein's fundamental contribution entailed opening the door to the recognition that it is a profound error to treat the activities of science as providing veridical descriptions of a foundational Real. More positively, Wittgenstein's contribution lies in his suggestion that science is the product of some of the same human actions that underlie the conceptual constructions of our form of life, or *Lebenswelt.* And, in this context Wittgenstein's concept of *language game* was a direct precursor to Kuhn's paradigm. Gadamer's contribution was a systematic demonstration that this move beyond objectivism and foundationalism did not necessitate a slide into absolute relativism (see Fig. 3.1). So feared by those committed to an empiricism and scientific realism. Hanson's (1958) analysis of the history of the physical sciences was significantly influenced by Toulmin (1953) and by the later Wittgenstein (1958), and it provided the necessary prerequisites for Kuhn's introduction of paradigms. On the basis of his analysis of the history of the physical sciences, Hanson drew three powerful conclusions about the actual practice of the physical sciences as distinct from the classical language game of neopositivism in which they were being described. These conclusions themselves provided a blueprint for the new relational methodology. The conclusions were that: (a) there is no absolute demarcation in the physical sciences between interpretative theory and observation or between interpretative theory and facts or data; a notion that was captured in his now-famous aphorism, *all data are theory laden*; (b) scientific explanation consists of the discovery of patterns, as well as the discovery of causes (see also Toulmin, 1953, 1961); and (c) the fundamental logic of science is neither a split-off deductive logic nor a split-off inductive logic, but rather it is an abductive (retroductive) logic.

Interpretation and Observation

Hanson's (1958) first conclusion, that *all data are theory laden*, became the core principle of the new relational methodology and of the very notion of paradigms as disciplinary matrix. The idea here of an indissociable complementarity, a dialectic, between interpretation and observation heals the split between subject and object, body and mind, created by Descartes (1969). And this complementarity destroys any possibility of a foundationalism, objectivism, atomism, or reductionism. It moves science to a constructivist stance in which knower and known merge and to a holistic stance in which the identities of objects and events derive from the relational context in which they are embedded. Wholes define parts and parts define wholes. In this context interpretation and observation, along with other fundamental bipolar concepts, cease being competing alternatives or exclusive dichotomies. Interpretation and observation are relational concepts not names of split-off natural entities. They are relational concepts in the same sense that Hegel's (1807) master–slave dialectic entails relational concepts, where it is impossible to define freedom without reference to constraints or to define constraints without reference to freedom. In the new methodology, interpretation becomes a necessary feature of the total scientific process, not something to be shunned as in neopositivism, nor treated as a peripheral convention in conventionalism. The core challenge here is a determination of exactly how interpretation – this sociological component – enters science. Kuhn's answer, and those who followed on, was to be that interpretation primarily enters as background ideas of a research program (Lakatos 1978a, 1978b) or research tradition (Laudan, 1977). These often remain silently in the background, but they provide the nondissolvable conceptual context and rich source of metaphors for current and future empirical puzzle solving.

Causality and Dynamic Patterns

Hanson's (1958) second conclusion that pattern (Aristotle's formal and final explanation) and cause (Aristotle's efficient and material cause) (Ross, 1959) have always operated jointly as explanations in the physical sciences, subverts the split empiricist narrative of a clear-cut line of demarcation between the natural and social sciences (see also Kuhn, 1977b). Prior to Hanson (see also Toulmin, 1953) the neopositivist and conventionalist narrative had been that *explanation* involved only causal attribution and was limited to the natural sciences. According to empiricists and conventionalists dynamic patterns – sometimes called *principles of intelligibility* in that they make the object of inquiry intelligible and give reasons for the nature and functioning of the object (Randall, 1960; Taylor, 1995) – such as kinship structures or mental structures were held to provide *understanding* and were considered features the social sciences. Hanson's analysis demonstrated that understanding and explanation need not be dichotomous, competing, alternative language games. Dynamic pattern explanation, which can entail intention and reasons, and causal explanation, which entails necessary and sufficient conditions, here become, as they were with Aristotle (Randall, 1960), relational concepts. Explanation then, defined as *intelligible ordering* (Hanson, 1958), becomes the superordinate concept that joins dynamic patterns and cause.

The challenge within the developing relational methodology was to establish a justifiable coordination of the two modes of explanation. von Wright (1971) presented a richly detailed and complex effort in this direction and Ricoeur (1984, 1991) later built upon and expanded this effort. Both focused on this intelligible ordering in the social sciences. Von Wright and Ricoeur each suggested that the coordination be made along the lines of an internal–external dimension. With respect to psychology, for example, *Intemal* here refers to the domain of the psychological person-agent or psychological action system. *Extemal* refers to movements or states. Following a critical distinction made earlier by Anscombe (1957), any given behavior can be considered internal under one description and external under another description. Thus, any specific behavior may be, as von Wright states, "intentionalistically [sic] understood as being an action or otherwise aiming at an achievement, or … as a 'purely natural' event, i.e., in the last resort muscular activity (p. 128)."

Within this framework, causal explanations, understood as Humean causes – defined by the logical independence or contingency relations between cause and effect account for external movements and states. Dynamic pattern explanations (i.e., action, action systems, intention, reason) account for the meaning of an act. Two things should be clear here: First, it is only within this sort of relational definition of explanation that the cognitive revolution in psychology could have possibly been allowed to bring meaning back into scientific psychology. Second, as dynamic patterns are necessarily inferential, their acceptance as a necessary feature of science provides another base for the introduction of a conceptual pattern called a *paradigms*.

Abduction/Transcendental Argument

Hanson's (1958) third conclusion was that neither split-off induction nor split-off deduction constitutes the fundamental logic of science. Each of these enters the operation of science, but Hanson argued that the overarching logic of scientific activity is *abduction*. Abduction (also called *retroduction*) was originally described by the pragmatist philosopher Charles Sanders Pierce (1992). In a contemporary version, this logic is defined as *inference to the best explanation* (Fumerton, 1993; Harman, 1965; Lipton, 2004). A form of abduction termed the *transcendental argument* was introduced by Kant (1781), and recently elaborated by Charles (1995; see also Grayling, 1993; Hundert, 1989).

Abduction or the transcendental argument operates by arranging the observation under consideration and all *background ideas* (i.e., the paradigm and theoretical terms) as a relational complementarity. The coordination of the two is explored by asking what, given the background ideas, must necessarily be assumed in order to have that observation. The inference to, or interpretation of, what must, in the context of background ideas, necessarily be assumed then constitutes the explanation of the phenomenon. This explanation can then be assessed empirically as an hypothesis to ensure its empirical validity (i.e., its empirical support and scope of application). An important relational feature of this logic is that it assumes the form of the *hermeneutic circle* (Gadamer, 1989) by moving from the phenomenological level (the commonsense object) to explanation and back in an ever-widening cycle. The difference between this and the previously described hypothetical-deductive explanation of empiricism is that in abduction, all background ideas (the disciplinary matrix or paradigm and theoretical terms), constitute a necessary feature of the process and the abductive explanations themselves become a part of the ever-widening corpus of background ideas, as shown in Fig. 3.2.

The basic logic of abduction operates as follows: (a) Step 1 entails the description of some highly reliable phenomenological – including background ideas – observation (i.e., 0 is the case); (b) for Step 2, with 0 as the explanandum, an *inference or interpretation* is made to a dynamic pattern explanation (E) resulting in the conditional proposition "If E is the case, then 0 is expected"; (c) Step 3 entails the conclusion that E is indeed the case. Thus, the logical form of the argument is:

1. 0 (phenomenological observation) is the case.
2. If E (action-pattern explanation) is the case, then 0 is expected.
3. Therefore, E is the case.

Abductive inference is illustrated in virtually any psychological work that assumes a centrality of emotional, motivational, or cognitive mental organization (i.e., system). Russell (1996), for example, has discussed the significance of abduction to the area of cognition. Kuczynski & Daly (2003), explore the importance of abduction in generating theory in parent-child relations. Chomsky's (1957) work on language and Piaget's (e.g., 1967) work in cognitive development are particularly rich in abductive inference. Consider as an illustration of the process the following example drawn from Piaget: (1) There is the phenomenal observation (O) that it is the case that a

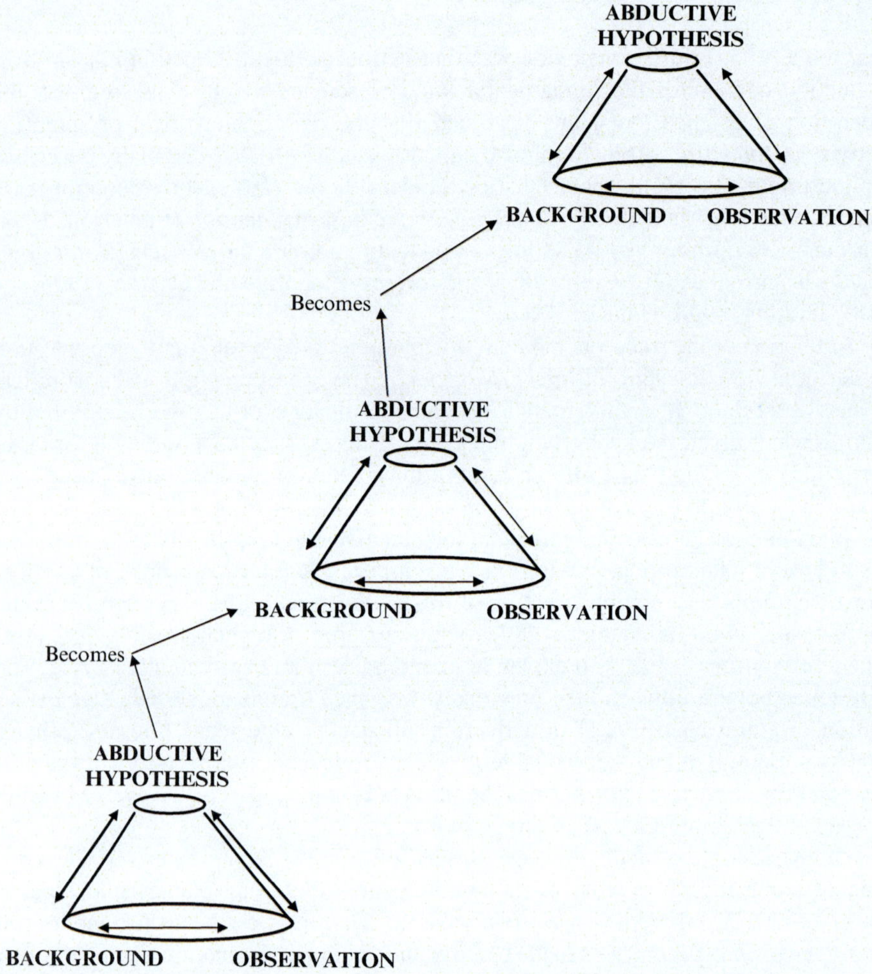

Fig. 3.2 The abductive process

certain group of people (children around 7 years of age) understands that concepts maintain the same quantity despite changes in qualitative appearances (i.e., conservation). (2) Given the relational background ideas that constitute Piaget's disciplinary matrix or paradigm, Piaget makes the abductive inference that the explanation of this observation (E) is that a certain type of action system, having specified features including reversibility (i.e., concrete operations), must be available to these people. This forms the conditional statement "If (E) concrete operational structure, then (O) conservation, is expected." (3) Given (O), the conclusion is, "Therefore, concrete operational structure explains the understanding of conservation."

As Fumerton (1993) has pointed out, it is obvious that, if the conditional in Step 2 is read as *material implication*, the argument would be hopeless because it would then describe the fallacy of the affirmed consequent (i.e., it would be *viciously* circular).

Quite correctly Fumerton recognizes that the "If … then" relation asserts some other sort of connection. Specifically, the connection is one that logicians refer to as *meaning relevance* between E and 0, where relevance is defined in terms of the intelligibility of the relation between E and 0 (Overton, 1990).

It is also the case that there must be criteria established that would allow one to choose among alternative Es, the best E. This is no major hurdle, however, because many of the criteria for theory or explanation selection that have been available under the traditional science language game can, with profit, be used here. These criteria include the scope of the explanation, the explanation's depth, coherence, and logical consistency, the extent to which the explanation reduces the proportion of unsolved to solved conceptual or empirical problems in a domain (Laudan, 1977), and, last but not least, the explanation's empirical support and empirical fruitfulness. Note here that scope, empirical support, and fruitfulness themselves bring the circle back to the observational world and thus keep the cycle open. Dynamic-pattern explanation, entailing the total research program (e.g., relationism, relational developmental systems), in fact, determines what will count as further observations, and the empirical task is to go into the world to discover whether these observations can be found. Thus, the cycle continually moves from commonsense observations and background presuppositions to dynamic-pattern explanations, returning then to more highly refined observations and back again to explanation. Explanation is, indeed, circular but it is not viciously circular.

Abductive logic or the transcendental argument is designed to explain the necessary conditions for an event's occurrence or what have been called *How possible?* questions (von Wright, 1971). Rozeboom (1997) has provided a richly detailed operational analysis of the abductive process along with practical advice on statistical and research strategies associated with the process. Given abductive research answers to the issue of necessary conditions, predictive analyses of individual variability of the dynamic pattern under consideration can then proceed in classical antecedent – consequent terms and their associated statistical models. Thus, again, pattern explanation and causal explanation enter in an indissociable complementarity.

As with Hanson's two earlier conclusions, the analysis of science's abductive character, with its inherently interpretative sociological component of incorporating background ideas into the very center of science, established firmer grounds for Kuhn's notion that paradigms, as disciplinary matrix, enter science as a part of the warp and woof of the scientific process.

In summary to this point, the revolution that was occurring in psychological science and the revolution that was occurring in science broadly became the background or grounding for the emergence of Kuhn's description of the nature of scientific revolutions and the nature and place of paradigms in science. The years that followed were years of critique, consolidation, and elaboration. Critiques were of two kinds, sympathetic and unsympathetic. The unsympathetic critiques generally arose in the context of a broader cultural revolution that was becoming influential at the time; that of skeptical postmodern thought (Overton, 1994a). Postmodern thought pursues the broad aim of exploring the definition of individual freedom. In this quest, the skeptical approach attacks all forms that might offer constraint and hence, limit the potential of absolute freedom, but absolute freedom implies absolute subjectivism and

Table 3.1 Fundamental antinomies – either/or dichotomies

Subject	Object
Mind	Body
Biology	Person
Culture	Biology
Person	Culture
Person	Situation
Intrapsychic	Interpersonal
Nature	Nurture
Stability	Change
Expressive	Instrumental
Variation	Transformation
Reason	Emotion
Form	Matter
Universal	Particular
Transcendent	Immanent
Analysis	Synthesis
Unity	Diversity

absolute relativism (see Table 3.1 on the relativistic abyss that opens when fails to make the leap from empiricism-conventionalism to a relational methodology), and these are intolerable in any empirical science (i.e., organized body of knowledge that whenever possible tests its propositions through empirical observations) (see Overton, 1984). When Kuhn and others, who were transforming the epistemological and ontological base of science, were mistakenly identified with this movement the traditional neopositivists and conventionalists attacked them as being antiscientific relativists (see Fig. 3.1 "Relativism"). One of the sharpest of critics was Popper (1970) who argued that "Kuhn's logic is the logic of *historical relativism* (p. 55)," and who discussed, contra Kuhn, "why I am not a relativist. I do believe in 'absolute' or 'objective' truth (p. 56)." What Popper and other critics (e.g., Shapere, 1964) failed to recognize was that Kuhn's position – while being an attack on the absolute of foundationalism and the notion of objective truth – does not lead to a subjectivism, nor an absolute relativism. For Kuhn (1970, *Postscript – 1969*) a paradigm was the product not of a subjective individual, but of a *scientific community,* and the relativism involved, as Latour (1993, 2004) later discussed, would not be absolute, but a relative relativism termed *relationism*.

Kuhn's more sympathetic critics such as Lakatos (1978a, 1978b) and Laudan (1977) also elaborated and extended the intrinsic role of paradigms in science. Their criticism of Juhn focused most specifically on Kuhn's notion of scientific change through a *Gestalt switch* based on the buildup of empirical anomalies (empirical falsifications). For Lakatos and Laudan, the *Gestalt switch* introduced an irrational component into science and, while they incorporated the basic notion of paradigms into their own research programs (Lakatos) and research traditions (Laudan), they also argued for more rational mechanisms of change.

Returning to my own participation in the exploration of scientific paradigms, following our own first analysis of the role of paradigms and models in science, psychology broadly, and developmental science specifically (Reese & Overton,

1970), Hayne Reese and I continued an elaboration in a series of joint publications (Overton & Reese, 1973 – exploring methodological implications of paradigms and models; Overton & Reese, 1976 – analysis of paradigms in understanding man–environment relations; Overton & Reese, 1981 – on paradigms and understanding psychological issues of stability–change and continuity–discontinuity). These were followed by several single authored and collaborative co-authored publications that continued to analyze the impact of alternative paradigms on psychological theory and research (Overton, 1984 – exploring the impact of Lakatos and Laudan's elaborations and extensions of Kuhn's work; Overton, examining the role of alternative paradigms in the construction of developmental theory; Overton & Horowitz, 1991 – exploring implications of alternative paradigms for theory and research in developmental psychopathology; Overton, 1994b – examining the ways in which paradigms form alternative contexts for the understanding of meaning; Overton, 1994a – exploring a Kuhnian revolution in the understanding of the concepts of change, cognition, and embodiment).

Beginning around 1994 – led by insights of people such as Charles Taylor (1995), Hilary Putnam (1987, 1990), Hans Georg Gadamer (1989, 1993), Bruno Latour (1993), Steven Toulmin (1990), Georg Henrik von Wright (1971), and Paul Ricoeur (1984, 1991), I began to find it useful to incorporate the earlier Kuhn–Lakatos–Laudan perspective into the concept of *metatheory*.

Scientific Paradigm as Metatheory

Metatheories transcend (i.e., *meta*) theories and methods in the sense that they define the context in which theoretical concepts and specific methods are constructed. A *metatheory* is a coherent set of interlocking rules, principles, or story (narrative) that, like the Kuhnian disciplinary matrix, *both describes and prescribes* (i.e., is descriptive and normative) what is acceptable and unacceptable as theoretical concepts and as methodological procedures in a domain of inquiry. The primary function of metatheory is to provide a rich source of concepts out of which theories and methods emerge. Metatheories ground, constrain, and sustain theories and methods. For example, one metatheory may prescribe that no *mental* concepts (e.g., *mind*) may enter theory, and that all change must be understood as strictly additive (i.e., no emergence, no gaps, strict continuity), and hence will be measured by additive statistical techniques. This is a description of some features of early behaviorism. Another metatheory may prescribe that mind is an essential feature of the system under consideration, that the system operates holistically, that novel features emerge, and that nonadditive statistical techniques are a welcome feature of any methodological toolbox. This is a description of some metatheoretical features of what was earlier described as a relational developmental system. Metatheoretical assumptions also serve as guidelines that help to avoid conceptual confusions. Take, for example, the word *stage*. In a metatheory that allows discontinuity of change and emergence, *stage* will be a theoretical concept referring to a particular level of organization of the system; in a metatheory that allows only continuity, if *stage* is

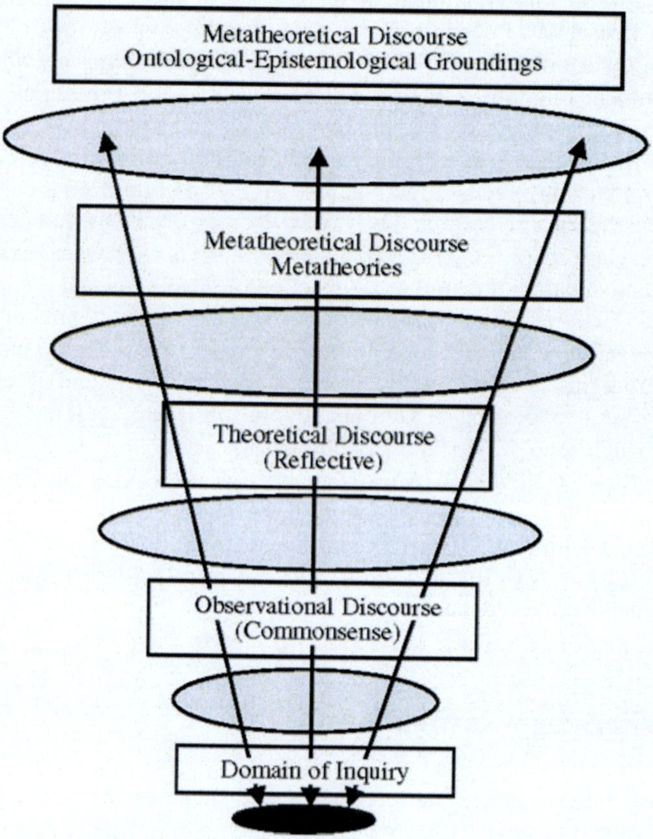

Fig. 3.3 Levels of scientific discourse

used at all – it will be a simple descriptive summary statement of a group of behaviors (e.g., the stage of adolescence), but never as a theoretical concept.

Importantly, metatheories are structured as a hierarchical organization consisting of several levels, defined in terms of increasing scope at each higher level. Each lower level is conceptually nested within the higher levels. To avoid confusions of the type that occurred with Kuhn's early definition of paradigm it is critical to maintain clear distinctions of discourse levels among an observational level, a theoretical level, and several levels of metatheory (see Fig. 3.3). Theories and methods refer directly to the empirical world, whereas metatheories refer to the theories and methods themselves. The most concrete and circumscribed level of discourse is the *observational level.* This is one's current commonsense level of conceptualizing the nature of objects and events in the world. For example, one does not need a professional degree to describe a person as thinking, wishing, desiring, feeling, acting, intending, etc. This observational, commonsense, or folk level of discourse has a sense of immediacy and concreteness, but when reflected on, it is often unclear, muddy, and ambiguous. It is the reflection on folk understanding that moves the level of discourse to a *reflective level,* which is the beginning of theoretical discourse.

Science itself has been defined in terms of the reflective criticism of commonsense observations (Pepper, 1942) leading to an organized body of knowledge. At this level, reflection is about organizing, refining, and reformulating observational understandings in a broader, more coherent, and more abstract field. At the *theoretical reflective level,* concepts are *about* the observational level and these range from informal hunches and empirically testable hypotheses to highly refined theories. Relatively refined theories that may themselves be narrow or broad. My own area of study (Overton & Dick, 2007), which at the observational level concerns the development of human reasoning from childhood to the late adult years, provides an example. Beginning in the 1970s and operating in the context of the generation and observational testing of abductive hypotheses (Fig. 3.2) which emerged from a relational and embodied developmental systems metatheoretical background, my colleagues and I have constructed a developmental theory of reasoning currently termed the *Competence < – – > Procedural Processing* theory (Overton & Ricco, 2010; Ricco and Overton, 2011). In brief outline form, this theory explains the development of logical reasoning in terms of the development of dual systems of processing. Competence refers to the development through action in the world of a *system of mental logic*. Procedures refer to the development through action in the world of a *system of real-time processing* that engages, overrides, facilitates the expression of the competence system. This is a cognitive developmental theory that is broadly compatible with theories of greater scope such as Piaget's and Werner's theories.

The metatheoretical level itself operates above, and functions as a contextual grounding for, the theoretical level. At the metatheoretical level, reflective thought is about basic concepts that, as mentioned earlier, form the contextual frame for the theoretical and observational levels. And here, to make matters a bit more complicated, it is further possible to discriminate levels of metatheory. Consider for example, the following: At the basic level of metatheory *Competence < – – > Procedural Processing theory* is framed by the metatheoretical concepts of *embodied action* (Overton, 2007a). The human organism is conceptualized as an active agent engaged in a world of sociocultural objects, functioning and changing through its intentional actions in this world. This metatheory is itself currently being termed an *enaction paradigm* by a number of scholars exploring implications of the paradigm's concepts for a broad range of psychological phenomena (Stewart, Gapenne, & Di Paolo, 2010). Some *dynamic systems* approaches (Overton 2007b; van Geert, 2003; van Geert & Steenbeck, 2005) are also variants of the embodied action metatheory; these focus on the active agent as active *system*. The same is true for Fischer and colleagues (e.g., Mascolo & Fischer, 2010) dynamic action systems. And similarly, *dialectic transactional* positions (e.g., Kuczynski & Parkin, 2009) represent a variant that explores the active agent in transaction with others. While embodied action, enaction, *dynamic action systems*, and dialectic transaction do, indeed, constitute a *family of metatheories,* they find their family structure in the context of a broader metatheory, which I earlier referred to as a *relational developmental system perspective*. In this context the embodied, enactive, dynamic system, transactional organism is further contextualized as a holistic (relational) complex, self-creating, self-organizing adaptive system. And, finally, relational developmental systems are framed by concepts of *relationism*, a metatheory that operates at the pinnacle of the metatheoretical hierarchy.

Metatheories operating at the pinnacle of this metatheory – theory – observation hierarchy are termed *worldviews* (Overton, 1984). These are metaphysical (i.e., about the nature, origin, and structure of the world) and are composed of coherent and interlocking sets of *epistemological* (i.e., issues of knowledge and knowing) and ontological (i.e., issues about the Real) principles. Because these function as the core principles of any research program (Lakatos, 1978a, 1978b) or research tradition (Laudan, 1977) they point to the future and as the prospective component of this essay, these now become the focus.

Worldviews as Metatheories: Paradigms

The early publications of my colleagues and myself concerned implications of scientific paradigms for theory and research in psychology generally and developmental science specifically, and we focused on three worldviews described by Steven Pepper (1942): The *mechanistic, organismic, and contextualist*. The broad single theme that ran through these writings was that behaviorism, the computational theory of mind, information processing, neopositivism, and conventionalism were formulated in the context of mechanistic metatheoretical concepts while theories such as those of Piaget (1967), Werner (1948), Erikson (1968), Vygotsky (1978), and the newly emerging methodology described by Kuhn, Lakatos, and Laudan were constructed within the organismic paradigm. Contextualism had an unstable ambiguity – as acknowledged by Pepper – sometimes seeming to mix with mechanism and sometimes with organicism (Witherington, 2007).

A change in the focus of the publications began early in the 1990s (Overton & Horowitz, 1991). Primarily through reading object relations theory (e.g., Bowlby, 1958; Fairbairn, 1952; Mitchell, 1988; Sullivan, 1953; Winnicott, 1965, 1971), along with second order cybernetics (Hayles, 1999; Maturana & Poerksen, 2004; Maturana & Varela, 1988), and especially through discussions with my friend and colleague Harvey Horowitz, I began to formulate the idea of broad metaphysical level metatheories that are best characterized along a *split–relational* axis (Overton, 1994a, 1994b), and form the grounding for both the relational developmental systems metatheory in psychology and relational methodology in science broadly.

Split Metatheory

The historical basis of *Split Metatheory* lies in Descartes splitting of mind–body, subject–object, and his invention of the concept *foundationalism* (1969). The metatheory entails several basic defining principles, including *splitting, foundationalism, atomism,* and *objectivism*. *Splitting* is the separation of components of a whole into mutually exclusive pure forms that are taken to describe basic elements. But, in order to split one must accept the principles of *foundationalism, atomism,* and *objectivism*. These metatheoretical propositions refer to the ontological conception that there is ultimately a rock bottom unchanging nature of the Real (the foundation of foundationalism); that this rock bottom is composed of elements – *pure forms* – (the atoms of atomism) that preserve their identity regardless of context; and that objects of knowledge exist independently of the mind of the knower (objectivism). A corollary principle here is the proposition that all complexity is *simple complexity* in the sense that any whole is taken to be a linear additive combination of its elements.

Splitting, foundationalism, and atomism are principles of decomposition; breaking an aggregate down to its smallest pieces, to its bedrock – the objective Real. This process also goes by other names including *reductionism* and *the analytic attitude* (Overton, 2002). Split metatheory requires another principle to reassemble or recompose the whole. This is the principle of *unidirectional and linear (additive) associative or causal sequences.* The elements must be related either according to their contiguity in space and time, or according to simple efficient cause–effect sequences that proceed in a single direction (Bunge, 1962; Overton & Reese, 1973). In fact, split metatheory admits no determination other than individual efficient causes, or these individual causes operating as a conjunctive (i.e., additive) plurality. That is, no truly reciprocal or circular causality is admitted (Bunge, 1962; Overton & Reese, 1973).

All fundamental antinomies in psychology emerge from a split metatheoretical context (see Table 3.1). The individual–social or individual–collective or person–social antinomy, for example, represents all behavior and action as the additive product of elementary bedrock pure forms identified as person and sociocultural. Arising from this splitting, behavior is understood as an aggregate composed of these two pure forms, and the question becomes one of the *primacy or privilege* of one or the other. Among many possible, one brief example of this strategy is found in a sociocultural article by Cole and Wertsch (1996). This article begins by acknowledging – on the basis of several direct quotes by Jean Piaget, a traditional villain of both socioculturalist and social constructivists, who is often inaccurately accused of privileging the person – that Piaget "did not deny the co-equal role of the social world in the construction of knowledge" (p. 251). However, these authors then switch the ground of the issue from the social world specifically to culture mediation entailed by the social world and argue, both in their heading ("The Primacy of Cultural Mediation," p. 251) and in text, that culture is, in fact, to be privileged:

> Social origins take on a special importance in Vygotsky's theories … . For Vygotsky and cultural historical theorists more generally, the *social world does have primacy over the individual* (p. 353, emphasis added).

Similar examples can be found throughout psychology with respect to the various antinomies (see Overton, 2006).

Relational Metatheory – Relationism

Relational metatheory finds its historical origins in Aristotle's insistence that form and matter cannot be separated into two discrete elements, and later in Kant's attempt to reconcile empiricism and rationalism, and Hegel's elaboration of dialectical logic. The aim of relational metatheory is to heal the fundamental antimonies and provide concepts that are inclusive and that more adequately ground science generally and psychology specifically. In an analysis of the historical failures of classical split metatheory, as well as the emptiness of its seeming rival –postmodern thought – Bruno Latour (1993, 2004) proposed a move away from the extremes of Cartesian splits to a center or *middle kingdom* position where entities and ideas are represented, not as pure forms, but as forms that flow across fuzzy boundaries. This is a movement toward what Latour terms *relationism,* a metatheoretical space where foundations are groundings, not bedrocks of certainty, and analysis is about creating categories, not about cutting nature at its joints. Relational metatheory or relationism builds on Latour's proposal.

Relationism is a worldview formed as a principled synthesis of Pepper's (1942) organicism and contextualism (for details see Overton, 2007a; Overton & Ennis, 2006a, 2006b). As a worldview it is composed of a coherent set of intertwined ontological and epistemological principles. The ontology of relationism offers a Real based on *process-substance* rather than a split-off substance (Bickhard 2008). This is the ontology of what Gadamer (1989) argues to be the movement of *to and fro* and what has been sometimes defined as an ontology of *Becoming* (Allport, 1955; Overton, 1991). It includes process, activity, change, emergence, and necessary organization as fundamental defining categories, but it does not exclude categories of substance, stability, fixity, additivity, and contingent organization.

The epistemology of relationism is, first and foremost, a *relatively inclusive* epistemology, involving both knowing and known as equal and indissociable complementary processes in the construction, acquisition, and growth of knowledge. It is *relatively* inclusive, because *inclusion* itself – much like Hegel's master–slave dialectic – can be grasped only in relation to its complement *exclusion.* Thus, just as *freedom* must be identified in the context of *constraint*, *inclusion* must be identified in the context of *exclusion*. Relational epistemology specifically excludes Cartesian dualistic ways of knowing, because Cartesian epistemology trades on absolute exclusivity; it constitutes a *nothing but* epistemology. For the same reason, relationalism rejects both the mechanistic worldview and a strict contextualist interpretation of the contextualist worldview (Overton, 2007a; Witherington (2007).

Epistemologically relationism begins by clearing the *nothing but splitting*, *foundationalism*, *atomism,* and *objectivism* from the field of play and in so doing it moves toward transforming antinomies into co-equal, indissociable complementarities. In the relational frame, fixed elements are replaced by contextually defined parts.

In place of the rejected splitting, foundationalism and atomism, relationism installs *holism* as the overarching epistemological first principle. Building from the base of holism, relational metatheory moves to specific principles that define the relations among parts and the relations of parts to wholes. In other words, relational metatheory articulates principles of analysis and synthesis necessary for any scientific inquiry. These principles are: (1) *The Identity of Opposites*, (2) *The Opposites of Identity*, and (3) *The Synthesis of Wholes*.

Holism

Holism is the principle that the identities of objects and events derive from the relational context in which they are embedded. Wholes define parts and parts define wholes. The classic example is the relation of components of a sentence. Patterns of letters form words and particular organizations of words form sentences. Clearly, the meaning of the sentence depends on its individual words (parts define whole). At the same time, the meaning of the words is often defined by the meaning of the sentence (wholes define parts). Consider the word meanings in the following sentences: (1) The *party leaders* were *split* on the *platform*; (2) The *disc jockey* discovered a *black rock star*; and (3) The *pitcher* was *driven home* on a *sacrifice fly*. The meaning of the sentence is obviously determined by the meaning of the words, but the meaning of each *italicized* word is determined by context of the sentence it is in. Parts determine wholes, and wholes determine their parts (Gilbert & Sarkar, 2000).

Holistically, the whole is not an aggregate of discrete elements but an organized system of parts, each part being defined by its relations to other parts and to the whole. Complexity in this context is *organized complexity* (Luhmann, 1995; von Bertalanffy, 1968a, 1968b), in that the whole is not decomposable into elements arranged in additive linear sequences of cause–effect relations (Overton & Reese, 1973). In the context of holism, principles of splitting, foundationalism, and atomism are, by definition, rejected as meaningless approaches to analysis, and fundamental antimonies are similarly rejected as false dichotomies. In an effort to avoid *standard* (i.e., neopositivistic) misunderstandings here, it must be strongly emphasized that *nondecomposability does not mean that analysis itself is rejected*. It means that *analysis of parts must occur in the context of the parts' functioning in the whole.* The *context-free specifications* of any object, event, or process – whether it be a gene, cell, neuron, the architecture of mind, or culture – is illegitimate within a holistic system.

Although holism is central to relationism, holism does not in itself offer a detailed program for resolving the many dualisms that have framed scientific knowing and knowledge. A complete relational program requires principles according to which the individual identity of each concept of a formerly dichotomous pair is maintained while simultaneously it is affirmed that each concept constitutes, and is constituted by, the other. This understanding is accomplished by considering identity and differences as two *moments of analysis.* The first moment is based on the principle of the *identity* of opposites; the second moment is based on the principle of the *opposites* of identity.

The Identity of Opposites

The principle of the identity of opposites establishes the *identity among parts* of a whole by casting them, not as exclusive contradictions as in the split epistemology but, as differentiated polarities (i.e., coequals) of a unified (i.e., indissociable) inclusive matrix – as a *relation*. As differentiations, each pole is defined recursively; each pole defines and is defined by its opposite. In this identity moment of analysis, the

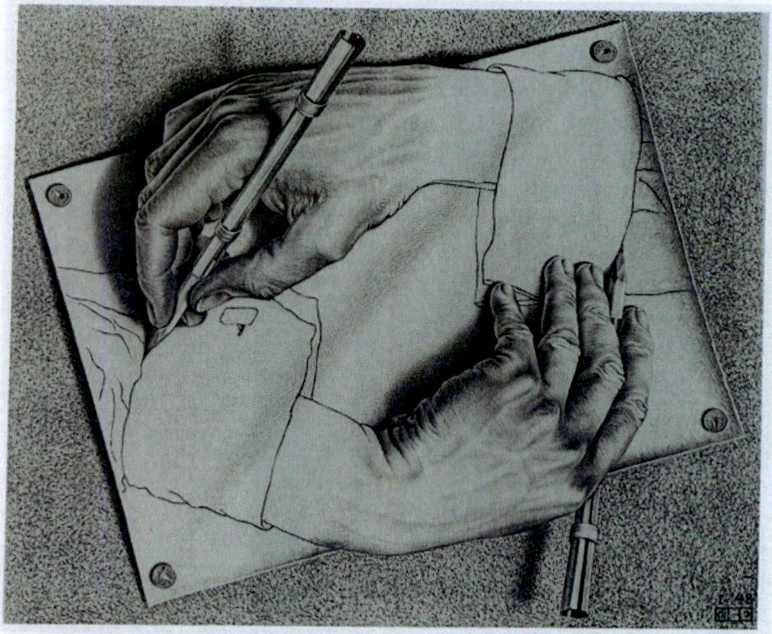

Fig. 3.4 M.C. Escher's "Drawing Hands" © 2009 The M.C. Escher Company-Holland. All rights reserved. www.mcescher.com

law of contradiction is suspended and each category contains and, in fact, *is* its opposite. Further – and centrally – as a differentiation, this moment pertains to character, origin, and outcomes. The character of any contemporary behavior, for example, is 100% nature because it is 100% nurture; 100% biology because it is 100% culture. There is no origin to this behavior that was some other percentage – regardless of whether we climb back into the womb, back into the cell, back into the genome, or back into the DNA – nor can there be a later behavior that will be a different percentage. Similarly, any action is both expressive and communicative/instrumental.

There are a number of ways to illustrate this principle, but a particularly clear illustration is found in considering the famous ink sketch by M. C. Escher titled *Drawing Hands*. As shown in Fig. 3.4, a left and a right hand assume a relational posture according to which each is simultaneously drawing and being drawn by the other. In this matrix, each hand is identical – thus coequal and indissociable – with the other in the sense of each drawing and each being drawn. This is a moment of analysis in which the law of contradiction (i.e., not the case that A = not A) is relaxed and identity (i.e., A = not A) reigns. In this identity moment of analysis, pure forms collapse and categories flow into each other. Here each category contains, and is, its opposite. As a consequence, there is a broad inclusivity established among categories. If we think of *inclusion* and *exclusion* as different moments that occur when we observe a reversible figure (e.g., a Necker cube or the classic vase-women illusion), then in this identity moment we observe only *inclusion*. In the next (opposite) moment of analysis, the figures will reverse, and there we will again see exclusivity as the hands appear as opposites and complementarities. Within the present identity moment of analysis, it is a useful exercise to write on each hand one of the bipolar terms of a traditionally split dualisms (e.g., biology and culture) and to explore the resulting effect. This exercise is more than merely an illustration of a familiar bidirectionality of cause and effects. The exercise makes tangible the central feature of the relational metatheory; seemingly dichotomous ideas that are often thought of as competing alternatives can, in fact, enter into inquiry as coequal and indissociable. It also concretizes the meaning of any truly nonadditive reciprocal determination (Overton & Reese, 1973) and any *circular causality* (see Witherington, 2011) in a way that simple bidirectionality cannot.

If inquiry concerning, for example, person, culture, and behavior is undertaken according to the principle of identity of opposites then various constraints are imposed, as constraints are imposed as they would be by any metatheory. An important example of such a constraint is that behavior, traits, styles, and so forth cannot be thought of as being decomposable into the independent and additive pure forms of biology and culture. Thus, the notion occasionally put forth by some sociocultural or social constructivist approaches, that society and culture occupy a privileged position in scientific explanation, is simply a conceptual confusion in the context of relational metatheory. Similarly, to conceive of the person as the additive (including additive statistical interactions) combination of nature and nurture is a conceptual confusion.

If the principle of the identity of opposites introduces constraints, it also opens possibilities. One of these is the recognition that – to paraphrase Searle (1992) – the fact that a behavior implicates activity of the biological system does not imply that it does not implicate activity of the cultural system, and the fact that the behavior implicates activity of the cultural system does not imply that it does not implicate activity of the biological system. In other words, the identity of opposites establishes the metatheoretical rationale for the theoretical position that biology and culture (like culture and person, biology and person, etc.) operate in a truly *interpenetrating relational* manner.

The justification for the claim that a law of logic (e.g., the law of contradiction) can reasonably both be applied and relaxed depending on the context of inquiry requires a recognition that the laws of logic themselves are not immutable and not immune to background ideas. In some metatheoretical background traditions, the laws of logic are understood as immutable realities given either by a world split off from the human mind or by a prewired mind split-off from the world. However, in the background tradition currently under discussion, the traditional laws of logic are themselves ideas that have been constructed through the reciprocal action of human minds and world. The laws of logic are simply pictures that have been drawn or stories that have been told. They may be good pictures or good stories in the sense of bringing a certain quality of order into our lives, but nevertheless, they are still pictures or stories, and it is possible that other pictures will serve us even better in some circumstances. Wittgenstein (1953/1958), whose later works focused on the importance of background or what we are calling metatheoretical concepts, made this point quite clearly when he discussed another law of logic – the law of the excluded middle – as being one possible *picture* of the world among many possible pictures.

> The law of the excluded middle says here: It must either look like this, or like that. So it really … says nothing at all, but gives us a picture … And this picture *seems* to determine what we have to do and how—but it does not do so. … Here saying "There is no third possibility" … expresses our inability to turn our eyes away from this picture: a picture which looks as if it must already contain both the problem and its solution, while all the time we *feel* that it is not so (para. 352).

The Opposites of Identity

Although the identity of opposites sets constraints and opens possibilities, it does not in itself set a positive agenda for empirical scientific inquiry. The limitation of the identity moment of analysis is that, in establishing a flow of categories of one into the other, a stable base for inquiry that was provided by bedrock *atoms* of the split metatheory is eliminated. In the split approach no relativity entered the picture; all was absolute. Reestablishing a *stable base* – not an absolute fixity, nor an absolute relativity, but a *relative relativity* (Latour, 1993) – within relational metatheory requires moving to a second moment of analysis. This is the oppositional moment, where the figure reverses and the moment becomes dominated by a *relational exclusivity*. Thus, in this opposite moment of analysis, it becomes clear that despite the earlier identity, Escher's sketch does illustrate both a *right* hand and a *left* hand. In this moment, the law of contradiction is reasserted and categories again exclude each other. As a consequence of this exclusion, parts exhibit *unique* identities that differentiate each from the other. These unique differential qualities are stable within any holistic system and, thus, may form relatively stable platforms for empirical inquiry. The platforms created according to the principle of the opposites of identity become *standpoints, points-of-view,* or *lines-of-sight,* in recognition that they do not reflect absolute foundations (Latour, 1993, 2004). They may also be considered under the common rubric *levels of analysis,* when these are not understood as bedrock foundations. Again considering Escher's sketch, when left hand as left hand and right as right are each the focus of attention, it then becomes quite clear that, were they large enough, one could stand on either hand and examine the structures and functions of that hand, as well as its relation to the other hand (i.e., the *co-actions* of parts). Thus, to return to the nature–nurture example, although explicitly recognizing that any behavior is both 100% biology and 100% culture, alternative points of view permit the scientist to analyze the acts of the person from a *biological* or from a *cultural standpoint*. Biology and culture no longer constitute competing alternative explanations; rather, they are two points of view on an object of inquiry that has been created by, and will be fully understood only through, multiple viewpoints. More generally, the unity that constitutes the psychological organism and its development becomes discovered only in the diversity of multiple interrelated lines of sight.

The Synthesis of Wholes

Engaging fundamental bipolar concepts as relatively stable standpoints opens the way, and takes an important first step, toward establishing a broad stable base for empirical inquiry within a relational metatheory. However, this solution is incomplete as it omits a key relational component, the relation of parts to the whole. The oppositional quality of the bipolar pairs reminds us that their contradictory nature still remains, and still requires a resolution. Furthermore, the resolution of this tension cannot be found in the split approach of reduction to a bedrock absolute reality. Rather, the relational approach to a resolution is to move away from the extremes to the center and above the conflict, and to there discover a novel system that will coordinate the two conflicting systems. This is the principle of the *synthesis of wholes*, and this synthesis itself will constitute another standpoint.

At this point, the Escher sketch fails as a graphic representation. Although *Drawing Hands* illustrates the identities and the opposites, and although it shows a middle space between the two, it does not describe a coordination of the two. In fact, the synthesis for this sketch is an unseen hand that has drawn the drawing hands and is being drawn by these hands. The synthesis of interest for the general metatheory would be a system that is a coordination of the most universal bipolarity one can imagine. Arguably, there are several candidates for this level of generality, but the polarity between matter or nature, on the one hand, and society, on the other, is sufficient for present purposes (Latour, 1993).

Matter and society represent systems that stand in an identity of opposites. To say that an object is a social or cultural object in no way denies that it is matter; to say that an object is matter in no way denies that it is social or cultural. And further, the object can be analyzed from either a social-cultural or a physical standpoint. The question for synthesis becomes the question of what system will coordinate these two systems. Arguably, the answer is that it is *life* or living systems that represent the coordination of matter and society. Because our specific focus of inquiry is the psychological subject, we can reframe this matter–society polarity back into a nature–nurture polarity of *biology* (matter) and *culture* (society). In the context of psychology, then, as an illustration, if we again write *biology* on one and *culture* on the other Escher hand, and question what system represents the coordination of these systems, it is life, the human organism, the *person* (Fig. 3.5a). That is, the person is the relational synthesis of biological and sociocultural processes.

At the synthesis, then, a standpoint coordinates and resolves the tension between the other two components of the relation. This provides a particularly broad and stable base for launching empirical inquiry. A *person standpoint* opens the way for the empirical investigation of universal dimensions of psychological structure–function relations (e.g., processes of perception, thought, emotions, values), the particular variations associated with these wholes, their individual differences, and their development across the life span. Because universal and particular are themselves relational concepts, no question can arise here about whether the focus on universal processes excludes the particular; it clearly does not as we already know from the

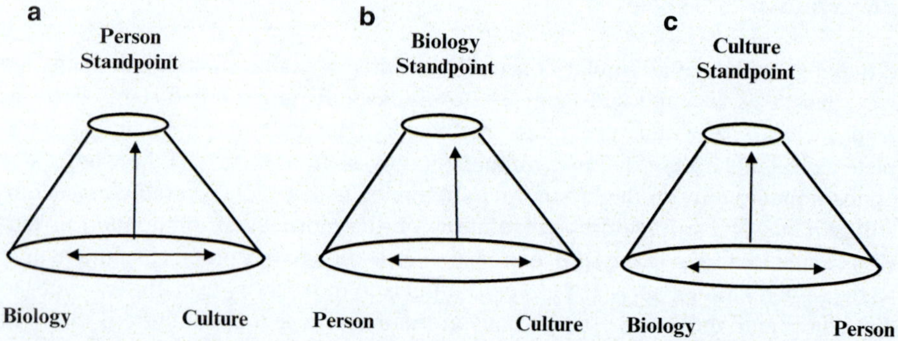

Fig. 3.5 The synthesis of relational standpoints in psychological inquiry: Person, biology, culture

earlier discussion of relations. The fact that a process is viewed from a universal standpoint in no way suggests that it is not situated and contextualized; the fact that it is viewed from an individual standpoint in no way denies its universality.

It is important to recognize that one standpoint of synthesis is relative to other synthesis standpoints. *Life and Society* are coordinated by *Matter*. As a consequence if we are considering the scientific domain termed psychology, *biology* represents a standpoint as the synthesis of person and culture (see Fig. 3.5b). The implication of this is that a relational biological approach to psychological processes investigates the biological conditions and settings of psychological structure–function relations and the actions they express. This exploration is quite different from split foundationalist approaches to biological inquiry that assume an atomistic and reductionistic stance toward the object of study. Neurobiologist Antonio Damasio's (1994, 1999) work on the brain–body basis of a psychological self and emotions is an excellent illustration of this biological relational standpoint. In the context of his standpoint, Damasio (1994) is empathic that:

> A task that faces neuroscientists today is to consider the neurobiology *supporting* adaptive supraregulations [e.g., the psychological subjective experience of self] … I am not attempting to reduce social phenomena to biological phenomena, but rather to discuss the powerful connection between them. … Realizing that there are biological mechanisms behind the most sublime human behavior does not imply a simplistic reduction to the nuts and bolts of neurobiology [emphasis added] (pp. 124–125).

A similar illustration comes from the Nobel laureate neurobiologist Gerald Edelman's (1992; 2006) work on the brain–body base of consciousness:

> I hope to show that the kind of reductionism that doomed the thinkers of the Enlightenment is confuted by evidence that has emerged both from modern neuroscience and from modern physics. … To reduce a theory of an individual's behavior to a theory of molecular interactions is simply silly, a point made clear when one considers how many different levels of physical, biological, and social interactions must be put into place before higher order consciousness emerges (Edelman, 1992, p. 166).

A third synthesis standpoint recognizes that *Person and Matter* are coordinated by *Society*, and again granting that our domain of scientific interest is psychological inquiry about psychological processes, then *culture* or *sociocultural* represents a standpoint as the synthesis of *person* and *biology* (see Fig. 3.5c). Thus, a relational cultural approach to psychological processes explores the cultural conditions and settings of psychological structure–function relations. From this *cultural standpoint,* the focus is on cultural differences in the context of psychological functions as complementary to the person standpoint's focus on psychological functions in the context of cultural differences.

This standpoint is illustrated by *cultural psychology* or *developmentally oriented cultural psychology*. However, not all cultural psychologies are consistent with relational metatheory. When, for example, a cultural psychology makes the social constructivist assertion that social discourse is "prior to and constitutive of the world" (Miller, 1996, p. 99), it becomes clear that this form of cultural psychology has been framed by split foundationalist background ideas. Similarly, when sociocultural claims are made about the "primacy of social forces," or claims arise suggesting that "mediational means" (i.e., instrumental-communicative acts) constitute the *necessary* focus of psychological interest (e.g., see Wertsch, 1991), the shadow of split foundationalist metatheoretical principles is clearly in evidence.

Valsiner (1998) gives one illustration of a relational, developmentally oriented *cultural standpoint* in his examination of the "social nature of human psychology." Focusing on the "social nature" of the person, Valsiner stresses the importance of avoiding the temptation of trying to reduce person processes to social processes. To this end, he explicitly distinguishes between the *dualisms* of split foundationalist metatheory and *dualities* of the relational stance he advocates.

Recently, Mistry and Wu (2010) have offered an explicitly relational sociocultural perspective on how children from diverse cultural backgrounds negotiate across cultures in developing their identities:

> Although the conceptual model was developed based on our interpretation of sociocultural theory, it is highly consistent with contemporary perspectives in developmental psychology that eschew dualisms (Lerner, 2006; Overton, 2006), such as the separation of individual and culture. We suggest that our conceptual model exemplifies a developmentally oriented embodied action metatheory (Overton, 2006) in which embodiment represents the interpenetrating relations between person, biology, and culture (p. 22).

Carpendale and Lewis (2010) further illustrate the relational posture of person and sociocultural points of view in the development of social knowledge (see also Carpendale & Müller, 2004).

When the three points of synthesis – biology, person, and socioculture – are cast as a unity of interpenetrating co-acting parts, there emerges what Greenberg and Partridge (2010) describe as a *biopsychosocial* model. In their tripartite relational developmental systems approach, each part interpenetrates and *co-constructs* the other or *co-evolves* with the other. Development begins from a relatively undifferentiated biosocial action matrix, and through co-constructive interpenetrating co-actions, the biological, the cultural, and the psychological or person part systems emerge, differentiate, and continue their interpenetrating co-construction, moving through levels of increased complexity toward developmental ends. This tripartite model stands in contrast to a *co-constructive biocultural* approach with its clear implication that the psychological system is explained by, driven by, and reducible to the *co-evolution* of two pure forms termed the *biological system* and the *cultural system* (Baltes, Lindenberger, & Staudinger, 2006).

As a final note concerning syntheses and the view from the center, it needs to be recognized that a relational metatheory is not limited to three syntheses. For example, *Discourse* or *Semiotics* may also be taken as a synthesis of *Person* and *Culture* (Latour, 1993). In this case, biology and person are conflated, and the biological/person and culture represents the opposites of identity that are coordinated by discourse.

Conclusions

This chapter has reviewed the history and current status of scientific paradigms. Scientific paradigms are coherent interlocking sets of principles that function in nested hierarchies ranging from narrow relatively concrete models to broad abstract worldviews. Paradigms, which we also refer to as *metatheories*, introduce a sociological dimension into science. They provide concepts that ground, constrain, and sustain scientific theory and methodology, and they are necessary indissociable components of any domain of scientific inquiry. While this judgment was controversial during the middle of the twentieth century, it has now become relatively commonplace. It is found among introductions to the philosophy of science (Godfrey-Smith, 2003), as well as among the discussions of eminent scientists such as Stephen Hawking:

> *There is no picture- or theory-independent concept of reality* … . We will adopt a view that we will call model-dependent realism: The idea that a physical theory or world picture is a model … and a set of rules that connect the model to observations. This provides a framework with which to interpret modern science (Hawking & Mlodinow, 2010, pp. 42–43; emphasis in the original).

Until the 1990s the nature and impact of the mechanistic, organismic, and contextualist worldview metatheories were prevalent in the psychological literature. But during this time, and earlier, there were also voices calling for the introduction of a more relational approach to inquiry. These voices included Wm James and John Dewey, and the early object relations theorists (e.g., Fairbarin, Winnocott). More recently an appeal for a movement towards a relational paradigm has been found across several disciplines including physics (Smolin, 1997; "Twentieth century physics represents a partial triumph of this relational view over the older Newtonian conception of nature[p. 19]."); anthropology (Ingold, 2000; "How can one hope to grasp the continuity of the life process through a mode of thought that can only countenance the organic world already shattered into a myriad of fragments? … What we need, instead, is a quite different way of thinking about organisms and their environments. I call this 'relational thinking' [p. 295]."); biology (Robert, 2004; "To understand the relationship between genotype and phenotype, we must transcend the dichotomy between them…[p. 66]."); social psychology (Good, 2007; "The relational nature of the ecological approach would thus entail a shift away from a focus on just the individual as the object of study and unit of analysis .... The growing interest in cognition as embodied and embedded is seen, nevertheless, as guaranteeing the continuing relevance of the ecological approach, with its emphasis on the reciprocity of perception and action and its relational ontology [p 268]."); and science studies (Latour, 2004; "Their [the sciences] work consists precisely in inventing through the intermediary of instruments and the artifice of the laboratory, the *displacement of point of view* .... They make it possible to shift viewpoint constantly by means of experiments, instruments, models, and theories .... Such is their particular form of relativism – that is, *relationism* [emphasis added] [p. 137].").

Relationism, as a scientific worldview paradigm composed of a coherent set of four interlocking principles (holism, identity of opposites, opposites of identity, synthesis of wholes), is an answer to these appeals. In the psychological and developmental psychological sciences, relationism has become the contextual frame for building contemporary metatheories of a narrower scope including *relational developmental systems*, and the narrower yet, *embodied action* (or *enaction*), *dynamic systems*, *dynamic action systems* and *dialectical transactional* paradigms, each of which represent the organism as a relational spontaneously active, complex, self-creating (autopoetic), self-organizing, self-regulating and embodied, adaptive system that functions and develops epigenetically through co-acting with a world of sociocultural objects. These narrower paradigms, in turn, have generated novel theoretical and methodological contributions (e.g., Bandura, 2006; Lerner, 2011; Marshall, 2009; Mistry, 2011; Müller & Newman, 2008; Overton & Ricco, 2010; Raeff, 2011 on theory; Granic & Hollenstein, 2006; Nesselroade & Molenaar, 2010; Molenaar & Campbell, 2009 on methods) along with significant empirical advances (e.g., Bub, Masson, & Cree, 2008; Chao & Martin, 2000; Demetriou, Mouyi & Spanoudis, 2010; Engel, 2010; Garbarini & Adenzato, 2004; Jackson & Decety, 2004; Kurtines et al., 2008; Lerner, von Eye, Lerner, Lewin-Bizan, & Bowers, 2010; Lewis, 2010; Liben, 2008; Lickliter, 2006; Mascalo & Fischer, 2010; Mistry & Wu, 2010; Mounoud, Duscherer, Moy, & Perraudin, 2007; Ricco & Overton, 2011; Santostefano, 2010; Tucker & Ellis, 2001, 2004). In the light of the productivity of the nested hierarchy of relational paradigms, it seems reasonable to suggest that as paradigms have become recognized as integral features of scientific research traditions, relationism offers a fruitful perspective for the future construction of scientific knowledge, especially the fields of psychological and developmental psychological science.

References

Allport, G. (1955). *Becoming*. New Haven: Yale University Press.

Anscombe, G. E. M. (1957). *Intention*. Oxford: Basil Blackwell.

Baldwin, J. M. (1895). *Mental development in the child and the race: Methods and process*. New York: Macmillan.

Baltes, P. B., Lindenberger, U., & Staudinger, U. M. (2006). Lifespan theory in developmental psychology. In R. M. Lerner & R. M. Lerner (Eds.), *Theoretical models of human development* (Handbook of child psychology 6th ed., Vol. 1, pp. 569–664). Hoboken, NJ: Wiley.

Bandura, A. (2006). Toward a psychology of human agency. *Perspectives on Psychological Science, 1*, 164–180.
Bernstein, R. J. (1983). *Beyond objectivism and relativism: Science, hermeneutics, and praxis*. Philadelphia: University of Pennsylvania Press.
Bickhard, M. H. (2008). Are you social? The ontological and developmental emergence of the person. In U. Mueller, J. I. M. Carpendale, N. Budwig, & B. Sokol (Eds.), *Social life and social knowledge: Toward a process account of development* (pp. 17–42). New York: Taylor & Francis Group.
Bowlby, J. (1958). The nature of the child's tie to his mother. *International Journal of Psychoanalysis, 39*, 350–373.
Bruner, J. (1990). *Acts of meaning*. Cambridge, MA: Harvard University Press.
Bruner, J., Goodnow, J., & Austin, A. (1956). *A study of thinking*. New York: Wiley.
Bruner, J., & Postman, L. (1949). On the perception of incongruity: A paradigm. *Journal of Personality, 18*, 206–223.
Bub, D. N., Masson, M. E. J., & Cree, G. S. (2008). Evocation of functional and volumetric gestural knowledge by objects and words. *Cognition, 106*, 27–58.
Bunge, M. (1962). *Causality: The place of the causal principle in modern science*. New York: The World Publishing.
Carpendale, J. I. M., & Lewis, C. (2010). The development of social understanding: A relational perspective. In W. F. Overton (Ed.), *Cognition, biology, and methods across the lifespan* (Handbook of life-span development, Vol. 1, pp. 584–627). Hoboken, NJ: Wiley.
Carpendale, J. I. M., & Müller, U. (2004). Social interaction and the development of rationality and morality: An introduction. In J. I. M. Carpendale & U. Müller (Eds.), *Social interaction and the development of knowledge* (pp. 1–18). Mahwah, NJ: Lawrence Erlbaum Associates.
Cassirer, E. (1951). *The philosophy of the enlightenment*. Boston, MA: Beacon.
Chao, L. L., & Martin, A. (2000). Representation of manipulable man-made objects in the dorsal stream. *Neuroimage, 12*, 478–484.
Chomsky, N. (1957). *Syntatic Structures*. Netherlands: Mouton.
Chomsky, N. (1959). Review of verbal behavior. *Language, 35*, 26–58.
Cole, M., & Wertsch, J. V. (1996). Beyond the individual-social antinomy in discussions of Piaget and Vygotsky. *Human Development, 39*, 250–256.
Descartes, R. (1969). *The philosophical works of Descartes*. 2 vols (Haldane, E. S. and Ross, G. R. T., Trans.). Cambridge: Cambridge University Press.
Damasio, A. (1994). *Descartes' error: Emotion, reason, and the human brain*. New York: Avon Books, Inc.
Damasio, A. (1999). *The feeling of what happens: Body and emotion in the making of consciousness*. New York: Harcourt Brace.
Demetriou, A., Mouyi, A., & Spanoudis, G. (2010). The development of mental processing. In W. F. Overton (Ed.), *Cognition, biology, and methods across the lifespan* (Handbook of life-span development, Vol. 1, pp. 306–345). Hoboken, NJ: Wiley.
Dewey, J. (1925). *Experience and nature*. La Salle, IL: Open Court.
Dray, W. H. (1957). *Laws and explanation in history*. Oxford University Press.
Edelman, G. M. (1992). *Bright air, brilliant fire: On the matter of the mind*. New York: Basic Books.
Edelman, G. M. (2006). *Second nature: Brain science and human knowledge*. New Haven: Yale University Press.
Engel, A. K. (2010). Directive minds: How dynamics shapes cognition. In J. Stewart, O. Gapenne, & E. A. Di Paolo (Eds.), *Enaction: Toward a new paradigm in for cognitive science* (pp. 219–243). Cambridge, MA: MIT.
Erikson, E. H. (1968). *Identity youth and crisis*. New York: W. W. Norton & Company.
Fairbairn, W. R. D. (1952). *An object-relations theory of the personality*. New York: Basic Books.
Frankel, C. (1957). Explanation and interpretation in history. *Philosophy of science, 24*, 137–155.
Fumerton, R. (1993). Inference to the best explanation. In J. Dancy & E. Sosa (Eds.), *A companion to epistemology* (pp. 207–209). Cambridge, MA: Basil Blackwell Inc.

Gadamer, H. G. (1960/1989). *Truth and method* (Weinsheimer, J. Marshall, D., Trans.). (2nd Revised ed.). New York: The Crossroad.

Gadamer, H. G. (1993). *Reason in the age of science*. Cambridge, MA: The MIT Press.

Garbarini, F., & Adenzato, M. (2004). At the root of embodied cognition: Cognitive science meets neurophysiology. *Brain and Cognition, 56*, 100–106.

Gilbert, S. F. & Sarkar S. (2000). Embracing complexity: Organicism for the 21st century. *Developmental Dynamics*. 219. 1–9

Godfrey-Smith, P. (2003). *Theory and reality: An introduction to the philosophy of science*. Chicago: The University of Chicago Press.

Good, J. M. M. (2007). The affordances for social psychology of the ecological approach to social knowing. *Theory and Psychology, 17*, 265–295.

Granic, I., & Hollenstein, T. (2006). A survey of dynamic systems methods for developmental psychopathology. In D. Cicchetti & D. J. Cohen (Eds.), *Developmental psychopathology* (2nd ed., Vol. 1, pp. 889–929). New York: Wiley.

Grayling, A. C. (1993). Transcendental arguments. In J. Dancy & E. Sosa (Eds.), *A companion to epistemology* (pp. 506–509). Cambridge, MA: Basil Blackwell Inc.

Greenberg, G., & Partridge, T. (2010). Biology, evolution, and psychological development. In W. F. Overton (Ed.), *Cognition, biology, and methods across the lifespan* (Handbook of life-span development, Vol. 1, pp. 115–148). Hoboken, NJ: Wiley.

Habermas, J. (1984). The theory of communicative action. *Reason and the rationalization of society*. (translated by Thomas McCarthy) (Vol 1). Boston: Beacon

Hanson, N. R. (1958). *Patterns of discovery*. London and New York: Cambridge University Press.

Hanson, N. R. (1970). Is there a logic of scientific discovery? In B. A. Brody (Ed.), *Readings in the philosophy of science* (pp. 620–633). Englewood Cliffs, NJ: Prentice-Hall, Inc.

Harman, G. H. (1965). Inference to the best explanation. *Philosophical Review, 74*, 88–95.

Hawking, S., & Mlodinow, L. (2010). *The grand design*. New York: Bantham Books.

Hayles, N. K. (1999). *How we became posthuman: Virtual bodies in cybernetics, literature, and informatics*. Chicago: The University of Chicago Press.

Hegel, G. W. F. (1807). *Phenomenology of spirit*. New York: Oxford University Press. (trans. A. V. Miller).

Hegel, G. W. F. (1830). *Hegel's logic: Being part one of the encyclopedia of the philosophical sciences*. New York: Oxford University Press. (trans. W. Wallace).

Heidegger, M. (1962). *Being and time*. New York: Harper and Row. (translated by J. Macquarrie and E. Robinson)

Hull, C. L. (1943). *Principles of behavior*. New York: Appleton-Century.

Hundert, E. M. (1989). *Philosophy, psychiatry and neuroscience: Three approaches to the mind*. New York: Oxford University Press.

Ingold, T. (2000). Evolving skills. In H. Rose & S. Rose (Eds.), *Alas, poor Darwin: Arguments against evolutionary psychology* (pp. 273–297). New York: Harmony Books.

Jackson, P. L., & Decety, J. (2004). Motor cognition: A new paradigm to study self–other interactions. *Current Opinion in Neurobiology, 14*, 259–263.

James, W. (1975). *Pragmatism and the meaning of truth*. Cambridge, MA: Harvard University Press.

Kant, I. (1781). *Critique of pure reason* (trans. F. Max Muller). New York: Anchor Books edition, 1966.

Kuczynski, L., & Daly, K. (2003). Qualitative methods as inductive (theory generating) research: psychological and sociological approaches (pp. 373–392). In L. Kuczynski (Ed.), *Handbook of Dynamics in Parent-Child Relations*. Thousand Oaks, CA: Sage.

Kuczynski, L., & Parkin, C. M. (2009). Pursuing a dialectical perspective on transaction: A social relational theory of micro family processes. In A. Sameroff (Ed.), *The transactional model of development: How children and contexts shape each other*. Washington, DC: American Psychological Association.

Kuhn, T. S. (1962). *The structure of scientific revolutions*. Chicago, IL: University of Chicago Press.

Kuhn, T. S. (1970). *The structure of scientific revolutions* (2nd ed.). Chicago, IL: University of Chicago Press.

Kuhn, T. S. (1977a). *The essential tension: Selected studies in the scientific tradition and change.* Chicago, IL: University of Chicago Press.

Kuhn, T. S. (1977b). Concepts of cause in the development of physics. In T. S. Kuhn (Ed.), *The essential tension: Selected studies in the scientific tradition and change* (pp. 21–30). Chicago, IL: University of Chicago Press.

Kurtines, W. M., Ferrer-Wreder, L., Berman, S. L., Lorente, C. C., Silverman, W. K., & Montgomery, M. J. (2008). Promoting positive youth development: New directions in developmental theory, methods, and research. *Journal of Adolescent Research, 23*, (Whole Issue), PP. 233–378.

Lakatos, I. (1978a). *The methodology of scientific research programmes: Philosophical papers* (Vol. 1). New York: Cambridge University Press.

Lakatos, I. (1978b). *Mathematics, science and epistemology: Philosophical papers* (Vol. 2). New York: Cambridge University Press.

Lakatos, I., & Musgrave, A. (Eds.). (1970). *Criticism and the growth of knowledge.* New York: Cambridge university press.

Laudan, L. (1977). *Progress and its problems: Towards a theory of scientific growth.* Berkeley: University of California Press.

Laudan, L. (1984). *Science and values: The aims of science and their role in scientific debate.* Los Angeles: The University of California Press.

Laudan, L. (1996). *Beyond positivism and relativism: Theory, method, and evidence.* Boulder, CO: Westview.

Latour, B. (1993). *We have never been modern.* Cambridge, MA: Harvard University Press.

Latour, B. (2004). *Politics of nature.* Cambridge, MS: Harvard University Press.

Lerner, R. M. (2006). Developmental science, developmental systems, and contemporary theories of human development. In R. M. Lerner, W. Damon, & R. M. Lerner (Eds.), *Handbook of child psychology* (Theoretical models of human development 6th ed., Vol. 1, pp. 1–17). Hoboken, NJ: Wiley.

Lerner, R. M. (2011). Structure and process in relational developmental systems theories: A commentary on contemporary changes in the understanding of developmental change across the life-span. *Human Development, 54*, 34–43.

Lerner, R. M., & Overton, W. F. (2008). Exemplifying the integrations of the relational developmental system: Synthesizing theory, research, and application to promote positive development and social justice. *Journal of Adolescent Research, 23*, 245–255.

Lerner, R. M., von Eye, A., Lerner, J. V., Lewin-Bizan, S., & Bowers, E. P. (2010). Special issue introduction: The meaning and measurement of thriving: A view of the issues. *Journal of Youth and Adolescence, 39*(7), 707–719.

Lewis, M. (2010). The emergence of consciousness and its role in human development. In W. F. Overton (Ed.), *Cognition, biology, and methods across the lifespan* (Handbook of life-span development, Vol. 1, pp. 628–670). Hoboken, NJ: Wiley.

Liben, L. (2008). Embodiment and children's understanding of the real and represented world. In W. F. Overton, U. Müller, & J. L. Newman (Eds.), *Developmental perspectives on embodiment and consciousness* (pp. 191–224). New York, NY: Taylor & Francis.

Lickliter, R. (2006). Developmental systems and psychological science. In B. D. Midgley & E. K. Morris (Eds.), *Modern Perspectives on J. R. Kantor and Interbehaviorism* (pp. 173–194). Reno, NV: Context Press.

Lipton, P. (2004). *Inference to the best explanation* (2nd ed.). New York: Routledge.

Luhmann, N. (1995). *Social systems.* Stanford, CA: Stanford University Press.

Marshall, P. J. (2009). Relating psychology and neuroscience. *Perspectives on psychological science, 4*, 113–125.

Mascalo, M. F., & Fischer, K. W. (2010). The dynamic development of thinking, feeling, and acting over the life span. In W. F. Overton (Ed.), *Cognition, biology, and methods across the lifespan* (Handbook of life-span development, Vol. 1, pp. 149–194). Hoboken, NJ: Wiley.

Masterman, M. (1970). The nature of a paradigm. In I. Lakatos & A. Musgrave, A. (eds.). *Criticism and the growth of knowledge (pp 59–90)*. New York: Cambridge university press.
Maturana, H. R., & Poerksen, B. (2004). *From being to doing: The origins of the biology of cognition*. Heidelberg: Carl-Auer.
Maturana, H. R. & Varela (1988). *The tree of knowledge: The biological roots of human understanding*. Boston: New Science Library, Shambhala.
Merleau-Ponty, M. (1962). *Phenomenology of perception (translation by Colin Smith)*. London: Routledge and Kegan Paul.
Merleau-Ponty, M. (1963). *The structure of behavior*. (translated by Alden Fisher). Boston: Beacon.
Miller, J. G. (1996). Theoretical issues in cultural psychology. In J. W. Berry, Y. H. Poortinga, & J. Pandey (Eds.), *Handbook of cross-cultural psychology: Theory and method* (pp. 85–128). Boston: Allyn and Bacon.
Miller, G. (2003). The Cognitive Revolution: A historical perspective. *Trends in Cognitive Science, 7*, 141–144.
Mistry, J. (2011). Integrative theoretical perspectives: Nature and process of development. *Human Development, 54*, 44–48.
Mistry, J., & Wu, J. (2010). Navigating cultural worlds and negotiating identities: A conceptual model. *Human Development, 53*, 5–25.
Mitchell, S. A. (1988). *Relational concepts in psychoanalysis: An integration*. Cambridge, MA: Harvard University Press.
Molenaar, P. C. M., & Campbell, C. G. (2009). The new person-specific paradigm in psychology. *Current directions in psychological science, 18*(2), 112–117.
Mounoud, P., Duscherer, K., Moy, G., & Perraudin, S. (2007). The influence of action perception on object recognition: A developmental study. *Developmental Science, 10*, 836–852.
Müller, U., & Newman, J. L. (2008). The body in action: Perspectives on embodiment and development. In W. F. Overton, U. Müller, & J. L. Newman (Eds.), *Developmental perspectives on embodiment and consciousness* (pp. 313–342). New York, NY: Taylor & Francis.
Nesselroade, J. R., & Molenaar, P. C. M. (2010). Emphasizing intraindividual variability in the study of development over the life span. In W. F. Overton (Ed.), *Cognition, biology, and methods across the lifespan* (Handbook of life-span development, Vol. 1, pp. 30–54). Hoboken, NJ: Wiley.
Overton, W. F. (1984). World views and their influence on psychological theory and research: Kuhn-Lakatos-Laudan. In H. W. Reese (Ed.), *Advances in child development and behavior* (pp, Vol. 18, pp. 191–226). New York: Academic Press.
Overton, W. F. (1990). Competence and procedures: Constraints on the development of logical reasoning. In W. F. Overton (Ed.), *Reasoning, necessity, and logic: Developmental perspectives* (pp. 1–32). Hillsdale, NJ: Erlbaum.
Overton, W. F. (1991). The structure of developmental theory. In H. W. Reese (Ed.), *Advances in child development and behavior* ((pp, Vol. 23, pp. 1–37). New York: Academic Press.
Overton, W. F. (1994a). The arrow of time and cycles of time: Concepts of change, cognition, and embodiment. *Psychological Inquiry, 5*, 215–237.
Overton, W. F. (1994b). Contexts of meaning: The computational and the embodied mind. In W. F. Overton & D. S. Palermo (Eds.), *The nature and ontogenesis of meaning* (pp. 1–18). Hillsdale, NJ: Erlbaum.
Overton, W. F. (2002). Understanding, explanation, and reductionism: Finding a cure for cartesian anxiety. In L. Smith & T. Brown (Eds.), *Reductionism* (pp. 29–51). Mahwah, NJ: Lawrence Erlbaum Associates.
Overton, W. F. (2006). Developmental psychology: Philosophy, concepts, and methodology. In R. M. Lerner (Ed.), *Theoretical models of human development* (Handbook of child psychology 6th ed., Vol. 1, pp. 18–88). New York: Wiley.
Overton, W. F. (2007a). A coherent metatheory for dynamic systems: Relational Organicism-Contextualism. *Human Development, 50*, 154–159.

Overton, W. F. (2007b). Embodiment from a relational perspective. In W. F. Overton, U. Mueller & J. L. Newman (Eds.), (pp. 1–18). *Developmental Perspective on Embodiment and Consciousness*. Hillsdale, NJ: Erbaum Associates

Overton, W. F. (2010). Life-span development: Concepts and issues. In W. F. Overton (Ed.), *Cognition, biology, and methods across the lifespan* (Handbook of life-span development, Vol. 1, pp. 1–29). Hoboken, NJ: Wiley.

Overton, W. F., & Dick, A. S. (2007). A competence-procedural and developmental approach to logical reasoning. In M. J. Roberts (Ed.) *Integrating the mind: Domain general versus domain specific processes in higher cognition*. New York: Psychology Press.

Overton, W. F., & Ennis, M. (2006a). Cognitive-developmental and behavior-analytic theories: Evolving into complementarity. *Human Development, 49*, 143–172.

Overton, W. F., & Ennis, M. (2006b). Relationism, ontology, and other concerns. *Human Development, 49*, 180–183.

Overton, W. F., & Horowitz, H. (1991). Developmental psychopathology: Differentiations and integrations. In D. Cicchetti & S. Toth (Eds.), *Rochester symposium on developmental psychopathology* (Vol. 3, pp. 1–41). Rochester, NY: University of Rochester Press.

Overton, W. F., & Reese, H. W. (1973). Models of development: Methodological implications. In J. R. Nesselroade & H. W. Reese (Eds.), *Life-span developmental psychology; Methodological issues* (pp. 65–86). New York: Academic Press.

Overton, W. F., & Reese, H. W. (1976). General models for man-environment relations. In H. McGurk (Ed.), *Ecological factors in human development*. The Netherlands: North Holland.

Overton, W. F., & Reese, H. W. (1981). Conceptual prerequisites for an understanding of stability-change and continuity-discontinuity. *International Journal of Behavioral Development, 4*, 99–123.

Overton, W. F., & Ricco, R. B. (2010). Dual-systems and the development of reasoning: Competence-procedural systems. In *Wiley interdisciplinary reviews: Cognitive science*. New York: Wiley.

Piaget, J. (1952). *The origins of intelligence in children*. New York: W. W. Norton & Company.

Piaget, J. (1954). *The construction of reality in the child*. New York: Basic Books, Inc.

Piaget, J. (1967). *Six psychological studies*. New York: Random House.

Peirce, C. S. (1992). *Reasoning and the logic of things: The Cambridge conference lectures of 1898*. Cambridge, MA: Harvard University Press.

Pepper, S. (1942). *World Hypotheses*. Los Angeles: University of California Press.

Popper, K. (1959). *The logic of scientific discovery*. London: Hutchinson.

Popper, K. (1963). *Conjectures and Refutations*. London: Routledge & Kegan Paul.

Popper, K. (1970). Normal science and its dangers. In I. Lakatos & A. Musgrave (Eds.), *Criticism and the growth of knowledge* (pp. 51–58). Cambridge university press: New York.

Putnam, H. (1983). *Realism and reason: Philosophical papers* (Vol. 3). New York: Cambridge University Press.

Putnam, H. (1987). *The many faces of realism*. Cambridge, England: Cambridge University Press.

Putnam, H. (1990). *Realism with a human face*. Cambridge, MA: Harvard University Press.

Quine, W. V. (1953). *From a logical point of view*. Cambridge, MA: Harvard University Press.

Raeff, C. (2011). Distinguishing between development and change. *Human Development, 54*, 4–33.

Reese, H. W., & Overton, W. F. (1970). Models of development and theories of development. In L. R. Goulet & P. B. Baltes (Eds.), *Life-span developmental psychology: Research and theory* (pp. 115–145). New York: Academic Press.

Randall, J. H. (1960). *Aristotle*. New York: Columbia University Press.

Ricco, R. & Overton W. F. (2011). Dual systems competence ←→ Procedural processing: A relational developmental systems approach to reasoning. *Developmental Review*. DOI: 10.1016/j.dr.2011.07.005.

Ricoeur, P. (1984). *Time and narrative* (Vol. 1. (translated by K McLalughlin and D. Pellauer)). Chicago: The University of Chicago Press.

Ricoeur, P. (1991). *From text to action: Essays in hermeneutics, II* (translated by K. Blamey & J. B. Thompson). Evanston, Ill: Northwestern University Press.

Robert, J. S. (2004). *Embryology, epigenesis, and evolution: Taking development seriously*. Cambridge: Cambridge University Press.
Ross, W. D. (1959). *Aristotle*. Cleveland: The World Publishing Company: A Meridian Book.
Rozeboom, W. W. (1997). Good science is abductive, not hypothetico-deductive. In L. L. Harlow, S. A. Mulaik, & J. H. Steiger (Eds.), *What if there were no significance tests?* (pp. 335–391). Mahwah, NJ: Lawrence Erlbaum.
Russell, J. (1996). *Agency: Its role in mental development*. Erlbaum: Taylor and Francis.
Santostefano, S. (2010). Developmental psychopathology – Self, embodiment, meaning: A holistic-systems perspective. In W. F. Overton (Ed.), *Cognition, biology, and methods across the lifespan* (Handbook of life-span development 2010th ed., Vol. 1, pp. 792–836). Hoboken, NJ: Wiley.
Schlick, M. (1991). Positivism and realism. In R. Boyd, P. Gasper, & J. D. Trout (Eds.), *The philosophy of science* (pp. 37–55). Cambridge, MA: The MIT Press.
Searle, J. R. (1972). Chomsky's revolution in Linguistics. *The New York Review of Books* (June 29).
Searle, J. (1992). *The rediscovery of the mind*. Cambridge, MA: The MIT Press.
Shapere, D. (1964). The structure of scientific revolutions. *Philosophical Review, 73*, 383–394.
Skinner, B. F. (1953). *Science and human behavior*. New York: Macmillan.
Skinner, B. F. (1957). *Verbal behavior*. East Norwalk, CT: Appleton-Century-Crofts.
Smolin, L. (1997). *The life of the cosmos*. New York: Oxford University Press.
Spence, K. W. (1956). *1956*. Behavior theory and conditioning. New Haven: Yale University Press.
Spiker, C. C. (1970). An extension of Hull-Spence discrimination learning theory. *Psychological Review, 77*, 496–515.
Stern, W. (1938). *General psychology: From the personalistic standpoint*. New York: The Macmillan Company.
Stewart, J., Gapenne, O., & Di Paolo, E. A. (2010) (Eds.), *Enaction: Toward a new paradigm in for cognitive science*. Cambridge, MA: The MIT Press.
Sullivan, H. S. (1953). *The interpersonal theory of psychiatry*. New York: W.W. Norton.
Taylor, C. (1964). *The explanation of behavior*. New York: The Humanities Press.
Taylor, C. (1995). *Philosophical arguments*. Cambridge, MA: Harvard University Press.
Toulmin, S. (1953). *The philosophy of science*. New York: Harper & Row.
Toulmin, S. (1961). *Foresight and understanding*. New York: Harper & Row.
Toulmin, S. (1990). *Cosmopolis: The hidden agenda of modernity*. Chicago: The University of Chicago Press.
Tucker, M., & Ellis, R. (2001). The potentiation of grasp types during visual object categorization. *Visual Cognition, 8*, 769–800.
Tucker, M., & Ellis, R. (2004). Action priming by briefly presented objects. *Acta Psychologica, 116*, 185–203.
Valsiner, J. (1998). *The guided mind: A sociogenetic approach to personality*. Cambridge, MA: Harvard University Press.
van Geert, P. (2003). Dynamic systems approaches and modeling of developmental proceses. In J. Valsiner & K. J. Connolly (Eds.), *Handbook of developmental psychology* (pp. 640–672). London: Sage.
van Geert, P., & Steenbeck, H. (2005). Explaining after by before: Basic aspects of a dynamic systems approach to the study of development. *Developmental Review, 25*, 408–442.
Vauclair, J., & Perret, P. (2003). The cognitive revolution in Europe: Taking the developmental perspective seriously. *Trends in Cognitive Science, 7*, 284–285.
von Bertalanffy, L. (1968a). *General system theory*. New York: George Braziller.
von Bertalanffy, L. (1968b). *Organismic psychology and systems theory*. Barre, MA.: Barre.
von Wright, G. H. (1971). *Explanation and understanding*. Ithaca, NY: Cornell University Press.
Vygotsky, L. S. (1978). *Mind in society: The development of higher psychological processes*. Cambridge, MA: Harvard University Press.
Werner, H. (1948). *Comparative psychology of mental development*. New York: International Universities Press. (Originally published 1940).

Werner, H. (1957). The concept of development from a comparative and organismic point of view. In D. B. Harris (Ed.), *The concept of development: An issue in the study of human behavior* (pp. 125–148). Minneapolis: University of Minnesota Press.
Werner, H., & Kaplan, B. (1963). *Symbol formation*. New York: Wiley.
Wertsch, J. V. (1991). *Voices of the mind: A sociocultural approach to mediated action*. Cambridge, MA: Harvard University Press.
Winch, P. (1958). *The idea of a social science and its relation to philosophy*. London: Routledge and Kegan Paul.
Winnicott, D. W. (1965). *The maturational process and the facilitating environment*. New York: International Universities Press.
Winnicott, D. W. (1971). *Playing and reality*. New York: Basic Books.
Witherington, D. C. (2007). The dynamic systems approach as metatheory for developmental psychology. *Human Development, 50*, 127–153.
Witherington, D. C. (2011). Taking emergence seriously: The centrality of circular causality for Dynamic Systems approaches to development. *Human Development, 54*, 66–92.
Wittgenstein, L. (1958/1953). *Philosophical investigations* (Anscombe, G. E. M., Trans.). (3rd ed.). Englewood Cliffs, NJ: Prentice Hall.

Part II
Paradigms in the Arts and Social Sciences

Chapter 4
Anthropology

Luciano L'Abate

I am not an anthropologist in any way, shape, form, or fashion. This is why I could not muster enough influence and traction to recruit or seduce a real, qualified, honest-to-goodness anthropologist to collaborate with an appropriate chapter in this volume. However, I did not think that this volume would be complete without a chapter in this discipline and other disciplines originally overlooked, such as economics (Chap. 6). Consequently, I had to make do, with the help of Karen Viars, to find whatever I could by cross-indexing paradigms within anthropological paradigms and theories.

Then, I tried to make sense of whatever information I gathered. I hope readers will forgive me for trying not to leave any stone unturned in editing this volume and adding a chapter in a discipline completely unknown to me. If real, honest-to-goodness anthropologists will find that this chapter is the work of an ignorant beginner, they will be completely correct. This ignorance is readily visible in my inability to distinguish between mainstream anthropology from second or third rate tangential, nonmainstream publications. Furthermore, there is no way this short chapter could review and include the vast literature on the role of paradigms in theory construction in anthropology. This chapter should be considered as a brief introduction, albeit incomplete and, therefore inadequate, of representative studies within the self-allotted 20 pages typed single space, rather than an overall review about the role of paradigms in anthropology (Moore & Mathews, 2001).

Anthropology, just like any other social science discipline, is composed of various specializations, as would be found in complex organizations. Wulf (2004), for instance, presented the paradigm of the "historical and cultural anthropology of education," providing an overview of its principles and major research perspectives. The author distinguished his concept of historical and cultural anthropology from other previous approaches (philosophical, historical, and cultural anthropologies).

L. L'Abate (✉)
Department of Psychology, Georgia State University, University Plaza, Atlanta, GA 30303, USA
e-mail: llabate3@bellsouth.net

L. L'Abate (ed.), *Paradigms in Theory Construction*, DOI 10.1007/978-1-4614-0914-4_4,

He gave an example of this anthropological approach in three important research areas: the human body, mimetic learning, and social agency performativity in rituals. Therefore, this introduction will allow to outline this chapter according to various types of anthropologies beyond those listed by Wulf (2004).

In line with Wulf's (2004) suggestions, Gremillion (2005) argued that scholarly interest in body size has increased in concert with recent efforts to shape and assess bodies in particular ways within industrialized social contexts. Attending to both overt and covert references to Eurocentric body projects, Gremillion's chapter reviews literature in anthropology, sociology, and cultural studies that addresses the cultural politics of body size in various parts of the world. The chapter begins with a discussion of biocultural paradigms, which accept certain biomedical categories even when challenging or reconfiguring their hegemonic power. Next is a survey of works analyzing body size within "non-Western" groups as well as European and North American subgroups. These studies often employ culturally powerful "Western" constructs as foils, an approach that risks cultural marginalization. The analysis then turns to the extensive literature that unpacks dominant Euro-American body practices and discourses. Here, diverse perspectives on several key concerns in sociocultural anthropology are considered; concepts of culture and power, theories of the body and embodiment, and understandings of human agency vary in instructive ways. The chapter concludes with a review of scholarship on postcolonial processes and representations that incorporated a critical perspective on Eurocentric preoccupations with body size.

Biological Anthropology

The foregoing discussion on body size is an apt introduction to biological anthropology. For instance, Ice (2005) argued that as the number of persons aged 65 and older is increasing dramatically in both developed and developing nations of the world, the health and well-being of elders has become a worldwide public health concern. Although older adults are now found in higher proportions across all cultures, the biology, behavior, and environment vary tremendously across older populations. Biomedical research largely follows a reductionist paradigm separating the domains of culture and biology. Even when health is examined in association with culture and behavior, biomedical researchers largely focus on static unidirectional associations instead of examining the dynamic multidirectional impact of culture, behavior, and the environment on physiology and ultimately health. Since aging and the processes of senescence clearly involve complex interactions among biological, environmental, and cultural domains, anthropologists with a biocultural and evolutionary perspective are well equipped to study variation in aging and senescence. While relatively few biological anthropologists have focused their attention on aging, a growing literature has demonstrated the utility of biocultural approaches to aging. This article serves as an introduction to a special issue that highlights the core of the biological anthropology of aging.

In line with Ice's conclusion about the importance of "biocultural approaches," Ingold (2004) argued that Darwinian accounts of human evolution are driven by a fundamental paradox. While asserting a principle of continuity between human populations and their evolutionary antecedents, they also posit a point of origin from which separate processes of culture and history rise up from a universal baseline of evolved human nature. But the belief in this nature rests on a confusion surrounding the notion of genetic information. Human capacities are not genetically prespecified but emerge within processes of ontogenetic development that are at once historical and evolutionary. Replacing the "population thinking" of the Darwinian paradigm with a "relational thinking" that focuses on the dynamics of developmental systems leads us to a new vision of anthropology, as a science of engagement in a relational world.

Flinn (2006) commented on Ingold's (2004) paradox and others who noted the flexible production of melanin in melanocytes in response to solar radiation. This latter paradigm, modeling culture – with mate choice is a specific component – is a compilation of flexible responses by individuals to specific environmental contingencies. This paradigm seems analogous to the biological concept of reaction norms and consistent with the basic premises of evolutionary psychology. Here Flinn (2006) pushed the evolutionary perspective beyond the concept of "evoked culture" as constrained response to variable environments guided by specialized psychological modules. Nonetheless, others reiterated the importance of this type of flexible evoked culture response to the material world. Awareness of these shortcomings, and the interest in a broader cross-cultural database advocated by others, have potential great strengths that will go far to advance our understanding of the nature of human psychology and its adaptive variations. This synthesis also portends renewed interdisciplinary efforts among anthropology, biology, and psychology across a broad range of research questions that may help resolve concerns. "Many psychologists have not been studying human nature, they have been studying the nature of educated, middle-class, young adult Westerners."

From an evolutionary viewpoint, current research in primatology employs the new paradigm of behavioral ecology and sociobiology for analysis and interpretation of variation in behavior and ecology (Rodman, 1999). Grouping and group sizes of primates are explained with reference to effects of predation, defense of resources, and female defense against male infanticide. Primates avoid close consanguineous mating, usually by dispersal of males from the birthplace, though in bonobos and chimpanzees males are philopatric. In many primates, nepotistic relations among females are explained by kin selection operating on the philopatric sex. Sexual dimorphism, dominance hierarchies, intrasexual competition, and particularly infanticide by males are best explained by the action of sexual selection. Comparative studies of primates indicate that the large brains of the genus Homo (enlarged cerebral cortex) evolved after bipedalism and human dental characters and probably depended on high-quality diets. Although dominant themes of current anthropology are not compatible with the epistemology, theory, or methodology of primate research and interpretation, primate studies fit easily within the future of anthropology as a four-field evolutionary study of the origins of humans and human nature.

Cognitive Anthropology

Tadajewski and Wagner-Tsukamoto (2006) introduced a new qualitative method that is theoretically underpinned by cognitive anthropology. This research strategy is introduced to further advance the understanding of complex green consumer behavior – in this case: life cycle analysis. These authors examined the contextual aspects of problem-solving behavior of green, environmentally concerned consumers. Cognitive anthropology develops a different, yet complementary, understanding of consumer cognition to a psychological approach. Through the concepts of practical thinking and bricolage, cognition and behavior are conceptualized on a contextual basis. Such an approach encourages a reassessment of how consumer research has traditionally conceptualized problem framing, information search, information processing, and related concepts. Tadajewski and Wagner-Tsukamoto (2006) drew upon in-depth, qualitative interviews with a wide range of green consumers from both the UK and Germany.

Findings provided some interesting clues regarding the nature of information search and information processing. In the sample, green consumers of the top clusters were able to see and retrieve life cycle information as it was offered by a shopping context and it was this context, as it is perceived by the bricoleur that ultimately limits information search and processing. Within the "objective" bounds of a choice context, skillful practical thinking and bricolage was shown in different degrees amongst the clusters, with considerable creativity shown in "seeing" life cycle information. Given that the research outlined in this article is monoparadigmatic, it is suggested that a future avenue for research in green consumer behavior would be the use of a multiple paradigm approach. These authors outlined a stepping approach to marketing communications directed towards the green, or potentially green consumer, suggesting that some form of community-based social marketing program might be a useful educational tool given the findings presented.

Cultural Anthropology

Culture is an extremely wide term that includes tangible and intangible mores, morals, and models. Clothing research, for instance, has attracted renewed interest in anthropology over the past two decades, experiencing a florescence that had been kept within bounds by reigning theoretical paradigms (Hansen, 2004). The works have been influenced by general explanatory shifts in anthropology, which inform disparate bodies of clothing research that otherwise have little unity. The most noticeable trend is a preoccupation with agency, practice, and performance that considers the dressed body as both subject in, and object of, dress practice. The turn to consumption as a site and process of meaning making is evident also in clothing research. Dress has been analyzed, by and large, as representing something else rather than something in its own right, although new efforts to reengage materiality suggest that this approach is changing. Little work has been done on clothing production issues, though some scholars examine the significance of dress in the context of the entire economic circuit and the unequal relationships between its actors.

Guillo (2007) raised some interesting questions about the relationship between cultural and biological evolution. He advocated memetics to construct a theory of culture, on the basis of an analogy with biological evolution. Such a neo-Darwinian theory of culture should be carefully distinguished from evolutionary psychology and cognitive anthropology, as it is not reductionist. More generally, it should not be identified with any of the theories of culture that are based on the hypothesis that the individual mind is active rather than passive when it adopts a given cultural trait. Although in some respects memetics is close to traditional social science paradigms, it forms a specific model. Despite its stimulating ideas and the interesting research it has encouraged, memetics is based on propositions and concepts – in particular that of the meme – which raise a number of difficulties. The nature of memes is uncertain and problematic. If they exist, their realm covers only some cultural phenomena. Finally, the attempt to prove the existence of memes encounters serious obstacles.

People's transitions between different cultural communities have been recorded in myths, legends, and chronicles since the early ages of human civilizations (Chirkov, 2009). Military crusades and occupation of foreign territories, enslavery and cross-tribal marriages, explorations and fortune seeking, trading and missionary work, education and professional work are the instances of such transitions. Encounters with strangers and following the diverse reciprocal influences of the involved parties have fascinated and attracted the interest of travelers, writers, and scientists. As a result, these cross-cultural transitions and encounters together with the processes that accompanied them on the levels of cultures, societies, groups, and individuals have become a focus of several social sciences: political science, sociology, social and cultural anthropology, and finally, cross-cultural and acculturation psychology. Acculturation psychology emerged in the second half of the last century within the frameworks, paradigms, and conceptual schemas of cross-cultural psychology. In doing their research, acculturation psychologists utilized definitions and conceptualizations of sociology and cultural anthropology. A creative scientific program of research has to encompass reflective and critical thinking about the existing paradigms and practices that guide this research.

In the last decade, and responding to the criticism of orientalism, anthropology has engaged in a self-critical practice, working toward a postcolonial perspective on science and an epistemological stance of partial and situated knowledge (De Clerck, 2010). In deaf studies, for instance, anthropological and sociological studies employing qualitative and ethnographic methods have introduced a paradigm shift. Concepts of deaf culture and deaf identity have been employed as political tools, contributing to the emancipation process of deaf people. However, recent anthropological studies in diverse local contexts indicate the cultural construction of these notions. From this viewpoint, deaf studies face a challenge to reflect on the notions of culture, emancipation, and education from a nonexclusive, noncolonial perspective. Deaf studies research in a global context needs to deal with cultural and linguistic diversity in human beings and academia. This calls for epistemological reflection and new research methods.

Influenced by the anthropology of sport, Thangaraj (2010) examined leisure spaces of sport as critical venues in the formation of South Asian American "self" and "selves" in opposition to, as well as in relation to, various "others." This author uses ethnographic research conducted with North American Indo-Pak Basketball leagues, primarily in Atlanta, Georgia, to demonstrate how South Asian American masculinity emerges in relation to ideas of "Blackness." In addition, participation in Indo-Pak Basketball for South Asian American men allows for creative practices that facilitate the collapse of "South Asianness" and "Americanness" (Maira, 2002) while contesting dominant paradigms of race and gender. Through analysis of Indo-Pak Basketball and related expressive practices, this author illustrates the processes of identity formation and consumption as critical to citizenship in the US national fabric. The author also argued that the subversive potential in basketball masculinity and its respective pleasures are limited by its relation to a "hegemonic masculinity" (Connell, 1995), as well as by the consumptive logic of late capitalism through which racialized, heterosexist, patriarchal formulations of community emerge.

Kuper (2003) commented on an article by Sousa (2003) who gave a good account of the fall of kinship, but the highly focused intellectualist perspective that he adopted leaves out too much. A broader context and a historical perspective are required if we are to make sense of what happened to kinship studies. We (anthropologists) have to look beyond anthropology. The modern orthodoxy on kinship in anthropology was part of a common paradigm of kinship, marriage, and the family within the social sciences. It collapsed, as the broader paradigm collapsed, in part because of internal problems – theoretical weaknesses, ethnographic contradictions – but more directly as a result of social and political developments in the outside world. As the critiques of kinship begin, in turn, to seem sterile and unpersuasive, there is room for new thinking. But my sketch of what happened to kinship theory suggests that internal pressures will not be enough in themselves to bring about a turn back to kinship studies.

If it is to attract interest, new thinking in this field will have to tell us all something about what is happening to kinship in our own societies and respond to concerns in the broader intellectual community. Kinship used to be described as what anthropologists do. Today, many might well say that it is what anthropologists do not do. One possible explanation is that the notion of kinship fell off anthropology's radar due to the criticisms raised by Needham and Schneider among others, which supposedly demonstrated that kinship is not a sound theoretical concept. Drawing inspiration from epidemiological approaches to cultural phenomena, in his article, Kuper (2003) aimed at enriching this explanation. Kinship became an unattractive theoretical concept in the subculture of anthropology not simply because of problems with kinship theory per se, but also on account of fundamental changes in the very conception of anthropological knowledge and the impact of these changes on the personal identity of anthropologists.

The fall of kinship studies supports the position (L'Abate & De Giacomo, 2003; L'Abate, 2005) that we cannot longer think of the traditional family as a rigid, "two parents with two children" system, as the reigning paradigm in family science. The variety of people who live under the same roof is such that the term "family" has lost any descriptive or explanatory meaning. Instead L'Abate suggested using the term "intimate relationships" defined by closeness, commitment, interdependence, and duration. Eventually, L'Abate gave up using monadic, nonrelational terms, such as personality and self, relying instead on a relational framework that would include intimate and nonintimate relationships (L'Abate, Cusinato, Maino, Colesso, & Scilletta, 2010).

The 1980s witnessed an increased interest among behavior analysts regarding a paradigm in cultural anthropology known as cultural materialism (Ward, Eastman, & Ninness, 2009). This perspective suggests that all behavior ultimately rests on the relationship between the natural environment and the methods used to obtain resources needed to maintain survival and a high standard of living, known as the mode of production. While useful, scientists in this area have overlooked one valuable resource – the laboratory. These authors manipulated the amount of resources dyads could harvest within blocks of five trials and across six conditions. Behavior consisted of harvesting and allocating resources. Token retention and a survival analogue were made contingent on resource sharing. Five of the seven dyads shared resources and no sharing occurred when participants could harvest sufficient resources to survive independently. Opportunities to integrate this research with optimal foraging research are discussed along with potential applications to real-world social issues.

At this critical historical juncture of environmental degradation, awareness of needed cultural change, and an emerging sustainability movement, attention to "re-enchantment" – the phenomena of sensory, emotional, and nonrational ways of connecting with the earth's living systems – strengthens our professional understanding as educators as well as our ability to contribute to institutional change (Barlett, 2008). Research among participants in the Piedmont Project – a faculty development program for sustainability across the curriculum at Emory University – and among sustainability leaders in higher education across North America shows that wonder, delight, awe, and meaning are linked to both personal and political spheres of action. The experience of re-enchantment can be understood in seven dimensions and provides restorative moments, fosters creativity for change, and supports a revised worldview. A stereoscopic paradigm that combines reason and re-enchantment will serve an anthropology that seeks to think, in Roy Rappaport's words, "not merely about the world but on behalf of the world."

As tourism is extending commoditization into every corner of the globe, poststructuralist approaches to the anthropology of tourism tend to focus on consumption, the ironies of cultural hybridization, and the instability or "virtuality" of identity and authenticity (McGuckin, 2005). While useful in the representation of highly particularized intersections of discourse and desiring bodies, poststructuralist discourses may tend to dematerialize political and economic processes with significant impacts on communities subjected to the tourist gaze. Using the author's fieldwork on the tourism industry in Dharamsala, India for context, he argued that by reemphasizing a focus on material production and class relations, and by transcending discourses of authenticity and virtuality through Marxist conceptions of alienation, an "engaged anthropology" of tourism can more usefully link the ironies of postmodern consumption with the inequalities that continue to be structured through capitalist production.

Health Anthropology

In health research, the concept of risk is used in several domains. It features in epidemiology and health promotion as well as in the sociology and anthropology of health and illness (Honkasalo, 2006). Important problems are embedded in risk as an analytic category, and some of these are widely discussed within the discipline. First, risk cuts out the experience of illness and replaces it with problematic objectifications. Second, there are a multitude of different meanings assigned to risk which make it problematic as an analytic category. Third, the question of uncertainty has, along with the discourse of risk, attained a solely normative meaning of threat or has become a residual category in anthropology and social science. Fourth, the individualizing character of the concept of risk eclipses the collective, historical, social, and political processes that contribute to the uncertainties of life. Departing from the methodological approach to "social suffering," this paper will argue through two ethnographic studies that normative and individualizing assumptions lead social studies to overlook the way that local actors, rather than narrowly controlling contingency, may engage in it and live with it as a normal experience of human life. Along with risk, there are variations in local attitudes to uncertainty and there are socially shared ways of accounting for and acting with it. A major challenge for social studies is to embrace the need for the development and implementation of a neutral, nonnormative language of contingency. The paradigm of social suffering enables a broader perspective to be brought to bear upon the lived experience of affliction and illness so as to expose a greater variation in human responses to the distress of everyday life.

Cheyney (2008) examined the processes and motivations involved when women in the United States choose to circumvent the dominant obstetric care paradigm by delivering at home with a group of care providers called direct-entry midwives. Using grounded theory, participant observation, and open-ended, semi-structured interviewing, Cheyney (2008) collected and analyzed homebirth narratives from a theoretical sample of women ($n=50$) in two research locales. Findings interpreted from the perspective of critical medical anthropology suggest that women who choose to birth at home negotiate fears associated with the "just in case something bad happens" argument that forms the foundation for hospital birth rationales through complex individual and social processes. These involve challenging established forms of authoritative knowledge, valuing alternative and more embodied or intuitive ways of knowing, and knowledge sharing through the informed consent process. Adherence to subjugated discourses combined with lived experiences of personal power and the cultivation of intimacy in the birthplace fuel homebirth not only as a minority social movement, but also as a form of systems-challenging praxis.

Almeida Filho (1985) reviewed the literature on the relationships between cultural change and psychopathology as demonstrated by Latin American researchers. Based on analysis of 13 epidemiological studies, it is shown that culture traditionally has been viewed by social psychiatrists as an independent variable associated with the prevalence of mental disorders. There are two basic approaches in this literature. The hypotheses of culture shock, acculturation stress, and cultural marginalization belong to the approach derived from anthropology, while the notions of urban stress, life change, social support, and goal-striving stress belong to the approach based on sociological explanations. The cultural change paradigm proposed by past scholars is the most complex paradigm proposed; its conceptual basis is criticized. Methodological problems involved in the evaluation of the culture–psychopathology link in contemporary social psychiatric research are discussed.

Depression will be the most common mental disorder by 2020, and it is also expected to be the second leading cause of disability, after cardiac diseases (Mosotho, Louw, Calitz, & Esterhuyse, 2008). Moreover, depression is likely to be a major public health burden in the future. This study evaluated the influences of culture on the symptoms of depression among Sesotho speakers. An evaluation of a sample of 100 participants diagnosed with depression was conducted, using the Psychiatric Interview Questionnaire. It was found that depression among Sesotho speakers is manifested in three areas: somatic symptoms, perceptual disturbances, and disturbances of the thought processes. Since it has become clear, on the basis of the investigation, that depression is a culturally diverse phenomenon, the authors also recommended that research in this regard should be conducted from a multidisciplinary perspective, so that other paradigms, including those of sociology and anthropology, can also be included.

Repeated drunkenness became a contradiction in terms of a healthy and rational lifestyle in the context of the Enlightenment in the eighteenth century (Schlimme & Emrich, 2004). Corresponding to the self-referential understanding of man as an independent and rational being, repeated drunkenness could no longer be understood as a conscious decision or a vice. The central point of the new paradigm of dependence was the installation of a special urge by a specific chronic chemical traumatization. But this original paradigm also required the acceptance of the human individual as a depth-structured being and shows influence by the Brownianism, therefore it is also an unspecific concept. This understanding corresponds with modern anthropology, in which the human being is understood dichotomously: on the one hand with consciousness and mind, on the other hand with unconsciousness, body, and the world around.

In psychiatric discourse, addiction is normally understood according to the original paradigm of "dependency" which claims lifelong and unchangeable dependency on the addictive substance or "psychotropic technique" (Schlimme, 2007). In the attempt to reduce damage or achieve abstinence, therapeutic interests and understanding of the human mind contradict this inflexibility. This contradiction in psychiatric understanding of addiction can be viewed as a dynamic part of the addiction process. A philosophical anthropology of addiction is presented in that light. Using the concept "the psychotropic technique achieves a superior position," both sides of the psychiatric understanding are combined and integrated. The addictive mind and its self-awareness can then be understood as a "fragile mono-identity." Additionally, Gilbert (1990) discussed the particular perspectives and methods that anthropologically produced data contribute to a biopsychosocial paradigm of alcohol use/pathology. Strategic uses of qualitative research approaches were suggested, and situations in which qualitatively produced data would be especially useful in the design and delivery of alcohol-related services were identified.

Lantéri-Laura (2005) compared psychiatry with psychoanalysis. To define psychiatry the author rejected the incompleteness of etymological research and favored the idea of a medical psychology that emerges when madness breaks its connections with the supernatural. Lantéri-Laura (2005) reviewed modern psychiatry starting from the end of the eighteenth century, distinguishing three periods each marked by the superiority of a particular paradigm in T. S. Kuhn's sense of the word: mental alienation in the singular (1793–1854), mental illnesses in the plural (1854–1926), psychological structures (1926–1977), and at the present time, postmodern psychiatry. The author rejected the idea of a philosophical history, which would somehow go beyond empirical and bibliographical research. He argued that twenty-first century psychiatry is implacably heterogeneous with very blurred and merely conventional limits. He studied its relationship with other disciplines, not only knowledge of the central nervous system, clinical neurology and neuropsychology, but also anthropology, criminology, and linguistics. Lantéri-Laura (2005) concluded that psychiatry and psychoanalysis have the same object, but see the object from different angles: the angles find their reflections in psychopathological theories; he maintained that psychoanalysis keeps a pertinence in the held of psychiatry connected with disorders of the central nervous system for even the demented, at least at the beginning of their disorders, maintaining both a conscious and unconscious mental life.

Qualitative research has been the historical bedrock of a number of social science disciplines, such as anthropology and history (Neale, Allen, & Coombes, 2005). However, interest in the approach has grown in recent decades as more researchers in basic disciplines and applied fields have adopted a qualitative paradigm. Within the addictions, qualitative studies have been occurring regularly since the 1920s. Over the years, qualitative methods have provided drug researchers with valuable techniques of scientific inquiry. In return, drug taking has provided qualitative investigators with an ideal arena in which to apply and develop their methods.

Currently, there are two challenges. First, those who are undertaking qualitative addiction research should have confidence in the scientific rigor and value of their methods and not hesitate in writing up their data for any journal that will reach their target audience. Second, addiction journals should adopt policies and practices that will potentially encourage more qualitative submissions. This might include a statement welcoming articles based on qualitative research, not being too prescriptive about the structure and length of papers, ensuring that qualitative work is always reviewed by those with qualitative expertise, commissioning publications from those known to be undertaking valuable qualitative investigations, and setting an approximate per annum quota of journal space for qualitative contributions.

Jenkins (1991) formulated an approach to cultural variation in social response to mental disorder based on an interactive conception of emotional atmosphere as outlined by H. S. Sullivan, E. Sapir, and G. Bateson. The main contribution of the expressed emotion (EE) paradigm is to identify criticism and emotional over-involvement as key domains. The kind of emotional material sought is not likely to be in the overt content of the interview, but may more likely be communicated through intended and unintended vocal characteristics of speech. EE must be developed further to be truly interactive; herein lies the complementary contribution of anthropology, to specify cultural conditions of emotional force based on variations in self, emotional experience, behavioral rules, and conceptions of mental disorder. Jenkins (1991) provided case examples of three families.

As "psy" clinicians, we are concerned with searching and finding the best possible solutions for our patients, in terms of restoration of equilibrium, harmony, and all of the ingredients contained in the current concept of total health (Betancourt, 2003). In order to overcome all dichotomies, reductionist stances, and a priories originated in the Cartesian paradigm in vogue during the medical and scientific education throughout the twentieth century, it is necessary to deepen our study of the philosophy of science, mind–brain relationships, the consideration of socio-history, art, and complex anthropology as crucial disciplines that could promote the upsurge of humanism and spirituality. Consciousness as a sui generis human phenomenon could be positioned at the center of our investigation. New journeys through territories neglected for centuries by modern science could be accomplished by positioning consciousness as an axis. Taking a formal account of the mind represented by nonordinary states of consciousness that underlie higher creativity manifestations, and applying an enlarged scientific methodology (that would include qualitative as well as quantitative research protocols) will contribute not only to the innovation of healing processes but also to human development, individually as well as planetary.

Historical Anthropology

The pioneering work of Margaret Mead's classic *Coming of Age in Samoa* is still the object of controversy and debate in anthropological circles. In 1928, Margaret Mead announced her stunning discovery of a culture in which the storm and stress of adolescence did not exist. The resulting book has since become a classic and the best-selling anthropology book of all time. Within the nature–nurture controversy that still divides scientists, Mead's evidence has long been a crucial "negative instance," an apparent proof of the sovereignty of culture over biology. In his book, Freeman (1983) presented startling but wholly convincing evidence that Mead's proof is false. On the basis of years of patient fieldwork and historical research, Freeman refuted Mead's characterization of Samoan society and adolescence point for point. Far from the relaxed transition to adulthood that Mead ascribed to permissive child-bearing and restrictive regulations against premarital sex. Freeman was a New Zealand anthropologist whose fieldwork disagreed with Mead's in regard to life for the Samoans. Her work favors nurture over nature as the most important factor in how children become adults in their societies, and he directly refutes her claims. Freeman has come under fire for not publishing his work until 5 years after Mead's death; in light of the fact that the two authors cannot have this argument themselves, other scholars are stepping in to debate what is called the "Mead-Freeman Controversy."

Shankman (2000), for instance, discussed Freeman's (1983) argument claiming that the central issue in the Mead-Freeman controversy is evolution. Freeman also claimed that Mead was wrong about Samoan sexual conduct because she was "hoaxed" by two young women telling lies. Freeman viewed Mead's text as not only misleading about Samoa but as a "sacred text" that promoted an antievolutionary paradigm among American cultural anthropologists. A review of Mead's writings on culture, biology, and evolution demonstrates that contrary to Freeman's claim, Mead favored an evolutionary approach throughout her career. Moreover, while Mead's book was a popular text and a bestseller, it was not a "sacred text" among anthropologists. Freeman's misrepresentation of the historical record concerning Mead's views and the place of Coming of Age in anthropology raised major questions about his scholarship.

The debate between Derek Freeman and the silenced ghost of Margaret Mead remains unintelligible to many readers because of the absence of historicist context for the early work of Mead (Murray & Darnell, 2000). Despite Freeman's apparent attention to historical documents, he made virtually no effort to situate Mead within Boasian anthropology or the interwar years more generally. Murray and Darnell (2000) provided such contexts in terms of the Boasian paradigm of the time and how Mead was understood by her contemporaries. This context dispels the black and white indictment of a disciplinary heroine with feet of clay and provides the baseline for a more balanced assessment.

Seventeen years after Freeman launched his bid to reform anthropology by refuting the ostensible key text of Boasian anthropology, the situation remains at stalemate, with Freeman claiming "ungainsayable" evidence for "closure," whereas his critics claim that his own refutation has been refuted (Caton, 2000). But no group of anthropologists espouses Freeman's "interactionist" paradigm, and no studies in anthropological method written since the commencement of the controversy recognize it as a significant contribution to the field. The current author argued that the Freeman reform of anthropology has no progeny because its basic game plan is not in alignment with the anthropological problem-set that he would reform. Several key examples of misalignment are discussed, including the hoaxing hypothesis.

In addition to the contribution by Margaret Mead, Gregory Bateson has exercised a lasting influence not only in anthropology but also in information science (Wardle, 1999). He was one of the last and most distinguished products of the school of anthropology that Haddon and Rivers created in Cambridge after the Torres Strait Expedition. Bateson used the years between WWI and WWII to create a theoretical approach that continued and deflected that of Haddon and Rivers. His major ethnography from this period, Naven, evidenced his complex academic positioning between the legacy of Rivers and the new paradigm emerging around Malinowski and

Radcliffe-Brown. After the WWII, Bateson's intellectual project emerged as even closer to Rivers' in both psychological and evolutionary dimensions. Bateson's influence is reflected directly in De Giacomo's Elementary Pragmatic Model (L'Abate & De Giacomo, 2003) and the information processing paradigm that is now present everywhere everyday in many cultures (see Chap. 18).

Linguistic Anthropology

Dialectics links anthropology and linguistics (Bartoli, 2007). Anthropology imported from structural linguistics the ability to reduce the complexity of data through models. In this respect, the most influential paradigm was levistraussian structuralism, which removed the dimension of use and practice. Starting from Malinowski's seminal intuitions about language in context, from the revolution of "the linguistic turn," and from a new attention paid to the political dimensions of language and behavior in the 1960s, new approaches have thrown new insights on language and culture. Ethnography has played a central role in that it has suggested ways of doing research within approaches deriving from sociolinguistics, ethnography of speaking, and conversation analysis. The outcome has been an interdisciplinary field which combines ethnographic concern for context with study of talk and considers the construction of social reality by social actors.

The Sapir-Whorf hypothesis has raised considerable controversy in the literatures of psychology and anthropology (Santa & Baker, 1975). Several misconceptions of the hypothesis are reviewed. In the present experiment with 37 undergraduates, the hypothesis was tested and supported in a visual reproduction paradigm. Participants were first given label training for a set of figures and were then asked to recall by drawing the shapes. Training with categorized labels resulted in a 25% improvement in recall when compared to a condition with nonword (paralog) labels. Even stronger evidence of linguistic influence on visual memory was obtained by examining the order of recall. The conceptual relationships among labels strongly influenced the sequence of reproductions. Language and culture, as conceptualized in traditional anthropology, may have an important influence on pain and brain–behavior relations (Fabrega, 1989). The paradigm case for the influence of language and culture on perception and cognition is stipulated in the Sapir–Whorf hypothesis. This hypothesis has been applied to phenomena "external" to the individual. The paradigm is applied to information the person retrieves from "inside" the body (i.e., "noxious" stimuli that register in consciousness as pain).

For some time now second language acquisition (SLA) research has been hampered by unhelpful debates between the "cognitivist" and "sociocultural" camps that have generated more acrimony than useful theory (Watson-Gegeo, 2004). Recent developments in second-generation cognitive science, first language acquisition studies, cognitive anthropology, and human development research, however, have opened the way for a new synthesis. This synthesis involves a reconsideration of mind, language, and epistemology, and a recognition that cognition originates in social interaction and is shaped by cultural and sociopolitical processes: these processes are central rather than incidental to cognitive development. Here Watson-Gegeo (2004) laid out the issues and argued for a language socialization paradigm for SLA that is consistent with and embracive of the new research.

Eastwood (1988) contended that the application of the scientific method to research in speech pathology may be less than adequate to meet the needs of the discipline. Speech pathology has failed to develop the core concepts or paradigms that inform the "hard" sciences, perhaps as a consequence of the unsuitability of experimental research approaches. Qualitative approaches to research allow for a more flexible and comprehensive method of investigation in clinical or research work and are employed by many other disciplines, such as sociology and anthropology, where the concern of the researcher is to investigate a complex, dynamic social setting or situation. It is concluded that an alteration of research orientation in speech pathology may provide clinicians and researchers with more appropriate and beneficial methodology.

Organizational Anthropology

Gregory (1983) described and critiqued organizational culture studies done in industrial settings, some of which were based on anthropological paradigms, including the structural–functional and configurationist holistic paradigms. Most studies failed to explore multiple native views. The author proposes a multicultural model for large organizations, and problems of cross-cultural contact are described. Native-view paradigms from anthropology, especially ethnoscience ethnography, are recommended for exploring multiple perspectives in detail. Data from the author's ongoing study of "Silicon Valley" technical professionals' native views demonstrates how ethno-science methods can be applied to the task of studying culture. About 75 professionals were interviewed using an open-ended format. Native views were primarily defined as cognitive systems.

Years later, Bate (1997) speculated on the nature of the "ethnographic quest," and on what might be gained from trying to put organization studies and anthropology back together again in some form or other. The question of whether social anthropology can contribute to organizational behavior and management research is explored. Ethnography can be defined as a particular type of method or fieldwork activity, a kind of intellectual effort or paradigm, and a narrative or rhetorical style. Reconciliation between organizational behavior and anthropology is discussed in relation to ethnographic studies of organizations. Different conceptions of ethnography, the texture of anthropological research, and qualities of good ethnography are discussed.

De Vries and Balazs (2005) addressed the subject of clinically oriented consultation in the workplace, offering an example of an intervention to illustrate the limitations of more traditional, rational forms of organizational consultation. The authors suggest that unconscious intrapersonal, interpersonal, and group-related dynamics have a serious impact on many decisions and policies in organizational life. The authors point out that both rational and irrational forces will affect life at work. Such "irrational" processes are a powerful force in explaining otherwise incomprehensible human motivations and actions. The authors then argue that the clinical approach to organizational consultation can make a significant, positive contribution in situations of problematic organizational transformation, dysfunctional leadership, collusive superior–subordinate relationships, destructive social defense mechanisms, ineffective intra- and intergroup relationships, and neurotic organizational culture. To help executives understand the dynamics of these irrational forces, the authors discuss a number of salient themes in contemporary psychoanalytic theory – a domain that includes contributions from dynamic psychiatry, developmental psychology, anthropology, neurophysiology, cognitive theory, family systems theory, and individual and group psychotherapy. The breadth of that grounding makes in-depth interpretations of organizational phenomena more powerful. The main parameters of the clinical paradigm are also reviewed in this article. In the final section, the authors deal with the role of the clinically oriented consultant, and discuss the impact of transferential and countertransferential processes in the client–consultant interface.

Psychological Anthropology

If anthropology is concerned with cultures and societies, psychological anthropology is concerned with how individuals function as influenced by their cultures. One could say that psychological anthropology is the lowest level of reduction achievable in that discipline. Schülein (1989), for instance, discussed symbiotic relationships between parents and children (primary symbioses) and between adults (secondary symbioses) within the theoretical framework of psychosocial development. In primitive, then in traditional, societies, symbiotic parent–child relationships were determined by economic, practical, and rational considerations of mutual need and usefulness. Under current societal conditions, primary and secondary symbioses are determined by complex dynamic, personal, affective, and irrational motivations. Psychoanalysis provides the best cognitive paradigm for the understanding of symbiotic relationships, based on the findings of modern anthropology on the structure of ontogenetic development expressed in psychosocial terms.

In another instance, Whitebook (2001) discussed the historical background of this article as a psychologically oriented anthropology, owing much to T. Hobbes's (1588–1679) view of the individual as isolated, instinct-driven, asocial and strategically oriented. The author presents a detailed critique of the intersubjective approaches of such authors as K. Apel (1973), J. Habermas (1967–1988), H. Joss (1980–1992) and, above all, A. Honneth (1985–2000), having as their central points of departure the diagnosis of a progressive disappearance of the work of the negatives in intersubjectivity paradigm.

A representative compendium about the psychology of cultural experience, Moore and Mathews (2001) included chapters about (1) theoretical and methodological approaches to the study of personal experience; (2) acquiring, modifying, and transmitting culture; (3) continuity and change in cultural experience; and (4) concluding with a comparative perspective that included cross-cultural studies in language and thought and comparative approaches in psychological anthropology.

Since gender was slightly touched in Moore and Mathews's work (one reference on gender roles on page 75), Singh (2007) critically reviewed the gender and development paradigms and proposed instead an alternative framework of research – identities of women. He contended that the gender and development paradigm is primarily guided by the tenets of Western feminisms and economic development. Singh (2007) also highlighted other limitations of the paradigm, including its pre-occupation with male–female inequalities, macro generalizations, and symbolic representation of women and a limited inclusion of local contexts. The identities of women framework proposes to address the limitations of the gender and development paradigm by studying women's conception of their environment and women's understanding of their relationship with these environments. The identities of women framework is informed by poststructuralist critique of feminism, cultural anthropology, and a sociopsychological approach to identities.

By the same token, Tsoukalas (2007) explored the microfoundations of group consciousness. In order to do that, it re-examines two well-established theoretical traditions within Western social science, the social cohesion paradigm within social psychology and the various mind–body theories within anthropology and sociology, hoping that this re-examination will have intellectual bearing on the issue of group consciousness and group action. As a result of this analysis it is suggested that the different ways a group enacts its ideology can help shed new light on many features of group life. A relatively new theory is therefore presented which can account for this relationship and also helps to organize and integrate many of the earlier findings that pertain to the issue of group consciousness and group action. This theoretical framework, originally developed by Harvey Whitehouse, postulates that divergent modes of encoding have differential social effects.

More specifically, there are two distinct modes of encoding, two distinct psychological ways of relating and acting within the framework of a group's ideology that are worth particular attention: the doctrinal mode of encoding and the imagistic mode of encoding. Their differences, manifested in the formal and informal behavior of groups, are many. Some of them, such as organizational structure, change patterns, leadership style, centralization–decentralization aspects, and institutional practice, are explored in depth. Whitehouse's cognitive theory is finally coupled with findings from sociology in order to furnish a fuller account of the social underpinnings of group-related phenomena. In particular, Tsoukalas (2007) used Mark Granovetter's network theory in order to show how interactional patterns can have an influence on issues of group consciousness. The two theoretical strands are fused into a dual model which shows great promise and merits further attention. In the final part of the article, examples are given in order to illustrate the real-life relevance of the stipulated model.

Social construction theory has offered a challenge to traditional anthropological models and has been responsible for the burst of innovative work in sexuality, both in anthropology and in other disciplines, since 1975 (Vance, 1991). The theoretical roots and implications of constructionist theory are explored. The intensifying competition between cultural influence and constructionist paradigms has been altered by the appearance of acquired immune deficiency syndrome (AIDS) and the subsequent increased support for research on sexuality. The expansion in funding threatens to strengthen essentialist models in biomedical contexts and cultural influence models in anthropology. The complexities and ambiguities inherent in the sexuality under study may both reveal the strengths of constructionist approaches and spur the development of research and theory in anthropology.

Anthropology and Religion

This is clearly an area that should contain more references that can be allowed within the 20-page limit. Smyers (2001), for instance, contended that anthropology still regards the experience of religion the same way it did when its interpretive paradigm was based on "scientific objectivity." To understand this situation, the work of C. G. Jung is helpful in two ways. First, by exploring how anthropology has dismissed Jung as a cultural universalist and/or mystic, often without an actual consideration of his writings, we see how he signifies what the field defines itself against. Second, Jung's empirical forays into the religious worldview provide us with both methodological and descriptive insights about that realm in which many of our informants (and even some anthropologists) live.

Velho (2004) commented on an article by Thomas J. Csordas that has had a stimulating effect on anthropological debates. Velho could testify to this stimulation for his own country, where "embodiment" has encountered an affinity with local practices. Now, 13 years after "Embodiment as a Paradigm for Anthropology," Csordas again surprised anthropologists. They are once again confronted with a major theoretical piece that will certainly give rise to much discussion and influence the future of the discipline. Although at first sight religion is not the main target in these discussions, the homology that exists between the general question of alterity and its application to religion, as Csordas proposes, may not be without consequence. One would like to endorse Csordas's point of not considering religion merely an individual experience, but One would suggest, following Latour to a certain extent, that the global upsurge of alterity could be seen as a lack of recognition of attachments in the name of a Western ideal of detachment.

Applications in Anthropology

Schensul (1985) argued that anthropology and ultimately all of science is a structure for and the result of the interactive processes between the scientist's action/practices and the natural/social world. An action/practice/theory interactive model is described that can improve the scientific nature of anthropological endeavors and more effectively keep up with the continuous change inherent in social reality. Two research cases – a mother–infant project and a health research project – are described to illustrate how this research paradigm can operate.

As early as more than 20 years ago, White and Truex (1988) discussed the use of computers in the analysis of fieldwork, texts, and socioculturally embedded systems, focusing on issues central to cultural, physical, archeological, and linguistic anthropology. Issues addressed included inferences drawn from texts compiled by computer, issues of psychological reality in the measurement of variability and stereotypy, modeling of cognition, and the use of computers to redefine paradigms. Applications of computers in anthropology are outlined as one way to help anthropology enter in the coming century.

Conclusion

It is clear to me and I hope to readers of this chapter will be able to see epistemological errors that do not distinguish clearly among paradigms, theories, models, and dimensions. These errors are still pervasive in anthropological thinking as they are in most social science disciplines. Furthermore, it seems that anthropology is still characterized by dialectical, qualitative methods, such as a semistructured interviews and participant observations. Only few studies used controlled observations. Whether this state of theoretical and methodological affairs is desirable and normative requires the judgment of anthropologists themselves. Nonetheless, suggestions by Damico, Simmons-Mackie, Oelschlaeger, Elman, and Armstrong (1999) provide basic, exemplary information about the rationale, design characteristics, strengths, and weaknesses of qualitative research methodologies that may be used within in anthropology and in social science disciplines.

With its long and well-established history in the fields of anthropology and sociology, the complex analytic paradigm of qualitative research does not favor a particular methodology over another (Damico et al., 1999). Objectives include taking a learning role, understanding procedural affairs, presenting a detailed view, focusing on the individual, and understanding the mundane. The strengths of qualitative research include its (1) design to study phenomena in natural settings; (2) open and relatively unstructured research designs; (3) use of the researcher as the key instrument for data collection; (4) design to collect descriptive data; (5) orientation to a more focused description than a broader one; (6) focus on the process of accomplishing social action; and (7) focus on participants' perspectives. Weaknesses include its (1) labor intensiveness; (2) experience-based learning; (3) operation based on diverse methodological assumptions; and (4) potential for abuse, through lack of rigor in data collection and the inability to verify findings.

References

Almeida Filho, N. (1985). Cultura e psicopatologia – Revisão da literatura epidemiológica Latino-Americana. *Jornal Brasileiro de Psiquiatria, 34*, 357–364.

Barlett, P. F. (2008). Reason and re-enchantment in cultural change: Sustainability in higher education. *Current Anthropology, 49*, 1077–1090.

Bartoli, V. (2007). Antropologie e linguistiche. Percorsi dialettici. *Rivista Di Psicolinguistica Applicata, 7*, 19–52.

Bate, S. P. (1997). Whatever happened to organizational anthropology? A review of the field of organizational ethnography and anthropological studies. *Human Relations, 50*, 1147–1175.

Betancourt, C. (2003). Conciencia y cambio terapéutico: La neurología del nirvana. *Psiquiatría, 19*, 35–40.

Caton, H. (2000). The Mead/Freeman controversy is over: A retrospect. *Journal of Youth and Adolescence, 29*, 587–605.

Cheyney, M. J. (2008). Homebirth as systems-challenging praxis: Knowledge, power, and intimacy in the birthplace. *Qualitative Health Research, 18*, 254–267.

Chirkov, V. (2009). Introduction to the special issue on critical acculturation psychology. *International Journal of Intercultural Relations, 33*, 87–93.

Connell, R. W. (1995). *Masculinities*. Berkeley, CA: University of California Press.

Damico, J. S., Simmons-Mackie, N., Oelschlaeger, M., Elman, R., & Armstrong, E. (1999). Qualitative methods in aphasia research: Basic issues. *Aphasiology, 13*, 651–665.

De Clerck, G. M. (2010). Deaf epistemologies as a critique and alternative to the practice of science: An anthropological perspective. *American Annals of the Deaf, 154*, 435–446.

De Vries, M., & Balazs, K. (2005). Organizations as optical illusions: A clinical perspective on organizational consultation. *Organizational Dynamics, 34*, 1–17.

Eastwood, J. (1988). Qualitative research: An additional research methodology for speech pathology? *The British Journal of Disorders of Communication, 23*, 171–184.

Fabrega, H. (1989). Language, culture and the neurobiology of pain: A theoretical exploration. *Behavioural Neurology, 2*, 235–260.

Flinn, M. V. (2006). Cross-cultural universals and variations: The evolutionary paradox of informational novelty. *Psychological Inquiry, 17*, 118–123.

Freeman, D. (1983). *Margaret Mead and Samoa: The making and unmaking of an anthro-pological myth*. Boston, MA: Harvard University Press.

Gilbert, M. (1990). The anthropologist as alcohologist: Qualitative perspectives and methods in alcohol research. *The International Journal of the Addictions, 25*(2A), 127–148.

Gregory, K. L. (1983). Native-view paradigms: Multiple cultures and culture conflicts in organizations. *Administrative Science Quarterly, 28*, 359–376.

Gremillion, H. (2005). The cultural politics of body size. *Annual Review of Anthropology, 34*, 13–32.

Guillo, D. (2007). Is cultural evolution analogous to biological evolution? A critical review of memetics. *Intellectica, 46–47*(2–3), 49–68.

Hansen, K. (2004). The world in dress: Anthropological perspectives on clothing, fashion, and culture. *Annual Review of Anthropology, 33*, 369–392.

Honkasalo, M. (2006). Fragilities in life and death: Engaging in uncertainty in modern society. *Health, Risk & Society, 8*, 27–41.

Ice, G. H. (2005). Biological anthropology and aging. *Journal of Cross-Cultural Gerontology, 20*, 87–90.

Ingold, T. (2004). Beyond biology and culture. The meaning of evolution in a relational world. *Social Anthropology/Anthropologie Sociale, 12*, 209–221.

Jenkins, J. H. (1991). Anthropology, expressed emotion, and schizophrenia. *Ethos, 19*, 387–431.

Kuper, A. (2003). What really happened to kinship and kinship studies. *Journal of Cognition and Culture, 3*, 329–335.

L'Abate, L. (2005). *Personality in intimate relationships: Socialization and psychopathology.* New York: Springer-Science.
L'Abate, L., Cusinato, M., Maino, E., Colesso, W., & Scilletta, C. (2010). *Relational competence theory: Research and mental health applications.* New York: Springer-Science.
L'Abate, L., & De Giacomo, P. (2003). *Intimate relationships and how to improve them: Integrating theoretical models with preventive and psychotherapeutic applications.* Newport: Praeger.
Lantéri-Laura, G. (2005). L'objet de la psychiatric et l'objet de la psychanalyse. *L'Évolution Psychiatrique, 70,* 31–45.
Maira, S.M. (2002). Desis in the house: *Indian American youth culture in New York City.* Philadelphia: Temple University Press.
McGuckin, E. (2005). Travelling paradigms: Marxism, post-structuralism and the uses of theory. *Anthropologica, 47,* 67–79.
Moore, C. C., & Mathews, H. F. (Eds.). (2001). *The psychology of cultural experience.* New York: Cambridge University Press.
Mosotho, N. L., Louw, D. A., Calitz, F. W., & Esterhuyse, K. F. (2008). Depression among Sesotho speakers in Mangaung, South Africa. *The African Journal of Psychiatry, 11,* 35–43.
Murray, S. O., & Darnell, R. (2000). Margaret Mead and paradigm shifts within anthropology during the 1920s. *Journal of Youth and Adolescence, 29,* 557–574.
Neale, J., Allen, D., & Coombes, L. (2005). Qualitative research methods within the addictions. *Addiction, 100,* 1584–1593.
Rodman, P. S. (1999). Whither primatology? The place of primates in contemporary anthro-pology. *Annual Review of Anthropology, 28,* 311–339.
Santa, J. L., & Baker, L. (1975). Linguistic influences on visual memory. *Memory and Cognition, 3,* 445–450.
Schensul, S. L. (1985). Science, theory, and application in anthropology. *The American Behavioral Scientist, 29,* 164–185.
Schlimme, J. (2007). Zur philosophischen anthropologie der sucht. *Der Nervenarzt, 78,* 74–80.
Schlimme, J., & Emrich, H. M. (2004). Das ursprüngliche paradigma der abhängigkeit – historische und kulturelle aspekte. *Psychiatrische Praxis, 31,* 78–82.
Schülein, J. A. (1989). Symbiotische beziehungen und gesellschaftliche entwicklung. *Psyche: Zeitschrift für Psychoanalyse und ihre Anwendungen, 43,* 1007–1028.
Shankman, P. (2000). Culture, biology, and evolution: The Mead-Freeman controversy revisited. *Journal of Youth and Adolescence, 29,* 539–556.
Singh, S. (2007). Deconstructing 'gender and development' for 'identities of women. *International Journal of Social Welfare, 16,* 100–109.
Smyers, K. A. (2001). Shaman/scientist: Jungian insights for the anthropological study of religion. *Ethos, 29,* 475–490.
Sousa, P. (2003). The fall of kinship: Towards an epidemiological explanation. *Journal of Cognition and Culture, 5,* 265–303.
Tadajewski, M., & Wagner-Tsukamoto, S. (2006). Anthropology and consumer research: Qualitative insights into green consumer behavior. *Qualitative Market Research: An International Journal, 9,* 8–25.
Thangaraj, S. (2010). Ballin' Indo-Pak style: Pleasures, desires, and expressive practices of 'South Asian American' masculinity. *International Review for the Sociology of Sport, 45,* 372–389.
Tsoukalas, I. (2007). Exploring the micro-foundations of group consciousness. *Culture and Psychology, 13,* 39–81.
Vance, C. S. (1991). Anthropology rediscovers sexuality: A theoretical comment. *Social Science & Medicine, 33,* 875–884.
Velho, O. (2004). 'Asymptote of the ineffable: Embodiment, alterity, and the theory of religion': Comment. *Current Anthropology, 45,* 180–181.
Ward, T. A., Eastman, R. L., & Ninness, C. (2009). An experimental analysis of cultural materialism: The effects of various modes of production on resource sharing. *Behavior and Social Issues, 18,* 1–23.

Wardle, H. (1999). Gregory Bateson's lost world: The anthropology of Haddon and Rivers continued and deflected. *Journal of the History of the Behavioral Sciences, 35*, 379–389.

Watson-Gegeo, K. (2004). Mind, language, and epistemology: Toward a language socialization paradigm for SLA. *The Modern Language Journal, 88*, 331–350.

White, D. R., & Truex, G. F. (1988). Anthropology and computing: The challenges of the 1990s. *Social Science Computer Review, 6*, 481–497.

Whitebook, J. (2001). Wechselseitige Anerkennung und die Arbeit des Negativen. *Psyche: Zeitschrift für Psychoanalyse und ihre Anwendungen, 55*, 755–789.

Wulf, C. (2004). Anthropologie historique. Nouvelles perspectives sur les fondements et les conditions de l'éducation. *Orientation Scolaire et Professionnelle, 33*, 571–586.

Chapter 5
Parallel Paradigms of Artists and Authors

Laura G. Sweeney

Visual artists and their creative writing paradigms exist to complement and inform each other just as the colors in artists' palettes impact and blend with each other. Moreover, visual arts and writing send an infinite number of messages to the ones who partake of the experience, leaving them forever transformed by the power art to convey information or be expressive. For instance, Andy Warhol's pop art paintings of Campbell's Soup Cans show that everyday items and even simple titles like "Campbell's Soup" can also become artistic concepts regardless of how mundane and anti-aesthetic. Even Edvard Munch communicated his unrequited love through highly linear and expressive paintings so much so they approached a calligraphic rather than exclusively symbolic quality, much like handwriting. Lewis wrote, "Most viewers feel more comfortable with artworks that refer to the world they know; many prefer art that tells a story" (416). Art expresses the mood, personality, and values of the artist in much the same way that writing exudes these valuable insights about the author. Nevertheless, regardless of how different most people consider the visual arts compared to the written medium, visual art paradigms interact and merge with paradigms in writing. Yet a conflict continues to exist between expressivity and the ability to convey useful information.

Underlying art paradigms are the foundational concepts of *art-for-art's-sake* and *art-for-commercial communication*, the latter of which tends toward practical expression. While the paradigms in visual arts are vast, they may be described generally as follows: (1) prehistoric-mythic; (2) classical-naturalist; (3) abstract-feeling; (4) minimalist; (5) constructivist; and (6) digital/post-modernist. Likewise, similar paradigms can be found in writing in terms of problems, research, and solutions. If prehistoric art focused on telling simple stories through symbols, then the oral storytelling tradition relied on metaphors and symbolism. Further, many of today's visual artists share elements of symbolism as these paradigms continue to reappear

L.G. Sweeney (✉)
Avant Garde Books, 301 Briarvista Way N.E.,
Atlanta, GA 30329, USA
e-mail: laurasweeneyworks@gmail.com

L. L'Abate (ed.), *Paradigms in Theory Construction*, DOI 10.1007/978-1-4614-0914-4_5,

in endless variations. Whereas a majority of artists produce works that evoke strong feelings, others simplify the shapes they create by resorting to symbolic minimalism. Post-modern art helps viewers embrace the possibility of open-ended answers. According to Bertens, "Postmodern art may even become 'non-art' in order to resist interpretation, and it is this avowed aversion that constitutes a definite break with Modernist art that invited, if not begged interpretation. Postmodernist art simply *is* and must be experienced" (Natoli & Hutcheon, 1993:32).

Similarly, numerous communicative and literary paradigms interrelate with the paradigms in visual arts. Literary paradigms include: (1) writing to inform; (2) writing to converse with feedback; (3) writing to persuade; (4) writing to entertain; (5) writing to heal; and (6) writing to express impromptu feelings through automatic writing. Naturally, there are numerous other purposes for writing, along with numerous literary genres that fall into the overlapping and overarching paradigms. Writing to inform entails sharing facts or taking notes, just as a realistic painter would share visual "reality." Writing to converse with feedback entails mutual communication such as in debate, much like the visual artist who invites the viewer to respond. Writing that entertains includes poetry, short stories, screenplays, and novels, much like comics and cartoons amuse their readers/viewers. One heals through writing done spontaneously in journals to think things out, much as in art therapy. Meanwhile, automatic writing entails writing one's thoughts spontaneously or automatically, often through poetic forms. Not surprisingly, automatic writing is also considered by some to be therapeutic whether or not derived from spiritual inspiration. Each of these paradigms and corresponding genres share themes that pervade the visual arts and writing.

The simplest paradigm is that of storytelling and realism combined in pictorial art and expressive writing. Such a storytelling narrative is manifested brilliantly in the form of the ecclesiastical archangel, a paradigm once literary and visual, which spans from the Middle Ages to modern times. In ecclesiastical art, content triumphs over abstraction through angelic and religious paintings such as those by Giotto. In the *Last Judgment*, the two archangels adjacent to Christ bear the emblem of the Knights of the Holy Sepulcher (Bellinati, 2003:138). This is among the most conspicuous realms of angels expressed by artists, which bears a hierarchy of three triads or spheres of angels that are as protective and consoling to their human dependents. Archangels, which belong to the third triad, were considered the closest to humankind and perpetually available to intercede in the event of disaster (Lucie-Smith, 2009:24). In medieval times, cathedrals and churches housed visual art that told stories, essentially making them the Middle Age equivalent to modern movies. Such stories were entertaining and provided assurance that humankind was equally beloved and protected by the Almighty.

In *Scenes from the Life of St. Francis: Receiving the Stigmata*, Giotto (1266–1336) utilized the frescoed painting medium to show how an angel provided the saint with consolation, protection, and empowerment. Visual art, much like writing and oral storytelling, taught humans valuable lessons of virtue while it offered consolation to those who felt innocent or victimized. Simply put, the figurative art of visual storytelling was quite literally a form of schooling for medieval communities at a time when oral storytelling, which originated from one-time pagans, was more prevalent. Such figurative art remains intertwined with the oral tradition that has been handed down to modern-day instructors who have taught lessons to various generations of students.

Today, Cindy Sherman (b. 1954) utilizes art as a vehicle for autobiography or dream-fantasy by drawing upon the female archetype. In contrast, Sherman re-imagines herself in photos in which she plays the roles of dramatic women in fairy tales or myths. Such archetypal, female imagery affects the style in which today's biographies and novels are written. In *Untitled* (1985), Sherman poses as an isolated murder victim whose throat was cut (possibly a metaphor for feminine silence) before she was isolated in the forest. Also apparent is the hint of a diagonal cross placed on the ground near where her head rests (recalling Christ and other angelic-archetypal art).

Art and writing paradigms exist as profound value systems that continue to generate new masterworks of literature and art. Since it could be too problematic to attempt to distinguish between an artistic paradigm, art theory, and artistic style, one might search instead for stylistic paradigms expressed in art models. Such models might arise in the form of architecture, sculpture, and three-dimensional expression. While researchers across disciplines explore models, theories and paradigms in other fields, historians refer to art and writing more often in terms of *style*, a term that deserves substantial attention.

According to the Merriam Webster Online Dictionary, style is "a distinctive manner of expression (as in writing or speech) (Merriam Webster Online Dictionary, n.d.)" Style is also defined as "a particular manner or technique by which something is done, created, or performed" (Encylopaedia Britannica, http://www.britannica.com/bps/dictionary?query=style&header_go=,2011). One might trivialize writing styles or having to develop a stylistic voice since the writer's voice has become the topic of so much discussion in writing classes. Despite the author's reluctance to focus on writing style, there is an ongoing dialogue in this area with experts readily making such classifications in terms of style and chronology.

Content arises from the imagination of the creator, who manifests inner visions to convey messages. These messages are sometimes created in the name of someone with something to sell (commercialism) or on behalf of the inner self (fine arts). In other words, form follows function while function also follows form in the confrontation between fine and applied arts. Most importantly, the *message* is central to all forms of art, be it fine or applied, written or visual, and it unites these separate entities under a single paradigmatic umbrella. Regardless of the paradigm into which specific art products fit, such products aim generally to convey messages and leave impressions through a variety of styles. "Written words also constitute art" (Hall & Citrenbaum, 2010:259). The products of people's drawing and painting pursuits communicate their beliefs in the same way graffiti suggests multicultural messages and meanings (Hall & Citrenbaum, 2010:259).

The elements of *propaganda* and *point-of-view*, both positive and negative, are inherent in all forms of artistic self-expression. During the Renaissance, churches employed artists to convey biblical messages in cathedrals and monasteries. Today, however, corporations have their own messages through which they sell products to consumers. Clearly, such messages may be either beneficial or detrimental. Consequently, Murphy notes that art has struggled for freedom from media influence and to be disassociated from all utility (Murphy, 2008:154). In fact, by the end of the nineteenth century, utopian groups took root to challenge commercialism while promoting the alternative aestheticism. Avant garde artists, after World War I, considered abstract art to be the embodiment of harmony (Pohl:337).

The shift from the so-called realistic to the abstract was indeed abrupt. Cusumariu loosely explains paradigm shifts in visual arts, in writing, and in sculpture as changes in style or authorship. Such shifts involve re-conceptualization across mediums:

> Transitions involving conceptual change can be cited in all forms of art in poetry, from rhyming to free verse and the prose-poem; in music, from song form to the more demanding counterpoint of the fugue to the structures of the classical style to the tone painting of romanticism to neoclassicism to the serialist abandonment of tonality to its return under minimalism … (Cusmariu, 2009:164).

James Joyce presented an estimable conceptual change to Homer's from the original Odysseus (ca. 700 BC) in his reinterpretation (1922) of Ulysseus, which exemplified an obvious paradigm shift in writing that shifted from the heroic classical to mundane realism. A similar, parallel process of deconstruction can be applied by most visual artists who, according to Cusmariu, begin working in the paradigm of the figurative tradition but later become more abstract. Whereas the initiate sculptor generally prefers realistic sculpture, the sculptor later tends toward simplification evidenced by Branchusi's Prometheus (1951), an abstract sculpture of the Greek god which reflects a paucity of human features (Cusmariu, 164). Whereas realism is closely linked to aesthetics, abstraction is not constrained to focus on beauty and thus provides the artist with the greatest of all freedoms. Consequently, stylistic growth from the tangible to the intangible continues to be evident in writing as it is in the visual arts.

French Impressionism fell within both the abstract and figurative traditions, also known in contemporary parlance as *paradigm value systems*. According to Kapos, the movement began in the spring of 1874 with impressionist exhibitions of works by Pissaro, Degas, Sisley, Cezanne, Monet, Renoir, and Manet. Ultimately, what started as a student revolt blossomed into one of the most acceptable and aesthetics-movements in art, thus giving credence to the necessity to record contemporary life freshly and as it happened in a manner that evoked cheerful conviviality and primarily celebrated pure sunlight (Kapos, 1991:29). The themes of impressionism, as essential to writers as they would be to visual artists, were pleasure, light, and happiness. As did numerous significant writers, impressionists broke with orthodox traditions since it was in breaking with tradition that creators became great. By painting directly from tubes of paint and, tending to mix their colors visually on canvas, impressionists taught viewers to record and appreciate every day as opposed to depicting, faithfully and consistently the archaic storytelling paradigms related to archangels and other legends. It is thus unsurprising that people today opt to decorate their homes with art that offers an impressionistic flair, scenes dabbled with brushstrokes and ad hoc color mixes, and brought to life with pointillist blends.

Impressionism is ultimately a light art form that carries intellectual undertones, similar in some respects to today's *Chick lit* writing for its liveliness.

The much more serious development of European Modernism began in about 1908 and remained until 1933 (Arnston, 2007:27). Artists such as Emilio Marinetti, Giacomo Balla, and Gino Severini illustrated the passage of *time* by painting progressive movements as shifts from one point to the next on canvas. This contrasted with the stationary, commercial approach known as *Plakatstil*, generally flat-color posters made popular by Lucian Bernhard during the early twentieth century. *Plakatstil* painting consisted of simple outlines, coloring, and shading that were filled in with flat colors in a manner that lent itself easily to printing to promote commercial products. *Plakatstil* (1906), German for poster style marked the first time that a flat graphic style was used in commercial art, and it is exemplary of art following function (Burrough & Mandiberg, 2009:82). Later, Duchamp's bicycle wheels and urinals turned ordinary commercial products into popular art, and Man Ray's mundane photograms followed suit (Arstson:27). Russian Constructivism, a structurally and architectonic approach to painting distinct from postmodern constructivism, began in Russia in 1917 (Arntson:28–29). Later movements, such as De Stijl and Bauhaus shared a liaison with functional design while proponents of Dada and Surrealism were antagonistic to being utilized in a practical manner. *Functionality-versus-nonfunctionality* seemed evident in all areas of visual arts and writing, although writing tended to be much more functional than did visual arts.

The Dada philosophy created by poets such as Tristan Tzara challenged all previous values and aestheticisms. "Dada was anti-aesthetic, anti-rational, and anti-idealistic" (Oxford University Press, 2011). In fact, Dadaism was a reaction to dehumanization and death associated with World War I trauma which was symbolized in Marcel Duchamp's Bicycle Wheel that shocked critics as it questioned the meaning of art. Dadaists also wrote poetry in a *stream of consciousness style* by expressing the state of being in a trance (Kawamura, 2005). Dada exposed the entanglement of art with social values in a manner that fostered greater inspiration for the arts under inspection. Art tended to cultivate visual aesthetic while intellectual expression had been previously neglected. Thus, Duchamp recognized the conceptual aspect of *ready-mades* that were anti-aesthetic (Judovitz and Duchamp, 2010:6–7). Tristan Tzara, also known as "Papa Dada" and the founder of the movement, initiated nihilistic Dada through the Dadaist manifesto in 1918 (Caws, 2004:239). Perhaps, one of the greatest roles played by Dada was that of provoking a sense of shock, which Brill called the "engagement with the senseless" and which prompted the condition of cause and effect. Art no longer needed to make sense in order to be meaningful (Brill, 2010:1–2). Tzara thus experimented with poems that came about by chance as they were cut randomly from newspapers. These included automatic writing in which letters were used to obliterate words by focusing more on their sounds than the meaning they conveyed.

In the first half of the twentieth century, De Stijl, Constructivism, and Bauhaus aligned with practical design to surpass *l'art pour l'art* aestheticism. showed readers that Tzara's Dadaist poems and Breton's automatic writing served as guides to how one might produce writing (Coulter-Smith, 2010). Both Dada and Surrealism broke progressively from practical design and architecture, which lead eventually toward the postmodern movement that informed each of the paradigms.

"Classical modernism evolved within an atmosphere of tension generated by a dichotomy between figurative and non-figurative art, realistic portrayal and abstraction" (Suckale, Wundram, Prater, Bauer, & Bauer, 2007:543). Kandinsky became the first totally abstract painter recognized as such. His group, the Fauves, liberated color in a forceful manner by making it the main emphasis of painting to be balanced in a specific manner and creating space by color arrangement rather than shading. So how did abstraction impact writing paradigms?

Abstraction in writing refers to abstract concepts. The author must conduct an intensive analysis of the subject just as painters analyze dimensionality, form, content, and even the picture plane. Many parallels can be found in the abstraction processes of writers and painters alike although there is no parallel writing movement short of automatism that corresponds with the transition from classical to abstract painting. Perhaps, surrealist writing comes closest to using the written word to see the world better from an alternative perspective. In *The Surrealist Mind*, Matthews wrote:

> The surrealist writer never turned to automatic writing in hopes of getting better. His ambition was to see better. The imagery liberated by automism was by no means exclusively visual in nature, of course. All the same, the examples one remembers most vividly frequently owe their effect (taking us by surprise) to their ability to make us see something not seen before (Matthews, 1991:43).

Thus, the surrealists' interest in writing led to "stream of consciousness" writing which, while it manifested the natural flow of ideas, it paid little or no attention to punctuation and grammar. Notable writers who adopted the approach successfully include Virginia Woolf and James Joyce (Claybourne, 2008:36).

Whereas, Matisse experimented with garish and typically discordant tones, Picasso (1881–1973) constructed figures with geometric shapes which revealed the subjects from various perspectives and represented the way an individual can see more than one view at a time as the eyes flit across the subject. In some respects, this cubist abstraction was more real than formal realism. In the early twentieth century, both Picasso and Braque altered the way the world would conceive objects with the invention of Cubism (Suckale et al., 2007:546). Following their deconstruction of shapes into geometric planes, a tremendous paradigm shift occurred between the extant static and the new moving version of art. Parallels in writing can be found in William Faulkner's *As I Lay Dying* in which the narratives of 15 characters join to produce a single storyline of visual culture.

Broadening Belief Systems

In his article entitled *Visual Culture in Arts Education*, Duncum explored the tension between the concept of fine art culture and popular contemporary visual culture (new applied arts) which might as well be considered two separate but merging paradigms. He discovered that visual culture pedagogy, which employs creative inquiry, was constrained by standards mandated by states and governments (68).

Visual culture then includes exemplars such as tattoos, graphic novels, avatar design, blogging, socializing on the web and digital filmmaking; however, Duncum conceived six extraordinarily broad definitions of visual culture which, he concluded, would benefit from striving to describe the concepts more concisely. Moreover, it may be that visual culture rests within the unconstrained paradigms of the *commercial constructivist* and *art-for-propaganda* realms; however, visual culture also covers multicultural civilization. Duncum (2009) eventually classified visual culture as its own paradigm, and although there is substantial truth in that conclusion, the author of this chapter considers visual culture as being linked to broadening belief systems.

As an example of automatism in design, Rauschenberg (1925–2008), the assemblage artist, claimed to be steered by the "random order" belief system. It appeared that the elements in his work had been placed together haphazardly, the culmination of spontaneous thought after prior planning. Yet, while Rauschenberg wanted to give the illusion of random order, much of that randomness was calculated and based on aesthetics. In "Should Love Come First?" it seemed that Rauschenberg understood the "risk of authorial self-exposure" (Katz:44). Such fears of revealing too much, combined with the ramblings of literature and visual art, pertained equally to both mediums, the arts of writing and painting-collage. In reference to Rauschenberg, Katz believes that "assemblage is the perfect medium for committing the perfect crime, the one where you do not get caught—caught in authority, intentionality, or worse still, self-revelation … ." (45).

Writing and assemblage have many commonalities, most evident among them the desire to communicate ideas such as hidden messages directed at specific groups. For instance, to those who saw only assembled rubbish and drippings, Rauschenberg's assemblage art seemed devoid of meaningful content. On the other hand, Rauschenberg's content transmits messages to a target audience which understands the messages aimed at a specific, alternative cultural group that embraces homosexuality. Further examples of art and communication were to be found in the secretive exchange between Rauschenberg and Jasper Johns.

Correspondence of a symbolic nature can occur through assemblage, much as it does in writing. According to Katz, "Assemblage represented a different kind of avant-garde authoriality, one associated not with resistance but with dissimulation, not with opposition, but with camouflage" (Katz, 2008). Likewise, "protest art" had its own particular message. According to Townsend, "[protest art] is Modernism's legacy of art as utopian social agent; it is the promise of the public sphere that constitutes a functioning democracy (8)." Art has taken upon itself the function of critiquing events through a culture of opposition. "Picasso's Guernica, that moving symbol of anti-war protest from an earlier age, remains one of the most potent images we have of the present conflict" (Townsend, 9). Protest art enlightens the minds of its viewers, an excellent example being the art made by Andy Warhol with his own photos (Townsend, 2007).

Not far from such protest art, feminist art has a direct connection to expressionism since such art expresses the innermost feelings, yearnings, and desires of women as they develop previously unheard voices. Most outstanding was the work by Judy

Chicago entitled "The Dinner Party" which also integrated a book filled with dazzling photos of the exhibition and linear drawings that expressed the emotions women feel, the personal pride for organic forms of the body. Highly controversial for uncloaking the scandalous issue of the "vulva" as a means of self-expression, this work certainly aroused reaction and commentary. Chicago's interest in three-dimensional plate painting grew out of her exposure to decorated China in an antique shop, which demonstrated clearly the strong connection between decorative and utilitarian arts. The companion book produced by Chicago reveals how artists make allusions to their creative forbearers. Specifically, Chicago alluded to O'Keeffe, Woolfe and others in her celebration of womanhood and the feminine paradigm. She also confirms how visual arts can evolve into a group project, as when contributors spent countless hours engaged in needlework for a "Dinner Party." Thus, "The needlework in the Dinner Party brings women's traditional way of operating into the main arena of society" (Chicago, 1979:228).

Marc Chagall (1887–1985), an artist who expressed himself through symbolism, flowing lines, and emotional color schemes, was in a category of his own, or so it seems on the surface. Yet, just underneath the surface of his playful paintings a somber tone is evident along with symbolism that reflects the world's darker side, which stemmed from him being scorned and then persecuted by the Nazis (Baal-Teshuba, 2008:7). Chagall's work crosses many boundaries and embodies the paradigm shift between religious arts, expressionist arts, symbolism, and even multicultural art. "Chagall's work is a tribute to the power that art has to transcend political realities and allow the artist to regain a lost world within memory and imagination" (Lewis:22). Surprisingly, Chagall's work, like that of Rauschenberg, attempts to express underlying messages that support minorities, namely the Jews in Chagall's work and homosexuals in Rauschenberg's work. Although the two artists' works, separated by time and style, appear very different on the surface, both were spontaneous and colorful purveyors of emotions. In contrast to Rauschenberg and Chagall, Judy Chicago opened up more freely and blatantly in her "Dinner Party" plates with vulva imagery because hers was a pure form of protest art conceived in the name of women.

In Chagall's "Triumph of Music" (1967), the artist confirmed that visual arts, storytelling, and even music exist in unison no matter how difficult the travails of life might be. Most predominant is the symbolic use of red that pervades the canvas, penetrated by light swirling lines that move in the direction of an implied spiral center. Mythological characters have been sketched out freely to imply dance and flight contemporaneously. Upon closer inspection, the viewer observes the violin, the bugle and other instruments being used by the swirling characters in an endless spiraling dance. An archangel prevails while the rest of the cast continues an eternal dance that tells many brief stories of people ignited by rhythm. It seems Chagall is alone in his good-humored style, yet, his works include a wide array of paradigmatic shifts.

Color plays a key role in some artistic paradigms and can help determine the mood an artist has chosen to illustrate. Chagall's highly expressive and bright colors express vivid emotionality. Likewise, various color schemes including monochromatic, analogous, triadic and complimentary, have been used to create visual harmony and unity.

One cannot deny the contrast between pastel colorings and monochromatic variations between cool and warm black tones. Colors undoubtedly correspond to the notion of imagery in writing, including simile and metaphor.

Visual artists and designers also employ language, via calligraphy and letter fonts of all shapes and sizes. Stuart Davis painted the word *Champion* across his pop landscapes including one painting entitled *Visa* (1951). Later, many conceptual artists would remove all other elements from their art, leaving only the lettering. Pentak and Roth remind us of something that is already clear: As human communication began, pictograms represented the single paradigm that united words and pictures (Pentak & Roth, 2004:156).

The visual arts and writing are forever intertwined. Chagall was considered as much a poet as an artist since his best friends were poets. Even today, many artists choose to express themselves through poetry and other forms of writing. Judy Chicago integrated explanations in her diaries and lists of exceptional females with her "Dinner Table" in praise of womanhood. Last but not least, Frida Kahlo, a Mexican representative of indigenous magnetism, created diaries that integrate both handwriting and art to express her passionate love for the exceptional painter and muralist, Diego Rivera (Kahlo, Fuentes, & Lowe, 1998). Linear arts, symbolism and writing unite timelessly with each element to enhance and influence the other, thus revealing the inner sentiments of Kahlo, a passionate woman, obsessed with romantic love. Here, a pole impales Kahlo's winged, headless torso; a leg is transformed into what seems to be a wooden crutch with directions for use printed to the right. Kahlo writes "apoyo 1" and "apoyo 2," literally meaning crutch one and crutch two. A not-so-elegant bird rests upon her headless torso. Ink fades through the other side of the page, clearly hinting at the passage of time.

When asked why Kahlo created so many self-portraits, she replied, "Because I am so often alone … because I am the subject I know best" (Frida Kahlo.com, 2011). In other works, Kahlo's symbolic paintings integrate words in the paintings as explanatory notes as to why she painted such subjects, most often portraits. Above the Self-portrait with Cropped Hair (1940), Kahlo wrote, "Look, if I loved you it was because of your hair. Now that you are without hair, I don't love you anymore" (MoMA Highlights (n.d.):181). Her estrangement from Rivera did not last very long as the two remarried. However, Kahlo's estrangement from life continued to be revealed through highly symbolic pictorial art that told stories, and through written words.

Julian Schnabel (1987) painted "L'Heroïne" (1987), one of a series of 14 cognitive paintings named after the American author, William Gaddis. He utilized the background, made of tarpaulins from military trucks, that contrasted with the shape of a cross. This work possesses elements of graffiti combined with military and spiritual symbolism (Suckale et al., 2007:680). The very large and stark hand-lettering of "L'Heroïne" interrupts the significant black space and stimulates cognitive reasoning.

Where does the graffiti end while visual art and writing assume control in a concerted effort? Are these two diverse mediums really so dissimilar? Poets and creative writers are akin to visual artists who paint on abandoned walls and artists who display their works in museums. Here, the common element is the desire to communicate intense, inner feelings by visual means. Graffiti tends to challenge members of the community who fear the writing insults public property; however, its basic goal is to give meaning to existence in a manner that stimulates both imagination and discussion.

Kant distinguished the role of the imagination between ordinary thought and perception: "The two elements are synthesized by an act of the imagination that constitutes them as a single experience—the experience of seeing a book" (Encyclopedia Britannica:8). Thus, there exists a shifting paradigm between creative aesthetics and the real world in which we live. Aesthetics represent the imaginary reconstruction of the world (Encyclopedia Britannica:8). Moreover, art and natural beauty, whether in writing or the visual arts, will arouse our emotions regardless of the paradigm from which they take form. Art to inform encourages participants to examine the world more closely; art to converse encourages constructive feedback; art to persuade takes participants to new realms, both real and imaginary; art to entertain helps participants relax; and art to heal reveals impromptu feelings. Meanwhile, automatism conveys subliminal feelings that cross all paradigms to interact and merge those that apply to both writing and visual arts.

If, indeed, arts creation is a stepwise process, those who write manuscripts and create visual arts may, conceptually, visit the shifting paradigms until the artist ultimately ends up in a place different from where the creative journey began. The investigation of the problem, the attendant research, and the eventual solution will take the creator down a path of spontaneous (automatic) creation that enhances what would otherwise be closed-minded rationalism. Thus, the product includes a variety of forces that unite to form novel art. With the emergence of photomontage and digital "paint" programs, it will be interesting to see how artists who also write will amalgamate mediums to embrace the overarching paradigms of being: (1) expressive and (2) practical-communicative.

Elucidating the Paradigms of Writing

Writing shares similar purposes with visual arts. One such purpose is to convey technical information while another, situated at the other end of the spectrum, is to express emotion. By extension, while the writing process is linear for some, it is a spiral for others. Writing typically consists of three primary stages: prewriting, writing, and rewriting. However, instructors today teach the writing process as a unidirectional, linear paradigm that includes prewriting, drafting, editing, revising and publishing. Regardless of which process the writer uses to achieve his or her intended product, by employing a nonlinear paradigm, the writer will discover that his belief system evolves as he progresses through the process of external research and inner contemplation. This organic growth process is coequal to the final product regardless of which paradigmatic view prevails. In fact, the processes of research and deliberation lead to change no matter how technical or utilitarian the written product might be. Ultimately, writers will tend to gravitate toward one of two overarching paradigms: (1) the *Informative Overarching Paradigm* or (2) the *Expressive Overarching Paradigm.*

In the book entitled *Instructional Development Paradigms*, Robert E. Horn (1997:697) asserts that structured writing is itself a paradigm because it possesses two of the essential qualities noted previously by Kuhn (1962, 1970): It attracts a large group of "adherents away from competing modes of scientific activity" and (2) leaves open-ended problems to be solved by supporters. Horn believes that since the overarching purpose of writing is to communicate technical material, there must be a better way to prepare writers to communicate effectively. In his observations, Horn poses a question that is supremely relevant to anyone working from within a utilitarian-informative paradigm; specifically, "What formats are optimum to enable users to make sense of the document as a whole?" (Horn:698) He finds sentences and images to be the two main components of effective communication. Visual markers, such as headings, enable readers to comprehend general concepts more effectively. Surprisingly, even a scientific theorist like Horn is able to find relevance in visual imagery even though his focus is on the utilitarian aspect of writing.

The reasons one writes or creates can include utilitarian underpinnings such as writing to persuade, entertain, communicate, and assert one's own existence. Each reason shifts and merges the notions of expressiveness and utilitarianism. Obviously, these paradigmatic movements influence productivity. Thus, no writer ever fully escapes the influence associated with the creative or communicative aspects of writing.

So how do paradigms ultimately enlighten writers and artists alike? In general, writers will demonstrate a preference for the metaphoric *Informative Overarching Paradigm* (utilitarian writing) or the *Expressive Overarching Paradigm* (expressive writing). The former paradigm encompasses genres including nonfiction, research, biography, autobiography, journalism, essays and even creative nonfiction genres, while the latter prevails over automatic writing, poetry, free writing, fictive novels, drama, graffiti, writing words integrated into the visual arts (such as scrawling across a canvas), and visual abstraction. Purely expressive writing may be as simple as pasting a single letter in an artistic collage or digital montage; however, since the symbol itself has larger implications, even writing of a purely technical nature cannot escape the influence of expressiveness.

Expressive writers create poetry in free verse and rhyming metrics. Fenton explains that "English poetry begins whenever we decide to say modern English language begins, and it extends as far as we decide to say that the English language extends" (2004:1). As Fenton expressed in *Writing for the Eye*, poets create an array of special shapes that fill the page to form concrete images or symbols:

> Any form of lettering that draws a particular attention to itself as lettering—the beautiful fonts of the Renaissance, the chunkiest sans serif scripts—may draw such attention to a sentence, a phrase, or a word, that it becomes an object of special aesthetic interest, and therefore something like a poem. (109)

These are the poems shaped like angels' wings, such as George Herbert's *Easter Wings* or a novice's poem created in the shape of a heart. Poems written to be seen on a page rather than to be heard rely tremendously on the negative space (background) as the poem itself becomes the foreground. Here, visual shape makes as much impact as the words themselves as such. Therefore, these shaped poems cannot be communicated adequately through auditory means alone.

Correspondingly, acrostic poems begin with vertical letters that spell words from top to bottom. More traditionally, poets count their syllables line-by-line to produce Shakespearean sonnets, villanelles and Haikus in very specific forms. Such metered writing has universal appeal, although sounds often prevail over clarity of meaning.

These types of creative writing contrast with writing that strives for clarity and is thus influenced by the informative paradigm. Compared to exclusively creative writing, Strunk and White hold that clarity is of utmost importance to the writer as the purpose of writing is utilitarian by nature and fulfills the purpose of the basic transmission of ideas. Writing with clarity produces journal articles and other examples of nonfiction. "But since writing is communication, clarity can only be a virtue. And although there is no substitute for merit in writing, clarity comes closest to being one" (Strunk & White, 2000:79). Perhaps, this overwhelming emphasis on clarity serves at least to mitigate the obscurity and symbolism (multiple meanings) found often in poetry. "Think of all the tragedies that are rooted in ambiguity, and be clear!" (Strunk & White:79).

In writing, one must determine whether to abide by the rules of grammar. Clearly, such rules are critical to the *Overarching Informative Paradigm* with its emphasis on utilitarianism. At the same time, free writing poets who choose not to adhere to those rules express themselves within the *Expressive Overarching Paradigm.* According to McCrimmon, a writer persuades by appealing to his readers through "force of character," by appealing to the "weight of the argument," and by "appealing to the emotions of the reader" (McCrimmon, Miller, & Salmon, 1974:307). Consequently, the writer has to sell himself in order to be persuasive.

Notably, romance novels are influenced by both paradigms since they teach lessons and empower readers by expressing the author's imagination. "According to statistics compiled by the Romance Writers of America, romance novels account for well over 50 percent of the mass market paperback fiction sold in the United States each year" (Michaels, 2007:5). It seems almost everyone needs a love story at some time, but it is important that authors let the love grow little-by-little as the plot unfolds (Michaels, 2007:43). Obviously, readers are attracted to these novels because of their typically happy endings and the optimism that prevails as each story ends. Interestingly, romance novels are usually accompanied by examples of semi-realistic visual cover art that depicts fictive protagonists. This nuance is characteristic of applied art (illustration), such that it serves the needs of writers and artists alike to express themselves fully within a utilitarian framework.

At the same time, a triangular paradigm is undeniably evident in popular fiction writing. To incite readers' interest, the protagonist victim, usually pitied by the audience, suffers as she experiences situations of jeopardy. A male character will likely become her savior. "However, there's always the danger of having a rescuer look like a fool for plunging in without enough thought" (Card:81). Thus, the victim must either be rescued or make the decision to be self-sacrificed. Readers will not empathize with the victim if she chooses martyrdom without a logical explanation; however, they will accept her sacrifice if it stems from worthwhile motives (Card:81). While it may seem that this scenario is a metaphor for art paradigms, it undoubtedly parallels the symbolism of the holy trinity in religious art.

This triadic interaction remains essential as the *Informative* and *Expressive Overarching Paradigms* deal constantly with *clarity*. In the confusing world of overarching paradigms, both writers and artists continue to provoke and to respond to discourse while evoking constant communication and criticism. Ultimately, while a single paradigm simply cannot prevail, each creator will ultimately adopt his or her personal paradigmatic slant.

Conclusion

Art and writing fall into broad utilitarian and nonutilitarian categories; the paradigms are broader than theories and styles. One must consider Western art versus multicultural art; female versus male art; realist versus abstract; and warm color schemes versus cool schemes. Ultimately, creativity is driven and fueled purposely by these paradigms.

References

Arnston, A. E. (2007). *Graphic design basics*. Boston: Wadsworth Cengage Learning.
Baal-Teshuba, J. (2008). *Marc Chagall*. Köln: Taschen. (Original work published 2003).
Bellinati, C. (2003). *Iconographic Atlas of Giotto's Chapel, 1300–1305*. Italy: Graphiche Vianello srl.
Brill, D. (2010). *Shock and the senseless in Dada and Fluxus*. Hanover, NH: Dartmouth College Press.
Burrough, X., & Mandiberg, M. (2009). *Digital Foundations [introduction to media design]: with the Adobe Creative Suite*. Berkeley, CA: New Riders.
Caws, M. A. (2004). *The Yale anthology of twentieth-century French poetry*. New Haven: Yale University Press.
Chicago, J. (1979). *The dinner party: A symbol of our heritage*. Garden City, NY: Anchor Press/Doubleday.
Claybourne, A. (2008). *Surrealism*. Chicago, IL: Heinemann Library.
Coulter-Smith, G. (2010). Evolutionary aesthetics: Rethinking the role of function in art and design. *Technoetic Arts: A Journal of Speculative Research, 8*(1), 85–91.
Cusmariu, A. (2009). Structure of an aesthetic revolution. *Journal of Visual Arts Practice, 8*(3), 163–179.
Duncum, P. (2009). Visual culture in art education. *Visual Arts Research, 35*(Summer), 64–75.
Hall, W. M., & Citrenbaum, G. (2010). *Intelligence analysis how to think in complex environments*. Westport, CT: Praeger Security International.
Horn, R. E. (1997). *Structured Writing as a Paradigm: Instructional development paradigms*. In Dills, C. R., & Romiszowski, A. J. (Eds.), Englewood Cliffs, NJ: Educational Technology 697–698.
Judovitz, D., & Duchamp, M. (2010). *Drawing on art: Duchamp and company*. Minneapolis: University of Minnesota Press.
Kahlo, F., Fuentes, C., & Lowe, S. M. (1998). *The diary of Frida Kahlo: An intimate self-portrait*. London: Bloomsbury.
Kapos, M. (1991). *The Impressionists, a retrospective*. Southport, CT.: Hugh Lauter Levin Associates.

Katz, J. (2008). Committing the perfect crime: sexuality, assemblage, and the postmodern turn in American art. *Art J, 67*(Spring), 39–53.

Kawamura, G. (2005). The art of nothing in the 20th century. *Washington State University Pullman, Washington*. Retrieved April 7, 2011, from http://www.wsu.edu/~delahoyd/20th/project1.html.

Kuhn, T. S. (1970). *The structure of scientific revolutions* (2nd ed.). Chicago: University of Chicago Press.

Lucie-Smith, E. (2009). *The glory of angels*. New York: Harper Collins.

Matthews, J. H. (1991). *The surrealist mind*. Selinsgrove, PA.: Susquehanna University Press.

McCrimmon, J. M., Miller, S., & Salmon, W. (1974). *Writing with a purpose* (5th ed.). Boston: Houghton Mifflin.

Merriam Webster Online Dictionary. (n.d.). *Merriam Webster*. Retrieved April 4, 2011, from www.merriam-webster.com/dictionary/style.

Michaels, L. (2007). *On writing romance: How to craft a novel that sells*. Cincinnati, OH: Writer's Digest Books.

MoMA Highlights. (n.d.). *MoMA|The Museum of Modern Art*. Retrieved May 31, 2011, from http://www.moma.org/collection/object.pho?object_id=78333.

Murphy, M. (2008). Pure art, pure desire: Changing definitions of l'art pour l'art from Kant to Gautier. *Studies in Romanticism, 47*, 147–160.

Natoli, J. P., & Hutcheon, L. (1993). *A postmodern reader*. Albany: State University of New York Press.

Oxford University Press. (n.d.). *Oxford art online*. Retrieved April 7, 2011, from http://oxfordart-online.com/public/page/themes/dadaandsurrealism.

Pentak, S., & Roth, R. (2004). *Color basics*. Belmont, CA: Thomson/Wadsworth.

Strunk, W., & White, E. B. (2000). *The elements of style* (4th ed.). New York: Longman.

Suckale, R., Wundram, M., Prater, A., Bauer, H., & Bauer, E. (2007). *Masterpieces of western art* (2nd ed.). Germany: Taschen.

Townsend, C. (2007). Protest art. *Art Mon, 303*, 207–210.

Chapter 6
Economics

Luciano L'Abate

I am not an economist in any way, shape, or form. This is why I could not muster enough influence or traction to recruit or seduce a real, qualified, honest-to-goodness economist to collaborate with one chapter in this volume. However, I felt that this volume would not be complete without a chapter in this discipline. Consequently, I had to make do, in complete humility and admitted ignorance, with the help of Karen Viars, with whatever I could find by cross-indexing paradigms with economics. Then I tried to make sense of whatever information I gathered through summaries that I reproduced here with complete bibliographic acknowledgments. Whether all these summaries should be within quotes because they are essentially downloaded with few cosmetic changes is a decision that I made after discussing with two colleagues who are versed in the ethics of citing published research. The reader should understand that perhaps the whole chapter should be within quotes.

I hope readers will forgive me for trying not to leave any stone unturned in editing this volume and adding a chapter in a discipline completely unknown to me. I cannot ever square my checkbook if it were not for the daily viewing of my online bank account. I hope that some real, honest-to-goodness economist will be able to correct and improve what I have attempted to do in this chapter out of complete ignorance.

Behavioral Economics

Choice is a central construct in behavioral economics, with choice research divided into choice of concurrent alternative reinforcers, which is conceptualized as relative reinforcing value, or choice of small immediate versus larger delayed rewards, usually

L. L'Abate (✉)
Department of Psychology, Georgia State University, University Plaza, Atlanta, GA 30303, USA
e-mail: llabate3@bellsouth.net

L. L'Abate (ed.), *Paradigms in Theory Construction*, DOI 10.1007/978-1-4614-0914-4_6,

of the same commodity, which is conceptualized as delay of gratification and delay discounting (Epstein, Salvy, Carr, Dearing, & Bickel, 2010). Relative reinforcing value, delay of gratification and delay discounting paradigms can be used to study obesity, which involves strong motivation to obtain and consume food reinforcers. Strong food reinforcement and difficulties in delay of gratification are risk factors for child weight gain, and both are related to individual differences in overweight/obesity. Delay discounting interacts with food reinforcement to predict energy intake. Epstein et al. provided a selective review of research on each of these areas, and argued that the division of choice into reinforcing value vs. delay discounting derived from an arbitrary definition based on the temporality of choices.

These authors presented a model that integrates reinforcing value and delay discounting approaches. Implications of this theoretical approach to better understand excess energy intake and obesity are discussed. The ratio of reinforcing value, i.e., pleasure/reward versus delay discounting, that is, avoidance of pain approaches is interesting and perhaps related at least conceptually to L'Abate's (2011a, 2011b) ratio of pleasures over hurt feelings.

Community Economics

Lutz (1988) showed that the Ifaluk understand fago as the compassion, love, and sadness one feels for a person who is weak and dependent. She argued that fago is compelling for the Ifaluk because it fits with the strategies they must use to survive the precarious material conditions of life on a coral atoll in the Pacific. Derné (1994) attempted to show how structural realities of marriage, family, and economy shape emotional paradigms of love in three cultures. The "mythic" American conception of love as a clear, enduring feeling for a unique, special person is no longer tenable. Derné argued that this notion is compelling for Americans because of the strategies they build around the institution of marriage. Derné (1994) compared these conceptions of love with those of urban, upper-caste, north Indian Hindu men. Hindu men often see love as a dangerous emotion that might tempt people away from their obligations to the larger joint families they value. The author argued that this image of love is compelling for Hindu men because it is consistent with the recurring dilemmas they face as they try to keep joint families together. Derné's (1994) main focus was on describing the correspondence between social structure and emotion culture, also examining how social structures create persistent dilemmas that shape emotion cultures and portraying emotion cultures as complex systems that contain competing paradigms.

Agency Economics

In an agentic relationship, one party acts on behalf of another. Kayaalp (1989), for instance, examined the concept of rational egoism through a review of the literature in economics and psychology. Both fields are studied to discern whether rational egoism can be successfully reconciled with some alternative postulates stemming from behavioristic and cognitive psychology, thereby creating a more relevant paradigm of economic choice. Rational egoism was reassessed and a critique was offered of some recent behaviorist remedies. Then behaviorist remedies were compared and contrasted with those offered by cognitive psychology. Kayaalp (1989) suggested that cognitive psychology may effectively help economic theory to acquire a more relevant character. It is curious that a concept that could not be more profoundly sociological does not have a niche in the sociological literature. This conclusion is supported by Chap. 8, where no reference was cited about agency theory.

In support of Kayaalp's position (1989), Shapiro (2005) began his review with the economics paradigm of agency theory, which cast a very long shadow over the social sciences. He then traced how these ideas diffuse to and are transformed (if at all) in the scholarship produced in business schools, political science, law, and sociology. Shapiro (2005) cut a swathe through the social fabric where agency relationships are especially prevalent and examined some of the institutions, roles, forms of social organization, deviance, and strategies of social control that deliver agency and respond to its vulnerabilities, and considered their impact. Finally, he suggested how sociology might make better use of and contribute to agency theory.

Communal Economics

Consumer research on gift-giving has been dominated by a paradigm that views human behavior as instrumental exchange (Belk & Coon, 1993). Results of qualitative research on dating gift-giving among a total of 110 college students revealed an alternative paradigm of gift-giving as an expression of agapic love. The economic exchange model is characterized by economic rationality, negative and balanced reciprocity, fear of dependence, and commoditization of partners. In the social exchange model, gifts may be seen as symbols of commitment, cues to compatibility, or extensions of self. The agapic paradigm, which encompasses not only romantic but brotherly, spiritual, and parental or familial love, is marked by emotion, expressiveness, and singularization of the recipient. It is suggested that agapic expressiveness is a needed addition to exchange instrumentalism for understanding gift-giving and perhaps for understanding consumer behavior in general.

To date, the marketing literature on gift-giving has focused on two approaches or paradigms – economic and relational exchange (Hollenbeck, Peters, & Zinkhan, 2006). This study adopted a different perspective, proposing a community paradigm to provide a holistic view of gift-giving. The data (based on 20 in-depth interviews and 2 group interviews) suggested that, on the Internet, social networks of relationships cohere into gift-giving communities that influence the purchasing of gifts. Findings about online gift-giving are presented according to three community themes: (a) shared rituals and traditions, (b) shared values, and (c) shared responsibilities.

Across two studies, Fetchenhauer and Dunning (2009) asked whether people trust too much or too little, relative to what an economic analysis would suggest. In the trust game paradigm, participants decided whether to hand money over to an anonymous individual who could either return more money back or keep all the money. Participants trusted too little, in that they grossly underestimated the proportion of their peers who would return money, prompting them to forgo profitable decisions to trust. However, participants also trusted too much. Given their high levels of cynicism and tolerance for risk, few should have handed money over, yet many still chose to trust. Possible explanations for this paradox of trusting "too little" yet "too much" are discussed.

Both Agency and Communal Economics models are included in Relational Competence Theory (RCT; L'Abate, Cusinato, Maino, Colesso, & Scilletta, 2010), as Model[4] for the ability to love and Model[5] for the ability to control and regulate self. Model[4] includes a communal aspect defined by how close (approach) or how distant (avoid) we are with and from each other. Model[5] includes an Agency aspect defined by how fast (discharge) or how slow (delay) we chose to approach or avoid intimates and nonintimates. While discharge is an objective measure derived from reaction time and temporal perspective, time discounting is an attribution and therefore not a direct measure of how fast or how slow one responds (Madden & Bickel, 2010). As Green and Myerson (2010) concluded in their chapter: "… what is abundantly clear is that discounting is not all of one piece" (p. 88).

These models interact redundantly with other models of RCT. Especially relevant to this chapter is a Triangle of Life Model[7] composed of Being or Presence, being available emotionally and instrumentally to self and intimates (expanded redundantly by Model[11] about the bestowal of Importance and Model[15] about Intimacy), Doing or Performance, defined by Information and Services, and Having or Production, defined by Money and Possessions. Being includes the Communal aspect. Doing includes the Agency aspect, and Having includes the economic aspect. Combining Performance and Production produces a super-ordinate model of Power (Guinote & Vescio, 2010). Another Model[12] covers priorities, a motivational view that includes goals needs and wants.

In view of that last RCT Model[12], Bierhoff and Fetchenhauer (2006) contrasted the general theoretical framework developed by Lindenberg (2006) with other explanations of prosocial and solidarity behavior (Fetchenhauer, Flashe, Buunk, & Lindenberg 2006). Lindenberg's framing theory posits three motivational processes (frames), apparently not known to many psychologists, that could integrate both psychological and economic theories: (1) a *normative frame* is appropriate for most people but may lead to relational uncertainty and vague norms; (2) a *gain frame* is efficient but it is sensitive to disturbances by costs and low degree of monitoring; and (3) an *hedonic frame* that feels good but it is sensitive to disturbances by own moods and unfriendliness by others. In particular, Bierhoff and Fetchenhauer (2006) discussed the meta-theoretical perspectives of neoclassical economics, classical (functionalistic) sociology, psychoanalysis, personality psychology, social psychology, and theories that have been developed within evolutionary biology and evolutionary psychology.

It might be interesting to relate conceptually Lindenberg's three frames to various models of RCT. For instance, the normative frame fits Model[7] by predicting that normality is a function of an appropriate balance of all three modalities of Being, Doing, and Having according to various stages of the life cycle. The hedonic frame could be related to Model[7] on the basis of seeking pleasure and avoiding pain through Doing, as well as Model[15] in the sense that intimacy is the sharing of joys, hurt feelings, and fears of being hurt, pleasure as the avoidance of pain (L'Abate 2011a, 2011b). The gain frame would fit well within the Modality of Having money and possessions.

Consumer Economics

A synthetic, culturally relative model for the diffusion of innovations in consumer behavior provides an account of how novel items of nonlocal origin are incorporated into the material culture inventory of Hausa-speaking peasants in Niger (Arnould, 1989). The model is composed of elements from the standard diffusion paradigm, world-systems theory, and economic and symbolic anthropology. Data from field work in Niger show that novel goods provide a medium through which alternative paradigms of consumer behavior and reality contend. Among the Hausa, a premarket model, a Western market-mediated model, and an Islamic ethnonationalist model compete for consumer affiliation.

Campbell (1991) discussed the upsurge of interest in the topic of consumption in both the social sciences and humanities. Some of the common factors underlying this new wave of research were identified, while the more significant developments in anthropology, sociology, economics, psychology, and history are reviewed. Campbell concluded that the preparadigmatic phase of the new wave is coming to an end and that critiques of the consumer society are now giving way to more objective and detailed studies. The continuing diversity of perspectives currently employed to study consumption suggests that no single paradigm has become widely accepted.

Although relationship marketing has developed into the prevailing marketing paradigm, it frequently encounters resistance from the demand side. Both management practitioners and academics indicate that at least some consumers show reactance against loyalty programs, i.e., against tactical instruments of relationship marketing. Nevertheless, relationship marketing has widely neglected reactance theory. Wendlandt and Schrader (2007) attempted to close this gap. Based on the fundamental principles of loyalty programs and reactance theory, Wendlandt and Schrader (2007) presented a set of hypotheses on the determinants and effects of situational consumer reactance against loyalty programs. This set tested these hypotheses on the basis of 388 face-to-face interviews with bookstore customers. These interviews included a between-subject manipulation on the reactance effect of economic, social-psychological, and contractual bonding potentials. To test the proposed hypotheses, the authors applied structural equation modeling. As expected, contractual bonds provoked reactance effects, while social-psychological bonds neither increased reactance, nor the perceived utility of the program. Economic bonds raised perceived utility up to a certain threshold level, from which the reactance effect dominated thereafter. As a consequence, a cautious and limited application of customer loyalty programs is advisable. The developed consumer reactance scale can help managers to evaluate the effects of planned or implemented customer retention measures. This is the first attempt to investigate situational reactance in a loyalty program setting.

Cultural Economics

Chen, Peng, and Saparito (2002) presented a cultural perspective on the assumption of opportunism in transaction cost economics (TEC). Researchers have criticized the TEC paradigm for over-generalizing the assumption of opportunism as human nature. The authors suggested that opportunistic propensity is affected by cultural prior conditioning of individualism–collectivism. Specifically, it was proposed that individualists have a higher opportunistic propensity in intra-group transactions, and collectivists in inter-group transactions. This cultural specification of opportunism is seen to help TEC to more effectively accommodate some criticisms and more realistically deal with problems of economic organization in today's global economy.

An assumption generally subscribed to in evolutionary economics is that new technological paradigms arise from advances is science and developments in technological knowledge (van den Ende, & Dolfsma, 2005). Further, demand only influences the selection among competing paradigms, and the course of the paradigm after its inception. In this article, the authors argued that this view needs to be qualified and modified. van den Ende and Dolfsma (2005) attempted to demonstrate that, in the history of computing technology in the twentieth century, a distinction can be made between periods in which either demand or knowledge development played the bigger role in shaping the technological paradigms. In the demand-enabled periods, new technological (sub-) paradigms in computing technology have emerged as well.

Economic interdependence of diverse and dissimilar cultures is the fruit of the global community phenomenon of the 1990s and the seed of twenty-first century political and economic transformation (Foster, 1997). Emerging from profound political change in Eastern Europe, Latin America and Africa, and coupled with a proliferation of market economies and commercial interpenetrations of hitherto closed societies, the international system has become a global village, where the rules of conduct of economic enterprise entail intricacies and sensitivities heretofore largely incongruous to mainstream organization theory. Interdependency implies relationships, and the pursuit of entwined cross-cultural economic relationships carries the social baggage and community traditions of the user and the used, buyer and seller, source and producer, native and alien.

Organization theorists have attempted to address the complexities of these interdependencies in disparate theoretical dialogue, much of which focuses on the development of organization theory based on a new dichotomy of the individual and the social ethic, basic tenets of individualism and collectivism which represent the cultural arena of global economics. But the ideal source for new organization theory may lie outside the customary domain of the organization theorists in the literature of sociology, ethics, and philosophy. This literature suggests that the communitarian ethic may provide a vehicle to traverse the dialectic, because it may be used to address the dualities of diversity and unity in both individualist and collectivist societies.

Using experimental paradigms from economics and social psychology, Savani, Markus, and Conner (2008) examined the cross-cultural applicability of three widely held assumptions about preference and choice: People (1) recruit or construct preferences to make choices; (2) choose according to their preferences; and (3) are motivated to express their preferences in their choices. In six studies, they compared how middle-class North American and Indian participants choose among consumer products. Participants in both contexts construct nonrandom preferences at similar speeds. Those in Indian contexts, however, are slower to make choices, less likely to choose according to their personal preferences, and less motivated to express their preferences in their choices. The authors inferred that the strong link between preferences and choices observed among North Americans is not a universal feature of human nature. Instead, this link reflects the disjoint model of agency, which prescribes that people should choose freely on the basis of their preferences. In contrast, Indian contexts reflect and promote a conjoint model of agency, according to which agency is responsive to the desires and expectations of important others and may require restraining one's preferences.

The purpose of Bose and Thomas' (2007) article was to link two key disciplines in finance and science in a way which is representative of the many challenges in the development of the knowledge economy. However, the valuation of intangibles remains a contentious issue in finance. Design/methodology/approach: In this article, the financial and economic issues pertaining to the valuation of intellectual capital are evaluated and addressed. An important aspect of raising capital is the ability to impute a fair value on the asset. The investigative process involved evaluating and assessing the appropriateness and efficiency of current models and finds them to be inadequate in yielding the true value of intellectual capital employed in knowledge-intensive firms. A new methodology for valuation is indeed required, and this value-driven valuation processes may produce significantly better understandings of the worth of intellectual capital in the knowledge economy. This article evaluated various methods that are currently used and recommended the development of a valuation process for new and evolving technologies.

Health Economics

Bickel, Hughes, DeGrandpre, and Higgins (1992) examined the effects of price manipulations on the consumption of concurrently available coffee and cigarettes. During fifteen 4-h sessions, coffee and cigarettes were concurrently available to seven healthy adult volunteers according to schedules of reinforcement. After consumption stabilized under 100 for both reinforcers, the response requirement for each reinforcer was varied separately, while the response requirement for the other reinforcer was kept at 100. Increasing the value decreased coffee and cigarette consumption to a similar degree. Also, as the price for cigarettes increased (and consumption decreased), coffee consumption decreased; however, as the price of coffee increased, cigarette consumption did not change. Findings illustrate the viability of using behavioral economics to examine drug self-administration in a choice paradigm.

Social differences in the role of habits in health-related behavior are explored within both sociology and economics (Lindbladh & Lyttkens, 2002). The corresponding theoretical perspectives are the *habitus theory*, the theory of individualization, and habits as rational decision rules. Sixteen thematically structured interviews were conducted with participants of different occupations. Three aspects of habits emerge: Association between habits and preferences, habits as a source of utility, and the relationship between habits and norms. It was found that people in lower social positions are more inclined to rely on habits and accordingly are less likely to change their behavior. These differences are reinforced as not only the disposition to maintain habits but also the tendency to conceive of the habitual as something good seems to be strengthened in lower social positions. Also, it was noted that the intensified individualization characterizing current society erodes the basis for habit-governed behavior, which may also contribute to social differences in well being. Finally, it was found that scientific dialog has enriched both scientific paradigms, and as a tentative hypothesis it is suggested that the traditional economic rational-actor model may be relatively less applicable to those with limited resources.

Effective advocates for people with intellectual and developmental disabilities must adopt strategies that are well founded, contemporary, and effective (Shaddock, 2003). In Western societies, there is a growing rejection of a strict economic rationalist agenda and an attempt to find a "third way" that recaptures some of the features of the welfare state. The context in which people with disabilities now seek to be heard and supported is one in which the focus is on economic, environmental, and social issues – the so-called "triple bottom line." This approach adopts the discourse of inclusion and pursues economic viability with a renewed commitment to equality in a civil society. The new paradigm also offers new opportunities for persons with disabilities and their supporters to demonstrate creativity.

Models or paradigms of disability are used to guide health care professionals' perceptions so that they can serve people with disabilities, enhance their futures, and facilitate the resources they need (Hubbard, 2004). Health care curricula, which in essence train students to make such decisions, are influenced by these models. The medical model, which locates disability within the individual, assumes the individual with a disability is a victim who must be cured or made more normal. The functional-limitation paradigm expands on the medical model, focusing on the interaction of physical or mental limitations with social and environmental factors. The economic model, based on the concept of employability, emphasizes a health-related inability (or limited ability) to work rather than physical functioning of the individual. The sociopolitical model views disability as a policy and civil rights issue. Health care professionals face a dilemma as the disability rights movement demands a shift in social power from the paternalistic view of the medical model to the autonomist view of the sociopolitical model. The question is asked if curricula are preparing our future health care professionals to distinguish how to view each situation and each individual through the lens of the appropriate model.

de Souza (2009) used the case study method to investigate the processes used by a local nongovernmental organization called the Society for People's Action for Development to organize sex workers in the slums of Bangalore, India, for HIV/AIDS prevention. The nongovernmental organization-facilitated HIV/AIDS program is based on the new paradigm of community organizing that encourages community participation and capacity building. Grounded in the culture-centered approach, this study documents the processes used to organize the women, while highlighting the role of communication in these processes. The study identified four primary processes used to mobilize the community, namely collectivization, community awareness and sensitization, capacity building, and providing legal education and support.

Each of these processes highlighted the importance of attending to the economic, social, and political realities that shape the health of women. The common thread linking these processes together is the notion of "voice." More specifically, each process serves as a catalyst to produce discursive practices that enable women to provide support to each other, increase awareness in the community about the problems that they face, build self-reliance through financial skills training and communication training, and defend their legal rights. In addition, the study suggests that the primary role of nongovernmental organizations should be the creation of "communicative spaces," which are discursive and material spaces within marginalized communities and mainstream society where cultural participants can identify problems (oftentimes beyond the realm of health), manage solutions to those problems, and advocate for health and social change.

Marketing Economics

Grönroos (1994) discussed the nature and sometimes negative consequences of the dominating marketing paradigm of today, marketing mix management, and furthermore discusses how modern research into, for example, industrial marketing and services marketing as well as customer relationship economics shows that another approach to marketing is required. This development is supported by evolving trends in business, such as strategic partnerships, alliances and networks. Grönroos (1994) suggested relationship marketing, based on relationship building and management, as one emerging new marketing paradigm of the future. He concluded that the simplicity of the marketing mix paradigm, with its four P model, has become a straitjacket, fostering toolbox thinking rather than an awareness that marketing is a multi-faceted social process, and notes that marketing theory and customers are the victims of today's mainstream marketing thinking. By using the notion of a marketing strategy continuum, Grönroos (1994) discussed a number of consequences of a relationship-type marketing strategy for the focus of marketing, pricing, quality management, internal marketing, and intraorganizational development. The author briefly commented on the possibility of developing a general marketing theory based on the relationship building and management approach.

Timmermann (1995) took a critical approach to the individualization thesis from the perspective of educational economics, investigating whether the individualization thesis can be used in the controversy about human capital theory, which presents a strict distinction between human beings and their labor power or labor potential as well as between the value of a human being and the value of the labor power of that human being. Timmermann (1995) concluded that the subject-related model of rational educational decisions underlying human capital theory can be rediscovered in the individualization thesis. He speculated about a phenomenon that is posed by developments in the economics and sociology of education during the 1970s and 1980s that seemed to counteract each other. The human capital paradigm dominated the economics of education thinking until the early 1970s and fawned upon the homo economics model of individual rational choice. That model came under severe criticism that pointed increasingly to the relevance for educational behavior of social background variables and of milieus located in the social structure of the society. As a result, debates in the economics of education became more open to arguments from the social science perspective. The basic hypothesis is that the theorem of individualization interferes in this dispute of paradigms – voluntary or nonvoluntarily – in favor of human capital theory by asserting the erosion of the power of the traditional social structures to stamp behavior: It strengthens the human capital theory way of looking at the world, and it relieves that theory from the pressure to legitimize itself.

Nida-Rümelin (1991) presented a thesis that a critique of decisions is not necessarily a critique of preferences. This thesis runs contrary to the fundamental assumption in economic theory (ET) that a critique of decisions will always be a critique of preferences, since decision behavior is a manifestation of preferences. If this thesis is correct, then the paradigm of instrumental rationality is in trouble. Preliminary remarks on the ET of rationality in general were presented, and the cooperation problem was discussed as a challenge to the ET. Attempts to "save" the ET were critiqued, and it was concluded that practical reason is concerned with actions and not with preferences.

The strategic management field has derived many of its theoretical concepts from other social science disciplines including economics, psychology, and sociology (McWilliams & Smart, 1993). Industrial organization (IO) economics and the structure-conduct-performance paradigm, in particular, provided many of the building blocks upon which strategy formulation was constructed. However, some researchers are now questioning this transfer of theory from IO economics to Strategic Management This paper aimed at investigating the possibility of a relatively uncritical transfer of theory from IO economics to Strategic Management and to assess whether this transfer of theory has led to inappropriate or costly generalizations.

Specifically, it is suggested that the structure-conduct-performance (S-C-P) paradigm of traditional IO economics has been transferred to and accepted in the strategic management discipline, and that the transfer has been pervasive and with little consideration of competing paradigms in the economics discipline. Further, it was proposed that the resulting emphasis on industry structure may be costly to strategy researchers and practitioner.

Social Economics

Zafirovski (1998) attempted to put the relationship between socio-economics and rational choice theory into proper perspective. This attempt was motivated by some recent confusions between the two. In particular, there was a tendency among the adherents of rational choice theory to subsume socio-economics and, for that matter, social science as a whole under their theory. On the other hand, many socio-economists or economic sociologists evinced some degree of lenience vis-à-vis rational choice theory as if the latter were fully compatible with their discipline. Zafirovski (1998) showed that the rational choice reduction of socio-economics and sister disciplines (e.g., economic sociology) to the "economic approach to human behavior" or to a utilitarian paradigm of socio-economic life, is theoretically and methodologically untenable, and thus that many socio-economists' generous attitude to rational choice theory is not entirely justified.

In this examination of the relationship between quality of life and the environment, Kilbourne (2006) argued that a broad framework within which the relationship can be established is required. The framework used is that of the dominant social paradigm (DSP) by Pirages and Ehrlich (1974) that contains political, economic, and technological institutions. These institutions determine both the quality of life and environmental constructs within any society. Kilbourne (2006) attempted to integrate these three areas in a single model that hypothesizes a particular relationship between the DSP, the environment, and quality of life.

Block and Kramer (2009) explored superstitious beliefs as a basis of product performance expectations and their impact on initial purchase likelihood and subsequent satisfaction. In doing so, they attempted to demonstrate instances when superstition-driven expectations cause consumers to make purchase decisions that run counter to economic rationality. In the first set of studies they found that Taiwanese consumers are relatively more likely to purchase a product with positive superstitious associations based on its "lucky" color, and are more likely to purchase and are willing to pay more money for a product with a smaller but "lucky" number of units contained in the package (e.g., eight tennis balls compared to ten). In contrast, consumers who do not hold such superstitious beliefs adhere to the more rational choice paradigm. Next, Block and Kramer (2009) showed that the differences in purchase likelihood are driven by superstition-based performance expectations. They further generalized these findings to product satisfaction, and find support for expectation disconfirmation sensitivity as a moderator of the effect.

Naylor (2004) review of Birch and Paul's (2003) book was framed within the debate of two philosophies: Economic rationalism, which relies on market forces, stresses productivity, and focuses on the bottom line of profits; and economic sustainability, which focuses on the triple bottom line of profits, people, and the planet. Coming from an economic sustainability paradigm, the authors argued that we must focus on quality of life to achieve balance between work and life, and between earning profits and moving the world into a sustainable environment. Their thesis was that "an organization's efficiency and profitability increase when the quality of life of its employees improves." In other words, employees who work fewer hours, spend more time with their families, and develop social relationships in their communities, are more productive workers. In order to achieve this balance, the authors called "… for a more generous, more compassionate and a more humane working life."

Neuroeconomics

Glimcher (2003) suggested a Carrtesian model of the brain and behavior. Descartes believed that all behaviors could be divided into two categories, the simple and the complex. It is argued here that Cartesian dualism operates from the fake premise that the reflex is able to describe behavior in the real world that animals inhabit. A mathematically rich cognitive theory could solve the most difficult problems that any environment could present, eliminating the need for dualism by eliminating the need for a reflex theory. Such a mathematically rigorous description of the neural processes that connect sensation and action has its roots in microeconomic theory. Economic theory allows physiologists to define both the optimal course of action that an animal might select and a mathematical route by which that optimal solution can be derived. The author outlined what an economics-based cognitive model might look like and how one would begin to test it empirically. The author also discussed the history of neuroscience and related questions about determinism, free will, and the stochastic nature of complex behavior.

Ortmann and Rydval (2004) reviewed Glimcher's (2003) book, in part history of neuroscience primer, in part philosophy of science reflections, in part a description of path-breaking neuroscience experiments and in part a proposal for a paradigm shift in neuroscience based on modern economic theory – is an amazing tour de force. The book has two parts. The first part of the book is an acquired taste. Some reviewers have lauded it and even considered it the major contribution of the book. Chapters 10–12 describe a series of experiments with humans and monkeys whose aim was to show that economic theory describes reasonably well not only human behavior but also the "behavior" of neurological circuits that produce the behavior. The book describes to some extent a reductionist program and is likely on these grounds not to sit well with those who like to conceptualize humans as more than stimulus-reaction machines, if rather noisy ones. The results of the author and his colleagues are indeed a very beautiful and fascinating demonstration that using game theory as a tool for analyzing the brain–behavior link is a viable and promising research strategy.

To make sound economic decisions, the brain needs to compute several different value-related signals. These include goal values that measure the predicted reward that results from the outcome generated by each of the actions under consideration, decision values that measure the net value of taking the different actions, and prediction errors that measure deviations from individuals' previous reward expectations. Hare, O'Doherty, Camerer, Schultz, and Range (2008) used functional magnetic resonance imaging and a novel decision-making paradigm to dissociate the neural basis of these three computations. Our results show that they are supported by different neural substrates: Goal values are correlated with activity in the medial orbitofrontal cortex, decision values are correlated with activity in the central orbitofrontal cortex, and prediction errors are correlated with activity in the ventral striatum.

Organization Economics

Throughout the growth period of Japanese manufactured exports, from the 1960s to the late 1980s, cultural arguments abounded to explain the influence of Japanese management practices and their role in Japan's competitiveness (Rafferty, 2001). The conclusions drawn were that Japanese organizations and management practices would not be transferable to foreign environments, since their cultural advantages could not be maintained. In addressing the question of intercultural transferability, this article presents a theoretical perspective on the issues of management practice as a means of identifying the causes and constraints inherent in the transfer of the Japanese model to the West and in particular to the UK. The theoretical propositions are underpinned with reference to empirical studies and hopefully identify areas for further research.

Kilbourne, Beckmann, Lewis, and Van Dam (2001) examined environmental attitudes of university students in England, Denmark, and the United States. A total of 386 participants from the three samples completed questionnaires assessing their positions on technology, politics, and economics; environmental attitudes and beliefs; and perceived amount of change necessary to alleviate environmental problems. Results indicate a significant relationship between attitudes toward technology, politics, and economics, the primary elements of the DSP, and environmental attitudes. Specifically, as beliefs in the elements of the DSP increase, the perception of the existence of environmental problems decrease. As a result of this decrease, perceived changes necessary to alleviate environmental problems also decrease. Results also indicate that there is a direct relationship between the DSP and perceived change. It is stated that the policy implications of this result suggest that what is needed is not only increased concern for the environment but also education about the DSP and its effect on the environment. Individuals may then break the cycles of technological advance, increased consumption, and reform politics, all of which appear to be complicit in environmental decline. Items used in the analysis and their statistics are appended.

Conceição and Heitor (2003) argued that innovation, a broad social and economic activity within the emerging learning societies, transcends any specific technology, even if revolutionary, and is tied to attitudes and behaviors oriented toward the exploitation of change by adding value. These authors analyzed the ongoing Portuguese path toward an innovative society, characterizing innovation in Portugal within the European context. Conceição and Heitor (2003) conceptualized "learning" and the process of knowledge accumulation as a framework to understand the new demands for being innovative. They concluded by suggesting elements for innovation policies for Portugal, arguing for the need to promote systems of innovation and competence building.

The family is a natural organization with its own specific roles, power hierarchy, status, and functionality. For example, David and Vicarelli (1989) analyzed data from studies on the family carried out in the 1970s and 1980s in an area of widespread productive decentralization in Italy. The role of the family was examined within this type of production context. The economic-territorial paradigm made it possible to identify the existence of a type of family with specific prerogatives in the economic and productive fields. Results underlined the full ability of the family to play an active economic role in society. Data are discussed in terms of utilization of public social and health services and different configurations of family strategies with respect to their economic-social position, life-cycle, and generation differences.

Hesterly, Liebeskind, and Zenger (1990) argued that organizational economics (OE) has made an important contribution to the understanding of the nature and purpose of organizations. The paradigm has expanded and redefined the concept of organization and has led to a resurgence of functionalist theory building. OE's value as a research tool is examined for (1) empirical consistency, (2) logical coherence and falsifiability, and (3) scope and parsimony. Organization theory can advance the usefulness of OE by refining existing theory, and expanding the paradigm's phenomenological scope.

Although previous research provided a foundation for developing, maintaining, and exiting relationships, the extant literature has yet to consider the influence that patterns of economic and social forces have in guiding the future of deteriorating relationships. To understand better and respond to relationships in decline and to salvage relationships that are destined to fail needlessly, Gassenheimer, Houston, and Davis (1998) used a political economic paradigm to identify symptoms of deteriorating relationships and provide a framework for combining relational forces that best guides relationship retention decisions. They propose a model based on the theoretical foundations of transactional cost analysis (TCA), social exchange, and distributive justice. Using relative dependence, interdependence, and mutual dependence to define the economic and social worth of the relationship, they incorporate dyadic patterns of behavior to illustrate similar and different interpretations and evaluations of fairness and the impact on relational outcomes. Managerial implications were discussed.

Godard (2002) drew on the new institutionalism in economics, sociology, and political studies in order to establish a foundation for analyzing how states shape employer human resource management and union relations. He then reviews and extends the available literature on this topic, establishing how, in addition to legal regulation, states help to shape the cognitive and normative rules that undergird employer decision processes, the social and economic environment within which employers act, and ultimately, the relations of authority constituting the employment relation itself and hence employer policy orientations. The article concluded with a discussion of the prospects for state policy initiatives in view of established employer paradigms, institutional logics, and state traditions, and identifies possibilities for further work in this area.

To get higher profits we need to work harder and longer. Right? Wrong! In this original book, Birch and Paul (2003) looked at the working lives of employees, managers and executives, and propose a new agenda for achieving a work/life balance. If companies get their relationships with employees and customers right, profits will follow as a matter of course. But too many organizations focus only on the short-term bottom line and are insensitive to the emotional and spiritual needs and dignity of employees. There is abundant evidence of mounting domestic dilemma as a consequence of industrial modernization, and social decay that remains unrelieved even as the political economy speeds along on paths uncontemplated in an ever-quickening metamorphosis.

As the communitarian self discovers her membership in society, so organizations and nations have discovered their membership in a global community. The communitarian self is embedded in society; organizations are likewise finding themselves implanted in societies with which they may not only be unfamiliar, but perhaps to which their customary goals and methods may be antithetical. This imbroglio fuels a demand for organization theory which addresses principles of mutual service, reciprocity, responsibility, and communion – concepts inherent to the communitarian ethic. One might frame the argument thusly: The inclusion of a communitarian perspective in organization theory provides an efficacious response on the part of American organizations to the challenges of international economic and political change, escalating intercultural competition and domestic social decay.

Psychological Economics

The integration of economics and psychology has created a vibrant and fruitful emerging field of study. Chapters in Frey and Stutzer (2007) took a broad view of the interface between these two disciplines, going beyond the usual focus on "behavioral economics." As documented in this volume, the influence of psychology on economics has been responsible for a view of human behavior that calls into question the assumption of complete rationality (and raises the possibility of altruistic acts), the acceptance of experiments as a valid method of economic research, and the idea that utility or well-being can be measured. Contributors, all leading researchers in the field, offered state-of-the-art discussions of such topics as prosocial behavior and the role of conditional cooperation and trust, happiness research as an empirical tool, the potential of neuroeconomics as a way to deepen understanding of individual decision making, and procedural utility as a concept that captures the well-being people derive directly from the processes and conditions leading to outcomes. Taken together, these chapters offered an assessment of where this new interdisciplinary field stands and what directions are most promising for future research, providing a useful guide for economists, psychologists, and social scientists.

Gagné (2008) reviewed Frey and Stutzer's (2007) edited book. The editors' goal in this book was to review some recent applications of psychological theories in the field of economics. Given economics' high reliance on agency theory, this book provides a fresh look into what standard economics has had a difficult time explaining: What economists have called "errors" in individual behavior and decision making. The gist of the book focuses on showing that these errors are in fact part of the human make-up, and the contributors use psychological theories and evidence to build a new economics paradigm. The book purports to question the assumption of the human as a rational being, to show the value of experimental methods in economics, and to make well-being a worthy outcome in economics. This book is of interest to economists who have an interest in microeconomics and want to break free from the narrow focus of agency theory. It is also of interest to psychologists who want to extend their work on prosocial behavior and well-being to include economic and institutional considerations as part of their accounts.

Socio-Economics

Etzioni (1986) contended that psychologists can formulate a new paradigm for the study of economic behavior. Economic behavior requires a distinct paradigm because rational-choice behavior is a distinct subject. The new paradigm of socio-economics must initially be parsimonious without being underdetermined. It must answer four key questions: How do people actually make choices when they aspire to act rationally? What is the role of emotions and values, next to intracognitive factors, in such decisions? What is the role of groups in setting individual choices? And, what is the role of society?

Sears (1991) discussed an alternative to the neoclassical economic paradigm. The new, more encompassing paradigm includes cognitive processes and the role of affect and values, individual processes and group structures, and symbolic factors. By the same token, Lamal (1992) tried to identify the major sources of misunderstanding of behavior analysis (BA) to demonstrate the similarities and differences between BA and socio-economics (SE). Both BA and SE (1) stress the necessity of taking the context of behavior into account and (2) reject the neoclassical economic paradigm. SE emphasizes the importance of hypothesized inner causes of behavior; BA does not. BA exalts experimental analysis; SE does not. SE is a distinct improvement over neoclassical economics.

Social capital concepts and theories have brought about an evolution of thinking in the field of economics and expanded the boundaries that traditionally separated economics and other social science disciplines (Storberg-Walker, 2009). Traditional (e.g., neoclassical) economics with its relatively under-socialized view of human nature – where individuals are seen as autonomous, rational decision makers – has had to develop new ways of understanding how "the social" influences affect economic behavior. Much of the theorizing around social capital has actually evolved from disciplines outside of economics, which has led to a truly multidisciplinary and evolving theoretical base. This article examines the most highly cited social capital theories, identifies their economic foundations, and analyzes how economic theories have been incorporated into the development of alternative social capital theories. It is then examined how heterodox economics might potentially influence HRD and suggest that a more nuanced understanding of the theoretical foundations of social capital will generate better and more diverse HRD theories and practice.

Conclusion

Bolacchi's (2008) chapter begins with the statement: "After more than a century of development, the social sciences, particularly economics, psychology, and sociology, face basic internal problems that each separately seems unable to solve." These three areas of the social sciences are discussed with an eye to how they might be integrated, thus producing a new paradigm. Whether that integration is possible and whether an overarching paradigm is desirable remains to be seen, as discussed in Chap. 23. Whether Bolacchi's attempt at integration is successful or not remains to be seen also. There is no question that each social science discipline is part of a largest whole. Whether the whole is better than its parts remains a desirable ideal or an actual reality depends on whether an overarching paradigm can perform that function.

References

Arnould, E. J. (1989). Toward a broadened theory of preference formation and the diffusion of innovations: Cases from Zinder Province, Niger Republic. *The Journal of Consumer Research, 16*, 239–267.

Bolacchi, G. (2008). A new paradigm for the integration of the social sciences. In N. K. Innis & N. K. Innis (Eds.), *Reflections on adaptive behavior: Essays in honor of J. E. R. Staddon* (pp. 315–353). Cambridge, MA: MIT.

Belk, R. W., & Coon, G. S. (1993). Gift giving as agapic love: An alternative to the exchange paradigm based on dating experiences. *The Journal of Consumer Research, 20*, 393–417.

Bickel, W. K., Hughes, J. R., DeGrandpre, R. J., & Higgins, S. T. (1992). Behavioral economics of drug self-administration: IV. The effects of response requirement on the consumption of and interaction between concurrently available coffee and cigarettes. *Psychopharmacology, 107*, 211–216.

Bierhoff, H., & Fetchenhauer, D. (2006). How to explain prosocial and solidarity behavior: A comparison of framing theory with related meta-theoretical paradigms. In D. Fetchenhauer, A. Flache, A. P. Buunk, S. Lindenberg, D. Fetchenhauer, A. Flache, & S. Lindenberg (Eds.), *Solidarity and prosocial behavior: An integration of sociological and psychological* (pp. 225–242). New York, NY: Springer Science.

Birch, C., & Paul, D. (2003). *Life and work: Challenging economic man*. Sydney: University of New South Wales Press.

Block, L., & Kramer, T. (2009). The effect of superstitious beliefs on performance expectations. *Journal of the Academy of Marketing Science, 37*, 161–169.

Bose, S., & Thomas, K. (2007). Valuation of intellectual capital in knowledge-based firms: The need for new methods in a changing economic paradigm. *Management Decision, 45*, 1484–1496.

Campbell, C. (1991). Consumption: The new wave of research in the humanities and social sciences. *Journal of Social Behavior and Personality, 6*, 57–74.

Chen, C. C., Peng, M. W., & Saparito, P. A. (2002). Individualism, collectivism, and opportunism: A cultural perspective on transaction cost economics. *Journal of Management, 28*, 567–583.

Conceição, P., & Heitor, M. V. (2003). Systems of innovation and competence building across diversity: Learning from the Portuguese path in the European context. In L. V. Shavinina & L. V. Shavinina (Eds.), *The international handbook on innovation* (pp. 945–975). New York: Elsevier Science.

David, P., & Vicarelli, G. (1989). Social and economic aspects of the family in an Italian peripheral area. *Marriage & Family Review, 14*, 145–157.

de Souza, R. (2009). Creating "communicative spaces": A case of NGO community organizing for HIV/AIDS prevention. *Health Communication, 24*, 692–702.

Derné, S. (1994). Structural realities, persistent dilemmas, and the construction of emotional paradigms: Love in three cultures. In W. M. Wentworth, J. Ryan, W. M. Wentworth, & J. Ryan (Eds.), *Social perspectives on emotion* (Vol. 2, pp. 281–308). New York: Elsevier Science/JAI.

Epstein, L. H., Salvy, S. J., Carr, K. A., Dearing, K. K., & Bickel, W. K. (2010). Food reinforcement, delay discounting and obesity. *Physiology and Behavior, 100*, 438–445.

Etzioni, A. (1986). Socio-economics: A proposal for a new interdisciplinary field. *Journal of Social Behavior and Personality, 1*, 475–482.

Fetchenhauer, D., & Dunning, D. (2009). Do people trust too much or too little? *Journal of Economic Psychology, 30*, 263–276.

Fetchenhauer, D., Flashe, A., Buunk, A. P., & Lindenberg, S. (Eds.). (2006). *Solidarity and prosocial behavior: An integration of sociological and psychological perspectives.* New York: Springer-Science.

Foster, J. (1997). Individuals, communities and organizations: A communitarian paradigm for cross-cultural enterprise. *Dissertation Abstracts International: Section A, 57*, 4532.

Frey, B., & Stutzer, A. (Eds.). (2007). *Economics and psychology: A promising new cross-disciplinary field.* Cambridge, MA: MIT Press.

Gagné, M. (2008). Review of 'Economics and psychology: A promising new cross-disciplinary field'. *Canadian Psychology/Psychologie Canadienne, 49*, 343–344.

Gassenheimer, J. B., Houston, F. S., & Davis, J. (1998). The role of economic value, social value, and perceptions of fairness in interorganizational relationship retention decisions. *Journal of the Academy of Marketing Science, 26*, 322–337.

Glimcher, P. W. (2003). *Decisions, uncertainty, and the brain: The science of neuroeconomics.* Cambridge, MA: MIT Press.

Godard, J. (2002). Institutional environments, employer practices, and states in liberal market economies. *Industrial Relations: A Journal of Economy and Society, 41*, 249–286.

Green, L., & Myerson, J. (2010). Experimental and correlational analyses of delay and probability discounting. In G. J. Madden & W. K. Bickel (Eds.), *Impulsivity: The behavioral and neurological science of discounting* (pp. 67–92). Washington, DC: American Psychological Association.

Grönroos, C. (1994). From marketing mix to relationship marketing: Towards a paradigm shift in marketing. *Management Decision, 32*, 4–20.

Guinote, A., & Vescio, T. K. (Eds.). (2010). *The social psychology of power.* New York: Guilford.

Hare, T. A., O'Doherty, J., Camerer, C. F., Schultz, W., & Range, A. (2008). Dissociating the role of the orbitofrontal cortex and the striatum in the computation of goal values and prediction errors. *The Journal of Neuroscience, 28*(22), 5623–5630.

Hesterly, W. S., Liebeskind, J., & Zenger, T. R. (1990). Organizational economics: An impending revolution in organization theory? *The Academy of Management Review, 15*, 402–420.

Hollenbeck, C. R., Peters, C., & Zinkhan, G. M. (2006). Gift giving: A community paradigm. *Psychology and Marketing, 23*, 573–595.

Hubbard, S. (2004). Disability studies and health care curriculum: The great divide. *Journal of Allied Health, 33*, 184–188.

Kayaalp, O. (1989). Reconciling economic postulates: Does 'adaptive' egoism satisfice? *Journal of Behavioral Economics, 18*, 289–306.

Kilbourne, W. E. (2006). The role of the dominant social paradigm in the quality of life/environmental interface. *Applied Research in Quality of Life, 1*, 39–61.

Kilbourne, W. E., Beckmann, S. C., Lewis, A., & Van Dam, Y. (2001). A multinational examination of the role of the dominant social paradigm in environmental attitudes of university students. *Environment and Behavior, 33*, 209–228.

L'Abate, L. (2011a). *Hurt feelings: Theory and research in intimate relationships*. New York: Cambridge University Press.

L'Abate, L. (2011b). *The seven sources of pleasure: Making way for the upside in the midst of modern demands*. Newport, CT: Praeger.

L'Abate, L., Cusinato, M., Maino, E., Colesso, W., & Scilletta, C. (2010). *Relational competence theory: Research and mental health applications*. New York: Springer-Science.

Lamal, P. A. (1992). Behavior analysis and socio-economics: Some similarities and differences. *The Journal of Socio-Economics, 21*, 173–180.

Lindbladh, E., & Lyttkens, C. (2002). Habit versus choice: The process of decision-making in health-related behaviour. *Social Science & Medicine, 55*, 451–465.

Lindenberg, S. (2006). Prosocial behavior, solidarity, and framing processes. In D. Fetchenhauer, A. Flache, A. P. Buunk, S. Lindenberg, D. Fetchenhauer, A. Flache, & S. Lindenberg (Eds.), *Solidarity and prosocial behavior: An integration of sociological and psychological perspectives* (pp. 23–44). New York, NY: Springer Science.

Lutz, C. (1988). *Unnatural emotions*. Chicago, IL: University of Chicago Press.

Madden, G. J., & Bickel, W. K. (Eds.). (2010). *Impulsivity: The behavioral and neurological science of discounting*. Washington, DC: American Psychological Association.

McWilliams, A., & Smart, D. L. (1993). Efficiency v. structure-conduct-performance: Implications for strategy research and practice. *Journal of Management, 19*, 63–78.

Naylor, L. A. (2004). Review of 'Life and work: Challenging economic man'. *Journal of Organizational Change Management, 17*, 228–231.

Nida-Rümelin, J. (1991). Practical reason or metapreferences? An undogmatic defense of Kantian morality. *Theory and Decision, 30*, 133–162.

Ortmann, A., & Rydval, O. (2004). Review of 'Decisions, uncertainty, and the brain. The science of neuroeconomics'. *Journal of Economic Psychology, 25*, 891–894.

Pirages, D. C., & Ehrlich, P. R. (1974). *Ark II: Social response to environmental imperatives*. San Francisco, CA: W.H. Freeman.

Rafferty, J. (2001). The synthesis of economic organization and the management of change: Intercultural transferability and the Japanese paradigm. *Management Decision, 39*, 415–420.

Savani, K., Markus, H., & Conner, A. L. (2008). Let your preference be your guide? Preferences and choices are more tightly linked for North Americans than for Indians. *Journal of Personality and Social Psychology, 95*, 861–876.

Sears, D. O. (1991). Socio-economics: Challenge to the neoclassical economic paradigm. *Psychological Science, 2*, 12–15.

Shaddock, A. J. (2003). People with disabilities in the era of the 'triple bottom line'. *Journal of Intellectual and Developmental Disability, 28*, 90–93.

Shapiro, S. P. (2005). Agency theory. *Annual Review of Sociology, 31*, 263–284.

Storberg-Walker, J. (2009). Heterodox economics, social capital, and HRD: Moving beyond the limits of the neoclassical paradigm. *Human Resource Development Review, 8*, 97–119.

Timmermann, D. (1995). Human capital theory and the individualization theorem. In G. Neubauer & K. Hurrelmann (Eds.), *Individualization in childhood and adolescence* (pp. 223–245). Berlin: Walter de Gruyter.

van den Ende, J., & Dolfsma, W. (2005). Technology-push, demand-pull and the shaping of technological paradigms – Patterns in the development of computing technology. *Journal of Evolutionary Economics, 15*, 83–99.

Wendlandt, M., & Schrader, U. (2007). Consumer reactance against loyalty programs. *Journal of Consumer Marketing, 24*, 293–304.

Zafirovski, M. (1998). Socio-economics and rational choice theory: Specification of their relations. *The Journal of SocioEconomics, 27*, 165–205.

Chapter 7
Education

Laura G. Sweeney

Understanding the hierarchy of paradigms, including how they influence theories and models, is the foundation for achievement in local and global education. However, paradigms in education cannot be discussed without first differentiating between a model, a theory, and a paradigm. All too often, educators confuse these three entities which belong to the hierarchy of education. More importantly, understanding the paradigm's hierarchical structure and its relationship to educational theory is essential to model building and broader educational change. For the most part, educators believe *erroneously* that these teaching components are interchangeable. Yet, theories and models differ from paradigms and are established to support a greater philosophical and archetypal framework, which is referred to as the paradigm. Unfortunately, the research suggests that these terms continue to be misused and inappropriated not only in psychology, but also in education, which concurs with L'Abate's (2009) observations that the terms have indeed been misused in various domains.

Paradigms, theories, and models are referred to frequently and inaccurately as simple examples of terminology. Moreover, the terms should be distinguished from each other so that researchers and theorists can appreciate that paradigms are conceptual constructions which control, or should control, the formulation of models and theories (L'Abate, 2009). Although conceptual paradigms are referred to inaccurately within other fields such as psychology or business, the author will focus in this chapter on three common paradigms rooted in the practical world of instruction. Further, some of the theories and models that relate to these paradigms will be discussed. These paradigms will be compared to education models and theories in an effort to determine whether paradigms are indeed overarching entities that are part of a pyramidal hierarchy or act as underlying foundations. Ultimately, it will be shown that paradigms are not necessarily situated at the top of the realm but can also

L.G. Sweeney (✉)
Avant Garde Books, 301 Briarvista Way N.E.,
Atlanta, GA 30329, USA
e-mail: laurasweeneyworks@gmail.com

L. L'Abate (ed.), *Paradigms in Theory Construction*, DOI 10.1007/978-1-4614-0914-4_7,

be found at the bottom of the inverted, solid pyramid. Thus, both are overarching and provide foundational elements.

According to Kuhn, paradigms "define scientific disciplines at certain points in time" and identify what is to be studied, the questions to be asked, and the nature of the questions along with the results (Experiment Resources & Shuttleworth, 2008). A paradigm shift is defined as a "series of peaceful interludes punctuated by intellectually violent revolutions" (Kuhn, 1962, p. 10). From the author's perspective, paradigms provide general frameworks, generalizations, and *colored lenses* that impact the body of principles, philosophies, and psychologies that seek to explain phenomena. Thus, it becomes apparent that both the pyramidal structures and frameworks function contemporaneously as paradigms.

Similarly, educators view the paradigm concept as a vast array of impressionist strokes that penetrate theory-building across the curriculum. Given that paradigms are more intuitive than theory and model-making, they will be more difficult to verify. Essentially, they bear comparison to the artist's spontaneous brushstrokes that add texture to an otherwise sparsely developed canvas. For instance, many individuals within an educational system share ideas about cooperative and collaborative inquiry, a spontaneous meeting in which people paint the educational canvas with ideas that originate in the larger community. This collaborative inquiry engages students of all ages in an innovative paradigm. Participants gather to answer questions they consider significant to their respective groups (Kasl & Yorks, 2010). The experiential nature of this participatory paradigm enables all involved in the process to contribute ideas that are very much like impressionist brushstrokes.

Paradigms lead to theories which are also known as a body of principles. These explain phenomena such as perennialism, idealism, realism, pragmatism, reconstructionism, and Protest Theory (including behaviorism, existentialism, and postmodernism). These philosophical education theories originated primarily in Europe and can be attributed to intellectuals such as Cornelius, Locke, Rosseau, Pestalozzi, Froebel, and Herbart (Pulliam & Van Patten, 2007). Moreover, each theory has unique dimensions, such as teaching methodologies and examinations, which are represented in various models. These models, also referred to as a structural designs and sets of plans, are dependent upon the larger theory they must support. In contrast, a theory represents an analysis of facts and the relationship they have with each other. Ultimately, a theory is a general, testable principle that is used to support a paradigm. Simply put, theories and models support the various conceptual constructions of broader paradigms while the paradigms themselves are infused with theory.

An educational philosophy does not become theory unless it includes the three dimensions of: (1) the role of the teacher; (2) the instruction methodology; and (3) evaluation methods. For example, the teacher's realist/essentialist role is to present the material in a "highly organized and very exact manner." Realist instruction methodology includes "lectures, demonstrations, sensory experiences, and teaching machines." Furthermore, realist/essentialist assessments must be "objective," which implies that the answers given are either correct or incorrect (Pulliam & Van Patten, 2007, p. 37). The realist/essentialist model that stems from the theory represents the systematic organization of the disciplines within the university's ordered body of knowledge. This theory conveys the need to immerse the learner in an environment rich in objects and practical experiences from which she will develop rationality. But still, what *exactly* is a theory?

According to Hillix and L'Abate (Chap. 1), "A theory is a conceptual framework constituted by models that allow direct empirical validation." Theories explain a category of observable facts, and one such theory is Transactional Distance Theory (TDT) developed by Moore in 1993. TDT is a global theory that has arisen in the midst of the distance education enterprise to clarify the meaning and effectiveness of distance learning and explain the relationship between teachers and learners separated by space. TDT is composed of models, such as the Transactional Distance Model adapted from Saba, and is populated with feedback loops that pass back and forth between teacher and student (Gokool-Ramdoo, 2008). Whereas this theory and others have been validated empirically, models such as the Transactional Distance Model differ since they simply support theories in a concrete, visual manner. For example, the Transactional Distance Model communicates effectively through an illustrative design to those who need to know that the theory entails continuous shifts of interaction between participants. This theory, along with its model, makes comprehension palatable to the learner.

Program Theory (Fear, 2007) comprises logic collected by an association or group and includes a hypothesis, convictions, and ideologies that solve problems. At first, one finds philosophies that have compelled the group to adhere to specific epistemological and ontological frameworks. A model results from these philosophies which categorize nonoverlapping ideas to facilitate their evaluation. This summative evaluation enables policy makers to see whether or not the policy is sufficient. Program theory thus enables researchers to understand the synthesis of models and paradigms. Of note, the box model arises from a logic paradigm. A logic model is subcategorized into the four components of: (1) inputs; (2) actions; (3) outputs; and (4) impact. Such self-contained box models suggest clearly that, by itself, the logic paradigm alone is flawed (Fear, 2007). Despite the limitations of Program Theory, however, it demonstrates clearly the symbiotic relationship between models, theories, and paradigms.

After researching the interconnectedness and use of paradigms, models, and theories in education, the author was dismayed to discover that these terms have been used interchangeably, apparently without regard to their fundamental differences. Under similar conditions and absent further clarification, scholars will continue to pursue research that reflects incorrectly the way these three concepts work together. In fact, the terms are neither interchangeable nor indistinguishable since their functions are remarkably diverse. While the number of paradigms, models, and theories is limited only by our imagination and innovativeness, one must consider that the paradigm embraces the other two components addressed here: Without paradigms, one has no theories, and without theories, conceptualists cannot build models.

Until now, research efforts have proven fruitless as scholars have used these forms sloppily and interchangeably for some time. In fact, the author noted that some models of the three terms place the hierarchy at different levels. For instance, the author noted on one occasion that the term "*model*" had been promoted above theory, a hierarchy that would compel one to question how a theory could be based upon a model. Instead, models should be based on theories. By extension, paradigms influence theories, and theories rightfully influence models. Of course, whether these forms are positioned at the top or bottom of the pyramidal structure is another question.

Here, the paradigm concept should be represented at the bottom – as its foundation – or at the top so that it may be recognized as an overarching concept. Alternatively, a paradigm might be represented by dots interspersed throughout the theories and models they influence. Such a representation would make it apparent that the concepts within the paradigm penetrate both models and theories concurrently. Because they are the stuff from which fruitful change is made, one would be unable to find hypothesis and structures without first having a proper understanding of paradigms. Regardless of where the paradigm is located on the diagram that illustrates its functions, the paradigm must always be present in the foundations that provide support for theories and models. For this reason, this chapter includes an illustration of three prevalent models of the paradigm as they work in conjunction with models and theories. First, an examination of three prevalent paradigms is in order.

Prevalent Education Paradigms

For discussion purposes, the author has coined names for the three prevailing paradigms that influence the hierarchy of theories and models:

1. Learning Standards
2. Consistent and Interconnected
3. Conceptual Constructivist

The three names help represent the unifying concepts that inform numerous education disciplines which range from the arts to the sciences. The theories tend to support these paradigms and others like them, and while the paradigms may have other names, they continue to exhibit paradigm shifts as society and knowledge change and grow. New paradigms such as these will continue to emerge as long as novel ideas are still being conceived. In short, they are interrelated, interconnected, and interdependent because they all influence one another and merge or shift into each other.

One example of paradigm shift has been described by Gokool-Ramdoo (2008) as the change from "organizational synergy to the transactional synergy" as distance-learning schools no longer focused on logistics but on making distance learning as excellent as its on-site counterpart learning (Gokool-Ramdoo, 2008, p. 3). Readers will certainly coin other names for the paradigm shifts that follow. However, for the purposes of discourse and classification, the author refers to three common paradigms under which one will find the historical, psychological, philosophical, and founding theories of education.

Learning Standards Paradigm

When educators work within the parameters of the Learning Standards Paradigm (LSP), their ultimate goal is to meet the specific benchmarks predetermined by the curriculum and to conduct testing to ensure that outcome. Such benchmarks vary according to disciplines, but testing goals remain the hallmark of achievement. From the LSP, countless theories have sought to describe and realize how students will pass tests and even surpass their peers locally and internationally. Yin Cheong and Mok (2008) coined the term "*first wave*" for a standards-based paradigm in which the curriculum mandates specific goals the learner must meet. It follows, then, that the effectiveness of the paradigm is determined according to whether or not learners meet those predetermined goals. According to Yin Cheong and Mok, the classroom must include specific factors, such as high standards for educational delivery, which encourage and enable students to meet those goals. A model classroom of that type promotes high expectations via effective classroom management techniques that emphasize rewards, teaching, feedback, extra help, flexibility, and an optimum curriculum. The underlying theory behind "*first wave*" presupposes that if all of the factors listed by Yin Cheong and Mok are present, a model classroom will ensue (Yin Cheong & Mok, 2008, p. 368). Assuming all the "necessary" components are stirred into the mixture, the final product should be optimum.

As a data-driven and strategy-based instructional program, the curriculum used by Learning Focused Schools is subsumed into the LSP along with many other models. As originated by Thompson (2008), teachers are expected to write essential questions that focus on the benchmarks and state standards that must be mastered in order to pass tests. The various strategies and techniques employed in Learning Focus Schools seem to be as important as the instruction itself. Teachers are required to work together as a team by utilizing consistent and interconnected strategies. This approach is intended to help schools meet their Annual Yearly Progress (AYP) requirements and compete successfully with rival schools for high passage rates. The utilization of graphic organizers and visuals paired with incessant questioning remains a positive reinforcement for students with a range of learning modalities. Therefore, the Learning Focused Turnaround/Transformation Model meets the needs of learners with a variety of learning styles due to its implementation of various graphic organizers, essential questions, and summarizing strategies. The model's creators boast that the model is driven by state standards, and in fact, the mapped curriculum model can be revised as needed to meet the ever-changing needs gauged through continuous evaluation. Professional development is administered regularly to demonstrate how to implement, sustain, monitor, and assess student data, the goal of which is to attain high test scores and increase student learning (Thompson, 2008).

In schools, the application of behaviorist theories stems from the prevalence of the LSP which supports test measures. According to Morrone and Tarr, "The behaviorist approach to learning is consistent with a transmission model in which the instructor is the possessor of knowledge. The knowledge is transmitted to the learner with little interaction between the instructor and the students" (Morrone & Tarr, 2005, p. 8).

Thus, reading long passages or watching videos allows students to absorb ideas without asking questions aloud. For instance, some online colleges require their students to perform college level work, such as reading long discourses by the professor. This is followed up with multiple-choice tests that reveal instantly whether or not the student can recall all of the information transmitted by the professor. Under these conditions, students are not given the opportunity to challenge the material by completing a subjective examination, although a discussion board may be available to entertain remarks and questions. If students fail to give the correct answers or if they have alternative perspectives, they must accept, according to the behaviorist approach, that they have erred. In fact, a graduate course offered by Chapman University evaluates student learning based solely on the results of a final, multiple-choice exam. The advantages of such a curriculum are that it saves the instructor valuable time that would otherwise be needed to grade exams, that the curriculum is efficient and that the final evaluation is objective.

In contrast, Shattuck believed that the "standards movement" is a deception because schools should not set standards without first developing the curriculum upon which they are based (Shattuck, 2005). Noddings questioned if a single set of standards should be created for all or if different standards should be prescribed for students who will never use algebra. Another issue Noddings mentioned is that universal standards may exist, among which reading is a Cardinal Principal (Noddings, 2007, pp. 208–209). Such standards should indeed reflect behavioral conduct and the level of proficiency before they are developed. Nevertheless, are those standards sufficient to ensure social interconnectedness?

Consistent and Interconnected Paradigm

The Consistent and Interconnected Paradigm (CIP) embodies the concept that what is taught must be consistent and interconnected throughout the various disciplines represented in the curriculum. All involved, teachers and learners alike, must work in unison to enhance each other's strengths. Because these paradigms often overlap each other, it should be no surprise that the theories employed by Learning-Focused Schools, 2010 fit well into this paradigm and the LSP. Consistency is maintained by presenting similar methods and materials to classes or by providing similar content. Of note, the concepts taught must not conflict unless there is sufficient logic to justify that divergence, and the team or educational leader must be aware of that divarication. In some instances, teachers work together to keep various disciplines in concert by teaching content across the curriculum or by integrating various subjects into a single lesson plan. An excellent example of this may be seen in art teachers and science teachers who collaborate to demonstrate the differences and similarities between mixing light pigments and paint pigments. Further, they utilize reading assignments intended to help students applying this knowledge to design theory. Perhaps a math teacher could expand these efforts by utilizing mathematical equations to determine color-mixing recipes.

The "*second wave*" paradigm described by Yin Cheong and Mok (2008) adheres to the Consistency and Interconnectedness paradigm since everyone involved is an investor in student success. As a "classroom for services stakeholders," the "second wave" is essentially an accountability paradigm in which education is conceptualized as a service to shareholders. Essential emphasis is placed on the school's ability to meet the needs of students and families. Employers and the community, and schools are expected to act as vessels of accountability and competition (Yin Cheong & Mok, 2008). In this second wave paradigm, classroom effectiveness is judged primarily on whether it meets the needs of the market. Test results ensure that learners are not only able to fulfill mundane business functions, but that they can also compete on a global level. Success entails satisfying the investors, primarily the taxpayers and businesses that stand to benefit from former students who move up the ranks of business (Yin Cheong & Mok, 2008). Above all, everyone in the "second wave" works together.

Collaborative/cooperative inquiry tends to be consistent and interconnected as everyone involved is at once a contributor and stakeholder. Using the artist's canvas metaphor for such cooperative investigation, each participant would add a unique brushstroke to the painting. According to Kasl and Yorks, "The epistemic principle posits that meaningful knowledge generation can grow only from the knowledge-maker's personal experience; this principal derives from the phenomenological assertion that one can best understand human experience by being inside that experience" (Kasl & Yorks, 2010, p. 317). In other words, each participant adds perspective to the group and contributes to answering the common question selected by the larger assembly. Afterwards, administrators seek to apply emergent theory to fit the needs of the collective community.

Paradigms of an *impressionistic* nature, such as collaborative/collective inquiry aim to benefit co-inquirers more so than the performance of the organization. At the same time, it is clear that a school's administration may seek to intervene in favor of meeting performance requirements rather than changing systems. The goal is that all members of the group, including the administrative body, become peers who work together to transform the core identity through emancipatory learning. The leader becomes a peer while the individual who initiates the inquiry becomes the observer. By extension, the participants become researchers. Such methods may well be applied to the various arts.

Arts-based methodologies, whether post-structural or performing, offer ways to address an array of nontraditional questions that disrupt qualitative and quantitative research paradigms. Authenticity and expressiveness govern arts practices in a manner that questions standardized learning, and these circumstances open pathways to new interpretations of past truths. By casting aside these past methodologies, learners make way for the very personal and independent conceptual constructivist domain (Rolling, 2010). Visual arts, music, dance, and automatic writing enable learners to engage free-flowing, spontaneous intuition, which generates greater analysis at higher levels. The practice of arts becomes a new teaching methodology capable of delving deep into the core of education, and it is from there that the Conceptual Constructivist Paradigm (CCP) emerges.

The Conceptual, Constructivist Paradigm

The CCP represents abstract reasoning at its most prolific and highest level. Educators encourage their students to develop abstract associations in the classroom learning environment. They guide students to comprehend underlying meanings and interpret them analytically in order to make fresh, constructivist discoveries. For instance, teaching students to appreciate a discipline may fall under the CCP if teachers convey the idea that learning must be meaningful, especially to the learner who conceptualizes and creates. Cheng and Mok's "*third wave*" paradigm falls into this category. According to Cheng, the "third wave" would be a new, flexible life-long paradigm in which students are free to study in accordance with their own schedules as they learn to think globally. Essentially, students learn to learn in an open and flexible environment (Yin Cheong & Mok, 2008, p. 376). Here, the goal is not only to retain knowledge, but also to advance it for the future. Learning must be both individualized and globalized through the integration of multiple intelligences. Computer-mediated instruction emphasizes constructivist education as do the "*third wave*" and the Conceptual Construction Paradigm.

Constructivism is a pedagogical orientation which conveys the argument that all knowledge is constructed rather than received. Its educational roots can be traced to Piaget, whose work was both structuralist and constructivist. According to the constructivists, students must seek their own ends in learning rather than receive messages from lectures. Active learning based on Piaget's "clinical method" is considered by constructivists to be superior to passive learning. Therefore, students must question why they are learning particular topics and why knowledge is essential in order to become empowered members of society (Noddings, 2007, pp. 129–131). Such is the focus of the Dynamic Assessment and Remediation Approach (DARA).

DARA employs dialogue and scaffolding techniques that are integrated into the assessment process. In short, it supports the theory that an interactive process will give educators a better understanding of student learning. While student reading is assessed continuously, students are encouraged to an even greater degree to think for themselves to find answers. The process-oriented DARA model entails testing, teaching, and re-testing while scaffolding throughout the journey. Unlike traditional testing methods, students are permitted to build and generate answers in a constructivist manner based on the comprehension they bring to the classroom. Whereas traditional teaching focuses on the product, DARA, as is the case with other constructivist approaches, focuses on a process that includes the dynamics of ability-level, culture, learning style, and environment. According to Macrine and Sabbatino (2008), "Too often, learners are not involved in their own evaluation" (2008, pp. 52–76). In contrast, DARA theory suggests that evaluation is of little use if educators and students do not build continuously on the results.

The DARA model is based upon two of Vygotsky's theories: (1) Language is for social communication (Vygotsky, 1985) and (2) that there is a zone of proximal development for higher intellectual purpose (Vygotsky, 1978, 1985) leading to socially mediated activities. Scaffolding is at the heart of the DARA model.

Further, the model and the theory fall readily into the Conceptual Construction Paradigm since students ask their own questions. Macrime and Sabbatino write, "Through a dialogical model, students are empowered with the knowledge, vocabulary, and labels to articulate the gist of a story, or what they should do when they don't know a word. Along with the teacher, students develop strategies in an authentic way to form generalizations, and develop a global gestalt of the processes involved in becoming a better reader" (2008, pp. 52–76).

Fortunately, today's computer-mediated communication stimulates critical thinking and constructivism, or the construction of meaning by students in the new model of content analysis. Buraphadeja and Dawson (2008) reviewed computer-mediated communication to gain a better understanding of whether this constructivist theory enabled students to think critically. This type of research is under the umbrella of the Conceptual Construction Paradigm since the researchers involved encourage their students to analyze and create as they learn. Computer-mediated curriculum theory is self-reflective and fosters an open-minded and cognitive presence. Critical thinking remains essential at the higher education level, which demands that one remain open-minded and able to make critical decisions in the changing contexts of learning. The theory holds simply that online scholars must advance learner–content, learner–instructor, and learner–learner interactions in an investigative community under the supervision of experienced instructors able to present online instruction which facilitates high-level communication. Essentially, the teacher or facilitator leads the class by encouraging students to ask and answer challenging questions. This approach is based on research into online instruction that utilizes Socratic questioning, which encourages learners to utilize higher levels of Bloom's taxonomy in their academic considerations (Buraphadeja & Dawson, 2008).

As recommended by De Freitas and Neumann, Synchronous Audiographic Conferencing (SAC) involves the presentation of live communication through various media by utilizing social presence, collaborative engagement, webinars, web casting, experiential learning, and problem solving in a community of inquiry (De Freitas & Neumann, 2009). SAC meets the needs of learners who would otherwise be isolated in an asynchronous environment without being exposed to the tutor. SAC benefits learners in nontraditional contexts worldwide and encourages them to work together in a context that stimulates deep discussion. The tools of the trade are the virtual whiteboard, testing, and multi-wave live audio transmission. In addition to other live interaction via technology, cues, emoticons, voting boxes, and instant feedback stimulate student participation. Additionally, instructors work concurrently with groups of varying sizes. SAC may also fall under the framework of the LSP since it includes online questionnaires and instant analysis of results. Above all, students integrate the social and global components to construct meaning via collaborative problem solving and dialogue.

TDT is a direct outgrowth of distance education. Over time, it has become a discrete academic discipline, the goal of which is to present education that surpasses on-site instruction. "Theory becomes important because it allows the exploration of more sophisticated issues that allow for more predictable generalizations" (Gokool-Ramdoo, 2008, p. 6). TDT entails transactional dialogue between the learner and the learning institution with a focus on individual learning and meta-cognition through reflective practice. Since there is more distance between the learner and tutor, the student assumes a more constructivist responsibility, which Saba (2007) refers to as the "feedback loop." As the participants proceed to generate constructivist meaning, various components are explored through TDT, such as cognitive, meta-cognitive, socio-economic, and affective issues, in addition to the constraints represented by geographical and social distance (Gokool-Ramdoo, 2008, p. 11). Thus, TDT readily falls into the Conceptual Construction Paradigm, sharing the global, communicative context with SAC. Perhaps TDT could readily be applied to distance writing activities in the psychology field where feedback is constant and a psychology-curriculum might emerge in a paradigm that integrates curriculum and psychology.

The model known as "Moore and Kearsley's Systems Model of Distance Education" (1966) stems directly from TDT. Various TDT components have been simplified and categorized into the well known, instructional design ADDIE model (analysis, design, development, implementation, and evaluation), with an additional E reflecting the learning environment context. Six rectangles are aligned on a vertical axis beginning with the *analysis* of student needs, organizations, theory, history, and philosophy in the first section. The second rectangle reflects the program's design, media, and evaluation. The model eventually notes the various delivery means, the learning environment, and evaluation which includes every available evaluative process (Gokool-Ramdoo, 2008, p. 12). This exemplary model reflects the values specific to the larger scale TDT while embracing the greater Conceptual Construction Paradigm described by the author of this paper.

Gokool-Ramdoo also proposed a model referred to as extending the applications of TDT. This model stems from Moore and Kearsley's model and TDT (2008, p. 13) and shows the cascading effect that occurs from one level to the next. Since quality assurance plays such a significant role in this model, it reflects the context and attitudes of the LSP and represents a paradigm shift between the Conceptual Construction Paradigm and the LSP. Simply put, this new model reflects the values of business accountability in addition to the positive effects of transactional distance within two paradigms. Perhaps what Gokool-Ramdoo has suggested is that two aspects of education, the teacher–student relationship and quality assurance, must meet business and national needs.

Music education theories typically fall under the CCP as music is born out of constructivist thought. Fernandez questioned what he explains as the music education paradigm in which students fear displeasing the teacher or being humiliated during performance practice. Some typical questions include: Did you practice? Have you memorized that movement? Music educators presume that their students seek to become noted performance artists; yet, the common assumption is that, unlike visual arts education which admits students at all levels, post-high school music education is reserved for the "talented" students. Music students are viewed as artistic geniuses or entertainers who are accountable only to themselves

(Gaztambide-Fernandez, 2010). As one might expect, however, perhaps, the goal in music is to construct or perform new compositions that emerge from the minds of learners. The standards of music-learning are reflected in the student's ability to perform the music accurately.

The ability to teach music, art, or literature for appreciation rests entirely on the foundation for the Conceptual Construction Paradigm. As models, portfolio creation and presentation fit into any of the three major paradigms, depending upon whether the portfolios are conceptual, meet specific standards, or are interconnected with the general curriculum. Portfolio creation models may be implemented to support any number of theories ranging from behaviorist theories to constructivist theories. In behaviorism, students are given rewards for having assembled an excellent portfolio. Conversely, since various modalities of knowledge expression are acceptable according to the Learning Styles Theory, student portfolio could be constructed kinesthetically (through collage) or visually with lines that evoke rhythm. Students create portfolios that are relevant to their own lives and illustrate the importance of student relevancy. Meanwhile, the content-based portfolio delivers, often in written form, the information the student has received during lectures.

One must look at the theories and their applicable models to discern how they relate to the greater paradigms within the hierarchy of education. Hierarchy is just as important in education as it is in psychology since understanding how the mind learns and how people progress through learning stages remains the foundation of teaching. Therefore, one must divide the many education paradigms into their primary components in order to learn how each relates to other paradigms. The paradigms are evident at the top of an overarching model, under which can be found the various theories embraced by the paradigms. Within the theories are models, all of which must be confirmed in the classroom, since without verification, they will become obsolete. Educators should not mingle paradigms, theories, and models haphazardly if these learning strategies are to remain relevant. According to L'Abate (2009), "Without hierarchy and order, there is confusion." Critically, the same conditions apply to education. Teaching strategies benefit from precise formulation to ensure they adhere to hierarchical vision, the dominant, overarching archetype under which theories and models are found. Theories must result in two- and three-dimensional models, or they will not be well defined.

Paradigms, theories, and models in education fulfill essential, although different, functions. Paradigms are the abstract constructions supported by theories because without paradigms, theories would not have conceptual focus. Conceptual focus is the concentrated effort to spotlight essential aspirations that are not necessarily goals. Paradigms unite theories under conceptual umbrellas where the various parts work together in organized unison rather than chaos. Models are the basic elements of theories in their simplest forms. A model lends a better understanding of the linear aspects of a theory, these components under the umbrella of the paradigm affecting positive outcomes in the classroom. Consistency and interconnectedness at all three levels essentially guarantee school effectiveness.

Educational Theories

For the sake of discussion in this chapter on education paradigms, theories and models, the author has selected five major categories of educational theories that fit the parameters of various paradigms. Of course, more than five categories of educational theories exist since the number of available theories continues to grow. Nevertheless, most educators who have completed training programs are familiar with theories that fall into the following general categories: (1) behaviorism theory; (2) theories related to student-relevancy, which hold that teachers should make learning relevant; (3) theories related to societal needs, which imply that schools must meet the needs of business and society; (4) learning style theories that cater to the various learning modalities; and (5) content-based theories in which the teacher is the students' purveyor of information. These categories have been simplified since it would be impossible to describe them in depth in this chapter. Additionally, each of the prevalent philosophies of education refers to theoretical constructs that fall into one or more of these five domains. Therefore, categorizing these theories provides readers a framework with which they can conduct the necessary comparisons and contrasts in order to understand the hierarchy, dependence, and independence of theories-in-use.

According to Edwards, theories and theories-in-use do not coincide entirely in a practical sense. Theories-in-use enable administration of the environment in which one works to include explaining, predicting, and controlling actions that become reproducible. Reflection-in-action aligns theories-in-use with representative learning outcomes and course content (Edwards, 2010). The role of theorists is to analyze practice at work or in school environments. Theories-in-use exist to be challenged in learning-centered environments through student feedback that may reveal incongruence and inconsistency. This outcome will result in redesigned curricula that mirror thoughtful education practice.

Grossman presented theories by Dreyfes and Dreyfes, also known interchangeably as models since they include diagrams and illustrations. Grossman (2008) examines the impact of theory on teaching and has constructed a model to illustrate this practice; however, no delineation of the differences in language usage is apparent when referring to either the theory or model in his work. Once again, the author finds that theories and models are terms used interchangeably in this attempt to determine how teachers learn while in the role of students.

Noddings notes that philosophers have typically analyzed and clarified theories and arguments in education. Today, however, philosophers create some theories and study education from philosophical perspectives (Noddings, 2007, pp. xiii–xiv). One of Socrates' most enduring theories holds that if the instructor asks questions, the student will reflect appropriately and respond. Dewey, for example, sought answers to questions regarding natural phenomena and objects in his environment. Dewey's theory was that students would grow when given the opportunity to pass through evolutionary experiences that lead to more education. According to his theory, education had to have an aim, which is what one refers to today as a goal. Students, he posited, had to launch their own objectives connected to prior experiences.

Dewey's theory included a model for solving problems. First, the problem appeared and confounded the learner who would resolve to establish a hypothesis. Next, the hypothesis would be tested. This would disclose consequences and results to be evaluated. The result was a generalization of knowledge; however, students could move through the learning model in the order they preferred (Noddings, 2007, p. 30). In short, Dewey's theorization was founded on the need to undergo experiences. Such problem-solving entails suffering the consequences of one's hypothesis. Educators can thank Dewey for many of the trial-and-error characteristics that are now essential to constructivism.

Student-Relevancy and Related Theories

Teaching-for-appreciation is the archetype for an effective education theory (or structural design), which enables educators to achieve the goals of the greater learning paradigm (Brophy, 2008). Such a beneficial theory, through which students realize that learning remains applicable to successful living, facilitates the creation of models of student engagement. Students must consider the three important aspects of learning; namely, how education will affect them now in the context of their peers, whether they will succeed in the learning environment, and what future benefits will accrue from their studies. Value in education comes not only from attaining knowledge, but also from continuing to build on it later in life. The integration of the cognitive and motivational aspects of learning helps students retain knowledge. Brophy believed that "activating schema networks" through the curriculum's purposes and goals offers the key to effective education. Here, if students activate schema networks that relate to personal plans, they will build readily on those branches. Ultimately, it is up to teachers to develop schema networks that are not too obvious to others. This may be done by giving explanations and utilizing cognitive modeling and scaffolding to develop an appreciation of schema connections. Brophy encourages teachers to return to the twentieth century educational values of explaining exactly why a particular task is being taught and addressing both the goals and intended outcomes (Brophy, 2008).

Pragmatism is a Student-Relevancy Theory as pragmatism focuses on teaching students to be capable problem solvers in flexible settings that are conducive to learning. Having emerged in the late nineteenth century, pragmatism entailed a curriculum of practicality based on science, evolution, change, and growth. Order and finality were no longer essential as change was embraced by proponents including Charles S. Pierce, William James, and John Dewey. Dewey, the founder of pragmatism, believed that the mind needed to be shaped by experience before it could develop ability to solve problems that arise continuously. Dewey asserted, "Theory, and – to some extent – practice, have advanced far enough to recognize that play-activity is an imaginative enterprise" (Dewey, 1916, p. 145). Although the primary goal of education was to solve problems, those problems changed continuously. Students needed to be prepared to solve new problems as they were presented to them. Dewey realized that humans were driven first to meet certain evolutionary needs. In fact, in *How We Think* (1910), Dewey wrote that learning must first be meaningful to the learner to ensure that he or she will grow through constant activities. Dewey favored child-centered environments of learning in which subjects were integrated across the curriculum, and his theories remain the mainstay in education today. Since predetermined goals change as education evolves, even the means used to the end vary as students change. Ultimately, however, learning remains relevant in effective schools (Pulliam & Van Patten, 2007, pp. 44–46).

Societal Needs Theories

Societal Needs Theories help meet the needs of the community as a whole while making the world a better place to live. These theories refer to society as a whole as opposed to the individual's "social" needs. These are not to be confounded with Maslow's Social Needs Theory. Instead, they propose that the curriculum should meet the objectives of a much larger group. For example, the reconstructionism theory encourages students to become social activists in nontraditional settings, even those without walls. In *Toward a Reconstructed Philosophy of Education* (1962), Brameld expressed the belief that the goal of education was to guarantee food, shelter, employment, and security to the learner and the learner's family. Unless humans were to reconstruct goals, there would be a proliferation of nuclear arms and other unfavorable outcomes. Knowledge was the power transmitted through education in these nontraditional settings without walls. Reconstructionists, also known as progressives, believed that the *whole child* learned best when provided with meals, health care, and even counseling. This led to social engineering efforts in schools in the form of searches for talent, racial integration, and busing. This also presaged the Elementary and Secondary Education Act, known more popularly as *No Child Left-Behind*, which seeks to empower at-risk and minority students (Pulliam & Van Patten, 2007, pp. 47–49).

Feminists, such as the philosopher Noddings (2007), rightfully theorize that schools should offer all students opportunities to develop skills in the areas of caretaking as opposed to developing skills that have been considered traditionally as masculine attributes. Whereas girls deserve a rich environment in science and mathematics, boys also need to learn to care for others, themselves, and the planet (Noddings, 2007, p. 216). Nurturing is not a domain reserved for women, and there are means by which this objective can be met for both sexes. Perhaps, the extant theories were conceived consciously to ignore the needs that must be met in an inclusive feminist paradigm.

In addition to working well in mixed ability groups, one theory suggests that working in diverse gender and cultural groups would enable students to perform better. Students learn better in cooperative learning groups with mixed-ability levels and ability groupings since such associations reflect miniature societies. "Many studies have shown that with appropriate organization, cooperative learning with other students can be more effective than independent learning" (Long, 2007, p. 142). Students developed better self-esteem and the ability to cope socially in these groups than they did in a competitive situation. In this manner, children with varying levels of ability can help one another which is particularly useful in the ESOL classroom.

Aranson et al. (1978) developed the idea of the "jigsaw" activity in which everyone is given a multifaceted task. Students explore the various tasks and eventually assemble them so that once the puzzle is solved, everyone shares the knowledge. In this manner, all learners support one another to the point that high ability level students tutor and assist those who are struggling. Such strategies based on theories which hold that students work well together are particularly useful in international and multicultural environments.

Behaviorist Theories

Behaviorism is dependent upon operant conditioning which is employed to generate specific and predetermined outcomes. Thanks to Skinner and Watson, rewards (and negative reinforcements) are used to teach competency in a variety of areas. However, external rewards seldom produce a desire among students to learn and achieve mastery. Skinner rewarded his subjects each time they achieved mastery in a specific task, the goal of which was to master one task at a time and build upon tasks. From these behaviorist theories, the notion of programmed instruction was conceived which involves the administration of multiple-choice tests and computer-assisted instruction along with a system of both rewards and punishments (Pulliam & Van Patten, 2007, pp. 68–69).

Behaviorist approaches are practical only for the management of poor classroom behavior. Learned problem behaviors are modified via the application of conditioning. Approaches such as flooding and exposing students to situations they consider fearful, enable students to overcome their classroom-based fears and anxieties while conditioning helps students overcome their improper behaviors (Long, 2007, pp. 272–273). It is essential to match positive reinforcements with good behaviors and ensure that extrinsic rewards do not create dependency.

Content-Based Theories

Perennialism is exemplary of Content-Based Theory or philosophy given that advocates believe in the value of learning from the classic works of literature, lectures, and discussions. Here, students reveal learning and knowledge through the production of essays. Perennialists tend to be grounded in Western approaches that could be applied to all students. Although perennialists believe in the separation of church and state, they insist that a universal sense of ethics must be instilled in learners. Perennialists value a curriculum based on specific subject matter that originated from Western culture. Classical literature should be taught through history, language, mathematics, logic, literature, and the humanities in much the same way it is taught in liberal arts colleges today (Pulliam & Van Patten, 2007, p. 43).

Perennialists share common features with the essentialists since both camps center their efforts on subjects, and both believe in nurturing the mind while developing essential values of work and morality. However, while perennialists focus on the three Rs and classical subjects including Latin and logic, the essentialists include six major domains in the secondary curriculum: English, math, science, history, foreign language, and geography. Nevertheless, both approaches advocate the "back-to-basics" approach (Ornstein, 2008, pp. 415–417).

In an organized approach, realists and essentialists employ lectures and demonstrations in addition to providing sensory experiences. Both camps hold that all teaching must come from the real world via the senses, which may be observed and studied. In order to understand the world, one must use reason in the examination of objects. Aristotle (384–322 BC), who was the first known realist, classified objects into categories and examined the physical nature of how something was made and how it was used ultimately in a larger framework. Today, realists share the belief that curricula should include intellectual disciplines that deal with the real, sensory world, and the way a teacher presents such a curriculum remains fundamental. According to the realist, Harry Broudy, author of *Building a Philosophy of Education* (1954), schooling would make the world a more habitable place (Pulliam & Van Patten, 2007, pp. 42–43).

Along with the perennialists and the realists, the idealists also fall into the realm of Content-Based Theories. Surprisingly, the theory that there exists a reality of abstract ideas, to include Plato's (427–347 BC) concept of mind, spirit and reasoning, holds that content may be transmitted readily to the learner through lectures, discussions, and requirements that students read and write essays. As did Plato, some educators today believe that spiritual aspects of education are real and may be even more important than the physical facets of teaching. Education is a search for ultimate truth, which becomes clear when learners acquire concepts and ideas that are superior to tangible matter. Idealist theories remain compatible with organized religions. For example, St. Augustine (354–430) embraced the *trivium*, which included logic, grammar, and rhetoric, in addition to the *quadrivium*, which included music, geometry, arithmetic, and astronomy. For idealists, the spiritual world and its truths, along with moral conduct, are all that ultimately exist and matter (Pulliam & Van Patten, 2007, pp. 39–41).

Idealism remains one of the oldest philosophies and is today seen as a theory which claims that students will become successful adults when they live up to the worldview of metaphysics. The notable proponents of this theory included Georg W.F. Hegel (1770–1831), Ralph Waldo Emerson (1802–1882), and Henry David Thoreau (1817–1862). Whereas Froebel's kindergarten was based upon idealist theories, the teachings of Hinduism and Buddhism fall into this realm as well

(Ornstein, 2008, p. 164). The commonality among these theorists is that they believe a single spiritual essence or universal mind is shared by all. The learner must recognize a priori ideas that have always been with him or her, a process enabled through lessons in deductive logic. The attainment of knowledge is tantamount to having recognized these underlying spiritual truths. The content consists of theology, philosophy, the arts (for inspiration), and the general curriculum.

Bentley (2010) called for utilizing postmodern techniques to deconstruct metanarratives of education theory and practice that have proven ineffective in providing quality education to handicapped and disabled students. In the postmodern paradigm, teachers must analyze and transform ineffective practices that limit the student's ability to learn through tangible life experiences, a process that runs counter to the more confined structure of behaviorism. Relative to what Bentley called the "mouse metaphor," students typically are boxed into structures that prevent them from encountering these experiences and must be released from those confines into the "unexpected, unpredictable" postmodern condition (Bentley, 2010).

Predominant Theories Today

Today's two prevailing theories include cognitive approach theory and operant conditioning. Conditioning involves forming associations between stimuli and responses, and classical conditioning is based on Pavlov's (1927) experimentation with dogs which learned to salivate on command. Skinner's operant conditioning is utilized to predict what will happen in response to an action. Although effective, such theory is applied primarily for behavior modification and is considered outdated with respect to general classroom settings that emphasize the retention of new information. Operant conditioning employs voluntary responses to positive reinforcement which help students modify their behavior.

Cognitive approaches include the internal representation of information to disclose changes in the student's development over time. Cognitive theory assumes that the student must process information before he or she can test inner models of the world that will apply to future performance (Long, 2007, p. 17). Such cognitive approach theories are represented effectively in two models labeled "The Structure of Memory" and "Long Term Memory" (Long, 2007, pp. 17–21).

Exemplary Models That Support Frameworks

Numerous exemplary models are available. Also known as intellectual pictures, these offer dimensions that support the framework of instructional design and are often represented as flow charts and timelines in academic textbooks and journals.

For visual thinkers, such models represent a tangible means to place specific instructions on paper or in digital media in a manner that will be understood by the practitioner. Models show steps for implementing a curriculum, a learning process, or the way theorists perceive the mind to work. Above all, models summarize concisely the steps to be taken to implement the curriculum.

One such model is Merrill's First Principles of Instruction, a problem-based model that engages students in four separate phases of learning (Merrill, unknown). Learners need to solve problems related to the real world, but such knowledge must be activated in order to stimulate new knowledge. Learning means that the student has applied what has been learned and integrated it into his or her environment.

One of the most fascinating models available to educators is the instructional design model, which appears in *Designing Effective Instruction* by Kemp, Morrison, Ross, and Kalman (2011). This model illustrates nine dimensions of instructional design showing how instructional design remains a continuous, ovular process comprised of nine elements that depend upon each other. According to Kemp, the

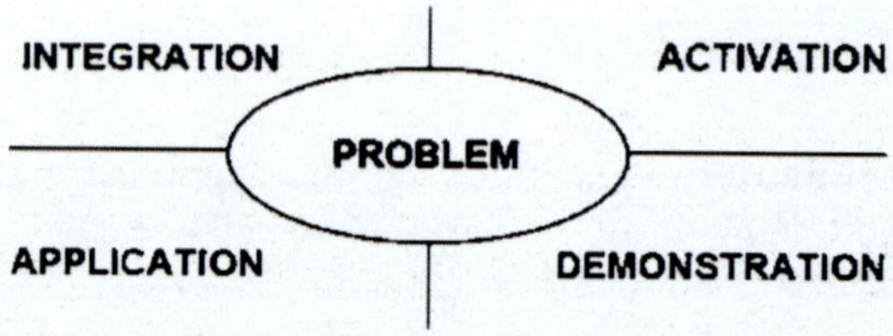

Merrill's First Principals of Instruction Diagram, Utah State University

instructional design model represents "how you apply instructional theory to create an effective lesson or unit" (Kemp et al., p. 383). In other words, models depend upon theories. He further categorized these theories into: (1) learning theory and (2) instructional theory as stimuli for models. When creating models to convey a plan, the number of elements in the model may vary widely based upon how the model will be applied. Of course, not all models can or should be circular in nature.

Models create images in readers' minds. Such models have three dimensions and may be envisioned or diagramed on the page with ease. Freeman (2005) identified two predominant models, one of which has been embraced by the United States' schools, and another much older model that dates to Socrates' times. The factory model presents mass public education as a means for assimilating immigrants and citizens alike in a well-designed production process. Students go from class to class at 1 h intervals in much the same way that products move down a factory assembly line. This factory model remains foundational in the prevailing cost-effective system (Freeman, 2005). Today, when leaders speak of school reform, they refer to ways in which schools can be made more *efficient* and *cost-effective* in order to increase output in the typical capitalist fashion. This model falls under the LSP.

Freeman contrasted the factory model with a much more ancient example he calls the "cultural womb," which dates to Plato's Academy and was employed during medieval times (Freeman, 2005). In this model, the student's culture is refined by exposure to classical subjects passed from one generation to the next. The cultural womb model connects well with the CCP in which students learn the material from a teacher who is philosophical in nature and who employs the Socratic approach.

Hierarchy in Education Paradigms

Education paradigms influence theory building and the creation of models by enabling instructors to envision the future through multicolored lenses. Research leads to the revelation that paradigms are also value systems that change and shift as do lenses when overlapped with others of different colors. Another metaphor that describes the paradigm is the color wheel system which mixes hues to produce new results. The color wheel gives us a better understanding of the hierarchical structure of the three entities of: model, theory, and paradigm, where each of the former is both separate and dependent upon the latter.

In the hierarchical structure, educational paradigms represent values that prevail over theory. According to literature, there are only a handful of popular paradigms that influence education and schools today, even though the number of paradigms that may emerge is unlimited. Clearly, there are more of these. For the sake of simplicity, the author categorized the major paradigms into three areas: (1) the LSP; (2) the CIP; and (3) the CCP. These paradigms are value systems of overarching philosophical movements that include humanism, perennialism, idealism, realism, pragmatism, reconstructionism, behaviorism, postmodernism, existentialism, and constructivism (the latter being the predominant paradigm/philosophy at present although it is influenced positively by other paradigm shifts). Earlier paradigms in education envisioned the child as a passive receptacle of information, but later paradigms promoted the teacher as a facilitator (Kuhn, 1962). Fortunately, educators can thank Piaget for cognitive approaches emphasizing the constructivist ways children and students will construct their own knowledge.

In this chapter, the author also explored definitions of the term, *theory*. According to Long, "Theories are general ways of looking at or understanding an area" (Long, 2007). Theories and models are supported exclusively by the larger philosophical and archetypal framework of the paradigm. Research has shown that these terms continue to be misused in education as they are in psychology. Findings coincide with L'Abate's observation that they are used inappropriately across the board. Hence, these three terms must be used appropriately to reflect the way conceptual paradigms (value systems) control the formulation of theories (hypothetical principals) and models (dimensional plans) in education and other domains.

Regardless of which colored paradigm-lenses educators wear, each of them creates models based on theories while the paradigm's relationship to educational theory is that of the overarching determinant of hypothesis-based theory. Theories do not exist without the visionary aspects of paradigms and overarching value systems shared by groups that will test them. Educational paradigms stimulate goal setting essential to generate hypotheses.

Finally, paradigms are indeed overarching entities that become part of a pyramidal hierarchy with an underlying foundation. Paradigms are simultaneously overarching and foundational, because they inspire theories and impact the models present in instructional design. But while models cannot exist without paradigms, a new model exists to explain how paradigms relate to theory creation and model building. The new model proposed here is triptych and appears in three formats rather than one static form: (1) Paradigms provide a foundation that supports the theories in the middle and the models at the top; (2) Paradigms also dominate theories and models, with theories exerting greater weight than models; (3) Paradigms are similar to the three primary colors of a the color wheel where secondary and tertiary colors represent theories and models in the same order. Like impressionist dots, paradigms penetrate all that educators do in schools.

Regardless of whether one envisions paradigms as being overarching, foundational, or all-pervading within all that one teaches, it is essential to understand that theories depend upon paradigms and are markedly different from them. Models, on the other hand, must be created to explicate theories so they can be put into practice. Furthermore, authors of texts related to education, educational leaders, and classroom teachers must consider these factors in curriculum building.

References

Aranson, E., Blaney, N., Stephen, C., Sikes, J., & Snapp, M. (1978). *The jigsaw classroom* (1st ed.). Beverly Hills, CA: Sage.

Bentley, J. K. C. (2010). Thoroughly postmodern Ralph: A commentary on scholar-practitionership. *Scholar-Practitioner Quarterly, 4*(1), 42–51.

Brophy, J. (2008). Developing students' appreciation for what is taught in school. *Educational Psychologist, 43*(3), 132–141.

Buraphadeja, V., & Dawson, K. (2008). Content analysis in computer-mediated communication: Analyzing models for assessing critical thinking through the lens of social constructivism. *The American Journal of Distance Education, 22*, 130–145.

De Freitas, S., & Neumann, T. (2009). Pedagogic strategies supporting the use of synchronous audiographic conferencing: A review of literature. *British Journal of Educational Technology, 40*(6), 980–998.

Dewey, J. (1916). *Democracy and education*. Lexington, Kentucky: Unabridged classic reprints, p. 145.

Edwards, P. M. (2010). Theories-in-use and reflection-in-action: Core principles for LIS education. *School of Information and Library Science, 51*(1), 18–22.

Experiment Resources & Shuttleworth, M. (2008). *What is a paradigm?* Retrieved August 8, 2010, from http://www.experiment-resources.com/what-is-a-paradigm.html

Fear, W. J. (2007). Programme evaluation theory: The next step toward a synthesis of logic models and organisational theory. *Journal of Multidisciplinary Evaluation, 4*(7), 13–24.

Freeman, R. (2005). *Competing models for public education*. Retrieved January 3, 2011, from http://www.commondreams.org/views05/0226-25.htm

Gaztambide-Fernandez, R. (2010). Wherefore the musicians? *Philos Music Education Review, 18*, 67–84.

Gokool-Ramdoo, S. (2008). Beyond the theoretical impasse: Extending the applications of Transactional Distance Theory. *International Review of Research in Open and Distance Learning, 9*(3), 1–15.

Grossman, P. (2008). Teaching development—experience and philosophy (Using the three Rs). *Teacher Education Quarterly,* (Spring), 155–169.

Kasl, E., & Yorks, L. (2010). Whose inquiry is this anyway? Money, power, reports, and collaborative inquiry. *Adult Education Quarterly, 60*(4), 315–338.

Kemp, J. E., Morrison, G. R., Ross, S. R., & Kalman, H. K. (2011). *Designing effective instruction* (4th ed.). Hoboken, NJ: John Wiley.

Kuhn, T. S. (1962). *The structure of scientific revolutions* (1st ed.). Chicago, IL: The University of Chicago Press.

L'Abate, L. (2009). Paradigms, theories, and models: Two hierarchical frameworks. In L. L'Abate, P. De Giacomo, M. Capitelli, & S. Longo (Eds.), *Science, mind and creativity: The Bari symposium* (pp. 107–122). New York: Nova Science.

Learning-Focused Schools (2010, November). *Learning-focus solutions that work.* Retrieved November 6, 2010, from http://www.learningfocused.com

Long, M. (2007). *The psychology of education* (6th ed.). Glasgow, UK: Bell & Bain.

Macrine, S. L., & Sabbatino, E. D. (2008). Dynamic assessment and remediation approach: Using the DARA approach to assist struggling readers. *Reading & Writing Quarterly, 24*, 52–76.

Merrill, M. D. (unknown). *First principles of instruction*. Retrieved January 3, 2011, from http://id2.usu.edu/Papers/5FirstPrinciples.pdf

Morrone, A. S., & Tarr, T. A. (2005). Theoretical eclecticism in the college classroom. *Innovative Higher Education, 30*(1), 7–20.

Noddings, N. (2007). *Philosophy of education* (2nd ed.). Boulder: Westview Press.

Ornstein, L. (2008). *Foundations of education* (1st ed.). Boston, MA: Houghton Mifflin.

Pavlov, I. (1927). *Conditioned reflexes*. Oxford: Milford.

Pulliam, J. D., & Van Patten, J. J. (2007). *History of education in America* (9th ed.). Upper Saddle Valley, NJ: Pearson.

Rolling, J. H. (2010). Paradigm analysis of arts-based research and implications for education. *Studies in Art Education, 51*(2), 102–112.

Shattuck, R. (2005). Shame of the schools. *New York Review of Books, 52*(6), http://www.nybooks.com/articles/archives/2005/apr/07/the-shame-of-the-schools/.

Thompson, M. (2008). *Learning-focused solutions that work!* Retrieved July 11, 2010, from http://www.learningfocused.com/index.php.

Yin Cheong, C., & Mok, M. M. C. (2008). What effective classroom? Towards a paradigm shift. *School Effectiveness and School Improvement, 19*(4), 365–385.

Vygotsky, L. S. (1978). *Mind in society: The development of higher psychological processes.* (M. Cole, V. John-Steiner, S. Scribner, & E. Souberman, Eds. & Trans.). Cambridge, MA: Harvard University Press. (Original work published in 1955).

Vygotsky, L. S. (1985). *Thought and language*. Cambridge, MA: MIT Press.

Chapter 8
Sociology

Alberto L'Abate

In this chapter I am transposing pages from a volume (L'Abate, A., in press) that deal in greater breadth and depth with themes that will be touched briefly here. Eventhough I am aware of the confusion proposed by the editor (See Chapter 1 of this volume), among paradigms, theories, and models, the major emphasis of this chapter does not concern itself with the hierarchical organisation among those three terms. Instead, this chapter will concern itself with delimiting the role of paradigms, as originally proposed by Masterman (1970). The major theme proposed here and suggested also by Hillix and L'Abate (Chapter 1 this volume), is the search for an unified paradigm within the field of sociology.

Various proposals will be evaluated: (1) Friedrichs (1972), who distinguished between prophetic and sacerdotal paradigms, with a revaluation of the first and going above and beyond the second; (2) Boudon (1980a, b), who distinguished between functionalism and neo-marxism and went above and beyond them with a third distinction called interactive relationships, which, however, in other sources, he defined as "methodological individualism"; (3) Eisenstadt and Curelaru (1976), who emphasised the actual importance of open systems; and (4) Morin (1984), who distinguished between a paradigm of classical science (CS), which he also defined as a simplification, and the paradigm he proposed about complexity, with a great deal of importance given to individual change and happenstance.

After reflecting on Ritzer's (1975, 1979) proposal who identified three major paradigms: (1) social facts (Durkheim); (2) social definition (Weber); and (3) behaviourism (Skinner) who accepts all of them and indicates how it is possible to move toward a unified paradigm, I arrive to a personally possible solution of the problem. I will analyse prevailing theories and models in sociology starting with Masterman's (1970) three different meanings or levels of what constitutes a paradigm: metatheoretical, sociological, and constructivist. After this analysis, I arrive at the conclusion

A. L'Abate
Via A. Mordini, 3, 50136, Florence, Italy
e-mail: labate.alberto@gmail.com

L. L'Abate (ed.), *Paradigms in Theory Construction*, DOI 10.1007/978-1-4614-0914-4_8,

that in the first metatheoretical level, there are, in my opinion, models of society in which the principle of uniqueness seems to be important.

This is a model of unstable equilibrium to overcome the dichotomy between consensus and conflict. In the second level of sociological theories, it's important to consider plurality rather than uniqueness especially starting from the bottom putting together single middle-range theories (Merton, 1949). This might be one way to arrive at a much wider level (with many doubts concerning their validation or falsification (Popper, 1968, 1970). According to Masterman, the true concept of a paradigm, supported also by Kuhn (1970) is constructivism, which I identify with methodology. At this level, according to me, neither uniqueness or plurality are worthy of consideration. Instead, I recommend emphasising four methodologies used in sociological research: casual, structural, functional, and process analyses. These four analyses are considered in greater detail in L'Abate (in press).

Paradigms in Sociology

The editor of the volume has shown that the main contribution to the development of the concept of "paradigm" came from T. Kuhn (1962), who saw the whole of scientific progress pass through the revolutionary change of paradigms. Kuhn defined the concept as a system of ideas which guide and organise a certain area of scientific investigation, making it readily communicable and modifiable within a community that shares the same language.

Other definitions of the concept of paradigm adopted by sociologists for their analyses. Giddens (1976, p. 166), for instance, defined it simply as "frame of meaning" and demonstrates its importance for research if we keep ourselves open to mutual confrontation. Ceresa, Mela, Pellegrini define it as "A world to give structure to, a group of questions and problems which at the same time represent an attempt to organise those problems in constructs which can be channelled into scientific terms". (1981, pp. 1,2). Finally, Ritzer gives a very complex, detailed definition: "A paradigm is a fundamental image of the subject matter within a science. It serves to define what should be studied, what questions should be asked, how they should be asked, and what rules should be followed in interpreting the answers obtained. The paradigm is the broadest unit of consensus within a science that serves to differentiate one scientific community (or subcommunity) from another. It subsu-mes, defines and inter-relates exemplars, theories, methods, and instruments that exist within it". (1975, p. 157). As we can see, the three definitions above are very different from each other, above all regarding the sphere they cover – the first is limited and vague, the last is very broad and specific. It is quite natural that starting from such a badly defined concept, incorporating different notions in reality, we have arrived at making the most disparate suggestions for its application in sociology. However, it would be well to have a glance at least at the foremost suggestions to see whether the concept can be used, or whether it is so confused and contorted that it should be eliminated from the language of sociological research, as Martindale (1979) and Cavallaro (1984) maintained. This will lead us to consider, too, the very lively debate among those who support the idea of having as a basic goal one unified paradigm to overcome the confusion created by the many paradigms used and the supporters of a multi-paradigmaticity of sociology.

Given that each proponent seems to have followed a different path directly inspired by Kuhn's book, but without explicit reference to each other, I shall follow a logical order moving from the simplest to the more complex proposals.

Simple Proposals

Among the simplest proposals I will refer to Friedrichs, Eisenstadt and Curelaru, and Boudon's. Friedrichs (1972) identified various paradigms, in particular a systemic one which puts the emphasis on integration within the system, and a conflictual one which emphasises disintegration and conflict. He holds that we have to accept both of them, thus supporting the multi-paradigmaticity of sociology. However, what lies at the core, in his estimation, is the concept of paradigm seen as an image the sociologist has of himself as an agent of scientific progress, and here he does not accept eclecticism as a simple summation. In this sector he identifies two other paradigms: the prophetic and the priestly.

The former sees the function of the sociologist as being not only that of understanding and interpreting the world, but also of transforming it, of making it better, more adequate for civil life together and for the results obtained by the research itself. So the sociologist is seen as an agent of social change. To Friedrichs this approach has been at the base of the birth and development of sociology in his country, the USA.

The latter paradigm is not so much preoccupied with changing the world, a task considered outside the sphere of the sociological world, as interpreting it on the basis of specific and accepted codes (like priests who limit themselves to celebrating a rite), established objectively (the *method* which is made equal to the rite) and usually borrowed from the more consolidated natural sciences. In this vision the sociologist is seen as a scientist who is "free from values". To Friedrichs this seems the latest development in North-American sociology. He does not hide his sympathy for the first of the two paradigms, considering it fundamental for a sociology that does not become estranged from the social milieu surrounding it.

Eisenstadt and Curelaru's proposal (1976) is slightly more complex: they identified three paradigms which they also defined as models:

Discrete paradigms: They focus on separate concrete entities, like ecological properties, size of groups, racial and psychological characteristics. Given the image of the world as an ensemble of isolated units, those who worked inside this paradigm had difficulty in dealing with relational matters, such as emergency, innovation, creativity.

Closed-system paradigms: Society was made up of separate, but interrelated, elements. Those who worked inside this paradigm tended to consider one element as dominating others (e.g. Marx and the importance of the economy).

Open-system paradigms: Focus on the internal, systemic, dynamic interconnections and continual processes of feedback among different components of the social system.

Eisenstadt and Curelaru's concluded that despite there not being a natural, linear movement from one paradigm to another, and that historically there has been a form of superimposition, there seems to be a long-term flow from the first type, across the second towards the third.

At a higher level of complexity, which comes close to, but in a certain way goes beyond those analysed below, are Boudon's proposals (1980a). He started from Lazarsfeld's statement that macrosociology is dominated by two theories: Marxism and Functionalism. Boudon maintains that it is more correct to consider them paradigms, not theories, and thinks of bringing the former back to that form of neo-Marxism which has been incorporated and realised in the work of many western sociologists. Furthermore, he considers that another paradigm exists and is fundamental, which he defines "interactive analysis", or better "analysis of the aggregations of actions", which distinguishes it better from interactionism. He believes this latter paradigm, which is used a great deal in economics and has illustrious precedents in the social field (e.g. in Rousseau's social contract), has been unjustly neglected by many contemporary sociologists.

According to Boudon even though there are considerable differences among them, some premisses that characterise the first two paradigms can be identified. They are:

Functionalism: The first premise is that nobody can impose their will on others through force and violence. The second is that what counts is not so much the internal imbalances in a society as much as a departure from the balance which has been reached among the various social parts that make up society may lead to a worsening of the situation for one or more of those who are worse off. Social changes, even if they can lead to an increase in internal inequalities, if they improve everyone's situation – both of those who are better off and those who are worse off – are seen as a passage from an unbalanced state to a balanced one. Indeed, according to those scholars who are associated with this paradigm a state of inequality is balanced if, and only if, a departure from this balance would worsen the situation for at least one of the parts (in an alternative version: "for those who are worse off") (1980a, p. 53).

Neo-Marxism: This paradigm's basic premise is that a class (the ruling one) has the power to impose its will on the class, or classes, that are ruled. Public ills such as inequality in educational opportunities, pollution, the wasting of natural resources, deterioration in the quality of life, crime etc., can be interpreted, therefore, by the mere fact that the profit gained from their existence by some members of the ruling class, thanks to their socio-economic role in society, is greater than the damage they suffer as members of a broader society.

Boudon held that Functionalism has not managed to explain the persistence of "social ills" in western society and that this has made the neo-Marxist paradigm popular among western sociologists. But he thinks that this second paradigm is limited by that strong premise that "a class (the ruling one) has the power to impose its will on the class, or classes, that are ruled" (1980a, p. 52).

Through two examples: inequality in educational opportunity in western society, and propaganda, Boudon shows how a third paradigm, which he calls an "analysis of the aggregation of actions", permits us to understand the negative results of our actions (which is not interpretable on the basis of the Functionalists' premisses) without, however, introducing that strong premise of neo-Marxism which, in his opinion, is not falsifiable.

This led Boudon to claim the importance of the third paradigm, which is based on a general methodological principle which he has defined "methodological individualism" in other books (Boudon, 1980b), that "macrosociological phenomena, even if they appear by definition on a social level, should be analysed as resulting from the aggregation of elementary actions" (1980a, p. 54). In presenting this paradigm Boudon has in fact presented a method for analysing social reality. This can help us when we are trying to overcome terminological confusion on the concepts of paradigms, theories and models.

Complex Proposals

Ceresa, Mela and Pellegrini (1981) identified seven main sociological paradigms:

Mechanic: Society is seen as a machine or a planetary system (celestial mechanism). Social actors are seen as atoms in mutual equilibrium. Its reference point as a scientific community (which, as we have seen, Kuhn placed at the core of the concept of paradigm) is social physics and the first systematic forms of sociology. It has a preferential relationship with organic and systemic paradigms, and with social cybernetics.

Organic: Society is seen as a living organism *sui generis*. Its reference point, as theoretic models, are the structural and evolutionary laws of living beings (differentiation, growth, pathology). Its scientific community is classical sociology, heavily influenced by bio-organics and Functionalism. It has a preferential relationship with mechanical and systemic paradigms (in particular socio-biology), and partially with the dialectic paradigm, corrected, however, by the second International and revisionism.

Dialectic: Man is seen as the subject who transforms natural material conditions. At the heart of historical processes it sees the struggle among social agents (class struggle). Its reference point as regards theoretic models are the laws governing the organisation of production (production methods, division of labour) and it sees the role of the economy as fundamental for the development of society. Its scientific community are intellectuals (economists, sociologists, etc.), politicians, militants and the proletarian avant-garde in general. It has preferential relationships with the organic, structuralist and social-action paradigms.

Structuralist: This paradigm's image is of social rapports governed by abstract systems of organisation, as in a linguistic system. Its theoretic models are the formal ones of regulation and invariance and the rules of transformation. It puts the emphasis on research into deep structures, unconscious ones, too. Its scientific community are linguists, psychologists, anthropologists and, in particular, the structural-functionalist trends in sociology. It has preferential relationships with the dialectic and systemic paradigms.

Systemic: The image is of an organised whole supplied with communication and information networks seen as a living organism placed within other whole-environments. Its theoretic models are the laws that explain the interrelationships among the parts and their conditions of equilibrium, the regulation of the system, its ends and evolutionary processes. Its scientific community is interdisciplinary, made up of scholars of mathematical and natural sciences who have tended to see the concept of "system" as valid in different fields, including social sciences. It has preferential relationships with the organic, mechanic, transactional (above all games theory) and structuralist paradigms.

Social action: The picture is of society as a process of interactions where the actors are engaged in defining their own position with respect to others, and where a fundamental role is played by the exchange of information and conversation. The theoretic model is the analysis of the subjective feeling the actor gives to his or her actions, models of acting and roles, and the creative work of the actor. The scientific community is classical sociology (above all Weber), North-American sociology, phenomenology and microsociology. It has preferential relationships with transactional, organic (functionalism), dialectic and systemic paradigms.

Transactional: The authors see this paradigm as being made up of three different schools, each one with a certain amount of autonomy. (A) *Drama*. Society is seen as a theatre. Its theoretic models are the characters, the scenes, the performance. It focuses on the study of expressiveness, face-to-face interaction, conduct, deference, the handling of the role and rituals. Its scientific community is microsociology following the tradition begun by Simmel. It has preferential relationships with the social-action paradigm. (B) *Strategy*. The image is of the rational construction of strategies for success. The theoretic models are the actors who play in games whose rules are to be found out in a climate of competition, with alliances and calculations about the efficacy of acting. Its scientific community is above all the scholars of the sociology of organisation. It has preferential relationships with the action and systemic paradigms. (C) *The market*. The picture is that of a market at the core of which is the negotiation of prices and exchange. The theoretic model is of the calculation of costs/benefits in social interaction (i.e. how much I earn or lose if I do something). Its scientific community is above all Behaviourism with its enormous clout in the North-American psychological tradition. It has preferential relationships with organic and partially with action (1981, pp. 89–90).

Ceresa, Mela and Pellegrini also posed the question of whether it is possible to arrive at only one paradigm and proposed some elements which could help a "new paradigmatic synthesis". These are in essence of the following:

1. Overcoming the dichotomy and opposition between theories and paradigms of social action (social action, transactional), which place man's activity and his capacity for social acting in the foreground, and "structural" paradigms (organic, mechanic, systemic, structuralist, dialectic, in part), which favour the functions of large systems (the social conditioning of human activity).
2. Going beyond the classical theories of stability and change which favour social equilibrium. The large systems can function also in conditions of great instability (see the concepts of negative and positive feedback and the complex rings of feedback).
3. Going beyond the dichotomous way of reasoning between "meaning understood subjectively" and "acquisition of meaning from the structure-system" (systems are not entities given a priori, but produced by the competent activity of subjects) (p. 102 ff).

Morin (1984, 2008), seems to have accepted the challenge to search for a new integrated approach, which he calls "the paradigm of complexity". He compares this paradigm with that of CS, which he calls 'the paradigm of simplification'. Here they are:

1. *Classical science* (CS) is based on principles of universality, so scientific knowledge is only general knowledge. Detail and the individual tend to be excluded as contingent or residual. The *paradigm of complexity* (from now on PC) considers this principle inadequate and highlights the fact that intelligibility begins from the detail and the individual.
2. CS, which is looking for general rules, tends to eliminate everything in relation with a single, specific event. PC needs to include history and the event in all its descriptions and explanations. A complex system can be understood only in reference to its history and evolution.
3. CS tends to reduce the knowledge of the parts to the knowledge of the elementary units (simple parts) which make them up. PC maintains it is impossible to know the parts without knowing the whole and vice versa.
4. CS tends to reduce the knowledge of organisation to the principle of order (constant, invariable laws). PC considers the principle of order only a part of the organisation.
5. CS is based on the principle of linear causality, superior and external to objects. PC is based on the complex interrelation of *mutual causality*, on inter-feedback, delays, mixes, deviations, re-orientation and on the principle of the endo-eso-causality of self-organisation.
6. CS considers order sovereign, absolute: there is a universal, impeccable determinism; casual events are due to ignorance. PC is based on a dialogic organisation-interaction between order and disorder; it is considered that casual events complete comprehensibility.

7. CS tends to isolate or separate the objects it studies from their environment. PC, on the contrary, tends to study its objects as distinct, but not disconnected, from their environment.
8. CS tends to separate an object from the subject that perceives it completely. The observer/experimenter's control is enough to exclude the knowing subject and guarantee objectivity. PC, on the contrary, tends to recognise the rapport between the observer and thinker and the object which is observed and thought.
9. CS tends to eliminate all the problems connected with the subject-observer from scientific knowledge. PC, however, holds that it is possible to formulate a scientific theory of the subject and carries out research with this aim in mind.
10. CS tends to eliminate being and existing through quantification and formalisation. PC aims at the introduction and recognition of the being and existing categories.
11. CS maintains that the concept of autonomy from everything is inconceivable. PC aims at the scientific recognition of the concept of autonomy.
12. CS is based on the absolute reliability of logic in stating theories (contradiction = error). PC is based on recognising the limits of logic in complex systems: contradictions and incoherence are signs of a deep, unknown sector, so PC tends to associate complementary, competitive, antagonistic concepts.
13. CS is based on mono-logic thought (clear, distinct ideas in generally accepted logic). PC, on the contrary, is based on dialogic thought, on thinking in macro-concepts and on the complementary linking of notions which may even be mutually antagonistic.

Ritzer (1975, 1979) also identified three paradigms considerably different from Eisenstadt-Curelaru and Boudon's. They are: *social facts*, *social definitions* and *behaviour*.

Social facts: The writer who contributed the most to the development of this paradigm is E. Durkheim. He tends to regard facts as things, objects outside of the individual who acts. This paradigm tends to place at the core of scientific research large-scale structures and institutions and their effects on the individual's thought and actions. From the methodological point of view those this paradigm appeals to, tend to use historical-comparative and questionnaire-interview methods more than any other. Theories which fall into this paradigm are structural-functionalism and the conflict and systemic schools.

Social definition: The writer who gave the main examples of research that those who follow this paradigm refer to is Max Weber. In this paradigm reality is seen as a social construction (not as an outside datum) and the greatest interest is on the way the actors define their social situation and on the effects these definitions have on the resulting actions and interactions. The methodology which distinguishes supporters of this paradigm is observation, even if they normally have recourse to the questionnaire-interview, too. The theories that can be included in this paradigm are: action theory, symbolic interactionism, phenomenology, ethnomethodology and existentialism.

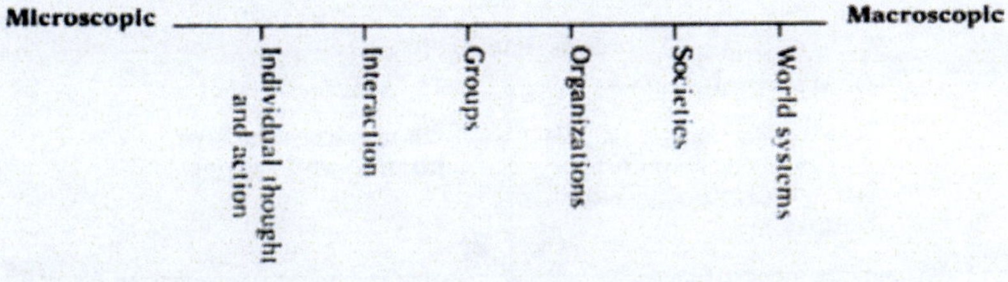

Fig. 8.1 Macroscopic–Microscopic continuum with key points identified (Ritzer, 1975)

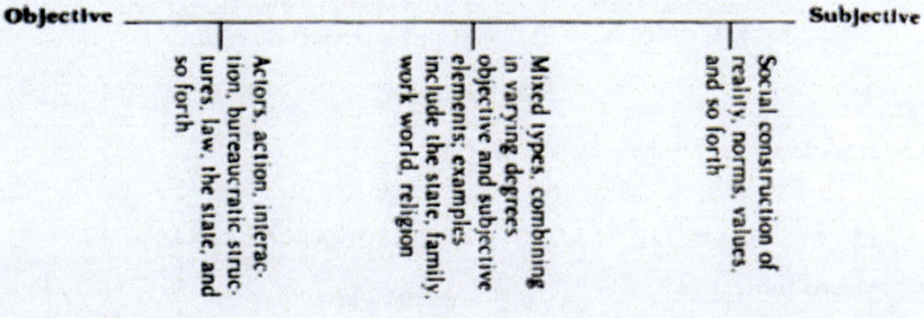

Fig. 8.2 The Objective-Subjective Continuum with mixed types identified (Ritzer, 1975)

Behaviorism: The clearest examples of this paradigm are Skinner's. The object of sociology for the social behaviourists is the unpredictable behaviour of individuals. Of particular interest are the rewards which draw out desirable behaviour, and punishments which impede or reduce undesirable behaviour. What distinguishes the supporters of this paradigm from the methodological point of view is the use of experiment, which is used more often than in others. The theories which can be included in it are behaviourist sociology and the exchange (or market) theory.

Ritzer poses the question of what to do, and how to do it, in order to move towards a better integration of different paradigms (while safeguarding their specificity and differences), and, taking certain pointers from G. Gurvitch, works on two continua of social reality: (a) the microscopic–macroscopic continuum; (b) the objective–subjective continuum. The former leads him to delineate figures 8.1 and 8.2.

The two continua overlap so that at each end of the micro–macro objective and subjective components can be differentiated. On the micro, individual level, for instance, there are the subjective mental processes of an actor and the objective models of action and interaction he/she engages in. The same is true of the macro. According to Ritzer, given that society is made up of both objective structures (like governments and laws) and subjective phenomena (like values or norms of behaviour), these levels cannot be disregarded if we want to understand reality in a valid way, and it is fundamental to understand the mutual interplay between micro-macro and object–subject. This leads to the formulation of diagram 8.3.

According to Ritzer an integrated sociological paradigm should deal with all four basic levels of social reality identified in the figure, and with their interrelations.

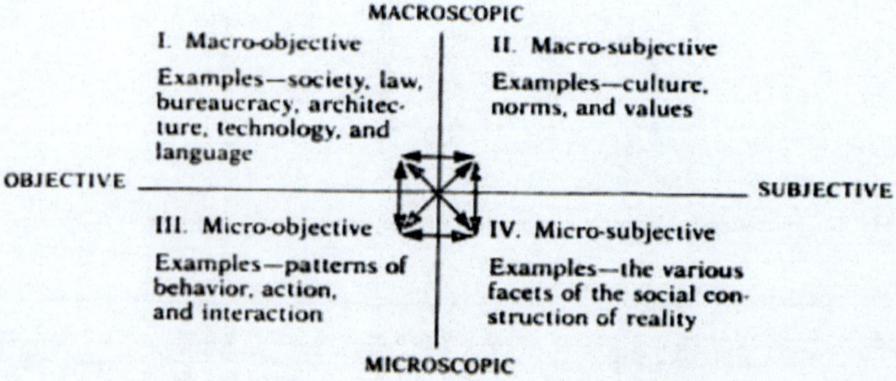

Fig. 8.3 Major levels of social reality (Ritzer, 1975)

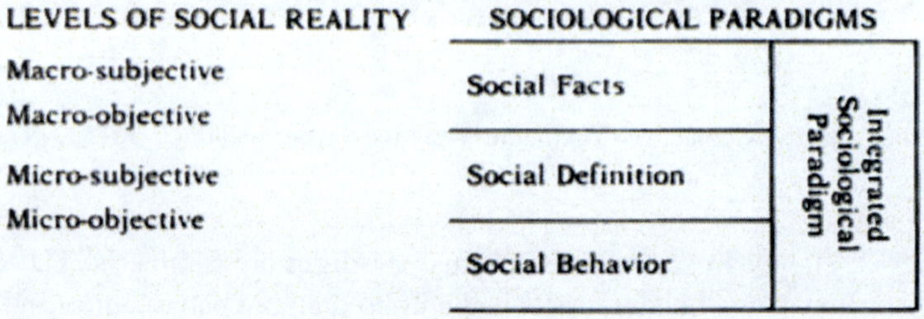

Fig. 8.4 Levels of social reality and major sociological paradigms (Ritzer, 1975)

But another question Ritzer regards as important is how the four levels of the diagram relate to the three paradigms outlined before. This led him to present another diagram in figure 8.4.

While existing paradigms slice the levels of social reality horizontally, an integrated paradigm slices them vertically. This makes clear why an integrated paradigm does not substitute the others. It deals with all levels, but it does not examine them with the degree of intensity that the others do: the others can, and have to, be used alternatively according to the question being asked. If they do not claim to explain social reality in all its complexity, they can be of great help. The core element of the new paradigm lies in the interrelation of the four basic levels of social reality, symbolised in the penultimate diagram by the interconnection of the arrows in the middle of the figure.

Ritzer made clear that this diagram does not represent the real world, which is always flowing and is not split into different levels, therefore it is not to be reified. But it may be useful in order to deal with the complexity of the real world, keeping in mind that it is the real world, and not the diagram, which is the subject of sociology. To do this all methods (questionnaires, interviews, experiments, observations, etc.) are valid, but there must always be a face to face with history, comparing different societies to understand the differences, and similarities, in how the various levels work for or against the others and vice versa. And we have to shun the idea that only one abstract theory can be developed which can explain social reality in all societies throughout history.

Ritzer concluded by saying that an integrated paradigm has much to gain from dialectical logic. "The dialectic is rooted in the real world of individual thought and action. It sees people producing larger social structures and those structures coming, in turn, to constrain and coerce actors. It therefore has a very clear image of the interrelationship between the microscopic and macroscopic realms. Yet it does not give primacy to one or the other; they are seen as dialectically related". (1979, p. 44).

Only One Paradigm or an Integration of various Paradigms?

The question we asked at the start of this article – whether it is better to move towards a single paradigm, as Morin suggests, or have an integrated understanding of the various paradigms, preserving sociology's multi-paradigmaticity, as Ritzer suggests – remains open. Further, if we accept the first suggestion we have to ask ourselves whether Boudon's paradigm of interactive analysis is more valid or Morin's PC; whether the conceptual confusion which is present in the definition of paradigm, as we have seen, makes it serviceable for sociological exploration in real depth or not, as Martindale (1979) and Cavallaro (1984) maintain. Finally, the last question we have to find an answer to is whether it is possible to draw valid methodological pointers from the comparison of different paradigms as has been attempted here. My opinion is that requires an analysis of two other concepts, theory and model of society, which are often mixed up with and considered a synonym of paradigm. Only at the end of this additional effort can we try to reply in a valid way to the above-mentioned questions. This is what I did in my work (L'Abate A., in press). I cannot repeat all my arguments in detail here. However, I shall limit myself to synthesise them as much as it is necessary to follow the logic of my arguments. In regard to theories, first I will analyse the elements that compose a theory (concepts, variables, theoretical propositions, models) . Then, I will expand in greater detail and depth the last concept through its three principal models of theory-making: assiomatic, casual and classificatory. I will conclude with three levels of theorixing, such as particular theories, of medium range and general range. The last classification will be used later on to reach some conclusions about the role of paradigms in sociological theories.

Still referring to the same source (L'Abate, A., in press), I will analyse different societal models used in sociological studies. This brings me to consider consensus and coersion models as viewed by Dahrendorf (1959) both as antagonists and as alternating with each other. I will also include other alternatives offered by various scholars (Alberoni, 1965; Coser, 1967; Feldmann & Moore, 1962; Pellicani, 1974) to try to reconcile and move above and beyond contradictions and gaps between these two models, tentatives that are very likely inadequate to solve problems found in those models. So much so, that eventually I will propose my own model.

At this juncture we are in a position to reply to the questions posed earlier – i.e. whether the concept of paradigm can be useful, even indispensable, for social research, whether it leads us astray, or is confusing. Also, whether we should move in the direction of a single paradigm or accept the eclecticism and multi-paradigmaticity of sociology itself.

We have seen how the three concepts (theories, paradigms, models) are often used as equivalents, or at least similar. For this reason different works can be found which speak of conflict theory, conflict model and also conflict paradigm, and so on. But this terminological confusion stops us from replying to the questions correctly.

A contribution to the clarification of concepts came from Masterman (1970). In her essay on "The Nature of a Paradigm", she identifies 21 different senses which Kuhn gave to the concept: (1) a universally recognised scientific achievement; (2) a myth; (3) a 'philosophy', or constellation of questions; (4) a textbook, or classic work; (5) an whole tradition, and in some sense, as a model; (6) a scientific achievement; (7) an analogy; (8) a successful metaphysical speculation; (9) an accepted device in common law; (10) a source of tools; (11) a standard illustration; (12) a device, or type of instrumentation; (13) an anomalous pack of cards; (14) as a machine-tool factory; (15) a gestalt figure which can be seen in two ways; (16) a set of political institutions; (17) a "standard" applied to quasi-metaphysics; (18) an organising principle which can govern perception itself; (19) a general epistemological viewpoint; (20) a new way of seeing, (21) something which defines a broad sweep of reality.

An analysis of things in common between these different definitions leads her to identify three main groups or areas of meaning:

1. *Metaphysical paradigms* (metatheoretical): a set of beliefs, a myth, a successful metaphysical speculation, a standard, a new way of seeing, an organising principle governing perception itself, a map, something which determines a large area of reality;
2. *Sociological paradigms*: a universally recognised scientific achievement, a concrete scientific achievement, a set of political institutions, an accepted judicial decision;
3. *Construct paradigms* (artefact): a textbook or classical work, supplying tools, instrumentation, an analogy, a gestalt figure, an anomalous pack of cards.

In illustrating these three different concepts of paradigm, Masterman points out how, according to Kuhn, the sociological paradigms precede theory and are different from it. This is why Kuhn needed a different term. For instance, the metaparadigm, the first group, is different from theory in that it is much wider and ideologically prior to it; it is a whole *weltanschaung* (conception of the world). The third paradigm is narrower than theory since it can be hardly theoretical at all, like a piece of scientific equipment. According to Masterman critics have analysed and focused their criticism on only the first of the area shown above, while she holds that "the fundamental one is the constructive sense of paradigm and not the metaphysical sense, or metaparadigm. It is only by using something artificial that brainteasers can be solved". (p. 143).

Masterman's clarification seems decisive to me. I think it would be wiser to use different terms for each of these three concepts. My suggestion is to use the concept "model of society" or *vision of the world* for the first group (metatheoretical), "sociological theory", or better still only "theory". It's a given that some useful theories may come from other social science disciplines, such as psychology, social psychology, economics, etc., for the second and keeping the term "paradigm" only to indicate the third concept, which I would define as "method of analysis".

This said, the problem is much easier to solve. I think we have to overcome the conflict between different approaches in the first level – "model of society" or *vision of the world*. We cannot continue to consider a human being as good or bad, and society as based on consensus or conflict. In my opinion the time has come to go beyond these ambiguities. We need to have a model which allows freedom to act, which considers people responsible for their actions, and sees conflict and consensus as fundamental for an understanding of both persistence and change.

We also need to study better (not theorise) how societies work and change. We have enough material to do so and to stop talking in an abstract way. In this case eclecticism and pluralism are not needed, what is needed is a unified, provisional model of society, che io definisco dell'equilibrio instabile. In questo modello integration and interdependence are seen as problematical and subject to significant changes. The organisation is seen as: "a set of conflicts, particularly of tensions between centripetal and centrifugal forces, which at the same time imposes and limits checks on the parts, which keeps the balance between their dependence and independence, and which separates and at the same time unites the parts" (Gouldner, 1959).

The situation is the complete opposite in regard to the other two groups: here pluralism is not only possible, but necessary. In the 'sociological theories' group (or better "theories", in order to remain open to contributions from the social sciences) we need to develop well-founded theories. For the third group, the paradigm group – or if we prefer, as I do, to call it the methodological group – here we have to follow Feyerabend (1970; 1975) and be pluralists. However, this does not mean accepting everything, especially what is new or original, as sometimes happens. We have to work hard to make clear the different methods of sociological analysis and codify their methodology better. I believe this is one of the most stimulating tasks we have.

A Methodological Discussion

This seems to me that we need to confirm completely the importance of using different terms for different concepts. We need to work on each of the three levels identified in different ways in the search for a unified model on the first level (without becoming dogmatic about it, and considering it provisional), by trying to be pluralistic on the other two levels (theories and methods). What has emerged confirms the importance of not limiting science to the sphere of the deductive method, but also of developing the inductive method and thus constructing theories from the bottom up proceeding along the following path: specific theory → middle-range theory-general theory and then starting again, but backwards this time. The return path would serve above all to activate experimental research – not only in the laboratory, but also in the field – which would allow us to get beyond the separation between "theory" and "practice", and make a valuable contribution to the development of that cognitive cycle which we have seen lies at the basis of the cognitive process and of our understanding of the world around us.

But before arriving at indications as to how we can carry out this task on a methodological level I feel it is important to have a look at what emerged on models and paradigms. First of all I feel it necessary to emphasise how the unstable equilibrium model finds a lot of support in the analysis of paradigms, in particular, in recognising that complex systems can work also in situations of great instability (Ceresa, Mela, & Pellegrini, 1981), and that order and disorder should not be seen as separate, opposite concepts, but rather as having mutual relationships, as dialogic, to use Morin's term, between order-disorder-organisation-order (Morin, 1984, 2008). Or in the importance and development of open paradigms (Eisenstadt and Curelaru, 1976) where conflict and consensus are mixed in relation to the ever-growing complexity of society.

From the examination of paradigms it seems to me that a series of methodological indications emerge – in support of my conclusions that we should consider methodological aspects as fundamental aspects of paradigms (as Masterman, 1970 maintains and Kuhn, 1970 recognises) – which we shall keep in mind and develop further (see L'Abate, A., in press).

It seems to me that the main methodological indications are:

1. Methodological individualism;
2. Complementarity between understanding and explanation;
3. Importance of subjectivity and objectivity;
4. Interrelation between the whole and the parts, but mutual autonomy.
5. Not only pluralistic, but also mutual, interconnected causality;
6. Order and disorder seen as complementary aspects of reality;
7. Science with conscience.

I would like to end this chapter by clarifying my opinion on methodological pluralism. I do not believe method should be eliminated, as Feyerabend (1975) does, but there are certainly a number of methods which must be identified. Apart from "causal analysis", which has been studied extensively and about which we can say that it has reached a good level of codification, other methods are at what we could call a "pre-elementary" level. I believe that at least three methods need to be worked out in depth: functional analysis, structural analysis, process analysis.

Formerly, I had mentioned four methods, placing systemic analysis among these three, but now as this work progressed, it seemed that it could be considered part of process analysis, given the importance of processes within systems. Each of these approaches, if we can call them so, in order to become a method could profit from the corresponding philosophy or perspective, but it also has to become autonomous and independent from it: functional analysis from functionalism, structural analysis from Marxism and structuralism, process analysis from interactionism, systemic analysis from structural-functionalism and the general theory of systems, too. This could be done by comparing studies that have used these different methods, as Merton (1949) began doing with functional analysis, trying to find pointers to codify each of these methods better.

Thus, all told, on the basis of this interpretation four main methods would exist (if we consider systemic and process analysis to be the same) which use, or can use, both qualitative and quantitative methods, even in different measures. This would bridge the gap between quantitative analysis being seen as conducted only, or almost exclusively on the level of causal analysis, and all the others as the domain of qualitative analysis.

To complete my arguments, I think it would be appropriate here to say something about the specific cognitive process in particular and research methods in general. Some scholars have criticised correctly at least in part the rigidity of the scientific method that risks to interfere with the attitude of 'creative uncertainty', which I believe the major fundamental orientation of any scientific researcher. However, I am of the opinion that those critics have not deals in depth with the limits of the so-called "scientific method".

The first of these limits is the underestimation of the inductive method to understand the world around us. It is common knowledge how most discoveries on the causes of cancer (for instance, the influence of certain manufactured substances) have not come from pre-existing theories or hypotheses or from large research and data-collecting centres around the globe (the so-called "cancer registers"), but from critical observation by G.P.s, who noted that in the area they were working the labourers in a certain factory got ill more frequently than others (Maccacaro, 1976).

This statement of fact is leading also epistemologists to have a close look at certain earlier theories of knowledge. David Hume's criticisms of the inductive method in "Enquiries Concerning Human Understanding and Concerning the Principles of Morals" (2007) is that the conclusions of inductive argumentation – contrary to deductive argumentation – have contents which are superior to those in the premisses. This provoked philosophical problems regarding the demonstrability of what has been defined "inductive support". This criticism has been carried forward above all by Popper (1968), who believes the inductive method is not provable and, therefore, unusable in a correct scientific procedure.

Bianca (1986), wrote:

> "In reality, Popper, even though he says science is a castle which can always be criticised, contradicts himself because he requires that scientific propositions, and inductive ones in particular, have absolute value in reality: he maintains, in effect, that through using his method of falsification it is possible to be sure of the truth of scientific propositions which would be valid in all conditions. This is something that has never been true, in particular because 'in all cases' can never be valid, not even after checks have been made on a wide and penetrating scale" (p. 24).

This and other criticisms are bringing about a re-evaluation of the inductive method. Indeed, according to Salmon (1969), a scholar of "logic", it is exactly the characteristic of having conclusions whose content goes beyond the premisses that "makes inductive argument necessary to support a great part of our knowledge" (p. 77). Bianca writes: "Truly, without induction man could not even survive because each experience he has would remain isolated from the others and his life would be a succession of discrete experiences. It would be unthinkable that each moment of man's being in the world would be always new: he would not be able to learn. Man, on the other hand, like animals, can survive precisely because he classifies his every experience, comparing past ones with present, and as they accumulate his present serves as a cognitive parameter for future experiences" (p. 20). Bianca then explained how by the introduction of assumptions (statements which are assumed, as though they had been formulated in reference to unexamined cases) the "logical leap" between single statements and general statements can be bridged, he writes "In reality there is no leap if induction is understood as a process of the simplifying kind founded on examined cases and cases that are accepted as though they had been examined" (p. 22).

As we see, induction has all the requisites for its legitimate inclusion as a valid scientific procedure. Now, let us have a look at the second criticism: the exclusion of practice. It is Galtung (1977) who emphasised how research should not be concluded with a written report, but only when the modification of the world around us as predicted by the results of the research has been achieved. And all scientists, above all biologists, know that scientific process necessarily goes by way of experimentation, not only of the theoretical sort, or shut inside specialised laboratories, but also the sort that modifies the world around us, or part of it, and keeps checking the results obtained where the intervention has taken place with those where there has been no intervention. This implies the inclusion of life itself in the cognitive process. This allows us to present the final diagram, which I produced, introducing some changes, based on a suggestion made by the French methodologist Loubet Del Bayle (1978), figure 8.5.

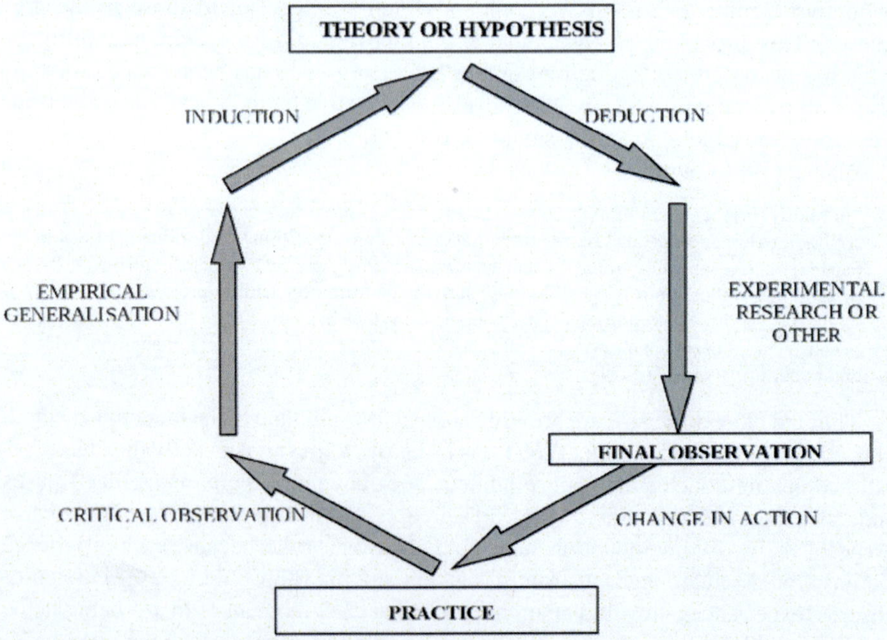

Fig. 8.5 Loubet del Bayle's Diagram (1978): the cognitive process

It appears to me that diagram 8.5 is more valid than other scientific methods, that view research according an exclusively deductive logic for instance, according to these phases Bailey (1978): (1) Choosing the problem and stating hypothesis; (2) Formulating research design; (3) Data collection; (4) Coding and analysing data; (5) Interpreting results; and back to point 1. This proposed model uses not only deductive logic but also inductive logic even though it introduces practice within the cognitive cycle. Scientific procedure can start both from the top, from the theory and/or hypothesis. From the right it can arrive at practice passing through deduction, research, final observations and the change in action (either total or partial, through experimental observation in the field). It can also start from the bottom, from practice itself, and through critical observation of this practice (which includes, or can include, research, as well) it can lead us to find empirical generalisations. These generalisations in turn can lead us to formulate hypotheses or theories through induction, or to confirm those already existing. Naturally, the cycle never ends and the hypotheses and theories which have been found and formulated undergo further research which may reinforce or invalidate them.

This diagram emphasises the importance of experimentation (in the field, not in the laboratory) as a tool passing through theory, research and action, and points out the limits of a type of sociology which restricts itself to working on "descriptive" and "analytical" levels, and is content to describe and explain, or interpret, the world around us. Stamler (1959), for example, speaks of four levels in epidemiology: descriptive, analytical, experimental and applied. Why should sociology and sociological research, limit itself to the first two levels, leaving the last two outside its cognitive cycle?

If this is true, from Comte's "to know in order to act", which favours knowledge over action, we have to go on to "act through knowing – know through acting", which recognises both practice and theory as the basis of knowledge, without favouring either, but above all which points out the importance of not limiting research to episodic moments, but to set up a valid system of information when we act which allows us keep what we do under control. This emphasises the importance of a systematic collection of data on specific problems we are working on (e.g. who is using the services, what needs they bring to the services, which needs cannot be solved by the service itself and requires intervention from elsewhere, what results we have, etc.). However, this should be done without making the data we collect excessive, which could risk making its collection a senseless matter of routine, allowing ourselves to carry out research, even episodic, on specific problems we want to respond to on the basis of initial, well-specified hypotheses.

As it draws to its end, our discussion should be taken back to the theory and the cognitive processes we have already spoken about, to induction and deduction. Given that we have maintained that both of these processes are important, specific, or particular theories are needed, which, once they are consolidated in intermediate theories, through induction and new studies, may lead us to construct vaster, general theories from the bottom up. And there is also a need for general theories from which we can formulate hypotheses, which are verifiable and still to verify, which can lead us to confirm or not to confirm them through a deductive process.

Conclusion

I would like to conclude this chapter by emphasising together with Turner (1978) that sociological theory finds itself in a provisional and initial state. According to Turner much of what is defined sociological theory is only a loose bundle of implicit assumptions, of inadequately defined concepts and of vague, logically disconnected propositions. Sometimes the premisses are defined explicitly and serve to inspire theoretical statements which contain well-defined concepts, but most sociological theory forms a verbal "image of society", rather than a rigorously constructed set of theoretical statements organised in a logically coherent format. Thus, in his opinion, a large part of so-called theory is in reality only a general orientation or perspective for looking at various aspects of social life, which may become scientific theories in the future if all goes well (p. 32). With this chapter and with the volume that expands and integrates it, I am trying to move sociology forward according to an more scientific and less temporary approach. However, only those readers and scholars who may refer to this approach will be able to judge whether I succeeded or failed in this goal.

References

Alberoni, F. (1965). Verso una teoria generale dell'azione sociale. *Studi di Sociologia, 3*, 194–210.

Bailey, K. (1978). *Methods of social research*. New York: The Free Press.

Bianca, M. (1986). Epistemologia. In M. Bianca (Ed.), *Temi di epistemologia, scienza e didattica delle scienze* (pp. 19–31). Quaderni del Centro di Documentazione: Provincia di Firenze.

Boudon, R. (1980a). The three basic paradigms of macro-sociology: Functionalism, neo-marxism and interaction analysis. In R. Boudon (Ed.), *The crisis in sociology, (whole charter 3)*. New York: Columbia University Press.

Boudon, R. (1980b). *The logic of social action: An introduction to social analysis*. Chicago, IL: University of Chicago Press.

Cavallaro, R. (1984). *Paradigma e sociologia: un dibattito insensato ? ISVI (Istituto Formazione e Ricerca Problemi Sociali dello Sviluppo)*. Italy: Catania.

Ceresa, P., Mela, A., & Pellegrini, M. (1981). *Una lettura della sociologia per paradigmi*. Torino: Celid.

Coser, L. (1967). *Continuities in the study of social conflict*. New York: The Free Press.

Dahrendorf, R. (1959). *Class and class conflict in industrial society*. Palo Alto, CA: Stanford University Press.

Eisenstadt, S., & Curelaru, M. (1976). *The form of sociology: Paradigm and crises*. New York: Wiley.

Feldmann, A., & Moore, W. (1962). Industrialization and industrialism: Convergence and differentiation. *Transactions of the Fifth World Congress*, Washington, DC, 2–8 Sept.

Feyerabend, P. (1970). Consolations for the specialist. In I. Lakatos & A. Musgrave (Eds.), *Criticism and the growth of knowledge* (pp. 197–230). Cambridge, UK: Cambridge University Press.

Feyerabend, P. (1975). *Against method: An outline of an anarchistic theory of knowledge*. London, UK: New Left Books.

Friedrichs, R. (1972). *A sociology of sociology*. New York: The Free Press.

Galtung, J. (1977). Empiricism, criticism, constructivism: Three Aspects of Scientific Activity. In J. Galtung (Ed.), *Methodology and ideology* (pp. 41–71). Copenhagen, Denmark: C. Ejlers.
Giddens, A. (1976). *New rules of sociological method*. Cambridge, UK: Polity Press.
Gouldner, A. (1959). Reciprocity and autonomy in functional theory. In L. Gross (Ed.), *Symposium on sociological theory* (pp. 241–270). New York: Harper, Row.
Hume, D. (2007). *An enquiry concerning human understanding and other writings*. S. Buckle (Ed.). Cambridge, UK: Cambridge University Press.
Kuhn, T. (1962). *The structure of scientific revolutions*. Chicago, IL: University of Chicago Press.
Kuhn, T. (1970). Reflections on my critics. In I. Lakatos & A. Musgrave (Eds.), *Criticism and the growth of knowledge* (pp. 231–278). Cambridge, UK: Cambridge University Press.
L'Abate, A., (in press). *Methods of analysis in social sciences and research for peace: An introduction*. Basel, Swiss: Transcend University Press.
Loubet del Bayle, J. (1978). *Introduction aux méthodes des sciences sociales*. Toulouse, France: Privat.
Maccacaro, G. (1976). Comunicazione. *Il registro tumori regionale: esperienze, prospettive e problemi* (pp.13-16; 64–65). Regione Toscana, Giunta Regionale, Dipartimento Sicurezza Sociale, Firenze.
Martindale, D. (1979). Ideologies, paradigms, and theories. In W. Snizek, E. R. Fuhrman, & M. K. Miller (Eds.), *Contemporary issues in theory and research: A metasociological perspective* (pp. 7–24). Westport, CT: Greenwood Press.
Masterman, M. (1970). The nature of a paradigm. In I. Lakatos & A. Musgrave (Eds.), *Criticism and the growth of knowledge* (pp. 59–90). Cambridge, UK: Cambridge University Press.
Merton, R. K. (1949). *Social Theory and social structure*. Glencoe, IL: The Free Press.
Morin, E. (1984). *Scienza con coscienza*. Milano, Italy: Angeli.
Morin, E. (2008). *On complexity (translation by Robin Postel)*. Cresskill, NJ: Hampton Press.
Pellicani, L. (1974). *Dinamica delle rivoluzioni*. Milano, Italy: Sugar Editore.
Popper, K. (1968). *The logic of scientific discovery*. New York: Harper & Row.
Popper, K. (1970). Normal science and its dangers. In I. Lakatos & A. Musgrave (Eds.), *Criticism and the growth of knowledge* (pp. 51–58). Cambridge, UK: Cambridge University Press.
Ritzer, G. (1975). Sociology: A multiple paradigm science. *American Sociologist, 10*, 156–167.
Ritzer, G. (1979). Toward an Integrated sociological paradigm. In W. Snizek, E. R. Fuhrman, & M. K. Miller (Eds.), *Contemporary issues in theory and research: A metasociological perspective* (pp. 25–46). Westport, CT: Greenwood Press.
Salmon, W. (1969). *Logica elementare*. Bologna, Italy: Il Mulino.
Stamler, J. (1959). The epidemiology of atherosclerotic coronary heart disease. *Postgraduate Medicine, 25*, 610–622.
Turner, J. (1978). *The structure of sociological theory*, revised edition. Chicago, IL: The Dorsey Press.

Part III
General-Integrative Paradigms in Psychology

Chapter 9
Biopsychosocial

Beatrice L. Wood

> *The predominant model of disease today is biomedical, and it leaves no room in its framework for the social, psychological and behavioral dimensions of illness. A biopsychosocial model is proposed that provides a blueprint for research, a framework for teaching, and a design for action in the real world of health care*
>
> (George Engel, 1977).

> *The success of a hypothesis, or its service to science, lies not simply in its perceived "truth," or power to displace, subsume or reduce a predecessor idea, but perhaps more in its ability to stimulate the research that will illuminate … bald suppositions and areas of vagueness*
>
> (William Glen, 1994).

The Biopsychosocial Model: Neither Fish Nor Fowl?

The BioPsychoSocial (BPS) model has been criticized and praised more than it has been explicitly applied in scientific inquiry. This article will neither criticize nor praise, but rather examine the nature of the beast. Is it fish or fowl? Paradigm or theory? Model or dimension? And, what impact has this beast had on the evolution, or revolution, of theory and practice in medicine and psychology?

B.L. Wood (✉)
Women and Children's Hospital of Buffalo, 219 Bryant Street, Buffalo, NY 14222, USA
e-mail: bwood@buffalo.edu

L. L'Abate (ed.), *Paradigms in Theory Construction*, DOI 10.1007/978-1-4614-0914-4_9,

Historical Background

The biomedical model, still the dominant framework guiding medical research, education, and clinical care today, focuses on the root cause of disease, where disease is "fully accounted for by deviations from the norm of measurable biological (somatic) variables" (Engel, 1977). George Engel's research and clinical practice, and that of others, convinced him that this model was no longer adequate to meet the scientific tasks and social responsibility of medicine. He aimed to revolutionize his profession by replacing the biomedical model with the "Biopsychosocial Model," a broader framework for understanding both physical and mental illness, which he believed would serve to improve medical science and treatment.

The biomedical model was seated in what has been referred to as the "classical" scientific paradigm which was first enunciated by Galileo and Descartes. This paradigm rests firmly upon the assumption that phenomena (for example, somatic diseases) can be fully accounted for by single causes or chains of causes (for example, biochemical or neurophysiological aberrations) that exist within that entity (the soma). Furthermore, the biomedical model also assumes that behavioral, psychological, or emotional deviations from the norm can be similarly reduced to such somatic aberrations. The reductionistic, from-the-bottom-up assumption underlying the classical scientific paradigm also dictates an analytic method of scientific inquiry. This method assumes that an entity under investigation can be resolved into parts constituting that entity, and that once explained these component explanations can be assembled into a complete understanding of the entity. Engel noted that the biomedical model had yielded substantial critical advances in medicine. However, Engel felt the model fell short. In his seminal 1977 Science paper he used polemic devices to try to unseat this model from its dominant position in medicine.

First Engel described the concept of "folk models (i.e., culturally derived belief systems) which have been applied to disease." These involve beliefs and explanations about disease, along with rules of conduct to rationalize treatment actions. "These folk models constitute socially adapted devices to resolve for the individual, as well as for the society in which the sick person lives, the crises and uncertainties surrounding the disease" (Engel, 1977, p. 130). Unlike scientific models, which are revised or abandoned when they fail to account for all of the data, folk models become dogma, and discrepant data are forced to fit the model or are discarded. Engel made the case that the biomedical model had become Western society's folk model. The evidence he cited for this was that the biomedical model requires that "all disease, including mental disease, be conceptualized in terms of derangements in underlying physical mechanisms" (Engel, p. 130). He further claimed that physicians' beliefs are molded and shaped by their professional education which reinforces the biomedical model without introducing the perspective of scientific critique.

Engle further traced the history of medicine, in which medicine became a scientific enterprise when physicians and other scientists developed a taxonomy, and applied scientific method to the understanding and treatment of disease. He traced the mind–body dualism prevalent in the current biomedical model back to the Church's early interdiction against the scientific study of the mind and behavior, which was based on the claim that mind and behavior had more to do with religion and the soul and therefore was in the purview of the Church. At the same time, the classic paradigm of Galileo, Newton, and Descartes, characterized by reductionist-analytic methods, relegated medicine to the study of the body as a machine, and disease was seen as the consequence of the breakdown of the machine. These historical factors lead to the ignoring (and ignorance) of the psychological, social, and emotional aspects of disease in medical scientific inquiry and practice.

Engel then presented a compelling empirical argument elucidating the inadequacies of a biomedical explanation for both diabetes and schizophrenia, showing the essential role of environment in the development of these diseases. He demonstrated that the biological defects are necessary but not sufficient explanations for these diseases. This final argument lead directly to his proposing an alternate model: the BPS model.

The BPS model was conceived within the context of the General Systems Theory (GST) paradigm. GST had emerged both in biology through Weiss' and von Bertalanffy's work (Drack, Apfalter, & Pouvreau, 2007) and in the social sciences (Bateson, 1972; Gray, Duhl, & Rizzo, 1969; Grinker & Hughs, 2011) during the formative years of Engel's theorizing, and no doubt it shaped the development of the BPS model. GST was a paradigm shift, away from linear causal assumptions deriving from classical physics and chemistry, towards assumptions of hierarchical organization of phenomena and non-linear or circular causal effects. Definitions of "system" range from the most general, e.g., "sets of elements standing in interrelation," (von Bertalanffy, 1969, p. 38) to highly specific, e.g., "the living body considered as made up of interdependent components forming a unified whole" (Princeton University, 2011). Focus in GST is on understanding phenomena "in context" of the organization of which they are a part. "In context" refers to the complex interrelations of the elements interconnected with the phenomena. The scientist must not only understand the properties of the elements comprising the context, but also the nature of their interrelation.

Another GST departure from "classical science" is the methodology. The classical paradigm of scientific inquiry required analytic procedures. An entity or phenomenon was to be understood by disassembling it into its component parts, with the expectation that understandings of the characteristics of each component could be compiled into an understanding of the whole entity. GST claims that the analytic approach violates the properties of organized complexity (i.e., systems), which include not only characteristics of the components of a system, but also the interrelations among them. Hence, analytic procedures are invalid. Assuming that disease (mental and physical) occurs in the context of the organized complexity of the body, the psyche and the social context, such reasoning converges with the notion of a BPS model of disease. In his 1977 paper, Engel stated his belief that in order for a medical science to have a complete understanding of disease, as the underpinning for rational treatment and health care, "it would need to incorporate the patient, the social context in which he lives, and the complementary system devised by society to deal with the disruptive effects of the illness, that is the physician role and the health care system" (Engel, 1977, p. 132). In short, medicine needed a BPS model.

Is the BPS Model a Scientific Model?

In the next segment of Engel's article the reader expects to find laid out a BPS model, with identified aspects of the biological, the psychological, and the social dimensions of the model. The GST reader would look for specified propositions about the nature of the interaction of these dimensions. Engel chose instead to pursue a polemic direction, sharpening his arguments with moral, ethical, and rational reasons as to why such a model was needed. In his 1977 paper Engel did identify GST as "a possible conceptual approach suitable for the BPS concept of disease" (Engel, 1977, p. 134). He stated that "since systems theory holds that all levels of organization are linked to each other in a hierarchical relationship, so that change in one affects change in the others, its adoption as a scientific approach should do much to mitigate the holist-reductionist dichotomy..." (Engel, p. 134). This is as close as Engel came to identifying a scientific model. Hence, because of its lack of specificity in identifying aspects of the levels and their inter-relation, the BPS "model" was clearly not presented as a scientific model, at least not in its original introduction to the scientific literature. Indeed, Engel, himself, referred to the 1977 article as a "critical" essay.

In his 1980 paper, presented on the occasion of his honorary lecture to the American Psychiatric Association, Engel went further in explicating the BPS model (Engel, 1980). He articulated meticulously and extensively the levels of organizational hierarchy in which he saw the individual embedded. The hierarchy extends from subatomic particles to the biosphere, with the "person" positioned midway, as the highest level of the organismic hierarchy and the lowest level of the social hierarchy (see Fig. 9.1). He further proposed, consistent with GST, that within the hierarchy as a continuum, each system is at the same time a component of a higher system, and is influenced by that higher system (see Fig. 9.2). Furthermore, he noted that the designation "system" bespeaks the "existence of a stable configuration in time and space" (Engel, p. 541) maintained not only by the "coordination of the component parts in some internal dynamic network but also by the characteristics of the larger system of which it is a component part" (Engel, p. 541). Finally he noted that the methods and rules that apply to each level of system are distinctive for that level of system and cannot necessarily be applied to other levels, and that different approaches are required to gain understanding of the collective order of each system level. He again asserted that given these propositions, this inter-related hierarchical continuum of systems cannot be understood by a reductionistic approach, i.e., as "merely as an assemblage (or reassemblage) of constituent parts" (Engel, p. 541). This is a fairly explicit set of assumptions about the kinds of theories or models which can be constructed within the BPS framework. But it does not specify variables or specific relations among the components, and does not contain the elements necessary to make specific testable hypotheses. This calls into question the BPS entity as a "model."

Critiques of the BPS model as a model are plentiful. Unfortunately most of them criticize Engel for not delivering what he indicated he was going to deliver rather than thoroughly examining the nature and impact of the BPS entity he called "model." Ironically the most vociferous, ad hominem, and weakly researched and

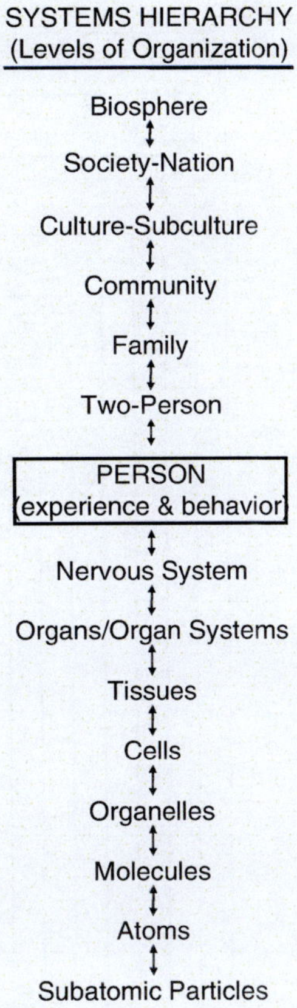

Fig. 9.1 Hierarchy of natural systems

poorly reasoned critiques come from within psychiatry. McLaren, an Australian psychiatrist, stated that "any value which may have derived from Engel's BPS model has been entirely fortuitous, because it is not a model at all" (McLaren, 1998, p. 87). McLaren based his critique on a scholarly analysis of the scientific meaning of "model" and came to the conclusion that the BPS notion was not a scientific model. McLaren went on, however, to critique the BPS model even as a "general theory," claiming that its stated reliance on GST to provide a conceptual and methodological framework is bankrupt because "once GST is extended outside its physical and mathematical basis, it becomes utterly banal" (McLaren, 1998, p. 89). He further stated, "It [GST] could never extend to a general methodology of human affairs, not the least because nobody has ever shown that the critical matter-energy transfer functions are applicable to the mind-body problem" (McLaren, 1998, p. 89). But he misunderstands a critical aspect of GST, which is that its theory and methodology rests firmly outside matter-energy functions. Rather, it is based in part on characterizing and explaining *patterns* of inter-relations, which is not limited in the way McLaren claims, and which is well suited to examining interactions across biological, psychological, and human systems (Bateson, 1972).

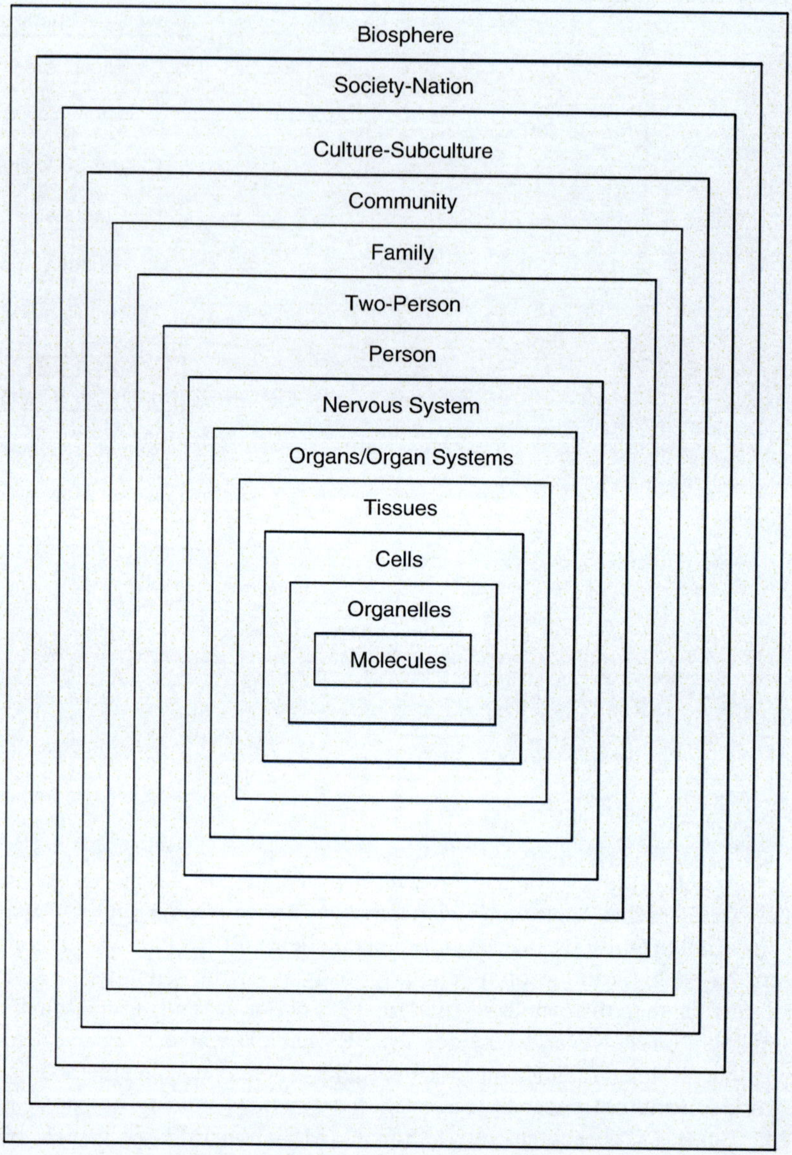

Fig. 9.2 Continuum of natural systems

Another vocal critic of the BPS model is Ghaemi, a psychiatrist from the US. Ghaemi claims that the BPS model is the conceptual status quo of contemporary psychiatry, but argues that while it has played an important role in combating psychiatric dogmatism, it has devolved into mere eclecticism (Ghaemi, 2009). It is highly questionable that the BPS model has become status quo in psychiatry. On the contrary, psychiatry, as judged in part by NIMH leadership and current education and practice, is highly biologically reductive in its science, education, and practice. After all psychiatry completed the "Decade of the Brain" (1990) in 2000 only to embark on the "Decade of the Genome" (McPherson et al., 2001). Both missions embrace and highly prioritize a reductionistic biomedical model. Both McLaren and Ghaemi claim that the BPS model is scientifically bankrupt, but they both woefully neglect the abundant scientific literature in support of BPS interactions in health and disease. (For reviews see Novack et al., 2007; Suls & Rothman, 2004).

The BPS "model," did give rise to several genuine and highly specified scientific BPS models. Such models will be sampled and illustrated later in this chapter. The difference between these and Engel's BPS "model" is that these models specify specific multi-level BPS variables, along with specific inter-relations among the levels, to explain a specific phenomenon. Engel's BPS notion was not specific in that way. Therefore it seems reasonable to conclude overall that the term "model" as applied to Engel's BPS model does *not* connote a scientific model. So if not a scientific model, is it a paradigm?

Is the Biopsychosocial Model a Paradigm?

A paradigm in Kuhn's inclusive sense encompassed everything necessary for practicing science, all the way from a set of metaphysical assumptions "at the top" through commitments to apparatus and experimental procedures "at the bottom." (See Hillix and Abate, Chap. 1) Clearly the BPS model does not meet this definition. Kuhn also stated that a mere "approach" to science is not sufficient, and that an approach is not a paradigm unless its accomplishments are impressive enough to attract the allegiance of essentially all of the practitioners of the branch of science involved. Lakatos (1970) refuted this, stating that it was not typical for a field of science to have only a single paradigm. Even allowing for multiple paradigms, the BPS model, at least at its inception, clearly did not meet the allegiance requirements for a paradigm.

Kuhn also asserted that if a paradigm is to be attractive enough, the procedures it advocates must be precise and clearly communicated, and its predictions must be clear and unequivocal. In his1980 paper Engel did specify by constraint some aspects of procedure. Namely he asserted that each system level required a different approach, and that the hierarchical continuum of systems was *not* amenable to reductionistic analytic approaches to explanation. He was clear and precise in laying out the hierarchical framework. He did not, however, specify what scientific approaches would be amenable. Furthermore, Engel's formulations were not of the sort that allowed for clear and unequivocal predictions. According to Kuhn, paradigms are universally recognized scientific achievements (in a given field) that for a time provide model problems and solutions for a community of practitioners [scientists] (Kuhn, 1996). It does not appear that the BPS model has achieved this status with regard to impact on scientific inquiry or by accruing a community of scientists who explicitly use the BPS model as its main framework.

Is the Biopsychosocial Model a "Paradigm Shift" in Its Early Beginnings?

On the other hand, Frankel, Quill, and McDaniel in their comprehensive and thoughtful book entitled "The Biopsychosocial Approach: Past, Present and Future," (Frankel, Quill, & McDaniel, 2003) propose that the BPS approach might be considered a paradigm "shift." They propose, following Kuhn, that a paradigm shift occurs when a scientist, or group of scientists, come to see the same facts or problems in a different light. They further assert that such changes proceed in science in two different ways, either by revolution, or by evolution, i.e., gradual change as noted by Laudan (1977). They give examples of each and make the case that the BPS approach embodies an evolutionary change.

As originally presented, and later followed up, the BPS model was most strongly advocated as an approach to looking at medical *practice*, not science, in a different way. Indeed more content in Engel's, 1977 paper was devoted to outlining a framework for a multi-level approach to diagnosis and treatment of patients than was devoted to the BPS implications for scientific inquiry. It appears to this reader that Engel had studied his Kuhn well, and determined that paradigm shifts came from socio-political efforts rather than scientific critique. Indeed, Engel disclosed his disenchantment with scientifically engendered overthrow of old theories in his 1977 paper. In his discourse Engel indicates that "despite there being a sizable body of knowledge based on clinical and experimental studies of man and animals which support biopsychosocial interactions,"… "the research remained unknown to the general medical public and education because of unremitting pressures to conform to scientific methodologies that were reductionistic in conception and inappropriate for the problems under study" (Engel, p. 134).

Hence it appears Engel turned his efforts to educating the educators of the next generation of physician-scientists. Apparently he took seriously Kuhn's notion that if young people could be attracted to a new paradigm, once stubborn devotees of the old paradigm eventually die off, the field will yield to a paradigm shift. Engel appeared to commit himself to use the BPS model to engender a socio-politically driven paradigm shift, but not one deriving from shifts occurring through scientific disproof of the biomedical paradigm. It is yet an open question as to whether the BPS model and Engel's socio-political advocacy ushered in a scientific paradigm shift.

The BPS as a Clinical Model or Approach

One potential resolution to the debate would be to specify the BPS model as a "clinical model" or approach. The BPS approach has had wide repercussions and influence in select areas of medical education and practice, and, as reasoned above, ultimately such effect may influence the direction, at least, of scientific inquiry.

Borrell-Carrio, Suchman, and Epstein, (2004) an internist, claims that the BPS model is both a philosophy of clinical care and a practical clinical guide. "Philosophically, it is a way of understanding how suffering, disease, and illness are affected by multiple levels of organization, from the societal to the molecular. At the practical level, it is a way of understanding the patient's subjective experience as an essential contributor to accurate diagnosis, health outcomes, and humane care" (Borrell-Carrio et al., p. 576). Most of the positive outcomes in medicine which claim to be a direct result of the BPS model explicitly identify it as an "approach," and concentrate on its clinical application. And these outcomes have been substantial.

A recent review claimed that the BPS model has become one of the main theoretical foundations for a number of reform curricula in undergraduate medical education (Berne, Maastricht, Harvard, Glasgow, Witten-Herdecke, etc.) (Adler, 2009). The authors noted that the integration of the BPS model into post graduate and clinical medicine has been less successful. However, integrated BPS care models are noteworthy, in pain clinics and medical-psychiatric units in the US, in psychosomatic units in Germany, as well as in departments of internal medicine using BPS assessment instruments in standard cases. Switzerland seems to be at the forefront. At the Medical School, University of Berne, a residency program in integrated BPS internal medicine has been successfully established (Adler, 2009). A curriculum in integrated BPS medicine for practicing physicians was established in 2002 in all five Swiss medical schools. The curriculum encompasses 360 h spaced over 3 years covering knowledge in psychosocial aspects, communication skills, and self-experience. At the end of this curriculum, the participants receive a diploma called Fähigkeitsausweis ("certificate of proficiency") (Adler, 2009).

The BPS model has been prominently applied to education and practice at the University of Rochester, where Engel was Professor from 1946 until his death in 1999. There is currently an innovative undergraduate medical education program at that University which incorporates the BPS approach as a conceptual, scientific, clinical, and educational framework. This program called, "the double helix curriculum" intertwines, like the DNA double helix, the basic sciences with clinical medicine across each of the 4 years. BPS integration is the hallmark of the curriculum (Frankel et al., 2003). In addition, one of the most prominent offspring of the BPS approach is the movement which was early identified as "Family System's Medicine" and is now referred to as "Family Systems Health." This movement, spearheaded and propagated by Susan McDaniel, Ph.D., Professor of Psychiatry and Family Medicine, had world-wide influence on the practice of medicine. Family Systems Health explicitly incorporates a family systems level into the BPS conceptualization of health and disease, and applies this new construct to clinical practice, education, and research.

Clearly the BPS approach has led to broad shifts in how medical practice is conceptualized and taught, at least in some parts of the world. However, what about medical science? It seems eminently fair to characterize Engel's model as a BPS Clinical Model, but let us return to the problem of how best to characterize the BPS notion in the scientific context.

Is the BPS Model a Theory?

According to the APA Dictionary of Psychology, a theory is a principle or a body of interrelated principles that purports to explain or predict a number of interrelated phenomena. In philosophy of science, a theory is a set of logically related explanatory hypotheses that are consistent with a body of empirical facts and that may suggest more empirical relationships. In its best use, theory signifies a systematic account of some field of study derived from a set of general propositions. (Hillix and Abate, Chap. 1) A "theory" rests conceptually and functionally somewhere between a paradigm and a model. According to our analyses above with regard to the BPS model's status as a scientific model, it appears that the BPS model also does not fit the criteria for the category "theory." What about as a "meta-theory?"

Is the BPS Model a Meta-Theory?

Feigl identifies a concept called "meta-theory" as a part of a paradigm (Feigl, 1970). A meta-theory is conceptualized as a philosophical component of a paradigm that determines how models and theories are to be constructed. Suls and Rothman state that, "no scientific field seems to advance without an implicit or explicit set of metatheoretical assumptions" (Suls & Rothman, 2004, p. 119). They refer to the BPS model as the meta-theory organizing research, practice, and policy for the field of health psychology, and they provide an extensive review of research emanating from this meta-theory.

The BPS model does seem to fit this definition given Engel's, 1980 presentation of the hierarchical continuum of the BPS model, along with its general system implications regarding problem conceptualization and associated constraints of scientific approach. Suls and Rothman conducted a critical review of the evolution of the BPS model and relevant research. They conclude that the BPS model is a meta-theory (Suls & Rothman, 2004). Meta-theories that encourage the development of models emphasize *deductive* approaches; those that ignore such development in favor of the direct study of relationships between empirical concepts take an *inductive* approach. (Hillix and Abate, Chap. 1) Engel's careful logical-deductive argument in support of the need for a BPS model (Engel, 1977) and his deductive analysis of the BPS model in his 1980 paper (Engel, 1980) suggest that as a meta-theory, the BPS model is deductive in nature. A meta-theory might also be analytic (encouraging breaking phenomena into small parts for study) or synthetic (encouraging the direct study of larger units). Clearly by proposing GST as an organizing paradigm for the methodology of the BPS model, Engel identifies the BPS model as synthetic. One pragmatic criteria for deciding whether the BPS model is well construed as a meta-theory is whether it shaped and inspired specific BPS scientific models. The following section will examine evidence for this criterion.

BPS Model as a Meta-Theory: Scientific Medical Research Consistent with the BPS Model

Novack and colleagues present several new fields of psychosomatic inquiry that are consistent with the BPS conceptualizations. They specifically present psychoneuroimmunology, psychoneuroendocrinology, autonomic psychophysiology as compelling examples (Novack et al., 2007). Most of the cited authors do not specifically identify the BPS model as a framework for their research. However, they may have been implicitly influenced by having been sensitized to BPS conceptualization.

Engel articulated the BPS model in 1977, heavily influenced by the science presented in the American Psychosomatic Society (APS) and by his colleagues at the APS. The early APS was a collection of PhD basic science researchers, and MD internists and psychiatrists. The membership was a rich crucible for searching for understanding and connections among the disciplines. Selye's stress research, the work of John Mason, and many other animal and human researchers highlighted the connections between the biological, psychological, and social spheres, and many researchers who presented at APS implicitly believed in a BPS formulation, even before it was explicitly articulated by Engel. Engel articulated in an elegant manner what he was seeing in the APS. However, as he noted in his 1977 paper, BPS psychosomatic science had not made much of an impact into the medical education world. Engel was president of the APS, so many researchers knew his work and ideas. At its core, psychosomatic research has always been about interconnections between disciplines, levels, and spheres of human functioning. It is not clear whether scientific researchers working subsequent to the publication of the 1977 article were influenced by the articulation of the model itself, but they may have been influenced by Engel's earlier work in psychosomatic medicine, and by the co-evolution of the BPS sensitivity within the psychosomatic research community (Dennis Novack, M.D., Past President of APS, personal communication).

BPS Meta-Theory as an Impetus for Paradigm Shift: Testable BPS Models

In addition to research findings that are consistent with certain aspects of a BPS Meta-Theory, there are specific BPS models that explicitly derive from this meta-theory.

The Interactive Biobehavioral Model (*IBM*): (Lindau, Laumann, Levinson, & Waite, 2003) The IBM embodies BPS core principles or assumptions: orientation to health, rather than illness, capacity to assess both illness and health outcomes; equal importance for biological, psychological, and social aspects of health; bi-directional causality and feedback; primacy of understanding health/illness in socio-relational context; interdependency of social and life course dynamics; the potential for factors to be assets or liabilities for health.

The IBM is applied to sexuality in diabetic couples in later stage life. The model allows for a complex, but testable, set of hypotheses about how physical impairments (e.g., obesity, frailty, and visual impairments) of elderly diabetics and their partners jointly affect the sexual health of the couple. Specifically, diabetics with intact sexuality and intimate relationships are hypothesized to exhibit better psychological, physical (glycemic control), and social functioning than similar diabetics with compromised sexuality. Individuals having diabetic partners with compromised sexuality will have poorer psychological, physical, and social functioning than those having healthy partners or those having diabetic partners with intact sexuality. Considering the reverse direction of causality, it is hypothesized that sexual health over time protects against, or slows, the development of Type II diabetes.

This model is specific, the variables are defined, and the relationships between the variables are specified with respect to the outcome variables of healthy sexuality and diabetes. Therefore, this model can be considered a specific model of the BPS Meta-Theory. The biological, psychological, and social levels are modeled specifically, and interactively, in a testable model of the bi-directional relations between sexuality and diabetes in intimate couple relationships. The model thus includes a specific dyadic relational (i.e., social) systemic level, and it focuses on health and illness on a continuum. Pilot data support this model (Lindau et al., 2003).

The paper describing the IBM is explicit that the IBM was inspired and informed by the BPS conceptualizations. It also expanded innovatively upon them, and most importantly it defined the variables and relationships among them in testable ways. Thus the IBM fulfills the criteria for a genuine scientific BPS model.

The Family System Genetic Illness Model (*FSGI*): (Rolland & Williams, 2005) This model addresses the psychosocial challenges of genomic conditions for patients and their families, and organizes this complex BPS landscape for clinical practice and research. One purpose of this model is to organize the examination of the relationship between individual and family dynamics as they relate to the psychosocial implications and challenges of genomic disorders, which are categorized *not* on their biological bases, but rather on clusters of their effect. The model clusters genomic disorders based on key characteristics of the psychosocial demands of these illnesses over time. Key disease variables include the likelihood of developing a disorder based on specific genetic mutations, overall clinical severity, timing of clinical onset in the life cycle, and whether effective treatment interventions exist to alter disease onset and/or progression. Thus the FSGI model is designed to be flexible and responsive to future discoveries in genomic research.

It is noteworthy, and an extension to Engel's BPS model, that the variables of primacy are not disease, or even biological aspects of the disease, e.g., life vs. death, or optimal biological management of the illness. Rather the variables of primacy are psychosocial, i.e., family relational well-being and individual psychological well-being of the patient and family members, as they influence and are influenced by the nature of the characteristics of the genomic disease. This model specifies the variables and their inter-relation in a manner that makes them testable by empirical research, and therefore the model is a genuine scientific model, despite the fact that it is heavily pragmatic and clinically oriented in its conception. It was directly and explicitly inspired by the BPS Meta-Theory.

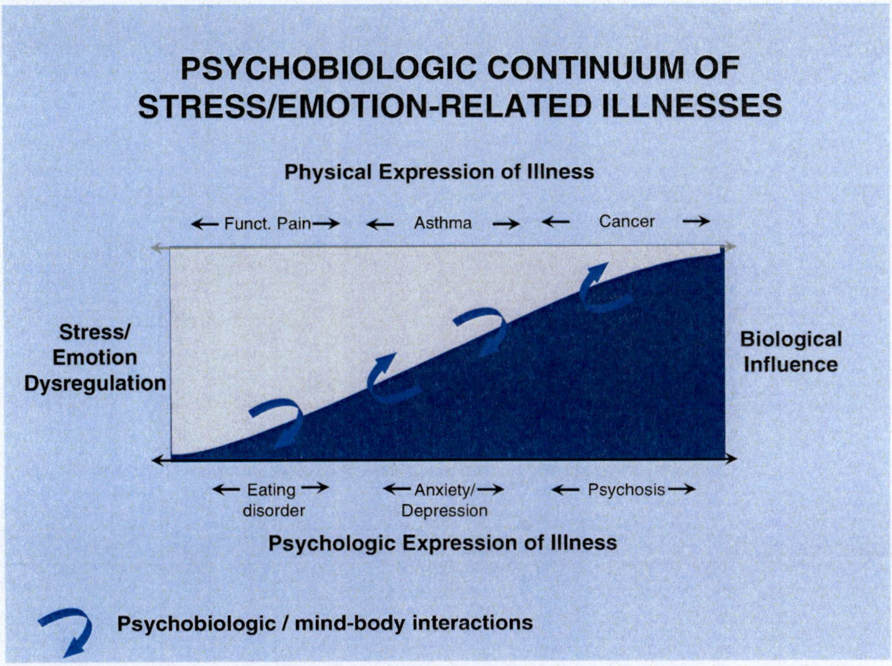

Fig. 9.3 Psychobiologic continuum of stress/emotion-related illnesses

The Biobehavioral Family Model (*BBFM*): (Wood et al., 2008) The BBFM is perhaps the most explicit BPS model, and one that has stimulated and been subjected to the most scientific research. It rests upon a meta-theoretical concept, which also is BPS inspired, call the "Mind-Body Continuum of Disease." This conceptualization features a continuum of psychological and physical disease which varies according to the relative proportions of psychosocial and biological influence on the disease process (see Fig. 9.3).

Both physical and psychological disease can be conceptualized in terms of a continuum of relative contribution of psychological and biological (i.e., mind–body) influence. The curved arrows in the diagram represent the interaction of psychosocial and biological factors. They represent mechanisms or pathways of BPS influence. The pathways are assumed to be bidirectional, such that having a disease elicits social and emotional stress, and social and emotional stress evokes the disease process. The BBFM specifies some of these pathways and illustrates how the family plays a pivotal role in health and illness.

The advantage of conceptualizing disease in this manner is that it organizes a more sophisticated approach to investigating factors influencing illness and it supports an integrated BPS approach to the research and clinical treatment of physically and psychologically manifested illness. Consistent with the biobehavioral continuum of disease, the BBFM is broadly applicable to any disorder, whether it be manifested physically, emotionally, or both.

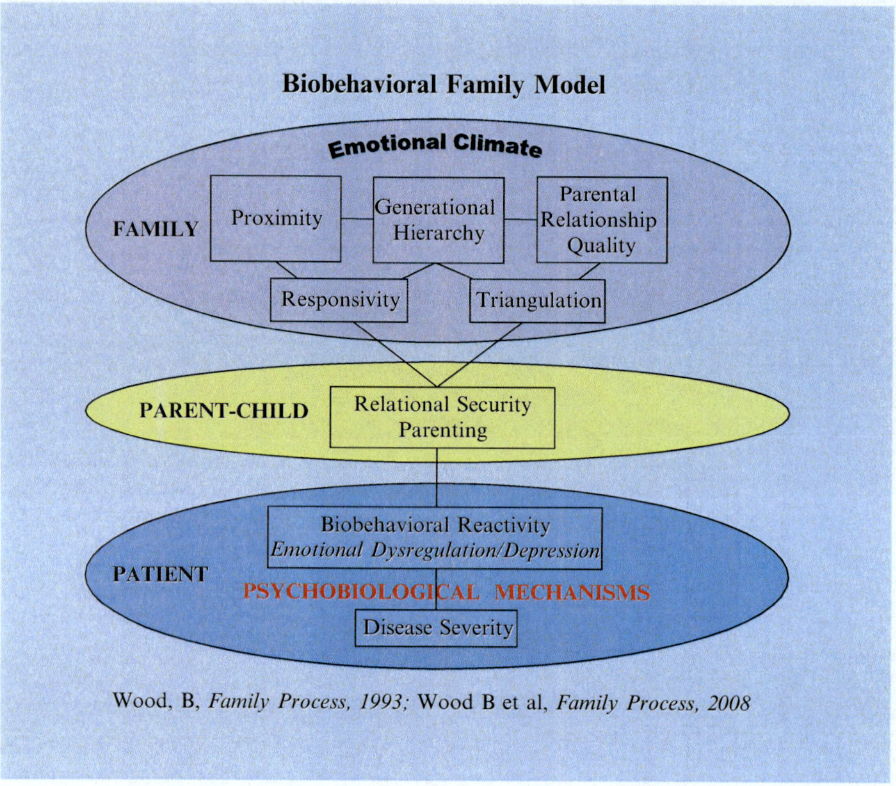

Fig. 9.4 Biobehavioral model (Wood et al., 2008)

The BBFM is one explicit model of pathways and mechanisms by which relational processes may influence both psychological and physical diseases (see Fig. 9.4).

Dimensions of the BBFM. The BBFM posits that family emotional climate, quality of parent–parent relations, parent–child relational security, and biobehavioral reactivity (emotion regulation/dysregulation) are processes that influence one another and collectively either buffer or potentiate psychobiological processes influencing disease activity in stress-related illnesses (Wood et al., 2008). Figure 9.4 illustrates the relational components of family emotional climate.

Family emotional climate refers to the overall intensity and valence of emotional exchange. It colors all aspects of family relationship, and therefore it is probably a key factor contributing to emotional status and outcomes in family members. A negative family emotional climate includes hostility, criticism, verbal attacks, etc., and is similar to the criticism construct of Expressed Emotion (EE) (Brown & Rutter, 1966). Positive aspects include warmth, affection, support, affirmation, etc. Family emotional climate is characterized by the intensity and balance of negative and positive emotional exchange among family members. This balance or imbalance can be construed as reflecting one aspect of family-level emotion regulation or dysregulation. *Proximity* is defined as the extent to which family members share personal space, private information, and emotions. *Generational hierarchy* refers to the extent to which caregivers are in charge of the children by providing nurturance and limits through strong parental alliance and absence of cross-generational coalitions. *Parental relationship quality* refers to interaction patterns which include mutual support, understanding and adaptive disagreement (respectful and resolving) vs. hostility, rejection, and conflict. Parental discord not only contributes to negative family emotional climate, it also has direct effects on children's emotional functioning, with depression mediating the link between parental conflict and asthma disease activity (Lim, Ballow, Miller, Wood, & Simmons, 2011). *Triangulation* refers to involving a child in the parental conflictual process in ways that render the child responsible, blamed, scapegoated, or in loyalty conflict. *Responsivity* refers to the extent to which family members are behaviorally, emotionally, and physiologically responsive to one another. Responsivity depends, in part, on the *biobehavioral* (that is, *emotional*) *reactivity* of each family member. Moderate levels of emotional/physiological responsivity allow for empathic response among family members. Extremely high levels of responsivity can exacerbate maladaptive emotional/physiological resonance in the family, possibly worsening psychologically influenced emotional or physical disorders. Extremely low levels of responsivity may be part of a general pattern of neglect or avoidance, leaving family members unbuffered from internal, familial, or environmental stressors (Wood et al., 2008).

Relational Security is a construct broader than, but related to, attachment. Parent–child attachment security is a key factor in the development of emotion regulation. Therefore, the BBFM proposes that *parent–child relational security* mediates and/or moderates the impact of stressful family process (or life events) on disease-related emotional and physiological dysregulation (Wood, Klebba, & Miller, 2000). Empirical finding support this hypothesis (Wood et al., 2006, 2008).

Biobehavioral reactivity is a pivotal construct in the BBFM, linking psychological and emotional processes to disease processes. It is conceptualized as the intensity and manner in which an individual family member responds physiologically, emotionally, and behaviorally to emotional stimuli (Jemerin & Boyce, 1992; Wood, 1993). Behavioral reactivity reflects the ability of the individual to regulate emotion. It is tightly linked to the constructs of emotion and physiological regulation and dysregulation. Depending on which physiological processes are activated or deactivated during patterns of emotion regulation or dysregulation (respectively) specific psychological or physical diseases can be influenced, when there is a pathogenic pathway.

Empirical validation of the BBFM. To date no single study has tested the BBFM in its entirety, but studies of children with inflammatory bowel disease (Wood et al., 1989), epilepsy (Langfitt, Wood, Brand, Brand, & Erba, 1999), Psychogenic Non-epileptic Seizure Attacks, PNEA (Wood, McDaniel, Burchfiel, & Erba, 1998), and asthma (Lim et al., 2011; Miller & Wood, 2003; Wood & Miller, 2005; Wood et al., 2000, 2006, 2008) report findings consistent with several links posited in the model.

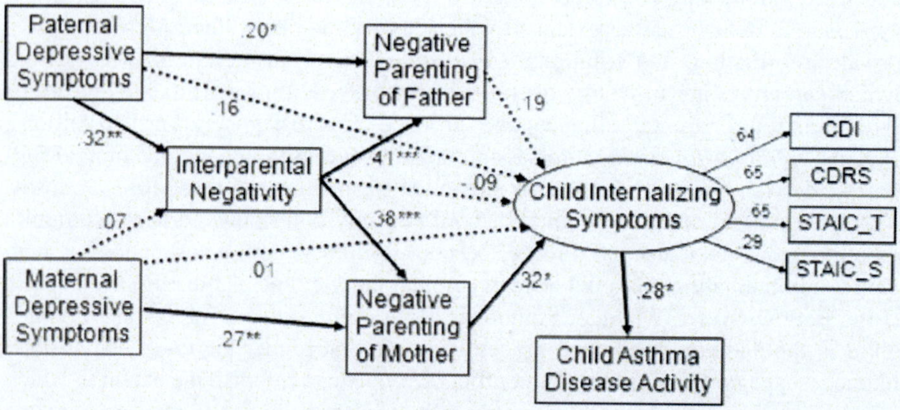

Fig. 9.5 Structural equation model showing the hypothesized direct and indirect pathways from parental depressive symptoms on child internalizing symptoms and asthma disease activity. $\chi 2(30, N=106)=36.18$, $p=0.20$, IFI=0.97, CFI=0.97, RMSEA=0.04.*$p<0.05$. **$p<0.01$. ***$p<0.001$

If complex BPS models are to be scientifically useful they need to be empirically testable in their full complexity. Recently a study by Wood and colleagues (Lim et al., 2011) tested all levels of the BBFM. Children with asthma ($N=106$, aged 7–17) participated with their fathers and mothers. Parental depressive symptoms were assessed by self-report. Inter-parental and parenting behaviors were observed and rated during family discussion tasks. Child emotional symptoms were assessed by self-report and by clinician interview and rating. Asthma disease activity was assessed according to National Heart Lung and Blood Institute guidelines. Results supported inter-parental negativity and negative parenting behavior as mediators linking parental depressive symptoms and the child's emotional and physical dysfunction. Furthermore, the child's depression predicted physical disease activity (Lim et al. 2011) (see Fig. 9.5).

Thus this research investigated and supported a family–psychological–biological interactive contribution to the child's asthma disease activity. Thus a BPS model was fully specified and tested in its entirety. This model may serve as a prototype for other models, specified as to illness type and family and individual psychological variables. Models like this are clear support of the BPS model as a Meta-theory. If more models are developed and more scientists become involved in this kind of modeling and research, then one might truly and fairly concede that the BPS model is not only a Meta-Theory, but also a true paradigm. It is unlikely that the BPS paradigm would actually replace the biomedical paradigm, because paradigms, theories, and models are likely to have specific value depending upon the phenomenon under consideration. Thus it is unlikely, and also perhaps undesirable for the BPS model to actually engender a full paradigm shift.

To summarize this point I quote Glen (1994):

> The success of a hypothesis, or its service to science, lies not simply in its perceived "truth," or power to displace, subsume or reduce a predecessor idea, but perhaps more in its ability to stimulate the research that will illuminate … bald suppositions and areas of vagueness.

Conclusion

Although Glen's point is well taken, it should not be inferred that there should be no organization in the process of scientific inquiry. For a hypothesis to be successful it is essential that it be embedded in the context of a model, theory, meta-theory, and paradigm. Unfortunately some aspiring scientists may not be actively cognizant of these levels, and therefore their hypotheses are less likely to be "successful." L'Abate (2009) has developed an elegantly organized conceptualization and categorization of relations among these levels of scientific proposition. If Engel had been schooled in such a framework, he likely would not have termed his BPS proposition a "model." Furthermore, although Engel clearly valued GST, he did not appear to understand it as a paradigm, nor did he have a clear sense of the hierarchical relationship of his proposition as being embedded within the GST paradigm. Rather he relegated GST to the task of "providing a methodological approach." Engel also seemed not to grasp the need for specificity in scientific modeling. He did not explicitly develop testable models, nor encourage that testable models embedded within his meta-theory be developed by others. Such lack of clarity is common in scientific practice, undoubtedly to its detriment. Perhaps Abate's categorization system of paradigm, metatheory, theory, and model will serve as a useful blueprint for more systematically articulated scientific inquiry (L'Abate).

References

Adler, R. H. (2009). Engel's biopsychosocial model is still relevant today. *Journal of Psychosomatic Research, 67*, 607–611.

Bateson, G. (1972). *Steps to an ecology of mind: Collected essays in anthropology, psychiatry, evolution, and epistemology*. Chandler: Chandler Publishing Company.

Borrell-Carrio, F., Suchman, A. L., & Epstein, R. M. (2004). The biopsychosocial model 25 years later: Principles, practice, and scientific inquiry. *Annals of Family Medicine, 2*, 576–582.

Brown, G. W., & Rutter, M. (1966). The measurement of family activities and relationships: A methodological study. *Human Relations, 19*, 241–263.

Drack, M., Apfalter, W., & Pouvreau, D. (2007). On the making of a system theory of life: Paul A Weiss and Ludwig von Bertalanffy's conceptual connection. [Review] [109 refs]. *Quarterly Review of Biology, 82*, 349–373.

Engel, G. L. (1977). The need for a new medical model: A challenge for biomedicine. *Science, 196*, 129–236.

Engel, G. L. (1980). The clinical application of the biopsychosocial model. *American Journal of Psychiatry, 137*, 535–544.

Feigl, H. (1970). The "orthodox" view of theories. Remarks in defense as well as critique. In M. Radner & S. Winokur (Eds.), *Analyses of theories and methods of physics and psychology*. Minneapolis: University of Minnesota Press.

Frankel, R. M., Quill, T. E., & McDaniel, S. H. (2003). *The biopsychosocial approach: Past, present, future*. Rochester: The University of Rochester Press.

Ghaemi, S. N. (2009). The rise and fall of the biopsychosocial model. *The British Journal of Psychiatry, 195*, 3–4.

Glen, W. (1994). *The mass-extinction debates: How science works in a crisis*. Stanford: Stanford University Press.

Gray, W., Duhl, F. J., & Rizzo, N. D. (1969). *General systems theory and psychiatry* (XXIIth ed.). Oxford: Little Brown.

Grinker, R. R., & Hughs, H. M. (2011). *Toward a unified theory of human behavior: An introduction to general systems theory* (2nd ed.). New York: Basic Books.
Jemerin, J. M., & Boyce, W. T. (1992). Cardiovascular markers of biobehavioral reactivity. *Developmental and Behavioral Pediatrics, 13*(1), 46–49.
Kuhn, T. S. (1996). *The structure of scientific revolutions* (3rd ed.). Chicago: University of Chicago Press.
L'Abate, L. (2009). Paradigms, theories, and models: Two hierarchical frameworks. In L. L'Abate (Ed.), *Science, mind and creativity: The Bari symposium* (pp. 107–122). New York: Nova Science Publishers.
Lakatos, I. (1970). Falsification and the methodology of scientific research programmes. In A. Musgrave & I. Lakatos (Eds.), *Criticism and the growth of knowledge* (pp. 91–195). New York: Cambridge University Press.
Langfitt, J. T., Wood, B. L., Brand, K. L., Brand, J., & Erba, G. (1999). Family interactions as targets for intervention to improve social adjustment after epilepsy surgery. *Epilepsia, 40*, 735–744.
Laudan, L. (1977). *Progress and its problems*. Berkeley: University of California Press.
Lim, J., Ballow, M., Miller, B. D., Wood, B. L., & Simmons, S. J. (2011). Effects of paternal and maternal depressive symptoms on child internalizing symptoms and asthma disease activity: Mediation by interparental negativity and parenting. *Journal of Family Psychology, 25*(1), 137–146.
Lindau, S. T., Laumann, E. O., Levinson, W., & Waite, L. J. (2003). Synthesis of scientific disciplines in pursuit of health: The interactive biopsychosocial model. *Perspectives in Biology and Medicine, 46*, S74.
McLaren, N. (1998). A critical review of the biopsychosocial model. *Australian & New Zealand Journal of Psychiatry, 32*, 86–92.
McPherson, J. D., Marra, M., Hillier, L., Waterston, R. H., Chinwalla, A., Wallis, J., et al. (2001). A physical map of the human genome. *Nature, 409*, 934–941.
Miller, B. D., & Wood, B. L. (2003). Emotions and family factors in childhood asthma: Psychobiologic mechanisms and pathways of effect. In E. S. Brown (Ed.), *Advances in psychosomatic medicine* (24th ed., pp. 131–160). Basel: Karger.
Novack, D. H., Cameron, O., Epel, E., Ader, R., Waldstein, S. R., Levenstein, S., et al. (2007). Psychosomatic medicine: The scientific foundation of the biopsychosocial model. *Academic Psychiatry, 31*, 388–401.
Princeton University. (2011). *About Wordnet. Wordnet [On-line]*. Retrieved, from March 15, 2011 http://wordnet.princeton.edu
Rolland, J. S., & Williams, J. K. (2005). Toward a biopsychosocial model for 21st-century genetics. *Family Process, 44*, 3–24.
Suls, J., & Rothman, A. (2004). Evolution of the biopsychosocial model: Prospects and challenges for health psychology. *Health Psychology, 23*, 119–125.
The Decade of the Brain. (1990). *House Joint Resolution 174 [On-line]*. Retrieved, from March 15, 2011 http://www.loc.gov/loc/brain/
von Bertalanffy, L. (1969). *General system theory; foundations, development, applications*. New York: G. Braziller.
Wood, B. L. (1993). Beyond the "psychosomatic family": A biobehavioral family model of pediatric illness. *Family Process, 32*, 261–278.
Wood, B. L., Klebba, K. B., & Miller, B. D. (2000). Evolving the biobehavioral family model: The fit of attachment. *Family Process, 39*, 319–344.
Wood, B. L., Lim, J., Miller, B. D., Cheah, P., Zwetsch, T., Ramesh, S., et al. (2008). Testing the biobehavioral family model in pediatric asthma: Pathways of effect. *Family Process, 47*, 21–40.
Wood, B. L., McDaniel, S. H., Burchfiel, K., & Erba, G. (1998). Factors distinguishing families of patients with psychogenic seizures from families of patients with epilepsy. *Epilepsia, 39*, 432–437.
Wood, B. L., & Miller, B. D. (2005). Families, health and illness: The search for mechanisms within a systems paradigm. In W. M. Pinsof & J. Lebow (Eds.), *Family psychology: The art of the science* (pp. 493–520). New York: Oxford Press.
Wood, B. L., Miller, B. D., Lim, J., Lillis, K., Ballow, M., Stern, T., et al. (2006). Family relational factors in pediatric depression and asthma: Pathways of effect. *Journal of the American Academy of Child and Adolescent Psychiatry, 45*, 1494–1502.
Wood, B., Watkins, J. B., Boyle, J. T., Nogueira, J., Zimand, E., & Carroll, L. (1989). The "psychosomatic family": A theoretical and empirical analysis. *Family Process, 28*, 399–417.

Chapter 10
Interbehaviorism

Mitch J. Fryling and Linda J. Hayes

Kantor was interested in a wholly naturalistic approach to scientific activity in general, and the science of psychology in particular. As a philosophy of science, interbehaviorism characterizes all scientific activities from a naturalistic foundation (Kantor, 1953; Kantor & Smith, 1975). As a scientific system interbehavioral psychology is an organized enterprise, coordinating all scientific activities through the detailed procedure of system building (Kantor, 1958). Kantor was a scholar of Renaissance proportions and his achievements were truly extraordinary. Unfortunately, his work and its implications for science and for the science of psychology in particular have often been overlooked or misunderstood. In this chapter we attempt to outline some of the unique features of both the philosophy of interbehaviorism and the scientific system of interbehavioral psychology. We will begin with an overview of interbehaviorism as a philosophy and conclude with the systemic details of interbehavioral psychology.

Interbehaviorism

As a naturalistic philosophy of science, Interbehaviorism requires the dismissal of all hypothetical constructs (Kantor, 1953; Kantor & Smith, 1975). As such, all constructions derived from a dualistic foundation, including contemporary substitutes for the mind (e.g., cognitive structures), are dismissed. The requirement that all

M.J. Fryling
Department of Applied Behavior Analysis, The Chicago School of Professional Psychology, 617 W. 7th Street, 8th Floor, Los Angeles, CA 90017, USA
e-mail: MFryling@thechicagoschool.edu

L.J. Hayes (✉)
Psychology Department/296, University of Nevada, Reno, Reno, NV 89557, USA
e-mail: lhayes@unr.edu

L. L'Abate (ed.), *Paradigms in Theory Construction*, DOI 10.1007/978-1-4614-0914-4_10,

hypothetical entities be dismissed is derived from a fundamental assumption about the philosophy of science: That all philosophical assumptions must be derived from contacts with confrontable events (Kantor, 1957). Along these lines, Kantor and Smith (1975, p. 411) remind us "Assumptions, whether informal references or technical postulates, are of course constructions and therefore should bear the closest relationship to the events with which the scientist interbehaves." In this sense the philosophy of Interbehaviorism differs from more traditional philosophical perspectives, wherein philosophical assumptions are permitted in the absence of any observational support (e.g., that the mind causes behavior). Moreover, from the perspective of Interbehaviorism the philosophy of science is itself a scientific enterprise, its workers observing natural events and developing constructs based upon those events (Kantor, 1953). Importantly, Kantor's philosophical system is a philosophy of *science*. At its most foundational level it involves assumptions or propositions pertaining to the domain of the sciences in general, and at its most specific level assumptions pertaining to individual scientific enterprises. As these issues are fundamental, our discussion of them will continue throughout the chapter.

In addition to articulating propositions for the philosophy of science, Kantor suggested that workers in the philosophical domain must also evaluate those assumptions as to their service to the enterprise. Specifically, Kantor (1969a, 1981) described the monitorial, coordinative, and semantic supervision roles that properly constructed assumptions play within a philosophical enterprise. The monitorial role primarily evaluates the development and validity of philosophical assumptions. In other words, monitorial evaluations ask questions pertaining to how various assumptions were developed, in addition to the extent to which they are consistent with other assumptions. Coordinative efforts focus on integrating various assumptions with other assumptions, both within and across individual scientific enterprises, assuring that the sciences are interrelated, whereby philosophical assumptions, and all subsequent works derived from them, demonstrate a general coherence. Finally, semantic supervision examines the way various terms are used within the sciences, again, both within and across individual scientific disciplines. For example, interbehaviorists have expressed concerns with the meanings of several terms in the domain of psychology, namely function, stimulus, and response (Kantor, 1970; Parrott, 1983b), owing largely to their having been borrowed from other disciplines.

Obviously, if the philosophical foundation of a scientific enterprise has not been explicitly articulated there is no way to examine it. Hence, failing to explicitly articulate philosophical assumptions not only decreases the likelihood that an adequate philosophy of science will emerge, but also increases the likelihood that a problematic philosophy will thrive (Kantor, 1969a). Put differently, failing to specifically articulate philosophical assumptions does not remove their powerful influence on scientific work. It simply assures that the influence will be a problematic one. It is for this reason that Kantor described an elaborate system building procedure, to which we now turn.

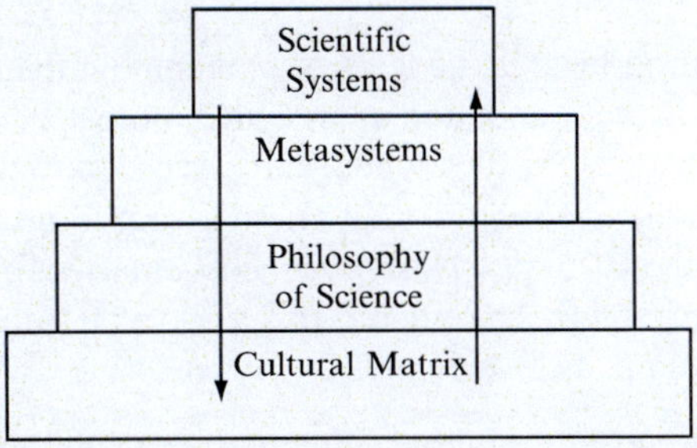

Fig. 10.1 Kantor's "Ziggurat of Science," adapted from Kantor and Smith (1975, p. 410)

System Building

The fact that the philosophy of science is viewed as a concrete scientific activity is only part of the uniqueness that characterizes Interbehaviorism. Closely related to this feature is the unique organizational character of this philosophy: Interbehaviorism is aimed at the systemization of all scientific activities (Kantor, 1958; Kantor & Smith, 1975, pp. 407–417). Interbehaviorists refer to such organizational efforts as *system building* (e.g., Clayton, Hayes, & Swain, 2005; Hayes & Thomas, 2004), as they pertain to the deliberate construction of scientific enterprises. In the following paragraphs we describe the "ziggurat of science and civilization," as it has been described by Kantor (1958, p. 63; Kantor & Smith, 1975, p. 410) (Fig. 10.1).

As shown in this figure, the cultural context is the most fundamental level, as all scientific activities are pursued within particular cultures. Like all other human enterprises, scientific work is carried out in the context of a specific culture whereby the intellectual and other practices of specific cultures always have a tremendous influence on scientific understandings (Kantor, 1953). Related to this, as the scientist is viewed as an active participant in all scientific works, the cultural background of the scientist cannot be ignored. For instance, a scientist's immersion in a dualistic culture necessarily impacts his or her work as a scientist. Importantly, the relationship between science and culture goes both ways. This is to say, while the culture exerts tremendous influence on the activity of the scientist, scientific activity can also exert tremendous influence on the culture (1953, pp. 49–50). Such as the case when the results of scientific research are infused into cultural institutions (e.g., medical and educational practices).

The fact that mentalistic and spiritistic assumptions are still present in the sciences exemplifies the powerful influence the culture has on science. That is, as the culture continues to embrace dualistic assumptions, such assumptions continue to make their way into scientific enterprises, and in particular, the science of psychology. Interbehaviorists explicitly acknowledge the inevitability of cultural sources of influence. Importantly, acknowledging influences from this source is not pursued to undermine objective scientific efforts, but rather to increase scientists' awareness of these influences whereby they may be minimized to the extent possible.

Just above the cultural level is the level of philosophy; that is, the philosophy or logic of science. This is the foundation upon which more specific meta-systems and individual scientific systems are developed, and serves to articulate the assumptions which are pertinent to all scientific works. These assumptions, termed "protopostulates" (Kantor & Smith, 1975, p. 411), serve a coordinative role. The following are the protopostulates of Interbehaviorism (taken from Kantor, 1958, pp. 64–65; Kantor & Smith, 1975, pp. 412–413):

1. Science is the enterprise of interbehaving with specific things and events which leads to a definite and precise orientation with respect to those things and events.
2. Scientific orientation concerns (a) the existence and identity of things and events or their components, and (b) the relationship between either the components of things and events, or between the various things and events themselves.
3. No science is concerned with existences or processes which transcend the boundaries of scientific enterprises. No scientific problem is concerned with a "reality" beyond confrontable events and their investigation.
4. Scientific orientation requires specialized instruments and methods, depending upon (a) the specific characteristics of the events interacted with and (b) the specific problems formulated about them.
5. Scientific interactions eventuate in protocols (records), hypotheses, theories, and laws.
6. Scientific construction – the formulation of (a) hypotheses and (b) theories and laws – must be derived from interbehavior with events and not imposed upon the events or scientific enterprise from nonscientific cultural sources.
7. Culture consists of the events and institutions (religion, art, economics, technology, social organization, and laws) of a specific group of people.
8. Scientific enterprises are evolutional; they develop in cultural situations as complex institutions. Scientific domains are cumulative and corrigible. They are completely free from all absolutes, ultimates, or universals.
9. Scientific enterprises can be and sometimes are autonomous and fundamental within a cultural complex. Only specific enterprises may cooperate and mutually influence each other with respect to basic investigational and interpretive procedures.
10. Applications of (a) scientific findings (records concerning events and their investigation) and (b) investigative results (laws and theories) may be localized within scientific enterprises. Such applications constitute the authentic basis for scientific prediction and control.

Just above the assumptions pertaining to the logic of science are the assumptions pertaining to the meta-system, in this case, of psychology. These assumptions pertain to the discipline in general, and in particular, to its relationships with other sciences. Moreover, meta-system postulates are based upon the aforementioned protopostulates of the philosophy of science. The meta-system postulates of Interbehavioral Psychology are as follows (Kantor, 1958, pp. 65–68; Kantor & Smith, 1975, pp. 413–415):

- Psychology is homogeneous with all other sciences.
- Psychology is a relatively independent science.

- An interbehavioral system of psychology departs from all traditional epistemological and ontological systems.
- A psychological system should achieve a comprehensive coverage of events, operations, and theory constructions.
- System construction requires adequate orientation with respect to systemological problems.
- A psychological system is not reducible to any other type.
- Psychological systems are relative and subject to continual corrective reformulation.

At the uppermost level are the assumptions pertaining to individual systems of particular scientific disciplines. In this case, the individual system is Interbehavioral Psychology, the assumptions pertaining to specific aspects of this subject-matter. The postulates for the science of Interbehavioral Psychology are outlined below (Kantor, 1958, pp. 77–82; Kantor & Smith, 1975, pp. 415–417).

- Psychological events consist of multifactor fields.
- Psychological events are interrelated with social as well as with biological and physical events.
- Psychological events are evolved from ecological interbehavior.
- Psychological events involve the participation of total organisms, not merely special organs or tissues.
- Psychological events are ontogenetic.
- Psychological interbehavior varies in specific details from other types of interbehavior.
- Psychological constructions are continuous with crude-data events.
- Psychological events consist of interrelated factors which do not admit internal or external determiners.

The systemic aspects of individual sciences warrant further comment, as their detailed articulation is a rather unique feature of Interbehavioral Psychology relative to other psychological approaches. As described by Kantor (1958, 1959), Interbehavioral Psychology is a thoroughly integrated, coordinated scientific system. This system is comprised a number of subsystems, including those of the investigative, interpretive, and applied domains. That these subsystems are viewed as coordinated participants within the larger system of Interbehavioral Psychology results in their being viewed as interrelated, and of equal importance to the enterprise as a whole. For example, because system assumptions are made explicit, investigations occurring within the system are more likely to relate directly to interpretation, which will, in turn, relate directly to application, and so on. That is, because there is an explicit effort toward organization and integration, the likelihood of productive relationships among the various subsystems is increased. This does not always seem to be the case in other psychological approaches, even in those relatively close to Interbehaviorism (e.g., Elliot, Morgan, Fuqua, Ehrhardt, & Poling, 2005). It is our view that this circumstance is owing to a relative lack of appreciation for systemic issues (Hayes, Dubuque, Fryling, & Prichard, 2009; Hayes & Thomas, 2004).

As discussed above, none of the sub-domains of Interbehavioral Psychology is held to have greater status or more importance than any other. This is not a widely held view. On the contrary, there is a tendency within the sciences for certain types of work to be valued more than other types. More specifically, investigative work, and particularly investigations of the applied sort tend to be most highly valued. Thus, it is not uncommon for work in the investigative and applied sub-domains to overshadow work in the other sub-domains, with the end result being an overall lack of coordination among workers within the discipline, and a general lack of appreciation for systemic issues. These sorts of outcomes are observed when a range of perspectives emerge within scientific disciplines, and especially when the work conducted under the auspices of one perspective has relatively little to do with the work conducted under the auspices of another. The enduring focus on system issues entailed in Interbehavioral Psychology decreases the likelihood of such a situation developing. In short, the philosophical foundation and the interpretive sub-domain are viewed as critical elements of the Interbehavioral Psychology, serving to sustain the coherence of the system as a whole.

The systemization of science is directly related to matters of validity, significance, and comprehensiveness, which are central to interbehavioral thinking (Kantor, 1958). Validity pertains to the extent to which sciences are internally consistent. In other words, validity is a within-system affair. Significance, on the other hand, pertains to the extent to which the assumptions of individual sciences are consistent with those of the larger scientific domain. In other words, significance is largely a between-systems affair. It is important to note that validity is a pre-requisite for significance; that is, a system cannot achieve significance if it is not internally consistent. Further, significance assumes validity, but validity does not assume significance. For example, some non-scientific psychological systems could potentially be valid (e.g., psychoanalysis), but not significant. Finally, from Kantor's perspective scientific systems must strive towards comprehensiveness, accounting for all of the phenomena that fall within the purview of the subject-matters. For example, a psychological system which provides an account of only certain types of psychological phenomena, such as relatively repetitive performances, may be said to lack comprehensiveness.

Finally, the interrelated nature of each of the four levels of scientific activity depicted above warrants further comment. The arrows are intended to highlight the fact that these levels are not discrete or independent but rather reflect a continuum, each exerting influence over the other.

Constructs and Events

Interbehaviorism attempts to make a thorough distinction between constructs and events throughout all aspects of scientific activity (Kantor, 1957; Smith, 2007). At the forefront of this distinction is the concern that constructs are often confused with events within the sciences, the resulting circumstance being that scientific work becomes further and further removed from the original events of interest. Kantor (1958, p. 73) acknowledges that the events of most interest to scientists, namely the naturally occurring or "crude events" (Kantor, p. 73) are never able to be studied in their pristine forms. Indeed, "Whenever the scientist brings an original event into context with scientific interests and activities (observation, experimentation) he endows it with properties additional to those it originally possesses" (Kantor, pp. 73–74).

We have recently elaborated on these issues in the context of psychological events in particular (Fryling & Hayes, 2009). While constructs and events are frequently confused throughout the domain of science, *psychological* events and constructs may be especially likely to be so confused. The increased likelihood of this confusion in the psychological domain is owing to both historical factors (i.e., the longstanding domination of the psychology domain by dualistic philosophy) and the unique features of psychological events themselves. In the following paragraphs we briefly comment on these unique features, and the practices that increase the likelihood of confusing events and constructs in the science of psychology.

First, psychological events may be relatively difficult to measure. Psychological events are functions obtaining between the stimulating action of environing things and the responding of organisms, and moreover, these functions are participants in multi-factored fields. This is to say, the psychological event is not comprised physical stimulus objects and biological organisms, but is rather comprised the *actions* of both organisms and stimulus objects. Because objects and organisms are relatively concrete relative to their actions, psychological events may be considered comparatively difficult to measure. In other words, it is difficult to measure a purely functional relationship, especially when the participants engaged in the relation have no substantive properties themselves. It is perhaps for this reason that psychological science has been given a "soft" status in the domain of the sciences, while other sciences, measuring more concrete or substantive things are said to be among the "hard" sciences, along with the implication that the latter are more "scientific" in nature. The unique features of psychological events have implications for both descriptive and investigative constructions, and it is to these that we now turn.

As we have stated, psychological events are particularly likely to be confused with constructs for a variety of reasons, and a number of common errors increase the likelihood of this happening. First, psychological events are often placed within the organism, as is the case when they are considered to be mental, cognitive, or neurological. A psychological event always takes place in an environmental setting, however, and the setting is a critical participant in all psychological events. Placing psychological events within the organism assures that the relationship between constructs and events is not continuous, as it is overlooked altogether.

Second, psychological events exist as a continuous stream of interaction; they are not discrete events. Thus, psychological events are not able to be constructed a priori, rather they depend upon the particular interests of investigators and their investigative situations. Third, psychological events always involve the whole organism, the implication being that they are always very complex happenings.

A psychological event might involve hearing, seeing, speaking, and writing, for example. One consequence of this characteristic is that particular aspects of the whole organism response tend to be emphasized for measurement purposes, or because they are assumed to be more important than other aspects, increasing the likelihood that the emphasized participant in the psychological event may be confused with the psychological event itself.

Fourth, the stimulating action of the environment is not always immediately apparent. Stimulation, as a factor in a psychological event, is explicitly distinguished from the source of stimulation. As such, a stimulus object may be home to a variety of stimulus functions, some of which may be substitutive in nature. The latter, owing to an organism's historical interaction with stimuli under conditions of their association, whereby the functions of absent stimuli may operate from present sources (Kantor, 1921, 1924), may be particularly difficult to observe. An unfortunate consequence of this difficulty has been a tendency to give "special powers" to the organism or to reduce the action of the whole to one or another of its parts. This problem is widespread throughout the general domain of psychology, and is directly attributable to the confusion between stimulus functions and stimulus objects (Kantor, 1947, 1970).

Lastly, psychological events become more and more elaborate over time. Psychological events are historical, whereby each and every psychological event entails its entire history up to that point (Hayes, 1992). For example, a number of historical factors are involved in psychological acts of remembering, and psychologists have long had a difficult time accounting for such interactions (e.g., Watkins, 1990).

Added to the difficulty of constructs in the sciences is that investigation requires an additional level of construction. Specifically, while descriptive constructs are employed to capture the subject-matter of a scientific enterprise, investigative constructs are derived from the former, and employed for experimental purposes. In Behavior Analysis, for example, the operant construct is frequently employed in investigative efforts. However, problems arise when investigative constructs are employed outside of the investigative domain and therein confused with the subject-matter of the enterprise in general. Furthermore, common investigative aims, such as prediction and control, often require additional constructions. Among them are dependency relations, and while such constructs may be perfectly suitable in the context of investigation, they are not descriptive of the subject-matter more generally. As a rule, investigation must always be interpreted in the context of the original event of interest. In other words, investigative constructs must be continuous with both descriptive constructs and the crude events of the scientist's original interest (see Kantor, 1958, p. 89; also see Hayes, Adams, & Dixon, 1997).

The emphasis on the distinction between constructs and events is a unique feature of Interbehaviorism, and is central to the interbehavioral conceptualization of the sciences more generally. In the following section we will provide an overview of the interbehavioral conceptualization of both disciplinary and interdisciplinary sciences, review practices that may compromise the value of scientific enterprises, and summarize the implications of the interbehavioral perspective for the future.

Scientific Enterprises

Disciplinary Sciences

From the perspective of Interbehaviorism science is better conceptualized in the plural form (Kantor, 1953, p. 5). In other words, there is not one science, but rather, many sciences. In this sense, scientific work is similar to other sorts of work, where there are many varieties of such. Scientific work is differentiated from other types of human activity, however. In particular, scientific work is distinguished by its seriousness and originality (Kantor, pp. 6–7), and aims to determine "(a) the existence or non-existence of certain things and events, and (b) the characteristics of such things when they do exist" (Kantor, p. 4). Importantly, it is the world of *nature* that sciences attempt to understand, and thus, all dualistic and spiritistic entities, which do not exist in the natural world are immediately dismissed from the scientific domain.

Nonetheless, the world of nature is a multi-factored constellation of happenings. This is to say, the natural world involves a multitude of interrelationships and thus the prospect of any one science developing an understanding of this constellation is an impossible undertaking. It is for this reason that there are a variety of sciences, all of which may contribute, in varying degrees, to a more adequate orientation to the natural world. The contributions of the various sciences, specifically the factors impacting their productivity and significance deserve further consideration.

Individual disciplinary sciences aim to understand a specific aspect of the natural constellation of happenings. As we have mentioned, the world of nature is comprised a multitude of interrelationships, and as such, disciplinary sciences must identify a specific relationship as their special object of study. Disciplinary sciences remain rather close one another, however, as their subject-matters are all isolated from the same matrix of natural happenings. Moreover, some disciplinary sciences focus on similar sets of events, and thus, some may be especially close to one another. Such as the case of cultural behavior, for example, where the disciplines of psychology, sociology, and anthropology are all focused on events of the cultural sort. The fact that the sciences are all derived from the same set of happenings, and moreover, that they may be especially close to one another at various times, relates to several important issues and problematic practices that may occur in the pursuit of interdisciplinary enterprises (see Hayes & Fryling, 2009b).

Given this perspective of disciplinary sciences, one may accept that sciences overlap with one another at various times, whereby one could imagine a circumstance where biologists, psychologists, sociologists, and anthropologists, for example, are all studying the same events. This circumstance is not a beneficial one, however. Each disciplinary science has the capacity to contribute to the overarching goal of understanding the world of nature only to the extent that it produces a *unique* piece of information. Thus, when various sciences examine the same event the result is redundancy, whereby scientific productivity is compromised. Given this, it is of the utmost importance for disciplinary sciences to differentiate themselves from one another, and nothing is more essential to achieve this differentiation than a precise definition of its special subject-matter. When the specific objects of study are clearly identified the likelihood of individual sciences overlapping and studying the same events is reduced. Thus, the value of disciplinary sciences depends upon the extent to which they have isolated a *unique* relationship as their special subject-matters from the complex multi-factored field that is the natural world. While this is a tall order in and of itself, there are other ways in which disciplinary sciences might compromise their contribution to the scientific domain and their subsequent participation in successful interdisciplinary enterprises.

Interdisciplinary Sciences

Our conceptualization of disciplinary sciences has direct bearing on our understanding of interdisciplinary enterprises. As scientific enterprises, interdisciplinary sciences are similar to disciplinary sciences with one important exception: the source from which their subject-matters are derived. While the subject-matters of individual disciplinary sciences are relationships derived from the natural world of happenings, the subject-matters of interdisciplinary sciences are the relations among the already existing relationships derived by the disciplinary sciences. That is, interdisciplinary sciences study relations among relations. Similar to disciplinary sciences, the value of an interdisciplinary science depends upon the extent to which a unique relationship is identified as its special subject-matter.

There are a number of additional concerns pertinent to interdisciplinary sciences, however. In particular, the value of interdisciplinary sciences depends upon the strength of the participating disciplinary sciences, and further, the manner in which the participating disciplinary sciences are constructed with respect to one another within the interdisciplinary system (Hayes & Fryling, 2009b). Indeed, a number of problematic circumstances may hinder the construction of effective interdisciplinary enterprises, and it is to these that we now turn.

First, it is not uncommon for interdisciplinary efforts to embody a collaborative structure, whereby two or more disciplinary sciences are working side-by-side. However, when a variety of disciplines work in parallel, the outcome is further knowledge about the respective subject-matters of the participating disciplines, and not genuine interdisciplinary knowledge. An interdisciplinary enterprise must identify a *unique* relationship if it is to produce new information, and thus, already existing disciplines, working side-by-side, does not constitute interdisciplinary science. Such efforts will surely produce new information, but not of the interdisciplinary sort.

Second, it is not uncommon for scientific boundary conditions to be compromised during interdisciplinary endeavors. Scientific boundary conditions are especially likely to be overlooked when they are poorly articulated to begin with. For example, if the boundary conditions between psychological and neighboring biological events are not explicitly articulated, boundary conditions are very likely to be compromised, perhaps even unknowingly. An interdisciplinary science cannot emerge from poorly systematized disciplines, and as such, this circumstance is certainly problematic. Not only does it prevent effective interdisciplinary sciences from developing, but it also compromises disciplinary productivity itself.

A similar compromise of disciplinary boundary conditions can occur even when boundary conditions are clearly drawn. Such as the case when the boundary conditions are expanded beyond the discipline, under the assumption that its subject-matter may be legitimately enlarged. For example, the discipline of psychology may import biological factors into its subject-matter. While it is possible that practices of this sort may engender a new scientific specialization, in which the unique contributions of psychology and biology are recognized. If the history of relations between these two disciplines is considered (Kantor, 1969b), however, the more likely outcome is that psychological events will be reduced to biological events or that the former will be held to be caused by the latter or some other unproductive relation between them will be constructed. In short, to the extent that disciplinary productivity depends upon a continued focus on a specific set of events, blurring disciplinary boundary conditions undermines the means by which productivity is achieved. Moreover, the collaboration of two unproductive disciplinary sciences does not bode well for the accumulation of new and different interdisciplinary understandings.

Boundary conditions may also be overlooked when the cumulative nature of scientific work is not appreciated. Boundary conditions may be overlooked because the boundaries between various scientific disciplines were arbitrarily drawn, and thus, may be assumed to be rearranged without problem. While it is true that the boundaries between various scientific enterprises were arbitrarily drawn, once articulated, continued acknowledgement of them is imperative. Scientific knowledge develops over time, and the productivity of a discipline depends upon the extent to which the same type of event is the object of focus for a sustained period of time. Thus, not only does the rearrangement of boundary conditions not constitute legitimate interdisciplinary science, but it even results in abandoning the disciplinary enterprise.

Overlooking the cumulative nature of scientific enterprises can also result in the misapplication of scientific products. It is not uncommon for scientific enterprises to borrow from other disciplines, most often laws and general principles, and then use those products to understand their own subject-matters. This situation seems to benefit both of the disciplinary sciences involved with the less developed science appearing to acquire more value or becoming more "scientific" by this practice while the more developed science appears to expand the scope of its laws and principles. Unfortunately, this practice overlooks the fact that scientific products are the result of particular sorts of scientific work, focused on unique events, and thus, apply only to those events. For example, the laws and principles developed in the science of biology are derived from *biological* work, and thus, apply to *biological* events. Borrowing biological products and applying them to psychological events implies that products are independent entities that can be applied to anything. At other times it can be a result of a more serious mistake, namely, the basis fallacy (Observer, 1969b), considered below. Either way, the result is certainly not interdisciplinary science, and again, even compromises the status of the participating disciplinary sciences.

The basis fallacy (Observer, 1969b), or reductionism more generally, can have a tremendous impact on the manner in which disciplinary sciences participate in interdisciplinary enterprises. As described above, the disciplinary sciences are all derived from the same matrix of natural happenings, and moreover, all sciences identify a unique aspect of this matrix for their special object of study. As such, all of the sciences are of equal value, as no one part of the natural world is considered to be more or less important than any other part. At the same time, some disciplines have been relatively more productive than others in many regards, and it is not uncommon for these disciplines to be considered more important for this reason. That one science may be considered more valuable on the basis of its relative productivity is not itself a problem, however, this hierarchical relationship is often assumed to mean that the less productive sciences require the support of the more productive sciences, or that they have their *basis* in them (Observer, 1968, 1969a). In this sense, the less productive science can be reduced to the more productive science, which can be further reduced to an even more productive science, and so on. Commonly, for example, sociological phenomena are assumed to be reducible to psychological events, psychological events to biological events, biological events to chemical events, and so on. This practice leads to problems in interdisciplinary endeavors. Most obviously, the discipline assumed to be the most fundamental quickly overshadows the discipline assumed to have its basis in it, whereby relationships among the participating disciplines are never examined.

As we have mentioned, failing to develop adequate disciplinary sciences, or to appreciate their value in interdisciplinary systems, prevents effective interdisciplinary sciences from emerging. Still, it seems likely that interdisciplinary sciences will continue to develop (see Hayes, 2001). The likelihood of productive interdisciplinary sciences developing may require specific planning on the part of all workers involved, however. The interbehavioral approach to scientific activity highlights important aspects of both disciplinary and interdisciplinary sciences, and embracing such a perspective increases the likelihood of effective interdisciplinary sciences developing in the future.

By now it is clear how the specific articulation of philosophical assumptions, and in particular, the construction of specific subject-matters, is central to the philosophy of science. In the following sections we elaborate on the implications of explicitly articulating the subject-matter for the science of psychology in particular. First, we will elaborate upon the distinction between biological and psychological events. This will be followed by a description of the psychological event, and a discussion of the implications of the psychological event for understanding complex behavior.

Distinguishing Between Biological and Psychological Events

Kantor explicitly distinguished psychological from biological events (Kantor & Smith, 1975, pp. 8–11; Pronko, 1980, pp. 9–11), as well as outlined the problems inevitable when the two were confused (Kantor, 1947). Kantor outlined seven features that served to distinguish psychological from biological events. First, psychological events are historical. Psychological events develop during the lifetime of the organism, through its interactions with its psychological environment, whereas biological events are present as soon as the biological structures are in place. Second, psychological events involve specificity, they are highly contextual. As described by Pronko (1980, p. 10), "Any foreign substance can cause the eye to secrete tears, but a German novel does not elicit universal responding from all humans." In other words, biological events are standardized across situations, whereas psychological events are dependent upon specific historical and situational circumstances. In this regard, psychological events are relatively more complex than biological events.

Third, psychological events are integrated happenings. For example, a child writing a sentence involves his gripping the pencil, pressing the pencil to the paper, and making specific movements with it that are organized into letters and words. Biological events do not integrate and develop into more elaborate events; rather they are relatively constant, discrete happenings. Fourth, psychological events are variable. Responding with respect to a specific situation changes as a result of experiences, and can do so rather quickly, as in the case of what is typically called "problem solving." We might first wiggle a key in a lock in particular way, and if that is not successful, we might try another action, we may push the key only part of the way in, turn it to the right, pull it out, push it all of the way in, and so on. The point here is that psychological events are variable, they shift and turn, and this same fact is not observed in the case of biological events which, again, are relatively similar from occasion to occasion of their occurrences.

Related to the fourth characteristic, psychological events can be modified. That is, we may develop a strong preference for a particular type of food, yet also develop an aversion to that very same food. At one point we might have a committed friendship with a particular person, and at another completely dislike that very same person. In both of these cases the psychological event was modified. Again, this does not seem to be the case with biological events. In other words, when wind blows in our eyes, we blink, and this does not change over time. Sixth, psychological events display inhibition. For example, a woman may be interested in a very expensive scarf, but purchase something less expensive. This example displays another unique aspect of psychological interbehavior, something which is not observed in biological happenings. Lastly, psychological events can be delayed. For example, we might make dinner plans with a friend, but not actually show up for the dinner until a much latter time (e.g., the following week). In this case, a current conversation initiates a psychological event which is not completed until a later date. Again, biological events are not suspended in this way; they begin and end in a relatively short period of time.

Our point in articulating the various differences between the events of the biological and psychological domains was not to suggest that events of the former are irrelevant to the latter. Obviously, biological events are participating factors in all events of the psychological type (Delprato, 2006): A psychological event cannot occur in the absence of a biological organism. Rather, in recognizing the differences between these two types of events, the importance of each in its own right may become more apparent. This recognition avoids the tendency to reduce psychological events to those of the biological domain, as well as eliminate the assumption of causal dependencies between them. We review the above distinctions not only because they represent an organized effort to distinguish biological and psychological events, but also because they highlight the unique features of psychological events unlike any other perspective attempts to do.

The Psychological Event

In the above sections we outlined the system postulates of interbehavioral psychology in the context of describing the more general philosophy of Interbehaviorism. In this section we provide an overview of the psychological event as it has been described by Kantor (e.g., 1958, p. 14, 1977, pp. 46–50). After doing so, we will elaborate upon some particularly interesting aspects of the psychological event, including their implications for conceptualizing complex human behavior.

From Kantor's perspective, psychological happenings involve a continuous stream of events. In other words, individuals begin interacting with their stimulating environments at the time of birth (or before) and this interaction is continuous throughout their lives. Psychological happenings involve the stimulating action of stimulus objects and the corresponding responding of organisms. Importantly, stimulation and responding are conceptualized as a unitary event in Kantor's system, that is, it is not a linear S→R interaction, but rather, one of Sf↔Rf. From this perspective responding cannot occur in the absence of stimulation, and likewise, stimulation cannot occur in the absence of responding. It is for this reason that Kantor prefers the term *inter*behavior, emphasizing the interactional nature of psychological happenings. Still, the Sf↔Rf function involves a number of factors (1958, 1977).

The fact that psychological activity is continuous necessitates its being constructed for descriptive purposes. Kantor's construction of this stream of interaction is the psychological event, or interbehavioral field.

$$\mathrm{PE} = C(k, \mathrm{sf}, \mathrm{rf}, \mathrm{st}, \mathrm{md}, \mathrm{hi}).$$

Here, the PE stands for psychological event. *C* represents the fact that the psychological event is an integrated whole; participants in it do not exist in isolation but are rather interrelated. The unique configuration of each and every psychological event is represented by the letter *k*. Psychological events are always unique, the exact same circumstance can never happen more than once, and the uniqueness of each event is central to interbehavioral thinking. The stimulus function of stimulus objects is represented by the term sf, and the responding of the organism by the term rf. Psychological events always happen in a complex interactional setting, and these participants are identified by the term sd. The medium of contact, md, refers to participating factors which facilitate the organism's contact with stimulating objects. For example, light might be considered a medium of contact that participates in a visual interaction with a stimulus. Finally, psychological events always involve historical factors, as indicated by the term hi.

Again, from Kantor's perspective all of these factors participate in a unitary, integrated psychological event. Moreover, that each of these factors is considered a participant renders all of them *equal* participants, whereby no factor may be considered more or less important than another. Consistent with this understanding, changing one factor involves changing the entire field, such that no factor may be considered to have independent, dependent, or causal status. Importantly, this does not negate the value of experimentation, but rather, questions the manner in which experimentation is traditionally interpreted.

Stimulus Substitution

A particularly important feature of Kantor's psychological system pertains to the distinction between stimulus objects and psychological stimulus functions (Kantor, 1924, pp. 47–48). The distinction between objects and functions allows for a coherent conceptualization of complex responding, where the stimulus function is often not apparent. Such as the case when one reminisces about old friends (e.g., see's them, thinks about them, hears their voices) in the absence of any apparent stimulus. Kantor (1921, 1922, 1924) has suggested that stimulus objects may acquire the functional properties of other objects when organisms interact with these objects under conditions of their spatio-temporal association. When various environing factors occur together in an individual's experience (e.g., a friend and a song), they may acquire the stimulational functions of one another, such that they may substitute for them in their absence. For example, a song which was frequently heard in the presence of a particular friend may "remind" you of that friend in her absence. When this happens we would say that the song substitutes for the friend, wherein you are interacting with the friend through the stimulating action of the song.

While far beyond the scope of the current chapter, it is important to note that Kantor's construction of stimulus substitution allows for a thoroughly naturalistic account of complex behavior, completing the analysis of such events in other systems (e.g., Radical Behaviorism), where this distinction has not been made explicitly (see Parrott, 1983b, 1983d, 1986). Indeed, a number of complex behaviors involve stimulus substitution (e.g., Delgado & Hayes, 2007; Dixon & Hayes, 1999; Fryling & Hayes, 2010; Fryling, Johnston, & Hayes, 2011; Hayes, 1991, 1998; Parrott, 1983c, 1984, 1987); and it is Kantor's explicit distinction between stimulus objects and stimulus functions that permits their naturalistic interpretation.

Implicit Responding

Implicit responding is how Kantor characterizes responding with respect to substitute stimulation (1924). In other words, when an individual thinks about a friend in the absence of the actual friend they are engaging in implicit responding.

This sort of responding can also be constructed as response substitution, as the psychological response is a substitute response, substituting for the response that would occur should the physical stimulus object itself be present. There are some interesting features of this sort of responding that warrant further comment. While a number of responses are determined by the physical properties of the objects interacted with, at least partially, implicit responses may have relatively little, or even nothing to do with the physical properties of the objects interacted with. Such as the case when one interacts with a liquid substance by calling it "coffee" in English speaking communities, for example. In this case the word "coffee" has nothing to do with the physical properties of the stimulus, and indeed, the same substance is referred to with other words in other languages. This sort of responding seems to be central to our understanding of cultural and verbal behavior, and to the distinction between social and cultural behavior (Hayes & Fryling, 2009b; Kantor, 1982; Parrott, 1983a, 1984).

Understanding private events. As we have mentioned, Kantor's conceptualization of the psychological event permits us to approach the most complex sorts of behavior. So powerful is this approach that it allows for the analysis of psychological events which are typically deemed private and unavailable for scientific analysis.

Of course, psychologists have long struggled with the topic of privacy in the analysis of behavior. In fact, so thoroughly engrained is the dualistic tradition that even the most stalwart behaviorists have struggled with the topic, as when radical behaviorists attempt to include "private events" in their system (Skinner, 1953, 1974). From our perspective, the need to include "private" events is the product of both historical dualism and the misconception of the psychological event itself. Moreover, suggesting that one day these "private events" will be made observable through progress in biology (Skinner, 1974) results in overlooking the psychological aspects of these happenings. From Kantor's perspective *all* psychological phenomena are observable in principle, and thus, events of these sorts cannot be relegated to biology. Instead, such events are conceptualized as exceedingly subtle psychological interactions requiring special observational practices. From this perspective, it is possible to observe what another person is thinking given particular historical circumstances and immediately present situational factors. Such as the case when an old couple speaks to one another less and less, because they "know what the other is thinking" through the observation of subtle psychological interactions of the substitutional sort (see Hayes, 1994; Hayes & Fryling, 2009a; Parrott, 1983d).

Psychological linguistics. Kantor (1936, 1977) also articulated a coherent, naturalistic approach to the topic of psychological linguistics, in which the activities of speakers and listeners are interrelated in special sorts of interbehavioral fields. For Kantor, the distinguishing characteristic of linguistic fields is bi-stimulation. In the case of what he calls referential interbehavior, stimulation arises from two sources simultaneously, one source being the object to which the speaker is adjusting, the other being the listener who is being oriented to that object by the speaker's action. Also constructed are linguistic events of the non-referential sort in which, again, two sources of stimulation are involved though in this case in serial fashion. More specifically, the stimulus products of a speaker's interaction with one source of stimulation provide a second source of stimulation by which the linguistic adjustment may be completed. While a thorough review of this aspect of Interbehavioral Psychology is far beyond the scope of the present chapter, it is worth mentioning, as it has engendered a significant body of empirical research (e.g., Bijou, Umbreit, Ghezzi, & Chao, 1986; Chiasson & Hayes, 1993), and continues to be the foundation for several research groups, particularly among the laboratories represented in the newly formed International Association of Interbehavioral Psychologists.

Kantor's construction of the psychological event permits a coherent, naturalistic approach to all of the events which fall within the purview of the subject-matter of psychology. Surely, this outcome is the product of a thoroughly articulated and fully systematized philosophical foundation.

Conclusion

Interbehaviorism is a fully systematized scientific philosophy, upon which a coherent, thoroughly naturalistic, and comprehensive science of psychology has been erected. Kantor's contributions to the scientific domain, in particular his articulation of detailed sets of postulates as multiple levels of scientific activity, his specifications for the coordination of the sub-domains of particular scientific enterprises, and the implications of these guidelines for effective interdisciplinary study are especially worthy of note by virtue of their rarity in contemporary treatises. Kantor's most important contribution to the science of psychology has been the elimination of all vestiges of dualism in his system. This, too, is a rarity. Indeed, it has been suggested that Kantor's work might provide a useful guide for other behaviorists as they attempt to construct more naturalistic conceptualizations of complex forms of human behavior (Morris, Higgins, & Bickel, 1982), and much of the work cited throughout this chapter has involved efforts to do just that.

References

Bijou, S. W., Umbreit, J., Ghezzi, P. M., & Chao, C. C. (1986). Psychological linguistics: A natural science approach to the study of language interactions. *The Analysis of Verbal Behavior, 4*, 23–29.

Chiasson, C. C., & Hayes, L. J. (1993). The effects of subtle differences between listenersand speakers on the referential speech of college freshman. *The Psychological Record, 43*, 13–24.

Clayton, M. C., Hayes, L. J., & Swain, M. A. (2005). The nature and value of scientific system building: The case of interbehaviorism. *The Psychological Record, 55*, 335–359.

Delgado, D., & Hayes, L. J. (2007). The acquisition of a conceptual repertoire: An analysis in terms of substitution of functions. *The Behavior Analyst Today, 8*, 307–315.

Delprato, D. J. (2006). Commentary on Lickliter. In B. D. Midgley & E. K. Morris (Eds.), *Modern perspectives on J. R. Kantor and interbehaviorism* (pp. 197–204). Reno: Conext Press.

Dixon, M. R., & Hayes, L. J. (1999). A behavioral analysis of dreaming. *The Psychological Record, 49*, 613–628.

Elliot, A. J., Morgan, K., Fuqua, R. W., Ehrhardt, K., & Poling, A. (2005). Self-and cross-citations in the Journal of Applied Behavior Analysis and the Journal of the Experimental Analysis of Behavior, 1993–2003. *Journal of Applied Behavior Analysis, 38*, 559–563.

Fryling, M. J., & Hayes, L. J. (2009). Psychological events and constructs: An alliance with Smith. *The Psychological Record, 59*, 133–142.
Fryling, M. J., & Hayes, L. J. (2010). An interbehavioral analysis of memory. *European Journal of Behavior Analysis, 11*, 53–68.
Fryling, M. J., Johnston, C., & Hayes, L. J. (2011). Understanding observational learning: An interbehavioral approach. *The Analysis of Verbal Behavior, 27*, 191–203.
Hayes, L. J. (1991). Substitution and reference. In L. J. Hayes & P. N. Chase (Eds.), *Dialogues on verbal behavior* (pp. 3–14). Reno: Context Press.
Hayes, L. J. (1992). The psychological present. *The Behavior Analyst, 15*, 139–145.
Hayes, L. J. (1994). Thinking. In S. C. Hayes, L. J. Hayes, M. Sato, & K. Ono (Eds.), *Behavior analysis of language and cognition* (pp. 149–164). Reno: Context Press.
Hayes, L. J. (1998). Remembering as a psychological event. *Journal of Theoretical and Philosophical Psychology, 18*, 135–143.
Hayes, L. J. (2001). Finding our place in a constructed future. In L. J. Hayes, J. Austin, R. Houmanfar, & M. C. Clayton (Eds.), *Organizational change* (pp. 349–372). Reno: Context Press.
Hayes, L. J., Adams, M. A., & Dixon, M. R. (1997). Causal constructs and conceptual confusions. *The Psychological Record, 46*, 97–111.
Hayes, L. J., Dubuque, E., Fryling, M. J., & Prichard, J. (2009). A behavioral systems analysis of behavior analysis as a scientific system. *Journal of Organizational Behavior Management, 29*, 315–332.
Hayes, L. J., & Fryling, M. J. (2009a). Overcoming the pseudo-problem of private events in the analysis of behavior. *Behavior and Philosophy, 37*, 39–57.
Hayes, L. J., & Fryling, M. J. (2009b). Toward an interdisciplinary science of culture. *The Psychological Record, 59*, 679–700.
Hayes, L. J., & Thomas, J. (2004). The nature and value of system building in behavior science. *The Behavior Analyst Today, 5*, 284–289.
Kantor, J. R. (1921). Association as a fundamental process of objective psychology. *The Psychological Review, 28*, 385–424.
Kantor, J. R. (1922). Memory: A triphase objective action. *Journal of Philosophy, 19*, 624–639.
Kantor, J. R. (1924). *Principles of psychology* (Vol. 1). Chicago: The Principia Press.
Kantor, J. R. (1936). *An objective psychology of grammar*. Bloomington: Indiana University Press.
Kantor, J. R. (1947). *Problems of physiological psychology*. Bloomington: The Principia Press.
Kantor, J. R. (1953). *Logic of modern science*. Chicago: The Principia Press.
Kantor, J. R. (1957). Events and constructs in the science of psychology: Philosophy – Banished and recalled. *The Psychological Record, 7*, 55–60.
Kantor, J. R. (1958). *Interbehavioral psychology*. Chicago: The Principia Press.
Kantor, J. R. (1959). *Interbehavioral psychology* (2nd ed.). Chicago: The Principia Press.
Kantor, J. R. (1969a). Scientific psychology and specious philosophy. *The Psychological Record, 19*, 15–27.
Kantor, J. R. (1969b). *The scientific evolution of psychology* (Vol. II). Chicago: The Principia Press.
Kantor, J. R. (1970). An analysis of the experimental analysis of behavior (TEAB). *Journal of the Experimental Analysis of Behavior, 13*, 101–108.
Kantor, J. R. (1977). *Psychological linguistics*. Chicago: The Principia Press.
Kantor, J. R. (1981). *Interbehavioral philosophy*. Chicago: The Principia Press.
Kantor, J. R. (1982). *Cultural psychology*. Chicago: The Principia Press.
Kantor, J. R., & Smith, N. W. (1975). *The science of psychology: An interbehavioral survey*. Chicago: The Principia Press.
Morris, E. K., Higgins, S. T., & Bickel, W. K. (1982). The influence of Kantor's interbehavioral psychology on behavior analysis. *The Behavior Analyst, 5*, 158–173.
Observer. (1968). Psychology: An interdisciplinary science. *The Psychological Record, 18*, 267–268.

Observer. (1969a). On the reduction of psychology to physics. *The Psychological Record, 19*, 515–518.

Observer. (1969b). The basis fallacy in psychology. *The Psychological Record, 19*, 645–648.

Parrott, L. J. (1983a). Defining social behavior: An exercise in scientific system building. *The Psychological Record, 33*, 533–550.

Parrott, L. J. (1983b). On the differences between Skinner's radical behaviorism and Kantor's interbehaviorism. *Mexican Journal of Behavior Analysis, 9*, 95–115.

Parrott, L. J. (1983c). Perspectives on knowing and knowledge. *The Psychological Record, 33*, 171–184.

Parrott, L. J. (1983d). Systemic foundations for the concept of "private events": A critique. In N. W. Smith, P. T. Mountjoy, & D. H. Ruben (Eds.), *Reassessment in psychology: The interbehavioral alternative* (pp. 251–268). Washington: University Press of America.

Parrott, L. J. (1984). Listening and understanding. *The Behavior Analyst, 7*, 29–39.

Parrott, L. J. (1986). On the role of postulation in the analysis of inapparent events. In H. W. Reese & L. J. Parrott (Eds.), *Behavior science: Philosophical, methodological, and empirical advances* (pp. 35–60). Hillsdale: Erlbaum Associates.

Parrott, L. J. (1987). Rule-governed behavior: An implicit analysis of reference. In S. Modgil & C. Modgil (Eds.), *B.F. Skinner: Consensus and controversy* (pp. 265–276). Barrcombe: Folmer.

Pronko, N. H. (1980). *Psychology from the standpoint of an interbehaviorist*. Monterey, CA: Brooks/Cole.

Skinner, B. F. (1953). *Science and human behavior*. New York: The Free Press.

Skinner, B. F. (1974). *About behaviorism*. New York: Knopf.

Smith, N. W. (2007). Events and constructs. *The Psychological Record, 57*, 169–186.

Watkins, M. J. (1990). Mediation and the obfuscation of memory. *American Psychologist, 45*, 328–335.

Chapter 11
Reflections on Personality Systematics and a Unified Clinical Science

Jeffrey J. Magnavita

In this chapter we will explore how developments in system theory has led to a new paradigm called personality systematics and suggest that a new trend toward unified clinical science is now within reach (Magnavita, 2006). It is suggested that knowledge of personality systematics has importance in understanding how human beings function and are led to states of dysfunction, as well as offering guidelines for clinical treatment or psychotherapeutics (Magnavita, 2005b). System theory (von Bertalanffy, 1968) is considered by many one of the most important paradigmatic shifts that occurred in science during the twentieth century. Prior to the development of system theory, scientists were mostly interested in understanding linear causation, which fails to account for the complexity in many open systems, which are being mutually shaped by both internal and internal elements and processes. This reciprocity represents a "simultaneous and mutually interdependent interaction among many components" (Capra, 1983, p. 267). Others have embarked on a search for a unified clinical science. Angyal (1941), who was ahead of his time in his theoretical modeling, proposed a basic framework for a unified paradigm, which he described as aptly:

> Of the total process of life a unified system of factors can be separated by abstraction. However, not every moment of the life process in organized into that system. The life process in its concrete form also contains factors alien to the system, or "random" from the point of the view of the system. The biological total process results from the interaction of system-determined (self-governed, autonomous) factors and factors which are alien to the system (governed from outside the system, heteronomous) (pp. 93–94).

A system contains elements, which interrelate with one another in a process of mutual influence. These feedback loops are essential to understanding system theory. Gleick (1987) wrote: "Nature forms patterns. Some are orderly in space but disorderly in time, others are orderly in time but disorderly in space." Clinical scientists

J.J. Magnavita (✉)
Glastonbury Medical Arts Center, 300 Hebron Ave. Suite 215,
Glastonbury, CT 06033, USA
e-mail: magnapsych@aol.com

L. L'Abate (ed.), *Paradigms in Theory Construction*, DOI 10.1007/978-1-4614-0914-4_11,

use constructs as pattern recognition tools. A simple positive feedback loop was apparent in a session I conducted while writing this chapter. A mother and daughter came in because of years of conflict evident as soon as they began speaking. As the conversation ensued mother began to criticize and blame her daughter for cutting her off for years. As mother continued, her daughter physically recoiled and became highly defensive. The more the daughter withdrew the angrier and more rejected the mother felt, which led to an increase in her blaming. In about 5 min a positive feedback loop was identified that was not producing the results they each said they wanted. This thin slice of their relationship was a fractal of what they had been re-enacting for years and unable to change. Each part of a system affects the state of the system. A fundamental assumption of system theory is that one cannot fully understand a complex system by isolating one part. One may indeed understand many important things when examining a part out of context but this alone is not sufficient for understanding the system in which it is embedded or how that system operates. For example, neuroscientists have deepened our understanding of how neurons transmit signals by releasing neurotransmitters in the synaptic gap but this knowledge cannot fully explain how we think. Thinking is an emergent phenomenon, which cannot be fully explained by the component parts. System theorists believe that there exist part–whole relationships whereby each part can impact the manner in which the entire system functions. In fact, a small perturbation in one part of a system can create massive change in the entire system. This tendency in complex systems to become chaotic with small changes in a component part was termed the "butterfly effect" (Lazzlo, 1996). It refers to the interconnectedness of complex systems. For example, a small increase in the ocean temperature in the Southern Hemisphere may impact weather patterns in North America. Mandlebrot (1977) wrote: "Random fluctuations and irregularities in ostensibly chaotic states may come to form not only complicated rhythms and patterns, but also demonstrate both recurrences and replicated design…here, the same shapes emerge from fluctuations time and time again, taking form sequentially on smaller and smaller scales" (p. 31).

The wave of revolutions in the Middle East can serve as an example. In one case, a tipping point was reached when, after being harassed by police, a street vendor immolated himself triggering a massive uprising against the regime in power. Our understanding of this type of nonlinear change has been advanced with the development of chaos theory. Termed the butterfly effect, it was suggested that a butterfly flapping his wings in China could create a tornado in South America. Weather patterns are good examples of how linear processes are inadequate to explain complex systems. By looking for causal relationships one may miss the patterns of complex interaction that lead a system from homeostasis to disequilibrium. There is a constant struggle between these two forces and change in humans requires states of disequilibrium as occurs in developmental transitions where old adaptive strategies must be eschewed for more adaptive ones.

Chaos or complexity theory a more recent iteration of system theory seeks to understand how complex systems operate incorporating the notion of fractals, tipping points, and self-organization, which are the ways that a system organizes and re-organizes itself at various levels of observation (Gleick, 1987). "Every system, whether a rock or an animal, tends above all else to keep itself in an ordered state" (Csikszentmihalyi, 1993, p. 20). When one looks at patterns in complex systems these tend to reveal a common organization called a fractal that can be seen at every level in the system. These fractals are common in nature. When looking every more microscopically at crystals one can see the same pattern evident at deeper structural levels. General system theory itself evolved as an understanding of complexity summarized by Gleick (1987):

> Nature forms patterns. Some are orderly in space but disorderly in time, others are orderly in time but disorderly in space. Some patterns are fractal, exhibiting structures self-similar in scale. Others give rise to steady states or oscillating ones. Pattern formation has become a branch of physics and of materials science, allowing scientists to model the aggregation of particles into clusters, the fractured spread of electric discharges, and the growth of crystals in ice and metal allows. The dynamics seem so basic – shapes changing in space and time – yet only now are the tools available to understand them (p. 308).

Complex systems can be viewed at various levels of abstraction allowing us to shift our perspective. As we narrow our lens our perspective becomes increasingly microscopic and when we widen our lens our perspective becomes broader. Each position gains and sacrifices something. A single perspective can never hope to capture the nature of a complex system and how it operates. Using multiple perspectives we call for a holographic-like image of the system. The human personality system can be viewed using a holographic (multiple views) perspective. The human personality system is a relational system, which is nested like Russian dolls (Bronfenbrenner, 1979): each level contained in the whole and interacting dynamically. Human beings exist at multiple levels from which personality and identity are emergent phenomena.

For heuristic purposes we can organize a complex system in a variety of ways, which allows clinical science to progress and is necessary. A lack of understanding of the embedded nature of any complex system can limit our understanding and lead us down false paths. For example, dysfunctional behavior can be viewed through a family lens by examining family structure, process, and communication, or a neurobiological lens by studying the actions of various neurotransmitters. We can see the complexity of the multiple levels of interactions when examining personality disorders which are subject to the multigenerational transmission process (Magnavita, 2000). This multigenerational transmission process may be explained by five factors:

1. Genetics (i.e., inherited differences in temperament, neurophysiology, and other variables).
2. Relationship dynamics within the family (e.g., variations in parenting style and skill).
3. Faulty learning (e.g., flawed ways of thinking acquired within the family).
4. Generalized familial pathology (i.e., global milieu effects that reflect broader dysfunctional processes among family members).
5. System propagation factors (e.g., familial roles and alliances, both hidden and overt, that maintain dysfunctional behavior patterns over time) (Bornstein, 2005, pp. 175–176).

The formation of these personality patterns involves a complex process, which includes a number of domain levels or system components interacting over the course of generations. Each component cannot be separated and require multiple perspectives. Each perspective is valid but neither is stand-alone. Like the analogy of the blindfolded men feeling different parts of an elephant and coming to very different conclusions about what they are feeling. Over reification of any one valid perspective within a system can be detrimental. For example, there has been a recent trend to find genetic and biological solutions to mental health disorders. Overemphasizing a narrow band perspective can result in an artificially closed system, which limits explanatory power. It is believed by many that mental disorders are the result of a neurochemical imbalance. The field of psychiatry has been in danger of being hijacked by an overemphasis on this neurobiological perspective. Examining the larger system one can see that the pharmaceutical industry, which profits from selling drugs to alleviate symptoms, may ignore that these symptoms are often the by-product of relational and societal dysfunction. Simplistic models sell drugs and some may profit but human functioning is much more complex.

System Theory and the Emergence of Family Therapy

The burgeoning family therapy movement quickly adopted a systemic paradigm during the later half of the twentieth century (Bray & Stanton, 2009). System theory offered family psychotherapists a new way to conceptualize the process that occurred relationally in family systems. Family therapy created a paradigmatic shift challenging the psychoanalytic system with its focus on individual psychotherapy and intrapsychic factors. Incorporating a systemic approach allowed individual functioning to be viewed, at least in part, as one expression of a dynamic system and how it is operating. An acting-out child could be conceptualized as a symptom bearer for a dysfunctional dyad or in a family where the hierarchy was disordered. System theory helped clinical science evolve from an intrapsychic perspective to a relational one. This important shift clearly embedded human functioning in the relational matrix that could now be understood with different methods such as communication process, linguistics, generational transmission, and many others.

Another important shift that this new systemic paradigm led to was in our understanding of psychopathology and how it should be treated. Instead of viewing psychopathology as existing within the individual, the landscape was dramatically changed and broadened to understanding the relationships that exist within a family, which was considered a system operating by its own unique principles, the patterns waiting to be identified. Family therapists eschewed the notion that psychopathology was inherent solely in the individual. This led to a theory about schizophrenia that was later refuted, called the double-bind theory which postulated that there are certain families that communicate in a manner which places vulnerable members in a double-bind where no matter how they respond they are wrong and the only solution is psychosis. Contemporary family psychology has its roots in and owes much to the contribution of system theory (Bray & Stanton, 2009). Psychology on the whole has not embraced system theory in the same way and continues to experience fragmentation as a result. A unified clinical science requires us to think systemically regardless of preferred perspectives.

Unified Clinical Science

The advancement of clinical science requires a unified approach. Human beings are highly complex organisms who are shaped and shape our ecology and culture. Human functioning cannot be separated from the total ecological system in which we operate and are intrinsically embedded. The self-organizing aspects of systems are survival adaptations, which allow us to survive and evolve meeting demands of our environment. During the first century of contemporary clinical science there were significant advances in our understanding of many of the component domains at various levels, accumulating evidence and building constructs essential to our understanding of human functioning. For example, cognitive scientists articulated and validated the concept of schema, which are essential tools of pattern recognition. Schemata are used to amalgamate the flood of information we are bombarded with, so that the most essential stimulus features can be ordered into recognizable patterns. We can then respond appropriately without experiencing overload and becoming incapacitated by more information than is needed to guide us through our lives. This advance in our understanding of schema led to the development of various forms of cognitive and cognitive-behavioral therapy. Cognitive theorists and researchers began mapping the belief systems in those individuals who were suffering from mental disorders such as anxiety and depression. Through this form of pattern recognition many common cognitive distortions were seen in those suffering from various disorders and new forms of restructuring cognitions were developed. Many other domains, which are part of the human system were also articulated and studied. Interpersonal theorists mapped dyadic processes and structures. Affective scientists continued the work of Darwin (1998/1872) in emphasizing the primacy of emotion as an adaptive evolutionarily adapted signaling system. The neurobiological domain was and continues to be explored and new knowledge garnered about how our brain, genetics, and physiology operate.

Even though there has been an explosion of scientific findings and advances in clinical science, the field has not fully embraced the systemic paradigm. As stated earlier, the uptake of systemic theory primarily resides with the specialized area of family therapy. References to systemic and complexity theory in general clinical science is still rare. There are some noteworthy exceptions such as developmental psychopathology, which as fully embraced systemic theory. The biopsychosocial model offered by Engel (1980) is another exception but even that has fallen short of embracing a systemic paradigm. Pilgrim (2002) believes that the resistance to the biopsychosocial model in medicine and mental health, which despite the scientific and ethical virtues have not been embraced and there has been a recrudescence of biological model as the primary force of psychology. As in the unified framework presented in this chapter, Stanton writes: "A systemic paradigm is helpful in psychology because it provides a framework for conceptualizing, assessing, treating, and researching human behavior" (p. 11).

Why systemic theory has not been more fully embraced is not clear. It may be that for various reasons scientists focus one area of specialization and study it with great effort but do not often raise their heads to look around. This may lead to fragmentation, which can limit scientific advances. Interdisciplinary collaboration is the exception and scientists especially those studying human behavior and psychotherapy tend to form tribes with their own specialized language and rituals. The lack of unity within the field of psychology is startling even though Staats (1983) and others more recently (Sternberg & Grigorenko, 2001) urged the field to move toward unification. We can see signs of this fragmentation when we look at a narrower slice of psychologists interested in psychotherapy. We can see how there are many different approaches to psychotherapy: probably over four hundred. Many of these approaches have their own constructs and technical language, and yet there are probably more commonalities than differences. Even though there has been a movement toward psychotherapy integration, by and large, the impact of this seems minimal. Looking at how the various adherents to different schools of psychotherapy advance their training one can make some assumptions. One can see that there are conferences and seminars attended by adherents to a particular school of psychotherapy who believe that this approach is where they should put their resources. This is probably not the most efficient way to train psychotherapists or advance the field. It leads to unnecessary turf wars and competition for limited resources. Unlike psychopharmacological approaches, psychotherapy does not have significant financial resources.

Research on Psychotherapy Revealing Principles

The research on psychotherapy is becoming more abundant and the findings show that the change factors responsible for a positive outcome may lie less in what approach is used and more in the alliance and nonspecific factors common to all psychotherapy such as expectancy, therapist belief in his or her approach, and client–therapist matching. Approximately 60,000 articles have been published on psychotherapy research in the last 30 years (Cooper, 2008). There is not an abundance of evidence that strongly shows that one approach is better than another. Although some researchers who argue the point contest this finding, the data from many meta-analytic studies seems to support the contention of psychotherapy equivalence. Researchers who use meta-analytic techniques can unknowingly bias their findings based on the studies, which are selected for review. It is a complicated situation. When we look at depression, for example, there are over 30,000 studies, which examine the effects of psychotherapeutic and psychopharmacological treatment. Clearly one can select a subset of studies to support just about anything. Studies that do not support a treatment effect tend not to get published and so many believe there is a strong bias that affects the research that is delivered to clinicians. What we are learning from the research that is applicable to our topic is that there are general principles of psychotherapy, which emerge from the literature and relationship factors, which seem essential to outcome. These unifying principles and processes maybe more important than the therapeutic approach that is being used.

Unifying Clinical Science with Personality Systematics

Personality can be depicted as a complex system, which is an emergent phenomenon as many believe is the case for consciousness. What this means is that personality does not exist as an entity nor does consciousness but most of us agree that human beings possess both personality and consciousness. Appel and Kim-Appwel (2010) describe the bases for a holistic unified model:

> Humans are indeed at least bio-cultural beings. We are neither biologically determined, nor tabula rasa, rasa, upon which culture is imposed. Rather, identity and personality organization emerge out of a jointly active and dynamic process (p. 273).

Personality systematics is the study of the interrelations among all the domains of the personality system from the micro-level to the macro-level. In other words, personality is not viewed as it was classically depicted as residing within the individual but is viewed as fundamentally relational. L'Abate (1986) used the term systematics in his in-depth volume on family systematics. In this volume he outlined many of the principles of systematics. By relational it is meant that personality is developed and represented in the relational matrix, which can be depicted at various levels going from the micro to the macroscopic perspective. We can divide the total ecological system, depicted as the nested dolls we discussed, at four levels starting with the microscopic and successively widening our perspective. At the intrapsychic level we are primarily concerned with what is occurring in the mind/brain system. At the interpersonal level, or dyadically, we depict what occurs between two people. At the triadic level we depict what occurs in dyads + 1. At the sociocultural and family level we see how all these levels interact in a process of mutual influence. Thus, personality systematics seeks to understand the relational processes and domains operative in the total ecological system and views personality as an emergent phenomenon of this complex system. Personality is a system, which includes genetic predispositions, individual patterns of adaptation, attachment style, interpersonal processes, family process, and sociocultural and political influences. Personality can be an organizing construct to unify psychology (Mayer, 2004, 2005). Other clinical theorists are developing unified models with many similarities to the author's unified model. Apple and Kim-Appwel (2010) use a Multipath Approach to Personality (MAP), which assumes that "personality and self are shaped by the combined forces of evolutionary, biological, situational, mental, as well as psycho-spiritual processes – all embedded in a temporal, socio-cultural, and developmental context" (p. 274). Peck (2007) also uses a multilevel system model with functionally nested structures. Others have applied chaos theory to clinical work in an attempt to understand the complexity of working with families (Butz, 1997).

Personality as an Operating System

Our personality is a complex system (Angyal, 1941; Block, 2002; Mayer, 1998), which can be equated to an operating system that runs a computer. A powerful computer with an antiquated operating system will be as slow as molasses. Our hardware is our physical self, body, and brain and our operating system is comprised of our attachment style, defenses, internalized beliefs, and schema about self and other all of which are in an endless process of iteration with others, family, institutions, and political systems. Individuals and groups can get caught in patterns of dysfunction, which are maintained by processes at any level of the system. A faulty neurobiological system from a brain injury resulting in TBI can affect the way in which the individual operates in all spheres of his or her life. We see this is evident in veterans returning from Iraq and Afghanistan whose injuries then impact their families and create a process of change that can result in much greater suffering. Sexual abuse and other forms of trauma can disrupt a system and create dysfunction that is maintained for years or even lifetimes. A perturbation in a system can create a dysynchrony, which reverberates generationally in family transmission processes (Magnavita, 2000).

A personality system tends to evolve with greater complexity over time through a process of differentiation and integration. As a system becomes more differentiated there is greater communication among the various parts. Csikszentmihalyi (1993) describes this process:

> *Differentiation* refers to the degree to which a system (i.e., an organ such as the brain, or an individual, a family, a corporation, a culture, or humanity as a whole) is composed of parts that differ in structure or function form one another. *Integration* refers to the extent to which the different parts communicate and enhance one another's goals. A system that is more differentiated and integrated than another is said to be *complex* (p. 156).

Personality systematics offers a personality-guided approach (Magnavita, 2005a) to treating the spectrum of mental disorders including personality dysfunction (Magnavita, 2010). Personality is seen as adaptive process to environmental and relational demands. When the system is stuck in a stable state and cannot profess through differentiation and integration, dysfunction may result which can result in a personality disorder. These disorders of personality are essentially dysfunctioning systems, which can be viewed at the intrapsychic-biological level, interpersonal-dyadic level, relational-triadic level, or sociocultural-familial level. Dysfunction is generally seen in multiple levels.

Advancing Clinical Science and Psychotherapeutics with the Unified Psychotherapy Project and *Psychotherapedia*™

In order to advance clinical science, the Unified Psychotherapy Project (UPP) (at website: unifiedpsychotherapyproject.org) was launched with the goal of establishing a forum for those interested in furthering unification. The four-level system described in this chapter was used to begin to organize the elements of clinical science and psychotherapeutics. A team of nationally known researcher clinicians were invited to serve as co-chairs overseeing the organization of the techniques of clinical science into an electronic database called psychotherapedia™. The need to catalogue all of the techniques of psychotherapy into a database seems essential for the field to advance. Using technology a wiki was established and is being populated with all the essential components of clinical science in an easily accessible format. It is hoped that once populated this will serve as an essential tool for clinicians and researchers who want to access this database.

This compendium of techniques catalogued in one database will allow us to begin to sequence techniques in various configurations and studying their effects. A unified framework does not eschew the validity of various approaches to treatment, which have been shown effective.

Summary

In this chapter a brief introduction to system theory and its application to personality systematics is presented and applied to the challenge of unifying clinical science and psychotherapeutics. There is a new wave evident in clinical science which seeks to unify our field but which requires a strong foundation in personality systematics, which is derived for system and complexity theory. Theoretical modeling allows us to develop what are hopefully useful paradigms to guide research and practice. Many components of previous models have been blended to create a stronger amalgam, which allows a meta-theoretical framework useful for guiding clinical treatment and practice.

References

Angyal, A. (1941). *Foundations for a science of personality*. New York: Oxford Press.

Appel, J., & Kim-Appwel, D. (2010). The multipath approach to personality: Towards a unified model of self. *Psychology, 1*, 273–281.

Butz, M.R. (1997). Chaos and complexity: Implications for psychologica theory and practice. New York: Taylor & Francis.

Block, J. (2002). *Personality as an affect-processing system: Toward and integrative theory*. Mahwah: Lawrence Erlbaum Associates, Publishers.

Bornstein, R. F. (2005). Psychodynamic theory and personality disorders. In S. Strack (Ed.), *Handbook of personology and psychopathology* (pp. 164–180). Hoboken: Wiley.

Bray, J. H., & Stanton, M. (2009). *The Wiley-Blackwell handbook of family psychology*. Chichester, West Sussex: Wiley-Blackwell.

Bronfenbrenner, U. (1979). *The ecology of human development: Experiments by nature and design*. Cambridge: Harvard University Press.

Capra, F. (1983). *The turning point*. Toronto: Bantam.

Cooper, M. (2008). *Essential research findings in counselling and psychotherapy*. Thousand Oaks: Sage Publications.

Csikszentmihalyi, M. (1993). *The evolving self: A psychology for the third millennium*. New York: HarperCollins.

Darwin, C. R. (1998). *The expression of the emotions in man and animal* (3rd ed.). New York: Oxford University Press. (Original work published 1872).
Engel, G. (1980). The clinical application of the biopsychosocial model. *American Journal of Psychiatry, 137*(5), 535–544.
Gleick, J. (1987). *Chaos: Making a new science*. New York: Viking/Penguin Books.
L'Abate, L. (1986). *Systematic family therapy*. New York: Brunner/Mazel.
Lazzlo, E. (1996). *The systems view of the works: A holistic vision for our time*. Cresskill: Hampton Press.
Magnavita, J. J. (2000). *Relational therapy for personality disorders*. Hoboken: Wiley.
Magnavita, J. J. (2005a). *Personality-guided relational therapy: A unified approach*. Washington: American Psychological Association.
Magnavita, J. J. (2005b). Systems theory foundations of personality, psychopathology, and psychotherapy. In S. Strack (Ed.), *Handbook of personality and psychopathology* (pp. 140–163). Hoboken: Wiley.
Magnavita, J. J. (2006). In the search of the unifying principles of psychotherapy: Conceptual, empirical, and clinical convergence. *American Psychologist, 61*, 882–892.
Magnavita, J. J. (2010). *Evidence-based treatment of personality dysfunction*. Washington: American Psychological Association.
Mandlebrot, B. (1977). *The fractal geometry of nature*. New York: Freeman.
Mayer, J. D. (1998). A systems framework for the field of personality. *Psychological Inquiry, 9*, 118–144.
Mayer, J. D. (2004). How does psychotherapy influence personality? A theoretical integration. *Journal of Clinical Psychology, 60*, 1291–1315.
Mayer, J. D. (2005). A tale of two visions: Can a new view of personality help integrate psychology? *American Psychology, 60*, 294–307.
Peck, S. (2007). TEMPEST in a gallimaufry: Applying multi-level systems theory to person-in-context research. *Journal of Personality, 75*(6), 1127–1156.
Pilgrim, D. (2002). The biopsychosocial model in anglo-American psychiatry: Past, present and future. *Journal of Mental Health, 11*(6), 585–594.
Staats, A. W. (1983). *Psychology's crisis of disunity: Philosophy and method for a unified science*. New York: Praeger.
Sternberg, R. J., & Grigorenko, E. L. (2001). Unified psychology. *American Psychologist, 56*(12), 1069–1079.
von Bertalanffy, L. (1968). *General system theory*. New York: Braziller.

Chapter 12
The Systemic Paradigm: The Intersubjective–Narrative Approach Versus the Relational–Generational One

Vittorio Cigoli and Eugenia Scabini

This purpose of this contribution is to present the distinctive features of two of the most important clinical–social approaches that follow from the systemic paradigm: the "intersubjective–narrative" approach, focused on principles and intervention modalities, and the "relational–generational" approach. Although originating from the same paradigm, each approach presents a history of research intervention and internal changes that should be recognized and appreciated as constituting two distinct approaches of thought and clinical intervention.

As we shall see, important differences already exist between the two approaches in their use of systemic theory. Such a critical rereading of the systemic paradigm is possible if a long period of time is encompassed, in our case about 60 years. In particular, when we write "systemic paradigm," instead, we refer to the clinical perspective within the action research.

Focusing on points in time that are too close to each other and on the immersion of clinical–social researchers in their own temporal period does not allow one to "see" the conceptual, methodological and empirical differences between the two approaches distinctly enough.

In particular, this contribution focuses on the use of the *relation* concept and on the *ethical* perspective that characterize the two approaches. In addition, it aims to shed light on the difference between the *clinical/systemic paradigm* (which is situated among the various research and clinical intervention paradigms) and family psychotherapy, which is a therapeutic genre.

Finally, the foundations of the *relational-symbolic* approach will be presented, an approach that we developed and used in more than 20 years of research on family relations in clinical as well as in psychosocial contexts.

V. Cigoli (✉)
Centro Ricerche Famiglia, Universita' Cattolica, Via Nirone 15, 20123, Milano, Italy
e-mail: vittorio.cigoli@unicatt.it

E. Scabini
School of Psychology, Catholic University of Milan, Largo Agostino Gemelli 1, Milano, Italy

L. L'Abate (ed.), *Paradigms in Theory Construction*, DOI 10.1007/978-1-4614-0914-4_12,

Systemic

We begin with Homer and, in particular, with the *Iliad*, a work that lies at the root of Western culture. Two features are of particular interest to us.

The first has to do with Homer's ability to take two viewpoints, those of both the Greeks and the Trojans; the second is that he attributes value to what men feel and experience in their relations with others. In fact, Homer reveals himself to be an acute connoisseur of the human soul and of the motives and aims that spur a person to act. Men for him are "those who eat bread" (wheat, vines, and olive trees are the sacred plants of Mediterranean culture). They thus have a close relationship with nature, i.e., the land and places, climate, social affiliations, and man's technological action ("technè") on nature itself. Using this terminology, we propose that man must be considered in his *context*.

We will attempt to follow Homer by taking the perspective of each of the two approaches that interest us, keeping in mind that both of them place importance on narratology and theatrical representation.

Let us begin with the "roots," with systemic theory. Literature on this topic is quite extensive and many authors have addressed its use in the area of family relations. Among these, we recall Wertheim (1973, 1975), Gurman and Kniskern (1981), Reiss (1971), Hoffman (1981), and Bertrando and Toffanetti (2000).

Here it is important to point out that systemic theory strives to use Kuhn's (1962) formulation, to be a new *paradigm*, i.e., a reorientation of conceptual and methodological vision to inspire research. This does not deny its evident indebtedness to Gestalt theory and to organicism. In short, the world one observes is always the same, but the "lenses" used to see it are different; even the instruments can be the same, but they are used in new and different ways.

It is well known that in 1930 von Bertalanffy and collaborators had already originated research on systems, but it is in the 1950s that the "Society for the Advancement of General Systems Research" was established. The goal, as was stated, is to create a new research paradigm able to formulate valid principles for all sciences beyond their specific nature. In particular, the intent is to enable the physical–natural sciences and the sciences of human action to meet and engage in dialogue with each other.

However, as Auerswald (1985) observed, the concept of *system*, the true fulcrum of Gregory Bateson, Norbert Wiener, and Ludwing Bertalanffy's thought, has been articulated in very different ways over the years, to the point of also becoming a source of epistemological confusion. On the other hand, the idea that the family can be view as a system is a legacy of the "family systems theory."

With respect to this confusion, Cigoli (1977) had already highlighted in 1977, after thoroughly studying contributions on this topic, that there are two strands within systemic theory. The first, which he defines as "ecological," is based mainly on cybernetic–communicational concepts that are applied to human interaction; the other, which he defines as "sociological–organizational," is based on the study of how groups and organizations function, taking their dynamics into consideration.

In the first case ("ecological"), especially through Bateson (1936–1958), negative feedback and homeostasis are recognized as explicative patterns useful for the study of social phenomena.

Cybernetics, as the science of control and communication, appeared to be the "good helmsman," able to expedite the cognitive jump with respect to behaviorism, but also to William James' introspectionism, both strands of psychology that are widespread in the North American context.

In the second case ("sociological–organizational"), general systems theory is construed more broadly and two crucial aspects of the theory are put into question. The first aspect has to do with "regulation by error" (feedback), which in groups and organizations assumes a more complex form that can be active (able to anticipate), passive (able to give rise to defensive barriers of an institutional nature), and ideological (able to modify ends and purposes). In other words, social organizations (including the family) present a different structure than biological organisms for which selection of an evolutionary character operates. As to "ends," a *reflective function* is present in human groups with respect to burdens and opportunities that the system presents, a function that is unknown in other self-regulating systems. As a result, cognitive and decisional processes are *indeterminate* and thus open to unpredictability.

Among the authors who have championed the sociological–organizational view, Buckley (1967) and Rapaport (1970–1971) stand out. Both emphasize that the aim of human systems is not homeostasis, but "ultrastability," i.e., the capacity to sustain profound changes in the dynamic relation between the system's interior and exterior. In particular, these authors expand the system's history and development. This approach received a vital infusion from Morin's foundational work on "method," whose conceptual and methodological reach is a source of inspiration for the clinical and psychosocial researcher. We later return to Morin's work addressing human identity and relational ethics (Morin, 2001, 2004).

The Two Approaches

Differences between the two approaches, even acknowledging shared principles (relationship between parts and whole, equifinality, regulation systems), have also had effects on intervention and clinical research. Cybernetics is prominent in the first approach. As Bertrando and Toffanetti (2000) affirm, the idea of system enters psychotherapy through cybernetics more than through General Systems Theory. We should not forget that for Jackson (1957) the family "is a cybernetic information system." For their part, Jackson and Weakland (1961) also underscored the presence of homeostatic mechanisms aimed at restoring the status quo in family systems. In turn, Bateson (1961), and later Selvini Palazzoli, Boscolo, Cecchin, and Prata (1975), considered the family a "biological system governed by feedback and activated by error."

It is probable that in adopting this cybernetic vision, clinical researchers were influenced by characteristics of redundancy and resistance to change in family groups presenting serious pathological forms. In any case, the underlying idea of the cybernetic approach is that the family tends to develop interactive schemes of a repetitive and lasting nature (patterns) that serve to *maintain equilibrium* when internal and external situations of acute stress present themselves. For this reason, there is a need to clinically intervene in an instructive–prescriptive way, whether through counter-paradox or powerful rituals, with the goal of creating new relational patterns.

It is interesting to note that this approach developed at the frontier of the American West, open towards the Asiatic world (the "new world"), and that it seeks to react to and go beyond behaviorism as well as introspectionism, both pillars of American scientific psychological thought. Information and communication levels (digital and analogic), "patterns that connect," thus become the principles around which research and clinical intervention are carried out.

Over the course of time, the cognitive leap of this approach arose precisely through the rejection of two cybernetic–systemic principles: that of the mind as a "black box," a legacy of behaviorism, and that of negative feedback understood as a homeostatic principle. In 1963, Maruyama had already introduced the concept of "second cybernetic," referring to the action carried out by positive feedback to create morphogenesis, but it is especially starting in the 1990s that the systemic approach references the physics of irreversible processes (Prigogine, 1961). This change entailed a profound revision of the relation between the client (family, couple, or individual) and the psychotherapist. The instructive–prescriptive model is abandoned together with the idea that clinicians possess prior knowledge of what constitutes health and illness (normality, abnormality). Rather, they are implicated in the exchange with the client system, are co-constructive of new relations, and also "irreverent" (Cecchin, Lane, & Ray, 1992), i.e., informal and open to the unexpected. Attention, then, is turned toward the modalities of *dialogue* (Bertrando speaks of the "dialogic therapist"), and to *narrative* models and their transformations (Bertrando & Arcelloni, 2006).

Therefore, there is a strong change of perspective: from the first *ecological–cybernetic* paradigm, to the development of a new approach more *intersubjective–narrative* oriented.

There is more, however. The "systemic approach" was to become a true *new clinical paradigm* and this dates to its "communicational" origins. More than the family relation itself, what interested the pioneers and successors of this "lineage" is the construction of a new clinical paradigm.

If we consider Palo Alto's "brief therapy model," we notice, in fact, that interest is focused above all on changing the client and his/her problem (Watzlawick, 1974), first through prescription of the symptoms and then through therapy centered on solving the problem (de Shazer, 1991).

Thus, as is the case with every clinical paradigm (psychoanalytic, cognitive, gestaltic) each with its own currents of thought and practice, over time the "systemics" wanted to become involved in every *therapeutic genre*: individuals, couples, families, groups. It is not by chance that "individual systemic psychotherapy" underwent substantial and important development (Boscolo & Bertrando, 1996; Ugazio, 1998; White & Epston, 1990) with attention focused on the semantics of communication and the construction of shared meanings in the therapeutic exchange.

The narrative metaphor and conversational techniques come to dominate in the clinical context (Anderson & Goolishian, 1992). We will not probe further into the creative aspects of various authors' contributions as this lies beyond our scope.

Instead, what interests us is reflecting on two foundational concepts: those of *relation* and *ethics*, which characterize this approach.

First, however, we must turn our attention to the other approach, the relational–generational one. Here clinicians have been concerned *essentially* with family psychotherapy. These are mostly therapists whose background is in psychoanalysis, of which, incidentally, they changed, even radically so, both the guiding principles and intervention techniques. Among these we find Nathan Ackerman, Ivan Boszormeniy-Nagy, James Framo, Carl Witaker, Donald Williamson, Norman Paul and, in Europe, Helm Stierlin and Maurizio Andolfi. Interestingly, this approach develops on the East coast of the United States and thus looks towards Europe (the "old world"). It is not by chance that the families of many of these clinical researchers originate in Europe. In this case, there is "superficial" interest in systemic theory. Some of its principles are accepted, but interest in intragenerational and intergenerational dynamics predominates. In this approach, one hears only a faint echo from afar of the passionate epistemological debates occurring within the ecological–cybernetic approach; an approach whose interest ranges from physics to biology (Maturana & Varela, 1980, 1984), particularly focused on the functioning and irreversibility of auto-organizational systems.

At the center of the relational–generational approach, we find a specific interest in family psychotherapy, a *therapeutic genre.* It would be wise for us to recognize the important contribution this approach has made to research and intervention with respect to family relations and its relational–generational connotation. This approach focuses on the subject as being deeply embedded in family dynamics. No other clinical paradigm has so richly contributed to psychotherapy and family counseling, both in key concepts and in therapeutic practice.

The relational–generational approach does not differ from behaviorism but rather, from psychoanalysis and, in particular, from the Freudian instinctual model and its individualistic closure. This does not alter that psychoanalytic thought and gestaltic–humanistic thought are the most frequent of interlocutors.

What do the two approaches have in common beyond their differences?

They share the idea that the unit of treatment does not refer to a single person and his/her interior world, whether in the psychoanalytic or the cognitive sense, but instead opens to the behavioral–intersubjective and cultural *context* in which he/she is rooted. In addition, there is a rejection of reductionism, be it biological or nosographic. However, the attention that both approaches give to semantic units that generate meaning has more importance. Indeed, there are those who privilege *dyadic* semantic units, as in the case of frequent referencing of attachment theory, and others who privilege semantic units that are *triangular* and at the intersection between triangles. We will return to this topic later.

However, our contribution seeks to go beyond commonality. As stated in the premise, it is our intention to probe the contrast between the two approaches by reflecting on how these approaches define the concept of relation and the ethical position taken.

If we survey the different clinical paradigms, we become aware of the various ways that *relation* is defined. *Relation* can be conceived as being inborn, as the outcome of evolution, or as being dialogical–cultural (Cigoli, 1997). So, in the case of the approach that developed in an intersubjective–narrative direction, relation is assumed to come from the interactive and conversational exchange between subjects, and manifests itself through patterns and carriers of meaning. It is, thus, a result, a "derivative" of the interactive exchange between subjects.

For the relational–generational approach, relation is conceived instead as a "first presupposition" (the foundation) of being human. In other words, relation transcends subjects in their exchange, being its constitutive matrix. Thus, subjects are immersed in the realm of the "relational," as if in a uterus, and it expresses itself in *bonds*.

We, thus, have two paths towards knowledge in the systemic arena. In the first path, clinicians focus on subjects interacting among themselves (therapist included) and attempt to grasp the *patterns* that connect them to the context. In the second path, clinicians focus on the *bond* (between generations, couples, siblings, communities), considered to be the *third party* among contracting parties who are weaving the dialogue. The quality of bonds is situated at various levels: the conscious and "measurable" bonds, and the symbolic bonds that also entail the presence of unconscious fantasies. The difference between the approaches can also be seen in the attention they give to the *ethical* perspective, specific to the world of action.

The intersubjective–narrative approach, especially in constructivist and constructionist currents, underlines the equal epistemic dignity of clients and therapists. The therapeutic journey is understood to be co-defined by the choices of those who participate in it. Moreover, it has a purpose: to increase the subjects' *spaces of freedom*, i.e., their possibility of choice and decision (von Foerster, 1994).

For its part, the generational–relational approach understands ethics to be *responsibility* vis-à-vis the other and human life in general, including that of future generations. Therefore, actions such as commitment to the bond with the other, the capacity for forgiveness, and the assumption of risk associated with reconciliation with the other are at the center of clinical work.

As can be seen in the first case, ethical value is attributed to the subject's interactions with the world (responsibility of choice and freedom). In the second case, ethical value is attributed to the bond between persons and generations (responsibility with respect to bonds).

In summary, we intended to highlight that there are two approaches to thinking and clinical practice within the systemic paradigm, one intersubjective–narrative–conversational and the other relational–generational, each with its own history and development. There have been various attempts to integrate the two approaches in the scientific literature (Gurman & Kniskern, 1981; Hoffman, 1981; Malagoli Togliatti, Angrisani, & Barone, 2003). In particular, researchers have tried to integrate the interactive–communicational strand with the psychodynamic–generational one, following from the systemic principle of connecting sciences and approaches. Their implicit aim is to offer a reading of systemic psychotherapy with one single history of development. We, however, believe it is the foundation on specific principles, conceptions, and development that makes it appropriate to differentiate between the two approaches.

In any case, a productive dialogue between approaches, even if present in a single paradigm, needs an acknowledgement of differences. Thus, we have highlighted the specificity of the two approaches present in the systemic paradigm with respect to the clinical encounter, their conception of relation, their ethical perspective, and their predilection for dialogue with other clinical paradigms. We have also underscored the difference between "therapeutic genre" (in our case, family clinical practice and psychotherapy) and "clinical paradigm."

We now present the relational–symbolic model, inscribed in the generational approach, that characterizes our clinical and psychosocial research on family bonds.

The Relational–Symbolic Model and Its Foundations

In the preceding section we elucidated the characteristics and history of the two most important clinical approaches within the systemic paradigm.

We now turn our attention to presenting the relational–symbolic model which guides research and intervention with a focus on family bonds in both clinical and psychosocial arenas. This model is inscribed in the *relational–generational* approach and was developed and refined starting in the 1990s at the Athenaeum Center for Family Studies and Research of the Catholic University of Milan.

What are the model's foundations with regard to conceptual, methodological, and empirical perspectives? We know that a variety of important research models can be found in the specialist literature. Among these we recall Luciano L'Abate's model focused on the Self in family relations (L'Abate, 1997; L'Abate, Cusinato, Maino, Colesso, & Scilletta, 2010). This model is founded on the guiding idea of the self's double competence: the ability to love and the ability to negotiate. Both capabilities (or their absence) have their foundation in the family relation. Recently, L'Abate talks about the ability to love and the ability to negotiate. We also recall David Olson and collaborators' (1986, 1989) "circumflex" model based on two specific dimensions (Cohesion and Adaptability); Beavers and Hampson (1990, 2000, 2003) model on "family competence" is focus on two dimensions: Competence/Health and Style (centripetal mixed centrifugal); Epstein and collaborators' (Epstein, Baldwin & Bishop, 1983; Epstein, Bishop, Ryan, Miller, & Keitner, 1993; Epstein, Bishop, & Baldwin, 2003) "McMaster Model of Family Functioning" developed several dimensions (problem solving, communication, roles, sensitivity and emotional involvement, and behavioral control); Kantor and Lehr's model (1975), later taken up by Constantine (1986, 1993), focused their attention on two dimensions (enmeshment vs. disengagement and chaos vs. rigidity) in order to distinguish the modalities of family functioning. Each of these models has its own properties and abilities to shed light on either the family system's overall functioning or on the self of members constituting the system.

Our model is *prototypical*, i.e., it is based on a procedure that calls for a standard of comparison (Carson, Butcher, & Coleman, 1988; L'Abate, 1997). It is founded on (or references) its own conception of system, has guiding ideas (pillars) and principles, and gives rise to a method for carrying out research and intervention.

Following Morin's (2001, 2004) contributions on method, we conceive the "family system" as a *unitas multiplex*. In it there is the *co-presence* of the whole ("generational intertwinement" and its qualities) and its parts (members and their beliefs, intentions and feelings). Cigoli (2000, 2006) also uses the strong metaphor of the "family body" to mean the "whole" and its qualities. Hence, it is difficult to capture the complexity of the family system.

If the conception of the family as "unitas multiplex" acts as the foundation, the pillars upon which the model rests are relation, transition, and generativity. Let us consider each of these in detail.

Relation is understood as the reciprocal bond between family members, as is distinctive of the relational–generational approach.

We can infer from the word's etymology ("re-ligo") that the bond has a sacred character and is thus situated at the origin of the person (and should be considered as a sign of the divine in the human being). To better understand the process of understanding the family, we need to distinguish the interactive from the relational levels; we clarified it in other contributions (Cigoli & Scabini, 2006; Scabini & Cigoli, 1998).

We follow an anthropological–philosophical point of reference in using the concept of *person* and not of the individual. The person is a subject in relation and is both generated and generating. S/he is thus subject to family history and to the culture of the times and is subjected to strong constraints from this standpoint: no one can choose from whom or when to be born. On the other hand, the person is also an active subject of a family's history and contributes to its construction.

The family relation is constituted by *bonds* between generations, in the couple, between siblings, and with the membership community, all of which have symbolic qualities and their own dynamics, as will be discussed later.

The second pillar is *transition*. The theme of transition has been an object of study in family research for some time. Haley (1973), in particular, had noticed that symptoms affecting one or more family members could be connected to the crisis that accompanies family transitions. Thus, symptoms are either the sign of the family's difficulty as a whole in facing a transformation of relations, or the right moment for clinical intervention. There are transitions that impose themselves with the force of an event (birth, adoption, marriage, divorce, death), and others that are more indistinct, such as adolescence and the condition of young-adulthood. It is not by chance that authors who have taken into consideration family transitions have been interested in the processes of coping that allow the family to manage situations of "stress" (Falicov, 1988; Walsh, 2008). The "stress and coping" model, widely disseminated in the scientific literature, should be distinguished from the relational–symbolic model that relies on the concepts of *crisis* (pain's factors) and *beneficial sources* (developmental/transformational factors).

We understand transition as a *critical passage* that allows one to capture the quality of family bonds, their "weak points," such as deficiencies and various forms of perversion, and strong points, such as the presence of reliable bonds and reciprocal acknowledgement between family members.

The crisis of transition always entails a *painful loss* that does not necessarily coincide only with the death of a family member. Freud (1914), in fact, emphasized that loss is "a class" that also has to do with ideals, values, the passages of life conditions, economic situations, and natural events.

L'Abate (1997), for his part, has emphasized that pain is a fundamental feeling often neglected in research (which privileges other terms, such as anguish, wound, threat) and that it is experienced for the most part in intimate interpersonal contexts.

Pain is not necessarily always visible; it can be denied even by the person who is experiencing it. In fact, as clinicians know well, pain can be projected onto others, as is the case with character disorders, or become encysted in the body, as in psychosomatic illnesses. Moreover, it can be cloaked by anger, shame, pride, and terror.

From this, it follows that pain management becomes crucial in family transitions. Indeed, we must not forget that family bonds are the place in which the expectation (and the *promise*) is precisely one of *sharing* joy and pain, as well as health and illness. It is here that the quality of the bonds themselves is measured. It also follows that the family should be considered a context of elevated relational ideality because one expects good to come from its bonds, even the family members' own awareness. As an example, consider divorce (a distinctive feature of Western culture for its frequency and expectation): from the point of view of transition, i.e., as a critical family passage, divorce signifies building contexts in which the bonds between parents and children, even with the pain of separation, nevertheless confirm their reliability.

The management of pain for the loss brought about by every transition can benefit from the presence of *rites of passage*. These exist in a variety of cultures but have become few and far between in Western culture in which greater value is attributed to individuals and to their expectations and rights. Today the lack of chorality, a typical aspect of ritual, is often addressed by putting one's individual condition online ("Facebook," "Messenger"). Meeting families or groups of families in a clinical setting also helps to restore chorality to the experience of pain, seeking ways to overcome it.

The third pillar of this model is the concept of *generativity*. Gender, generation, generativity have the same etymological matrix. Thus, addressing generativity and its implications means paying attention to that which occurs in the exchange between genders (male, female), between generations (prior, successive), and between family lines (paternal and maternal, paternal/maternal lineages).

The philosopher Nussbaum (1986) affirms that generating (an aspect that is typical of humankind which does not limit itself to merely reproducing) puts people at risk in the sense that it is above all by generating that one experiences the *fragility of the good*. This is to say that there is no action riskier than this. Many cultures protect themselves from the risk by preparing for the birth of a child with a set of precepts (customs, habits, roles, and prohibitions). In Western society, in which generating is conceived as the *choice* of single individuals, the anguish of this fragility clearly emerges: being a "good mother" or "good father" becomes a difficult challenge to face. Once the challenge is embraced, the scene becomes one in which technology rules: doctor, psychologist, educator, dietician, economist, and various trainers throng around the new mother, father, and baby.

In psychosocial and clinical research, it was Erikson (1982) and his followers (Kotre & Kotre, 1998; Snarey, 1998) who focused on generative commitment. In particular, they shed light on the relationship between generativity and a sense of mortality.

Apparently a paradox, it is the acceptance of one's own death (of one's limit) that causes the love of life to mature.

McAdams (2006) and de St Aubin & McAdams (1995), in turn, assert that generativity is the integration between aspects of self-expansion (creating something in one's own image) and care for the new generation. In particular, the authors do not consider generativity as a personality trait, but rather as a complex relational construct that integrates beliefs, desires, and behaviors whose purpose is caring for new generations. All in all, these authors focus on the individual in interaction with family and social contexts.

For our part, we address generativity as the *fruit of the exchange* between genders, generations, and lineages. We are mindful that the good conveyed in family relations is precious and fragile, which allows us to look with tenderness at family bonds. This does not alter the fact that we consider attentively various degenerative outcomes of the exchange that give rise to forms of mild to acute suffering in family members.

Having considered in what sense our model is "systemic" with reference to the concept of family as "unitas multiplex," and having illustrated the pillars that support it – the concepts of relation, transition, and generativity – we can now address the *principles* that guide research-intervention action. We thus turn our attention to organizing, symbolic, and dynamic principles, with a concluding note on methodology.

We will introduce the organizing principle with this question: what does the family organize? The family organizes relations: not every type of relation, but those that we define as primary. The family is a primary social form because it is at the origin of civilization in that it guarantees the generative process interconnecting the biological, genetic, psychic, and sociocultural realms. We know that it is through the ban on incest, even if manifested differently in different cultures, that the cultural order enters nature (Héritier, 1996; Lèvi-Strauss, 1969). But how can we characterize the family's identity structure that distinguishes it from other social groups and organizations? Referring especially to ethno-anthropology, we think that it is founded on *three differences*: that between genders, between generations and between lineages, and that its specific purpose, as we have already argued, is to *generate persons.* Precisely due to the co-presence of three differences that must be managed and to its generative purpose (the relational good that is constituted by the new generation), family organization is fundamentally *dramatic.*

The etymological root of "drama" is "dran," which means *action.* Family action is open to favorable outcomes from the point of view of generativity, but also to decidedly unfavorable ones. Rather than seeing the family either with its aspects of deficit (the "pathogenic family") or as resource (the "resilient family") – respectively, the glass half empty as opposed to half full – we consider its dramatic aspect. Thus, depending on the time period and culture, we ask ourselves: what are the challenges family bonds must face and how do they come to terms with them? Family organization is not "evolutionary," i.e., as anthropologists are well aware, it does not progress from primitive to evolved but can assume different forms both over time and in different cultures. These forms are variants of the organization of family bonds. This also implies that the *recognition* of who is part of the family and who is not can change. Here not only the children of unknown parents or those who are abandoned come to mind, but also who is deemed to be part of the extended family, and who is not. The new heterologous fertilization technologies and the "renting of uteruses" render problematic the identification of lines of ancestry and progeny, further complicating family recognition.

Let us consider the profound significance of the three differences. The difference between genders establishes the *personal limit* together with the specific identity of male and female (Kernberg, 1995); that between the generations establishes the *principle of hierarchy* and of the responsibility borne by previous generations towards successive ones; finally, that between lineages establishes the *necessity of encountering the other*, the stranger (exogamy). Indeed, in many cultures, including Western culture, there exists the fantasy of being able to generate alone (endogamic fantasy), which implies not opening oneself to the encounter with another who is not of the same group and "blood."

What we call "familyness," then, is constituted precisely by the ways in which the three differences are confronted (Scabini & Cigoli, 2000). In the case of families with homosexual parents, legally recognized in several countries, the difference of gender is weakened, but not eliminated. In any case, we are born from the encounter between the masculine and feminine, at least until there will be the possibility of cloning (Héritier, 1996).

From this it follows that psychosocial and clinical research on family bonds must be able to address this "triple difference" through techniques and instruments that are coherent with the objective.

We now come to the second principle, the symbolic one. The word symbol comes from the Greek "synballein," which means to connect separate parts. This was the case of the small medallion snapped in two that was placed on a chain around the necks of children abandoned by families due to serious economic hardship. When the mother returned to the orphanage to reclaim her child (if s/he was still alive), she would show the other half of the medallion as a sign of recognition.

For us, the symbolic matrix of family bonds is represented by basic qualities of an affective or ethical nature that transcend differences between cultures and that, when joined together, constitute the human essence of family bonds: it is what makes them species specific. To follow Brofenbrenner (1981), we can say that the family makes human beings as human. These qualities have to do with the bonds of kinship, the married couple, parents, and the community.

We are dealing here with the three symbolic qualities of trust, hope, and justice. These should be understood in a dialectic sense, i.e., as "coexisting" with their opposite (mistrust, hopelessness, injustice). Erikson (1968, 1982) and Bowlby (1969) locate trust and hope at the center of their thinking, and today these qualities are recognized as crucial factors in "close relationships" (Borawski, Ievers-Landis,

Lovegreen, & Trapl, 2003; Yuki, Maddux, Brewer, & Takemura, 2005). Meltzer and Harris (1983) also list instilling hope and generating trust among the basic affective functions carried out by the family relative to individual growth.

While trust needs proof and confirmation, hope (not by chance a virtue) is forward looking. As the logic of the Latin construction indicates, "I hope" is followed by the future infinitive. Trust and hope usually go hand in hand, but hope distinguishes itself by its ability to overcome an obstacle and to renew itself in the face of the crises traversed by bonds. We could also say that it is the expectation of fulfillment, in the sense that one *always* expects to experience trust and to receive justice in bonds. The loss of trust–hope in family bonds (which are primary bonds, as has been noted) entails serious consequences for individual family members as well as for "the family body" as a whole.

Let us now examine the third symbolic quality: *justice.* This was of particular interest to Boszormenyi-Nagy and Spark (1973, 1986) who spoke of "multipersonal justice," in the sense that every act of justice or injustice affects not only its recipient, but also the generational network. The authors distinguish between *distributive* justice, connected to what is inherited from prior generations, and *retributive* justice, which comes from the balance between giving and receiving in generational exchange. Family loyalty derives from justice, in the sense of respect or the lack thereof, towards obligations, but also from the presence or absence of openness to giving in the relationship with the other. Boszormenyi-Nagy and then Docommun-Nagy (2006) thus speak of a "relational ethics" from which trust and positive feelings derive in family bonds.

We accord particular value to the *equitable act*, which means an act that is able to restore order that has been attacked or seriously disturbed. This is "restorative justice" whose purpose is to repair the value of the bond. Repairing, forgiving, reconciling with the other are the consequent actions. Thus, within the concept of justice, as the thread that connects family bonds, we consider the notions of duty, gift, loyalty, and fairness.

In synthesis, the symbolic principle is like a latent underground current running through family bonds that thrives on the triangular and reciprocal relationship between hope, trust, and justice. As a consequence, psychosocial and clinical research that addresses family bonds must take into account this underground current through the use of appropriate techniques and instruments.

The third principle is the *dynamic* one and has to do with the exchange between family members. Another triangle comes into play here, constituted by the actions of *giving–receiving – reciprocating.* When we consider family bonds, our attention shifts to the world of action in addition to that of representation. We have various research traditions that examine the dynamic of the exchange: one is sociological and psychosocial, another generational psychodynamic, and a third is ethno-anthropological. The first sees the exchange between giving and receiving in utilitarian terms and is based on the value of equilibrium in the cost–benefit relationship. Developments in this tradition have distinguished between "close relationships" and relations of a commercial nature, however (Clark & Mills, 1993). In fact, in "close relationships" concern for the other's well-being and the value of commitment to relations are present.

The second tradition, to which we have already alluded (Boszormenyi-Nagy & Spark, 1986), conceives exchange as based on ethics and extends it to encompass generational time. The reciprocity of giving–receiving is thus broadened to include reciprocation–restitution, which goes beyond the asymmetry in the relationship between parents and children. Thus, when children generate in turn, this constitutes a form of restitution and of generational renewal.

The third tradition is of an ethno-anthropological (Benveniste, 1969; Godbout, 1992) and cultural psychological type (Vernant, 1983). The willingness to *give* is the central category in this approach. The gift is understood as the expression of an act of trust and hope towards the other. Godbout, in particular, stands on its head the position that sees obligation and indebtedness as primary in the intergenerational exchange and emphasizes the value of the gift's gratuitousness. From this comes the notion that the inability to give and the various forms of perversion of the gift (self-interested and instrumental gifts) constitute forms of relational pathology that have to do primarily with generational exchange.

For our part, we affirm the *ambiguous nature* of the dynamics of the exchange in family relations; indeed, both obligation-duty and the gift coexist in them. More precisely, we propose another triangle, the one represented by the actions of *giving* (bestowing, offering, doing something for another person), *receiving* (accepting, acknowledging), and *reciprocating* (doing something for another person who then gives in turn) in the protracted time of generational exchange. Gift and duty are factors of expansion of bonds. Receiving life is a gift and a debt that waits to be accepted and reciprocated, from generation to generation.

Thus, we look for the health and illness (psychopathology) of family bonds in the *three triangles* discussed here: the organizational, the symbolic and the dynamic, which intersect with one another to form a three-dimensional figure that conveys the depth of family bonds. The difference of gender–generation–lineage is dissected by the dynamic of giving–receiving–reciprocating, and both of these are traversed by the symbolic triangle that speaks of hope, trust, and justice.

The Model and Its Method

The value of a prototypical model, as this one is, lies in offering a research framework and methodology at the same time. In our own research, the family is not studied as the *context* for individual development and socialization (a theme of obvious importance) so much as the "unitas multiplex" characterized by the joint action of those who are part of it. The joint action of multiple family members and generations is thought of as the place where health illness is situated. Hence, we consider conflict and its management as "the soul of the relation."

As Beavers (1990) maintained, there is no single thread that leads to knowledge about family relations. On the other hand, each thread (in our case, the relational–symbolic model) must be able to respond to choices made (in our case, the "pillars" and "principles") and to search for their empirical and clinical evidence. This takes place precisely through method, which is the path to seeking such evidence.

We will not dwell here on various methodological problems that range from research aims to the treatment of data and information, passing through the choice of instruments (self-report, observational, "dialogic"). Rather, we are interested in highlighting the constraints on and opportunities for research. In our case, the *constraints* are the following: research must be based on family relations and not just on single family members, it must address transitions, and it must focus on aspects of generativity–degenerativity present in the generational exchange (outcome).

As for *opportunity*, we can attest that, when carried out in this way, research is able to gather valuable information and to intervene (action-research) in the exchange between genders, generations, and lineages, as well as in the gift–debt dynamic and the latent web of bonds represented by the crucial sentiments of trust, hope, and justice. Finally, such research has repercussions on the training of psychosocial and clinical practitioners.

Reiss and Oliveri (1991) offered a decisive contribution regarding the *relational nature* of findings and information. They observed that using instruments such as self-reports and questionnaires, the exchange that is accentuated is the one between the researcher and individuals (in our case, the single family members), while the use of observational and dialogic instruments highlights the direct exchange between family members. We call "dialogic" those instruments that engage all family members in direct dialogue with one another. In this case, the researcher does not limit him/herself to observing, but intervenes in order to probe contents and dynamics of the exchange. This is therefore a typical form of "therapeutic assessment" developed by Finn (2007).

Over the years, we have refined a set of analytical instruments and techniques that are coherent with the model in order to apprehend with increasing precision the family as *unitas multiplex*. Specifically, we have used self-report instruments that take into account the affective as well as ethical aspects of family relations and have created observational and dialogic instruments, such as the Family Sceno-Test (Margola, 2009), Family Joint Drawing (Gennari & Tamanza, in press), Family Life Space (Gozzoli & Tamanza, 1998), "The Double Moon" (Greco, 1999, 2006), and the Generational Clinical Interview (Cigoli & Tamanza, 2009). Furthermore, we have been mindful of the relational–symbolic dimension in the treatment of findings. In the case of findings obtained from self-report instruments, we have paid attention to the non-independence of family findings (Kenny & Judd, 1986), which forces the researcher to think in terms of families and not of individual family members.

In the case of observational and dialogic instruments, we have developed analytical grids for single family members' actions as well as for the members' joint action, taking into account the final product as much as the process of exchange.

There are numerous psychosocial and clinical studies pertaining to family transitions that reference the relational–symbolic model: becoming a couple, the transition to adulthood, the transition to adoptive parenthood, migrant families, the effects of an older person's serious illness on family relations, divorce and its generational effects, reconstituted families, and the psychopathology of the couple and family. The list of these researches would be longer. Some are cited in the bibliography.

To better delve into these studies, we refer you to the website of the *Athenaeum Centre for Family Research and Studies*.

A model is considered to be alive if it is self-reflective, and if it continually sharpens itself and delivers empirical and clinical evidence. What is more, a model is alive if it enters into constructive dialogue with other models, not accepting scientific reductionism, whether theoretical or methodological.

References

Anderson, H., & Goolishian, H. (1992). The client as the expert: A not knowing approach to therapy. In S. McNee & K. Gergen (Eds.), *Therapy as a social construction* (pp. 25–39). London: Sage Publications.

Auerswald, E. H. (1985). Thinking about thinking in family therapy. *Family Process, 24*, 1–12.

Bateson, G. (1936–1958). *Naven*. Stanford: Stanford University Press.

Bateson, G. (1961). *Perceval's narrative: A patient's account of his psychosis 1830–1832*. Stanford: Stanford University Press.

Beavers, W. R., & Hampson, R. B. (1990). *Successful families: Assessment and intervention*. New York: W. W. Norton.

Beavers, W. R., & Hampson, R. B. (2000). The Beavers systems model of family functioning. *Journal of Family Therapy, 22*, 128–143.

Beavers, W. R., & Hampson, R. B. (2003). Measuring family competence: The Beavers Systems Model. In F. Walsh (Ed.), *Normal family processes: Growing diversity and complexity* (2nd ed., pp. 549–580). New York: Guildford Press.

Benveniste, E. (1969). *Le vocabulaire des institutions européennes* [Dictionary of European institutions]. Paris: Les Editions de Minuit.

Bertrando, P., & Arcelloni, T. (2006). Hypotheses are dialogues. Sharing hypotheses with clients. *Journal of Family Therapy, 28*, 370–387.

Bertrando, P., & Toffanetti D. (2000). *Storia della terapia familiare. Le persone, le idee* [The history of family therapy. Persons and ideas]. Milano: Raffaello Cortina Editore.

Borawski, E. A., Ievers-Landis, C. E., Lovegreen, L. D., & Trapl, E. S. (2003). Parental monitoring, negotiated unsupervised time, and parental trust: The role of perceived parenting practices in adolescent health risk behaviors. *Journal of Adolescence and Health, 33*(2), 60–70.

Boscolo, L., & Bertrando, P. (1996). *Terapia sistemica individuale* [The individual systemic therapy]. Milano: Raffaello Cortina Editore.

Boszormenyi-Nagy, I., & Spark, G. (1973). *Invisible loyalties: Reciprocity in intergenerational family therapy*. New York: Harper & Row.

Boszormenyi-Nagy, I., & Spark, G. (1986). *Between give and take*. New York: Brunner/Mazel.

Bowlby, J. (1969). *Attachment and loss attachment: Vol. 1*. New York: Basic Books.

Brofenbrenner, U. (1981). *On making human beings human*. London: Sage Publication.

Buckley, W. F. (1967). *Sociology and model system theory*. London: Prentice-Hall.

Carson, R. C., Butcher, J. N., & Coleman, J. C. (1988). *Abnormal psychology and modern life*. Glenview: Scott Foresman & Company.

Cecchin, G., Lane, G., & Ray, W. A. (1992). *Irriverenza: Una strategia di sopravvivenza per i terapeuti* [The irriverence: Strategies for the therapists to survive]. Milano: Franco Angeli.

Cigoli, V. (1977). *Modelli di interazione familiare* [Interaction's family models]. Milano: Franco Angeli.

Cigoli, V. (1997). *Intrecci familiari. Realtà interiore e scenario relazionale* [Family plots. Inner reality and relational scenery]. Milano: Raffaello Cortina Editore.

Cigoli, V. (2000). *Il vello d'oro. Ricerche sul valore famiglia* [The golden fleece. Research on the value of the family]. Cinisello Balsamo: San Paolo.

Cigoli, V. (2006). *L'albero della discendenza. Clinica dei corpi familiari* [The tree of parentage. Clinic of family's bodies]. Milano: Franco Angeli.
Cigoli, V., & Scabini, E. (2006). *Family identity. Ties. Symbols, and transitions*. New York: Taylor.
Cigoli, V., & Tamanza, G. (2009). *L'intervista clinica generazionale* [The Clinical Generational Interview]. Milano: Raffaello Cortina Editore.
Clark, M. S., & Mills, J. (1993). The difference between communal and exchange relationships: What it is and is not. *Personality & Social Psychology Bulletin, 19*(6), 684–691.
Constantine, L. (1986). *Family paradigms: The practice of theory in family therapy*. New York: The Guildford Press.
Constantine, L. (1993). The structure of family paradigms: An analytical model of family variation. *Journal of Marital and Family Therapy, 19*, 39–70.
de Shazer, S. (1991). *Putting difference to work*. New York: Norton.
de St Aubin, E., & McAdams, D. P. (1995). The relations of generative concern and generative action to personality traits, satisfaction/happiness with life and ego development. *Journal of Adult Development, 2*, 99–112.
Docommun-Nagy, C. (2006). *Ces loyautés qui nous libèrent*. Paris: JC Lattès.
Epstein, N. B., Baldwin, L. M., & Bishop, D. S. (1983). The McMaster family assessment device. *Journal of Marital and Family Therapy, 9*(2), 171–180.
Epstein, N. B., Bishop, D. S., Ryan, C. E., Miller, I. W., & Keitner, G. I. (1993). The McMaster Model: View of healthy family functioning. In F. Walsh (Ed.), *Normal family processes* (2nd ed., pp. 138–160). New York: The Guilford Press.
Epstein, N. B., Bishop, D. S., & Baldwin, L. M. (2003). McMaster model of family functioning: A view of the normal family. In F. Walsh (Ed.), *Normal family processes. Growing diversity and complexity* (3rd ed.). New York: The Guilford Press.
Erikson, E. H. (1968). *Identity: Youth and crisis*. New York: Norton.
Erikson, E. H. (1982). *The life cycle completed*. New York: Norton.
Falicov, C. J. (1988). *Family sociology and family therapy contributions to the family development frame work: A comparative analysis and thoughts on future trends*. New York: The Guildford Press.
Finn, S. E. (2007). *In our clients' shoes: Theory and techniques of therapeutic assessment*. Mahwah: Lawrence Erlbaum Associates.
Freud, S. (1914). On narcissism: an introduction. *Standard Edition*, 14.
Gennari, M. L., & Tamanza, G. (in press). *Il disegno congiunto della famiglia. Uno strumento per l'analisi delle relazioni familiari* [The joint family drawing. An instrument for the assessment of the family relationships]. Milano: Franco Angeli.
Godbout, J. (1992). *L'esprit du don* [The spirit of the gift]. Montreal: Editions du Boreal.
Gozzoli, C., & Tamanza, G. (1998). *Il Family Life Space: L'analisi metrica del disegno* [Family life space: The drawing metric analysis]. Milano: Franco Angeli.
Greco, O. (1999). *La doppia luna. Test dei confini e delle appartenenze familiari* [The Double Moon: Test for family boundaries and belongings]. Milano: Franco Angeli.
Greco, O. (2006). *Il lavoro clinico con le famiglie complesse. Il test La doppia luna nella ricerca e nella terapia* [Clinical intervention with troublesome family. The use of Double Moon test in therapy and research]. Milano: Franco Angeli.
Gurman, A. S., & Kniskern, D. P. (Eds.). (1981). *Handbook of family therapy*. New York: Brunner/Mazel.
Haley, J. (1973). *Uncommon therapy. The psychiatric techniques of Milton Erickson*. New York: Norton.
Héritier, F. (1996). *Masculin, feminin. La pensée de la difference*. [Male and female. The thought about difference]. Paris: Edition Odile Jacob.
Hoffman, L. (1981). *Foundations of family therapy: A conceptual framework for systems change*. New York: Basic Books.
Jackson, D. (1957). The question of family homeostasis. *The Psychiatric Quarterly Supplement, 31*(Pt 1), 79–90.

Jackson, D., & Weakland, J. (1961). Conjoint family therapy: Some considerations on theory, technique and results. *Psychiatry, 24*(2), 30–45.
Kantor, D., & Lehr, W. (1975). *Inside the family*. San Francisco: Jossey-Bass.
Kenny, D. A., & Judd, C. M. (1986). Consequences of violating the independence assumption in analysis of variance. *Psychological Bulletin, 99*(3), 422–431.
Kernberg, O. (1995). *Love relations: Normality and pathology*. New Haven: Yale University Press.
Kotre, J., & Kotre, K. B. (1998). Intergenerationals buffers: "The damage stops here". In D. P. McAdams & E. de St Aubin (Eds.), *Generativity and adult development: How and why we care for the next generation* (pp. 367–389). Washington: American Psychological Association.
Kuhn, T. S. (1962). *The structure of scientific revolutions*. Chicago: University of Chicago Press.
L'Abate, L. (1997). *The self in the family*. Atlanta: Georgia State University.
L'Abate, L., Cusinato, M., Maino, E., Colesso, W., & Scilletta, C. (2010). *Relational competence theory*. New York: Springer.
Lèvi-Strauss, C. (1969). *The elementary structure of kinship*. Boston: Beacon.
Malagoli Togliatti, M., Angrisani, P., & Barone, M. (2003). *La psicoterapia con la coppia. Il modello integrato dei contratti. Teoria e pratica* [Psychotherapy with couples. Contracts integration models. Theory and practice]. Milano: Franco Angeli.
Margola, D. (2009). *Tecniche psicologiche di indagine clinica. Sceno-Test, FLS, La Doppia Luna, TAT* [Psychological techniques for clinical research. The Scheno-Test, FLS, The Double Moon, TAT]. Milano: Franco Angeli.
Maruyama, M. (1963). The second cybernetics: Deviation-amplifying mutual causal processes. *American Scientist, 5*(2), 164–179.
Maturana, H. R., & Varela, F. J. (1980). *Autopoiesis and cognition. The realization of the living*. Dordrecht: Reidel Publishing Company.
Maturana, H. R., & Varela, F. J. (1984). *El arbol del conocimiento* [The tree of knowledge]. Madrid: Editorial Debate.
McAdams, D. P. (2006). *The redemptive self: Stories Americans live by*. New York: Oxford University Press.
Meltzer, D., & Harris, M. (1983). *Child, family and community: A psycho-analytical model of learning process*. Paris: Organization for Economic Co-operation and Development.
Morin, E. (2001). *La Méthod. V. L'humanité de l'humanité. Tome 1. L'identité humain* [The Method. V. Humanity of humanity. Volume 1. The human identity]. Paris: Edition du Senil.
Morin, E. (2004). *La Méthod. VI. Etique* [The Method. VI. Ethics]. Paris: Edition du Senil.
Nussbaum, M. (1986). *The fragility of goodness. Luck and ethics in Greek tragedy and philosophy*. Cambridge: Cambridge University Press.
Olson, D. H. (1986). Circumplex model VII: Validation studies and FACES III. *Family Process, 2*(5), 337–351.
Olson, D. H. (1989). Circumplex model of family systems: Family assessment and intervention. In D. H. Olson, D. H. Sprenkle, & C. R. S. Russell (Eds.), *Circumplex model: Systemic assessment and treatment of families* (pp. 7–49). New York: Haworth Press.
Prigogine, I. (1961). *Thermodynamics of irreversible processes*. New York: Interscience.
Rapaport, E. A. (1970–1971). The resolution of a delusional hair fetish. *Psychoanalytic Review, 57*, 617–631.
Reiss, D. (1971). Varieties of consensual experience. *Family Process, 10*, 1–35.
Reiss, D., & Oliveri, M. E. (1991). The family's conception of accountability and competence: A new approach to the conceptualization and assessment of family stress. *Family Process, 30*, 193–213.
Scabini, E., & Cigoli, V. (1998). The role of theory in the study of family psychopathology. In L. L'Abate (Ed.), *Family psychopathology. The relational roots of dysfunctional behavior* (pp. 13–34). New York: Guilford Press.
Scabini, E., & Cigoli, V. (2000). *Il famigliare. Legami, simboli e transizioni* [The familyness. Ties, symbols, and transition]. Milano: Raffaello Cortina Editore.
Selvini Palazzoli M., Boscolo L., Cecchin G., & Prata G. (1975). *Paradosso e contro paradosso* [Paradox and counterparadox]. Milano: Feltrinelli.

Snarey, J. (1998). Ego development and the ethical voices of justice and care: An Eriksonian interpretation. In P. Westenberg & A. Blasi (Eds.), *Personality development: Theoretical, empirical, and clinical investigations of Loevinger's conception of ego development* (pp. 163–180). Mahwah: Lawrence Erlbaum Associates.

Ugazio, V. (1998). *Storie permesse, storie proibite. Polarità sematiche familiari e psicopatologia* [Allowed stories, prohibited stories. Family semantic polarities and psychopathology]. Torino: Bollati Boringhieri.

Vernant, P. (1983). *Myth and thought among the Greeks*. London: Routledge & Kegan.

von Bertalanffy, L. (1930). *Lebenswissenschaft und bildung*. Stenger: Erfurt.

von Foerster, H. (1994). Etica e cibernetica di secondo ordine [Ethics and second order cybernethics]. *Psicobiettivo, 14*(3), 47–57.

Walsh, F. (2008). *La resilienza familiare* [The family resilience]. Milano: Raffaello Cortina Editore.

Watzlawick, P. (1974). *Change*. New York: Norton.

Wertheim, E. S. (1973). Family unit therapy and the science and typology of family systems. *Family Process, 12*, 361–376.

Wertheim, E. S. (1975). The science and typology of family system II: Further theoretical and practical considerations. *Family Process, 14*, 285–309.

White, M., & Epston, D. (1990). *Narrative means to therapeutic ends*. New York: Norton.

Yuki, M., Maddux, W. W., Brewer, M. B., & Takemura, K. (2005). Cross-cultural differences in relationship and group-based trust. *Personality and Social Psychology Bullettin, 31*(1), 48–62.

Chapter 13
Constructivism

Alexander Riegler

Introduction

In one of his first papers, Thomas Kuhn (1959) addressed the "essential tension" implicit in scientific research, i.e., the contrast between convergent and divergent thinking. He considered both to be central to the advance of science. Convergent thinking is what scientists do in their daily "normal research projects," where the "scientist is not an innovator but a solver of puzzles, and the puzzles upon which he concentrates are just those which he believes can be both stated and solved within the existing scientific tradition" (*ibid.*, p. 234). The convergent mode is "neither intended nor likely to produce fundamental discoveries or revolutionary changes in scientific theory" (*ibid.*, p. 233). As he would describe in greater detail in *The Structure of Scientific Revolutions* (1970), this is because students are already discouraged from developing divergent-thinking abilities, partly because their education is based on textbooks, which "exhibit concrete problem solutions that the profession has come to accept as paradigms… Nothing could be better calculated to produce 'mental sets' or *Einstellungen*" (Kuhn, 1959, p. 229). Clearly, from this perspective, mental sets (or "mental inertia") play an important role in paradigms as they prevent the normal scientist from gazing beyond the limits of her paradigm. Kuhn also emphasized the importance of convergent thinking as "no part of science progressed very far or very rapidly before this convergent education and correspondingly convergent normal practice became possible" (*ibid.*, p. 237). However, Kuhn also recognized the divergent method because in order to assimilate new discoveries and theories "the scientist

A. Riegler (✉)
Centre for Logic and Philosophy of Science and Leo Apostel Center for Interdisciplinary Studies, Vrije Universiteit Brussel, Krijgskundestraat 33, 1160, Brussels, Belgium
e-mail: ariegler@vub.ac.be

L. L'Abate (ed.), *Paradigms in Theory Construction*, DOI 10.1007/978-1-4614-0914-4_13,

must usually rearrange the intellectual and manipulative equipment he has previously relied upon, discarding some elements of his prior belief" (*ibid.*, p. 226).

For psychologists, this interplay between convergent and divergent thinking – Kuhn's essential tension – is strongly reminiscent of Jean Piaget's notion of the ongoing attempts of the subject to restore her equilibration on the basis of assimilation (trying to fit the new puzzle pieces to the already existing pieces, or in Kuhn's view: "convergence thinking") and accommodation (trying to come up with new innovative ways of changing the game in the light of non-fitting pieces, or "divergent thinking," according to Kuhn). Is the continuous integration of new information into pre-existing structures, the transformation of existing structures, and the construction of new ones that are based on experience to accomplish assimilation what Kuhn had in mind? It leaves us with the question: How much constructivist (in the sense of Piaget) was Kuhn? In a series of papers several authors have also argued that Piaget and Kuhn have opposing viewpoints: while the former proposes continuous and cumulative progress, the latter emphasizes discontinuity (cf. Kitchener, 1985, 1987; Tsou, 2006; Burman, 2007). Also, Ernst von Glasersfeld, who founded radical constructivism (RC) in the 1970s, is clearly on Kuhn's side: "The history of scientific ideas shows all too blatantly that there has been no overall linear progression." (Glasersfeld, 2001, p. 32) Furthermore, according to Bird (2009), "Kuhn… denied any constructivist import to his remarks on world-change [but] acknowledge[d] a parallel with Kantian idealism." This, however, sounds odd because Kuhn claimed that "the proponents of competing paradigms practice their trades in different worlds… Practicing in different worlds, the two groups of scientists see different things when they look from the same point in the same direction" (Kuhn, 1970, p. 150)."[1]

Clearly, a statement like this sounds very constructivist, for constructivism expresses the idea that mental structures and operations are actively constructed by one's mind rather than passively acquired. "Constructing" means that there is a developmental path from some initial state, rather than a teleological progress towards some final state (Burman, 2007). It comes as no surprise then that different individuals take different paths resulting in different (momentary) states. Given these observations, is constructivism an over-arching perspective that accommodates both Kuhn's and Piaget's respective philosophies of science?

In this paper I will first describe the paradigm of constructivist approaches, in general, and radical constructivism, in particular. I will then investigate the notion of mental sets through the constructivist lens. The paper's goal is to show how central notions in Kuhn's theory can be interpreted in terms of constructivism and its emphasis on an alternative approach to knowledge.

[1] In his autobiography, von Glasersfeld (2010, p. 245) described how Kuhn grew indignant over von Glasersfeld's comment that Fodor's talk about "representations" was irresponsible if it did not add that they could never be representations of reality. So we can assume that in the above quote Kuhn interpreted "different worlds" as hypothetical or even fictitious worlds that are not corroborated by experiments whereas for von Glasersfeld these were different experiential worlds with no ontological connotation (Marco Bettoni, personal communication, 2011).

Constructivist Paradigms

Constructivism is not a homogenous paradigm. Various strands of empirical insights and philosophical reflections have led (and are still leading) to the formulations of a number of constructivisms. Vincent Kenny (2010) speaks of "ever branching new sub-disciplines." And even though many constructivist scholars have known each other personally, and have even been bound by ties of friendship, their respective approaches – including radical constructivism, social constructivism, constructionism, etc. – have remained at a certain distance from each other. So instead of speaking of *the* constructivism, I will use the term "constructivist approaches." It refers to the idea that the mental world – or the experienced reality – is actively constructed or "brought forward," and that the observer plays a major role in any theory.[2]

In some sense the variety of constructivist approaches seems to resemble that which Kuhn called "immature science" (Kuhn, 1970), which lacks consensus and where "much intellectual energy is put into arguing over the fundamentals with other schools instead of developing a research tradition" (Bird, 2009). However, there is a crucial difference. Constructivist approaches are not a discipline-bound endeavor but rather a horizontal "meta science" way of thinking that covers a variety of disciplines and interdisciplinary topics: neural networks, cognition, learning, living systems, organizations, architectural design, sociology, literary sciences, media sciences, and systemic family therapy (cf. Müller, 2010).

Looking closely at the differences among constructivist approaches one will recognize that they vary in function of how far they are willing to take the idea that reality is constructed. In the following section the attempt is made to classify constructivist approaches into two dimensions: (1) disciplinary paths to constructivisms and (2) dualistic versus non-dualistic approaches with regard to the dichotomy between mind and reality. For each of these categories I will paradigmatically discuss a typical – exemplary – proponent.

Paths to Constructivism

Phenomenological Constructivism: Ernst Mach

Some authors embrace a phenomenological perspective that considers perception to be the grouping of experiential complexes. For example, in the late nineteenth century, Ernst Mach, a precursor of constructivism,[3] claimed that "things" consist

[2] Cf. Maturana's "Everything said is said by an observer to another observer who can be himself or herself" (Maturana, 1978, p. 31) and von Foerster's "Objectivity is the delusion that observations could be made without an observer" (quoted in Glasersfeld, 1995, epigraph).

[3] It should be emphasized that this is the very same Ernst Mach whose name was used for the *Verein Ernst Mach*, which later became known as the Vienna Circle of the logical positivists. However surprising this "double life" might be, in his first publication on radical constructivism, von Glasersfeld (1974) already considered Mach (together with Percy Bridgman) an ally. They both neglected developmental aspects, which are crucial for constructivism.

of a functional assembly of sense elements, i.e., they are thought symbols for a sensational complex of relative stability (cf. also von Foerster's reference to eigen-behavior, see below). Consequently, the actual elements of the world are not things and bodies but rather sensations such as colors, sounds, pressures, spaces, and times as well as moods, emotions, and the will (Mach, 1959, ch. 1.2). However, this does not imply that the world is a mere sum of sensations. Rather, Mach defined the world in terms of "functional relations of the elements" (Mach, 1959, p. 363). These also include "psychical" entities such as the self, which are characterized as sensations that are related to each other in different ways. From this perspective, according to Mach, science is the "discovery of functional relations… the dependence of experiences on one another." (Mach, 1959, ch. 1.14) and the growth of knowledge is nothing more than the adaptation of thoughts to facts and "the discovery of logical relations via the accommodation of thoughts with each other" (Mach, 1992, p. 27). Indeed, it was on this basis that Mach developed the concept of the "economy of thoughts," which emphasizes the importance of compressing experiences into laws. He considered the object of science to be "to replace, or save, experiences, by the reproduction and anticipation of facts in thought. Memory is handier than experience, and often answers the same purpose" (Mach, 1960, p. 577).

For the non-dualist Mach, a (scientific) fact is a conscious sensation (Blackmore, 1972, p. 32) because experiential elements are always of the same sort:

> The antithesis between ego and world, between sensation (appearance) and thing … vanishes, and we have simply to deal with the connection of the elements *a b c* [the complex composed of volitions, memory-images, and the rest] *A B C* [complexes of colors, sounds, and so forth, commonly called bodies] *K L M* [the complex known as our own body] … This connection is nothing more or less than the combination of the above-mentioned elements with other similar elements (time and space). (Mach, 1959, ch. 1.7)

From his phenomenological perspective, according to which the world consists only of our sensations, knowledge does not refer to material entities but to sensations only.

Furthermore, Mach's perspective claims that the idea that experiences are the result of the effects of an external world extending into consciousness should be rejected. It is not bodies that produce sensations but complexes of (sensational) elements that make up bodies. Ultimately, "the world does not consist of mysterious entities which, by their interaction with […] the ego, produce sensations, which alone are accessible. For us, colors, sounds, spaces, times, etc. are provisionally the ultimate elements, whose given connection it is our business to investigate." (Mach, 1959, ch.1.13)[4]

[4] Some 100 years later, Francisco Varela shared Mach's strong emphasis on the first-person perspective. Having a complementary interest in phenomenology and neuroscience (rather than physics, as was the case with Mach), he developed *neurophenomenology* (Varela, 1996). It combines systems neuroscience with a pragmatic approach to becoming aware of our lived experience (Froese, Gould, & Barrett, 2011).

Biological Constructivism: Maturana et al.

Many variations of constructivism are founded in empirical (mostly biological) results rather than philosophical considerations. All these approaches emphasize: (1) the primacy of the cognitive system and (2) its organizational closure.

Building on a large number of insights into biological mechanisms, which he had collected in his long career as a neuroscientist, Rudolfo Llinás expressed the primacy of the cognitive system in very clear sentences: "[al]though the brain may use the senses to take in the richness of the world, it is not limited by those senses; it is capable of doing what it does without any sensory input whatsoever" (Llinás, 2001, p. 94). He goes on to claim that since the ability of the nervous system is to generate a sensory experience of any type, "we are basically dreaming machines that construct virtual models of the real world." (*ibid.*,). In his view, the mind is primarily a self-activating system, "one whose organization is geared toward the generation of intrinsic images," (*ibid.*, p. 57) which conveys weight to the idea that cognition acts independently of the environment. It merely requests confirmation for its ongoing dynamical functioning and otherwise works autonomously. In other words, the mind is organizationally closed, which implies that the mind must construct its reality and the entities it is populated with in the first place. Perturbations from the outside may, at best, modulate the dynamical construction process of the mind but may not determine it.

These were also the conclusions made by Heinz von Foerster, Humberto Maturana and Francisco Varela years before Llinás. Von Foerster picked up an old neurophysiological insight, the "principle of undifferentiated encoding," which is thought to highlight the peculiarity of the nervous system: "The response of a nerve cell does not encode the physical nature of the agents that caused its response. Encoded is only 'how much' at this point on my body, but not 'what'." (Foerster, 2003a, pp. 214–215). Together with the insight that a large majority of the synapses in the primary visual cortex (as well as in other areas such as the lateral geniculate nucleus of the thalamus) do not seem to be devoted to signals coming from the sensors (Sillito & Jones, 2002), the question arises: What is the brain, in its isolation, busy with? Maturana suggested that we can compare the situation of the mind to that of a pilot using instruments to fly a plane (or a submarine navigator piloting in the dark depths of an ocean). All he does is "manipulate certain internal relations of the plane in order to obtain a particular sequence of readings in a set of instruments" (Maturana, 1978, p. 42). Von Foerster suggested a solution based on the idea of (infinite) repetition, or *eigenbehavior*.[5] He pointed out that "what is referred to as 'objects' … in an observer-excluded (linear, open) epistemology, appears in an observer-included (circular, closed) epistemology as 'token for stable behavior'" (Foerster, 2003b, p. 261).[6] He argued these eigenbehaviors, or attractors, result from the recursion of accounting for the changes in an organism's sensations by its actions that in turn are

[5] A simple example often used by von Foerster himself is that of repeatedly applying the square root operator to the result of its own operations, which at its limit will always result in one irrespective of the initial number.

[6] In the terminology of complexity research it is called an "attractor."

described in terms of its sensations. Therefore, what appear to us as objects are equilibria that determine themselves through circular processes. They reside "exclusively in the subject's own experience of his or her sensorimotor coordination" (Foerster, 2003b, p. 266). This agrees with Piaget's (1954) insight that in the sensorimotor period children repeat certain actions over and over again in order to gain sensorimotor mastery.[7] Eventually they will also have to do this as scientists who are drilled during their educational period (see Feyerabend's criticism of stereotypical research below, and Burman, 2008).

The idea of circularity was also picked up by Humberto Maturana and Francisco Varela in their work on autopoietic systems (or "biology of cognition"). What began as an answer to the question, "What is life?" has become a fully encompassing explanatory network that includes living, cognition, languaging, and emotioning (cf. special issue of *Constructivist Foundations*, Riegler & Bunnell, 2011). At the focus of their work is the biological individual, who is a particular type of self-organizing system, a so-called "autopoietic system" (cf. Maturana & Varela, 1980). It obeys the following criteria: (1). Its components take part in the recursive production of the network of production of components that produced those components. (2). An entity exists in the space within which the components exist by determining the topology of the network of processes. (A system that does not fulfill these criteria is called allopoietic, e.g., it is a machine that serves a different purpose than maintaining its own organization.) Due to its circular organization, an autopoietic system is clearly "an inductive system and always functions in a predictive manner: what occurred once will occur again. Its organization (both genetic and otherwise) is conservative and repeats only that which works." (Maturana & Varela, 1980, pp. 26–27). It should therefore come as no surprise that science, carried out by autopoietic living systems, works inductively, too.

Many have criticized Maturana and Varela on the grounds of being self-contradictory due to their emphasis on biology: How can biology of cognition explain its own axioms, i.e., the biological a priori of cognition?[8] As will be pointed out below, the biological link can be considered superfluous, which makes the idea of autopoietic systems a formal rather than biological theory, thereby avoiding the criticism.

Dualism vs. Non-dualism

Dualistic Approaches: Cognitive Constructivism

Many constructivist approaches assume a dualistic relationship between constructed reality and mind-independent reality.[9] They maintain that constructed mental structures gradually adapt to the structures of the real world. Such a "cognitive

[7] Cf. Mach, "We have become accustomed to regarding an object as existing permanently." (Mach, 1970, p. 30).

[8] This criticism is also applied to evolutionary epistemology, cf. Riegler (2005b).

[9] In the German-speaking literature on constructivism, the distinction is often made between *wirklichkeit* (from the German "wirken", meaning "to have an effect on") – the world as the domain of our experience – and reality (from Latin "res" = thing) – the world as the domain of things in themselves.

constructivism" was championed by Jean Piaget. He suggested an interplay of assimilation (integration of experience) and accommodation (modification of the cognitive apparatus based on new experience to enable assimilation) that progressively leads to knowledge of reality: "Knowing reality means constructing systems of transformations that correspond, more or less adequately, to reality." (Piaget, 1970, p. 15).

Based on Piaget, Ulric Neisser (1976) defined perception as the "pickup of information" controlled by the mental structure that is constructed from earlier perceptions. He called this "schemata-controlled information pickup." It expresses the view that an individual's cognitive apparatus determines the way she looks at the environment. The apparatus constructs anticipations of what to expect and thus enables the organism to perceive the expected information. Without these anticipations the individual would not be able to see anything.

For example, a circle drawn in sand is perceived as a circle not because of sophisticated image processing in our head, which compresses the perceived trace into the mathematical concept of a circle, but due to the projection of a (mathematically ideal) circle onto sensory data and anticipating not too much of a difference (Riegler, 2006).[10]

The idea that perception and cognition in general are based on anticipation forces us to reconsider the information-processing paradigm, which portrays cognitive organisms as computers transforming information input into behavioral output. In this paradigm, organisms have to filter all the available information to find relevant features in order to control their behavior. This seems computationally implausible as the combinatorial variety is simply too large for biological brains and artificial computers alike. Very much in the sense of Karl Popper's (1979) "searchlight view of mind," the alternative would be "perceptive interaction on demand."

According to what I called the "constructivist-anticipatory principle" (e.g., Riegler, 1994, 2007), it is not all the available information from outside that is filtered for relevant issues in order to control the behavior of an organism. Instead, the cognitive system constructs cognitive structures in the first place and occasionally uses sensor signals to verify their validity (or as von Glasersfeld would say: viability). It may be compared with a relay race where the participants focus solely on their running, except for the short moments of coordination when they pass the baton to the next runner. One could describe the moments of coordination as "checkpoints" (Riegler, 1994, 2001a, 2007), where runners verify whether they are still on track so that the race can go on with the subsequent team member.

As a consequence of this, cognitive decisions are not taken in response to an environmental challenge but are a consequence of internal cognitive dynamics. Clearly, this emphasizes the key role of the cognitive apparatus in the process of reality construction, which is not simply the (passive) representation of a mind-independent reality. Therefore, cognition is not about information processing but rather about information generating.

---

[10] Similarly, George Kelly (1963) emphasized that a "person's processes are psychologically channelized by the ways in which he anticipates events."

The information-generating paradigm has great implications for science since it rejects the predominant positivist view (for example of the Vienna Circle) which considers observation (expressed, e.g., in protocol sentences describe sense-data) as the neutral arbiter between competing theories. From Kuhn's perspective, however, observation is influenced by prior beliefs and experiences (cf. Bird, 2009). This "theory-ladenness" was emphasized by Kuhn when he noted that a pendulum was seen by Aristotelians as body that "was simply falling with difficulty" while for Galileo it was a body "that almost succeeded in repeating the same motion over and over again ad infinitum" (Kuhn, 1970, p. 119). He concluded that "[t]wo scientists who perceive the same situation differently, but nevertheless use the same vocabulary to describe it, speak from incommensurable viewpoints" (Kuhn, 1970, p. 201).

Some 30 years before Kuhn, Ludwik Fleck (1935) had already emphasized the theory-ladenness of observation. What he described as "thought-style" determines not only the meanings of the concepts used by the scientists, but also the perception of the phenomena to be explained (Oberheim & Hoyningen-Huene, 2009).

The idea of "incommensurable viewpoints" can easily be accounted for by Neisser's perceptual cycle in which the mind "accepts information as it becomes available at sensory surfaces and is changed by that information; it directs movements and exploratory activities that make more information available, by which it is further modified" (Neisser, 1976, p. 55). In this view, reality is cognitively constructed on the basis of the mutual interplay between the mind and the "experiential data" it assimiliates into the existing network of schemata. For Piaget, too, understanding is a dynamical network process: in order to include phenomena it is necessary "to include them in a network of relations becoming increasingly remote from appearance and to insert appearance in a new reality elaborated by reason." (Piaget, 1954, p. 381). The emphasis on networks and their dynamical self-modification due to its accommodation to new experience makes it clear that constructivist accounts of reality construction and cognition cannot be based on static propositional knowledge but must – methodologically – include insights from the newly emerging network sciences (cf. chapter on "Towards a formal interpretation of radical constructivism" below).

Unfortunately, epistemologically speaking Piaget et al.'s cognitive constructivism is flawed as there are two problems with this (as von Glasersfeld calls it, "trivial constructivist") view. First, there is no way to prove the validity of our knowledge, i.e., whether we have constructed "an adequate representation of reality" (Piaget, 1954, p. 381), since all the means at our disposal to verify our knowledge are the very senses through which we gathered the empirical evidences for this knowledge in the first place (see von Glasersfeld's skeptical attitude below in the section, "Radical constructivism"). Therefore, we cannot be sure of the correctness of our beliefs about a mind-independent reality and consequently of our scientific theories. Secondly, notions such as observation, cognitive apparatus, and nervous signals are constructions, too. The situation is reminiscent of Wittgenstein's ladder: having it climbed we are obliged to throw it away, and with it any claim to be able to know the "true nature" of reality – whatever this denotes. Von Glasersfeld was very clear about this when he wrote that "those who merely speak of the construction of knowledge, but do not explicitly give up the notion that our conceptual constructions can or should in some way represent an independent, objective reality, are still caught up in the traditional theory of knowledge." (Glasersfeld, 1991b, p. 16). Therefore, any form of constructivism must necessarily strive for completeness with its claim to uncompromisingly apply the idea of constructions on all levels. Two such non-dualistic approaches, which try to dispense with the dichotomy between mind and reality, are reviewed next.

Non-dualistic Approaches: Theory of Cognitive Operators

While the dualistic position is intuitively easily comprehensible, it has been rejected by several authors on philosophical grounds. For example, it can be argued that the relationship between environment and sense organs resembles the constellation in scientific experiments where measurement devices gauge reality. While many intuitively assume that reality influences the reading on the measurement device, some assume that perceived patterns and regularities have to be regarded as invariants of inborn cognitive operators. This idea goes back at least to Immanuel Kant, who referred to the notions of space and time as being a priori and as indispensable for understanding raw sensory experience.

Kant's notion of the "Copernican Turn" addresses the question of how reality and knowledge relate to each other: does reality inform our knowledge or does knowledge inform reality? Konrad Lorenz's (1941) intention was to re-interpret Kant's a priori categories as *phylogenetically* acquired categories "fixed prior to individual experience." What is of interest to the constructivist, however, is the question how these (and other) categories occur in *individual* ontogeny. Olaf Diettrich's (2001) "theory of cognitive operators" is an attempt to shed light on this question. It starts with the claim that properties of scientific entities are defined as invariants of measurement devices. This means that certain notions in science cannot be defined independent of the measurement. Diettrich claims that this does not only apply to science but also to observations in general, i.e., they have to be considered the results of measurements of cognitive operators in the mind. In other words, observations are invariants of these cognitive measuring devices. Observations are therefore human specific in the sense that they do not represent independent ontological elements of an outside reality. Diettrich stresses that reality is simple and predictable "if and only if the way we describe the world is closely related to the way we act upon the world" (Diettrich, 2001, p. 277) and has nothing to do whether the world is simple and predictable *in itself*. For Diettrich this is even a way to account for the "unreasonable effectiveness of mathematics" because that which is perceived and the mathematical structures are invariants of the same mental operators (*ibid.*, p. 296), which means that any mind-independent reality is not part of the equation.

With regard to Kuhnian philosophy of science, we can identify two important implications and parallels:

1. Truth as the correlation between knowledge and reality cannot serve as an arbiter in assessing the former. According to Diettrich, this applies to the evaluation of scientific theories too. Lacking this arbiter, theory building must look for alternative criteria such as consistency. Similarly, Kuhn rejected the correspondence theory of truth. He claimed that scientific progress should not be equated to closer approximations to truth but to the weaker criterion of utility, i.e., that of puzzle-solving efficacy: "We… have to relinquish the notion… that changes of paradigm carry scientists… closer to the truth." (Kuhn, 1970, p. 170). Rather, scientists appeal to shared standards such as accuracy, consistency, scope, simplicity and fruitfulness (cf. Tsou, 2006, p. 216), or simply accuracy of predictability: "Theories are… to be evaluated in terms of such considerations as their effectiveness in matching predictions with the results of experiment and observation." (Kuhn, 2000, p. 209). This instrumentalist view in science is shared by Kuhn with von Glasersfeld (as pointed out in the next section).
2. The fact that different sets of cognitive operators bring forth different cognitive phenotypes makes it virtually impossible to communicate with beings equipped with those alternative operators. Variations among cognitive operators do not need to have biological roots – Diettrich mentioned the immense difficulties with alien races from outer space – but can also result from the ongoing ontogenesis of individuals resulting in a profound incommensurability among human beings. Similar to what Kuhn wrote, such incommensurabilities do not necessarily mean a gradation in consistency or efficiency among these sets of cognitive operators.

Non-dualistic Approaches: Radical Constructivism

A wide-spread objection to dualistic versions of constructivism starts with the question: Can the structures of the real world be compared with mental ones at all? Some are skeptical about this possibility as this would require a comparison independent of those senses through which the mental structures were constructed in the first place. For example, Ernst von Glasersfeld (1995), following this tradition of skepticism, criticized the realists' assumption that we can determine the truth content of our knowledge by comparing it with reality, as Wittgenstein's "in order to tell whether a picture is true or false we must compare it with reality" (Wittgenstein, 1922, Sect. 2.223) suggests. Von Glasersfeld vehemently objects to the possibility of comparing one's knowledge with reality as we would need to be able to stay outside our own knowledge (i.e., we would need to know that which we perceive before we perceive it).

Instead, von Glasersfeld proposes a different route to knowledge. Starting with defining reality as "a black box with which we can deal remarkably well" (Glasersfeld, 2007, p. 81) he maintained that knowledge about this black box is the result of trying to find regularities in its input–output behavior. So, in RC, the construction of reality is based on the recurrent extraction of repetitive patterns from the flow of experience. It is crucial to note that these regularities are not presented in the flow of experiences as such. Rather, repetition is created by the cognitive operation of assimilation (Glasersfeld, 2001, p. 42), i.e., the act of abstracting from details. The constructive aspect of knowledge was molded into the first principle of RC: "1. Knowledge is not passively received either through the senses or by way of communication, [rather] knowledge is actively built up by the cognizing subject." (Glasersfeld, 1991a, p. 233).

It is furthermore crucial to note that our apparent success in recognizing patterns says nothing about any ontological existence of these patterns. So even "if we posit causes for the sense data [...], this does in no way entail that these causes exist in the spatio-temporal or other relational structures into which we have coordinated them." (Glasersfeld, 2007, p. 82). Therefore knowledge can only fit reality, similarly to a picklock fitting a lock, rather than match reality in the sense of an iconic representation of it. The fit describes the capacity of the key rather than the property of the lock. His second principle reads thus as follows: "The function of cognition is adaptive, in the biological sense of the term, tending towards fit or viability; cognition serves the subject's organization of the experiential world, not the discovery of an objective ontological reality." (Glasersfeld, 1991a, p. 233).

Influenced by the writings of Piaget, for von Glasersfeld the construction of reality includes four steps (e.g., Glasersfeld, 1982): (1) Through (tedious) repetition, the subject organizes her sensorimotor experience into operational structures, i.e., three-part action schemata, which consist of an perceptual compound, an action, and an expected result.[11] The subject retains those schemata that lead to an equilibrium in the face of perturbations. Later on, perceptual compounds are externalized as objects. (2) By including sensory material from various source, perceptual compounds become multi-modal and the externalized objects "more real." (3) As soon as the subject uses schemata to construct other schemata, she is capable of "reflective abstraction" that allows the abstraction from these organizational structures of the sensorimotor content that led to their construction in the first place. It also makes it possible to reuse organizational structures in different contexts. (4) The final level of reality is reached as soon as the subject constructs herself as experiencer among others, which adds further ways of validating schemata, i.e., agreement and confirmation by others.

The continuing viability of operational structures in the face of further experiences gives rise to the belief in the mind-independent "existence" of the regularities they express and, consequently, of the objects into which they are projected. Therefore knowledge about the world is and can only be knowledge about our own experiential reality rather than an ontological mind-independent reality. Reality is a network of concepts that so far have proven to be viable in the light of the experiences of the subject because they have repeatedly served as a tool for successfully surmounting problems of life or for assimilating complexes of experiences (Glasersfeld, 1997, p. 47).

---

[11] The last part distinguishes these action schemata from stimulus-response schemata used for example by behaviorists.

These considerations led von Glasersfeld to the formulation of what he called "radical constructivism." It is crucial to understand "radical" not as "extremist" but rather as "thoroughly consistent" so that it avoids the criticism of being self-contradictory and circular. The radical constructivist paradigm rests on four pillars (Schmidt, 1993, p. 329): skepticism (as sketched above), instrumentalism (as "truth" is replaced by "viability," i.e., the utility of theories, mental models, and ideas), Kantian philosophy, and developmental psychology (in the tradition of Jean Piaget). Von Glasersfeld himself characterized RC as an "unconventional" (some say, "unintuitive") approach to the problem of knowledge and knowing because it maintains that "knowledge, no matter how it is defined, is in the heads of persons, and that the thinking subject has no alternative but to construct what he or she knows on the basis of his or her own experience." (Glasersfeld, 1995, p. 1). Similar to Mach's position, experience constitutes the only world we consciously live in. Experience, according to von Glasersfeld, "can be sorted into many kinds, such as things, self, others, and so on. But all kinds of experience are essentially subjective, and though I may find reasons to believe that my experience may not be unlike yours, I have no way of knowing that it is the same. The experience and interpretation of language are no exception." (Glasersfeld, 1995, p. 1). It is the insight that there cannot be any exceptions that makes von Glasersfeld constructivism "radical," and so avoiding the criticism of being self-contradictory (see also next section). Importantly, this is not to say that RC is "more true" than other philosophies. To claim that something is more true than something else means to neglect a basic principles of RC, namely the self-applicability of its findings. As von Glasersfeld (1991b, p. 13) puts it, "I would be contradicting one of the basic principles of my own theory if I were to claim that the constructivist approach proved a true description of an objective state of affairs…[I]ts values will depend mainly on its usefulness in our experimental world."

Towards a Formal Interpretation of Radical Constructivism

In the early 1980s, von Glasersfeld's approach was widely popularized in Watzlawick's book *The Invented Reality* (1984). Later, a group of German communication scientists around Siegfried J. Schmidt started to propagate a new amalgamated paradigm, developed from the philosophies of von Glasersfeld, von Foerster and Maturana and Varela, under the umbrella term of "radical constructivism" (Schmidt, 1987; cf. Scholl, 2010), which ultimately confused both proponents and opponents. In Riegler (2001b), I presented a reconciliatory interpretation of RC which also avoids criticism such as biological/psychological self-refutation. The interpretation is based on four formal principles.

(P1) describes RC as an approach focusing on organizationally closed systems, i.e., systems which can be characterized as networks with hierarchically arranged components of short characteristic path lengths between them and a high clustering coefficient. (P2) defines the agnostic perspective with regard to an "external/objective" reality: whether or not the world is amorphous is left to speculation. There is no need to assume external order parameters since order arises from within the system. (P3) emphasizes the circularity of the trains of thoughts, i.e., that experiential components are linked with each other thus forming a network of relations. Trains of thoughts can be described in terms of state cycles in the network. (P4) demands that reality construction in such systems is severely constrained rather than arbitrary. The limitation arises from inherent properties of the hierarchical network.

The last principle deserves a more in-depth discussion as it, on the one hand, responds to the criticism that RC is a self-contradictory doctrine and, on the other hand, serves a main goal of this paper, i.e., to interpret Kuhn's central notions of mental sets (which characterize the normal scientist) and incommensurability (arising from the operational closure of cognitive systems) from the perspective of constructivism.

Since in the "stream of consciousness" (James, 1890) sensations and experiences are made and linked to each other over the course of time, cognition is a *historistic* process and construction complexes are historistic collections in which experiences are positively or negatively related with each other. Consequently they form a *network of hierarchical interdependencies* (Riegler, 2001b). The components of such a network become mutually dependent: removing one component may change the context of another component. In this sense they impose constraints on each other. Consider the following analogy: by car, you can reach only those points that are connected to the road network; on foot, all the points (such as mountain peaks) in between can be accessed, but only if they are within walking distance. The basic component in both cases is the means of transportation that restricts the availability of reachable destinations. Free arbitrariness is not possible since different means of transportation have different degrees of flexibility and speed. This analogy may also be applied to the notion of incommensurability, which translates to the situation in which scientists from different paradigms are located at mutually unreachable positions because they use different (cognitive) vehicles. Similarly, the construction network envisaged by RC is also necessarily non-arbitrary. It follows the canalizations that result from the mutual interdependencies among constructive elements. Once a certain path is taken relating elements to each other in a particular manner, existing constructions are used as building blocks for further constructions. In other words, the dynamics of reality construction resemble a ratchet (Riegler, 2001a): the constructions run into "canalizations" (i.e., the radical reduction of freedom in future developments) or "constructivist entrenchments" which result from the requirement of assembling and fitting experiences.

Constructivist Entrenchment

In *Structure*, Kuhn argued that normal science was puzzle solving, i.e., that it is concerned with solving tricky problems within the confines of the currently valid paradigm. It is in particular this relationship between scientists and their paradigm which is interesting from a constructivist perspective. Kuhn (1970, p. 46) wrote: "Scientists work from models acquired through education and through subsequent exposure to the literature often without quite knowing or needing to know what characteristics have given these models the status of community paradigms." And it is such continuous repetitions of a particular methodical schema that inevitably define the scientists' capability (or "mental set") for problem solving. Long before Kuhn, José Ortega y Gasset (1929) pointed out that scientists work with available methods like a machine. To achieve a wealth of results, scientist can go about in a very pragmatic-instrumentalist fashion, which makes it superfluous to have a clear concept of the meaning and the foundations of these methods. It is the way in which many average savants contribute to the progress of science as they are locked in their laboratories. Amusingly, Ortega compares their situation to that of a donkey in its whim. And Wolfgang Stegmüller (1971) ridicules this dogmatism even more. He wrote that we should feel sorry for the average scientists since they are uncritical, narrow-minded dogmatists who want to educate students in the same way. This tight link between paradigm and dogmatism was characterized by Josef Mitterer (1994) with his neologism "paradogma." For others, such as Karl Popper, Kuhn's characterization went too far and he accused him of exaggeration. He wrote that "In my view the 'normal' scientist, as Kuhn describes him, is a person one ought to be sorry for" (Popper, 1970, p. 52).

Can this dispute be solved from the constructivist perspective?

As pointed out, in RC the hierarchical networks of relational dependencies lead to entrenchment. The older a relational element (sensation, experience) is, the deeper is its embeddedness. In other words, such elements are more difficult to remove than more recently added elements. They constitute what we refer to as "habits of thought" in common language. In philosophy, Wittgenstein (1953) associated them with the fact that humans are "profoundly enmeshed in philosophical – i.e., grammatical confusions", out of which they cannot be easily pulled. In cognitive psychology they are known as "set-effects" in (logical, mathematical, etc.) problem-solving tasks (Duncker, 1945; Luchins, 1942; see below). And in psychotherapy they turn out to lay at the root of social and family problems, which need "habitual reframing": "To reframe [...] means to change the conceptual and/or emotional setting or viewpoint in relation to which a situation is experienced and to place it in another frame which fits the 'facts' of the same concrete situation equally well or even better and thereby changes its entire meaning" (Watzlawick, Weakland, & Fisch, 1974, p. 95).

Levels of Constructivist Entrenchment

While both problem solving and social habits can be changed, the question arises of whether there are elements in the network of experiences which defy access and hence change?[12] Can we escape that which Paul Feyerabend addressed as stereotypical

[12] This is not to say that mentals sets *have* to be changed in an everyday context as they play a crucial role, e.g., in personal identity.

research schemata and their possible origins and enjoy the fruits of the divergent method as described in the introduction? He localized their roots in the cognitive development starting in early childhood: "From our very early days we learn to react to situations with the appropriate responses, linguistic or otherwise. The teaching procedures both shape the 'appearance', or 'phenomenon', and establish a firm connection with words, so that finally the phenomena seem to speak for themselves..." (Feyerabend, 1975, p. 72). In other words, he argued that starting from early on people (and scientists are no exception here) are subjected to an education that distinctly outlines both the way they have to view the world and the way they have to act in the world. Alternatives to these entrenchments are suppressed or referred to the realm of fantasy.

Annette Karmiloff-Smith (1992) identifies three levels in her theory of "representational redescription." The content of level E3 is consciously accessible and can be described verbally, such as the intellectual problems we find in science. E2 eludes verbal description. Here, expertise (cf. the difficulties in transferring the "knowledge" of a human expert to an expert system, Dreyfus & Dreyfus, 1988) and common sense (Varela, 1988) are located. Finally, E1 can be neither consciously accessed nor verbally described. Could our belief in the mind-independence of objects be the expression of the fact that we fail to access consciously their construction process?[13]

As pointed out above, von Foerster (2003b) defined objects as the result of sensorimotor attractors. This suggests that objects which were constructed in the network of experiences at a very early stage elude conscious access and hence the modification which we can apply to other more recent constructions such as social and intellectual entities. On Karmiloff-Smith's scale they are located on E1 and are inaccessible to the consciousness: this is commonly referred to as "childhood amnesia."[14] It is striking that once individuals start reasoning in language, they cannot reach older non-verbal (unconscious) sensorimotor sensations, as the experiment of Gabrielle Simcock and Harlene Hayne (2002) suggests. According to their findings, very young children's verbal descriptions of an event are "frozen in time, reflecting their verbal skill at the time of encoding, rather than at the time of the test." Usually we hold an outside reality responsible for the existence of "things." However, the different degrees of changeability of constructed mental complexes suggest that our belief in the existence of "things" results from constructing mutual relationships among sensations. Since this construction process eludes conscious access we are led to

[13] Von Glasersfeld admitted that we "build that world for the most part unaware, simply because we do not know how we do it" (Glasersfeld, 1984, p. 17). However, we claimed that this ignorance was quite unnecessary because "the operations by means of which we assemble our experiential world can be explored" (*ibid.*). He referred to Silvio Ceccato's notion of "consapevolezza operativa", i.e., to become aware of one's own mental operations, which can lead to different and perhaps better constructions.

[14] This notion refers to the phenomenon that we forget our experiences from our earliest childhood until the age of three or four.

assume that if humans cannot translate their preverbal (unconscious) memories into language, basic sensorimotor constructs made in that early period cannot be reasoned about and be claimed to be part of a mind-independent reality (Riegler, 2005a).

Orthogenesis of the Construction Process

The reality construction process has at least two *endogenous* properties, as discussed in the following section. It was perhaps Jean Piaget's most important contribution to psychology to show that as knowledge can only be acquired incrementally and that the individual builds on previous steps to reach the highest level of cognition, it therefore takes the shape of a hierarchical organization. Analogously, it can be claimed that the purpose of paradigms is to secure acquired scientific knowledge and to provide a base for extending developments in this knowledge, thus forming a hierarchical structure.

Herbert Simon (1969) provided a good argument for hierarchical organization being important. There are two insights to be drawn from his well-known watchmaker analogy, which shows that the hierarchical-modular way of constructing watches is superior to the non-modular way.[15]

1. As Simon (1969, p. 195) put it, "hierarchic systems will evolve far more quickly than nonhierarchical systems of comparable size." Kuhn has always emphasized that normal science is the most productive phase while in the periods of pre-science progress is small: one of the major advantages of convergent thinking as described in the introduction. Simon's argument confirms the idea that only in a system where you can firmly build on the previous insights of others can you increase your knowledge.
2. The hierarchical organization introduces a developmental direction that "is provided [...] by the stability of the complex forms, once these come into existence" (Simon, 1969, p. 203). Also for Piaget there is "orthogenesis" (Piaget, 1971), a directional tendency, in epistemic development toward epistemic subject and epistemic object (as well as in biological evolution towards an ideal equilibrium between organism and environment; Kitchener, 1987).

However, Kuhn's orthogenesis introduces a burden with respect to the number of available options which makes people incapable of (experientially and conceptually) perceiving concrete and abstract entities outside the habitual context. This phenomenon

[15] His watchmakers are building clocks consisting of n parts. Each time their work is interrupted at random moments of probability p an unfinished clock falls apart. For the watchmaker who tries to assemble each watch in one go, the probability of actually finishing one is $p_1=(1-p)^n$. However, for a watchmaker whose watch consists of stable subassemblies of k parts each, the probability of completing a watch is $p_2=(1-p)^k$. For example, for $n=1{,}000$ parts and probability $p=0.01$, the second watchmaker will produce watches 3,775 times faster than his colleague.

of "mind set" or "mechanization of thoughts" was described in the psychological literature long before Kuhn used the expression. Duncker (1945), for example, claimed that our thinking is canalized (or fixed) with respect to the way we have learned to deal with things.[16] In his experiments, subjects failed to use a box filled with tacks as an unusual yet effective support for candles by emptying it and tacking it to the wall. Luchins (1942) reported that subjects easily got stuck in an initially learned action procedure of applying a certain lengthy sequence of pouring water from one jug into another to measure out specific quantities of water. They failed to notice that for new tasks a simpler procedure would have yielded the same result. While such "mental sets" clearly reduce the cognitive efforts of learning something new, they prevent new, innovate ways from being found. As with the adage which says, "It ain't broke so don't fix it," their mind was set to the previously successful strategy.

In this sense, progress in terms of knowledge can be considered a trade-off between stability and acceleration, on the one hand, and becoming burdened, on the other. The long-term effects of this interplay can be seen in evolution, most notably in what happened after the "Cambrian explosion" some 600 million years ago (Gould, 1989). At that time a vast variety of new forms appeared in the animal kingdom but after a while development slowed down because the number of hierarchical interdependencies constantly increased up to the point at which no significant innovations were possible anymore.

The consequences are clear. During academic education scientists are subject to courses and seminars in which they are drilled in a certain way of solving problems. On the one hand, this consolidates paradigmatic thinking, which makes it possible to assign cognitive resources to new problems as entire branches in the search space can be pruned, thus leaving more time to concentrate on the unknown part. However, on the other hand, it disregards a (probably vast) amount of (probably better) alternatives.

In the context of Kuhnian revolutions, we can observe the same laws: each time a new discipline appears with a different set of paradigms, it has to start from scratch but is free to explore a much richer set of challenges and questions than later on when this pre-scientific phase has turned into the period of normal science with all its limiting entrenchments.

All these observations strongly suggest that the formation of scientific paradigms is an inherent property of the process of scientific knowledge construction.

Conclusion

We have seen that many notions in Kuhn's theory can be explained in constructivist terms. Two such terms, "mental sets" (as a crucial characteristic of the normal scientist working according to a paradigm) and "incommensurability" are at the focus of

[16] This is reminiscent of Piaget's claim that "all knowledge is tied to action and knowing an object or an event is to use it by assimilating it to an action scheme" (quoted in Glasersfeld, 1982, p. 613).

this chapter. After reviewing some exemplary cases of constructivism ranging from Mach's phenomenological constructivism to dualistic and non-dualistic versions we finally arrived at radical constructivism. One of the four defining core elements of this paradigm (i.e., that reality constructions are entrenched due to the hierarchical organization of the constructed knowledge) is of particular interest not only because it saves RC from being a self-contradictory biologism but also because it explains the advantages of paradigms and accounts for Kuhn's notion of incommensurability (being the unavoidable result of the different developmental paths which arise in entrenchment) as well as orthogenesis in Piaget's philosophy of science.

Acknowledgments I wish to express my gratitude to Marco Bettoni, Jeremy Burman and Armin Scholl for their helpful comments on a previous draft version of this article. Furthermore, I acknowledge the financial support from the Research Foundation - Flanders (FWO).

References

Bird, A. (2009). Thomas Kuhn. In E. N. Zalta (Ed.), *The Stanford encyclopedia of philosophy*. Available at http://plato.stanford.edu/archives/fall2009/entries/thomas-kuhn.

Blackmore, J. T. (1972). *Ernst Mach: His work, life, and influence*. Berkeley: University of California.

Burman, J. T. (2007). Piaget no "remedy" for Kuhn, but the two should be read together: Comment on Tsou's 'Piaget vs. Kuhn on Scientific Progress'. *Theory & Psychology, 17*, 721–732.

Burman, J. T. (2008). Experimenting in relation to Piaget: Education is a chaperoned process of adaptation. *Perspectives on Science, 16*(2), 160–195.

Diettrich, O. (2001). A physical approach to the construction of cognition and to cognitive evolution. *Foundations of Science, 6*(4), 273–341. Available at http://www.univie.ac.at/constructivism/pub/fos/pdf/diettrich.pdf.

Dreyfus, H. L., & Dreyfus, S. E. (1988). Making a mind versus modeling the brain: Artificial intelligence back at a branch-point. *Artificial Intelligence, 117*, 309–33.

Duncker, K. (1945). On problem solving. *Psychological Monographs, 58*(270), 1–112 (German original published in 1935).

Feyerabend, P. K. (1975). *Against method: Outline of an anarchistic theory of knowledge*. London: NLB.

Fleck, L. (1935). *Entstehung und Entwicklung einer wissenschaftlichen Tatsache*. Basel: Benno Swabe (English translation: Fleck, L. (1979) *Genesis and development of a scientific fact* (Edited by T. J. Trenn & R. K. Merton). Chicago: University of Chicago Press).

Foerster, H. von (2003a). On constructing a reality. In H. von Foerster, *Understanding understanding* (pp. 211–228) New York: Springer. (Originally published in F. E. Preiser (Ed.) (1973) *Environmental design research*, Vol. 2. Stroudberg: Dowden, Hutchinson & Ross, pp. 35–46).

Foerster, H. von (2003b). Objects: Tokens for (eigen-) behaviors. In H. von Foerster, *Understanding understanding* (pp. 261–271) New York: Springer. (Originally published in 1976).

Froese, T., Gould, C., & Barrett, A. (2011). Re-viewing from within: A commentary on first- and second-person methods in the science of consciousness. *Constructivist Foundations, 6*(2), 254–269. Available at http://www.univie.ac.at/constructivism/journal/6/2/254.froese).

Glasersfeld, E. von (1974). Piaget and the radical constructivist epistemology. In C. D. Smock & E. von Glasersfeld (Eds.) *Epistemology and education* (pp. 1–24) Athens: Follow Through Publications. (Reprinted in: Glasersfeld, E. von (2007) *Key works in radical constructivism* (edited by Marie Larochelle). Rotterdam: Sense, pp. 73–87. (Page numbers in the text refer to the reprint). Available at http://www.vonglasersfeld.com/034).

Glasersfeld, E. von (1982). An interpretation of Piaget's constructivism. *Revue Internationale de Philosophie 36*(4): 612–635. Available at http://www.vonglasersfeld.com/077.

Glasersfeld, E. von (1984). An introduction to radical constructivism. In P. Watzlawick (Ed.), *The invented reality: How do we know what we believe we know? Contributions to constructivism*. (pp. 17–40) New York: Norton. (Originally publish in German in 1981. Available at http://www.vonglasersfeld.com/070.1).

Glasersfeld, E. von (1991a). An exposition of constructivism: Why some like it radical. In G. J. Klir (Ed.) *Facets of system science*. (pp. 229–238) New York: Plenum. Available at http://www.vonglasersfeld.com/127.

Glasersfeld, E. von (1991b). Knowing without metaphysics: Aspects of the radical constructivist position. In F. Steier (Ed.) *Research and reflexivity (Inquiries into social construction)* (pp. 12–29). London: Sage. Available at http://www.vonglasersfeld.com/132.

Glasersfeld (1995). *Radical constructivism: A way of knowing and learning*. London: Falmer Press.

Glasersfeld, E. von (1997). Fiktion und Realität aus der Perspektive des radikalen Konstruktivismus. In E. von Glasersfeld, *Wege des Wissens* (pp. 45–61). Heidelberg: Carl Auer. Available at http://www.vonglasersfeld.com/133.

Glasersfeld, E. von (2001). The radical constructivist view of science. *Foundations of Science* 6 (1–3): 31–43. Available at http://www.univie.ac.at/constructivism/pub/fos/pdf/glasersfeld.pdf.

Glasersfeld, E. von (2007). Aspects of constructivism. Vico, Berkeley, Piaget. In von E. Glasersfeld (Ed.), *Key works in radical constructivism* (pp. 91–99). Rotterdam: Sense. (Originally published in Italian as: Glasersfeld, E. von (1992) Aspetti del costruttivismo: Vico, Berkeley, Piaget. In M. Ceruti (Ed.) *Evoluzione e conoscenza* (pp. 421–432). Bergamo: Lubrina. Available at: http://www.vonglasersfeld.com/139.2).

Glasersfeld, E. von (2010). *Partial memories: Sketches from an improbable life*. London: Imprint Academic.

Gould, S. J. (1989). *Wonderful life: The Burgess Shale and the nature of history*. New York: W.W. Norton & Company.

James, W. (1890). *The principles of psychology* (Vol 1). New York: Holt. (Reprinted in 1950 by Dover, New York).

Karmiloff-Smith, A. (1992). *Beyond modularity: A developmental perspective on cognitive science*. Cambridge: MIT Press.

Kenny V. (2010). Exile on mainstream. Constructivism in psychotherapy and suggestions from a Kellian perspective. *Constructivist Foundations, 6*(1), 65–76. Available at http://www.univie.ac.at/constructivism/journal/6/1/065.kenny.

Kitchener, R. F. (1985). Genetic epistemology, history of science and genetic psychology. *Synthese, 65*, 3–31.

Kitchener, R. F. (1987). Genetic epistemology, equilibration and the rationality of scientific change. *Studies In History and Philosophy of Science Part A, 18*(3), 339–366.

Kuhn, T.S. (1959). The essential tension. Tradition and innovation in scientific research. In C.W. Taylor (Ed.) *The third (1959) University of Utah research conference on the identification of scientific talent* (pp. 162–174). Salt Lake City: University of Utah Press. (Reprinted in: Kuhn, T. S. (1977). *The essential tension* (pp. 225–239). Chicago: University of Chicago Press).

Kuhn, T. S. (1970). *The structure of scientific revolutions* (2nd ed.). Chicago: University of Chicago Press.

Kuhn, T. S. (2000). *The road since Structure: Philosophical essays, 1970–1993, with an autobiographical interview* (Edited by J. Conant & J. Haugeland). Chicago: University of Chicago Press.

Llinás, R. R. (2001). *I of the vortex*. Cambridge: MIT Press.

Lorenz, K. (1941). Kant's Lehre vom Apriorischen im Lichte gegenwärtiger Biologie. *Blätter für Deutsche Philosophie* 15: 94–125. English translation: Lorenz, K. (1982). Kant's doctrine of the a priori in the light of contemporary biology. In H. C. Plotkin (Ed.), *Learning, development and culture* (pp. 121–143). Chichester: John Wiley.

Luchins, A. S. (1942). Mechanization in problem solving. *Psychological Monographs, 54*/248.

Mach, E. (1959). *The analysis of sensations* (3rd Ed.) (C. M. Williams & S. Waterlow, Trans.). New York: Dover Edition. (The German first edition was published in 1886).

Mach, E. (1960). *The science of mechanics: A critical and historical account of its development* (6th ed.). Chicago: Open Court. (Originally published as: Mach E. (1912) *Die Mechanik in ihrer Entwicklung, historisch-kritisch dargestellt* (7th ed.). Leipzig: Brockhaus).

Mach, E. (1970). The guiding principles of my scientific theory of knowledge and its reception by my contemporaries. In S. Toulmin (Ed.) *Physical reality* (pp. 44–53). New York: Harper. (German original published in 1910).

Mach, E. (1992). Ernst Mach. In J. Blackmore (Ed.) *Ernst Mach – A deeper look*. Dordrecht: Kluwer Academic Publishers. (Originally published in 1913).

Maturana, H. R. (1978). Biology of language: The epistemology of reality. In G. A. Miller & E. Lenneberg (Eds.) *Psychology and biology of language and thought: Essays in honor of Eric Lenneberg* (pp. 27–63). New York: Academic Press. Available at http://www.enolagaia.com/M78BoL.html.

Maturana, H. R. & Varela, F. J. (1980). *Autopoiesis and Cognition: The realization of the living*. Boston Studies in the Philosophy of Sciences (Vol. 42). Boston: Reidel.

Mitterer, J. (1994). *Das Jenseits der Philosophie*. Vienna: Edition Passagen.

Müller, K. H. (2010). The radical constructivist movement and its network formations. *Constructivist Foundations, 6*(1), 31–39. Available at http://www.univie.ac.at/constructivism/journal/6/1/031.mueller.

Neisser, U. (1976). *Cognition and reality*. San Francisco: W. H. Freeman.

Oberheim, E. & Hoyningen-Huene, P. (2009). The incommensurability of scientific theories. In E. N. Zalta (Ed.) *The Stanford encyclopedia of philosophy*. Available at http://plato.stanford.edu/entries/incommensurability/.

Ortega y Gasset, J. (1929). *La rebelión de las masas*. Madrid: Revista de Occidente. (English translation: Ortega y Gasset, J. (1994). *The revolt of the masses*. New York: W. W. Norton).

Piaget, J. (1954). *The construction of reality in the child*. New York: Ballantine. (French original published as: Piaget J. (1937). *La construction du réel chez l'enfant*. Neuchâtel: Délachaux & Niestlé).

Piaget, J. (1970). *Genetic epistemology* (E. Duckworth, Trans). New York: Columbia University Press.

Piaget, J. (1971). Biology and knowledge: An essay on the relations between organic regulations and cognitive processes (B. Walsh Trans.). Edinburgh: Edinburgh University Press. (Originally published in 1967).

Popper, K. R. (1970). Normal science and its dangers. In I. Lakatos & A. Musgrave (Eds.), *Criticism and the growth of knowledge* (pp. 51–58). New York: Cambridge University Press.

Popper, K. R. (1979). *Objective knowledge: An evolutionary approach*. 5th revised edition. Oxford: Clarendon Press.

Riegler, A. (1994). Constructivist artificial life: The constructivist-anticipatory principle and functional coupling. In J. Hopf (Ed.), *Proceedings of the 18th German conference on artificial intelligence (KI-94). Workshop on genetic algorithms within the framework of evolutionary computation*. (pp. 73–83). Max-Planck-Institute Report No. MPI-I-94–241.

Riegler, A. (2001a). The cognitive ratchet. The ratchet effect as a fundamental principle in evolution and cognition. *Cybernetics and Systems*, 32, 411–427. Available at http://www.univie.ac.at/constructivism/riegler/14.

Riegler, A. (2001b). Towards a radical constructivist understanding of science. *Foundations of Science*, special issue on "The impact of radical constructivism on science", 6(1–3), 1–30. Available at http://www.univie.ac.at/constructivism/riegler/20.

Riegler, A. (2005a). Constructive memory. *Kybernetes*, 34(1/2), 89–104. Available at http://www.univie.ac.at/constructivism/riegler/39.

Riegler, A. (2005b). Like cats and dogs: Radical constructivism and evolutionary epistemology. In: *Evolutionary epistemology, language and culture: A non-adaptationist, systems theoretical approach* (pp. 47–65). Dordrecht: Springer. Available at http://www.univie.ac.at/constructivism/riegler/36.

Riegler, A. (2006). Is a closed-loop discovery system feasible? In L. Magnani (Ed.), Computing and philosophy (pp. 141–149). Pavia: Associated International Academic Publishers. Available at http://www.univie.ac.at/constructivism/riegler/41.

Riegler, A. (2007). The radical constructivist dynamics of cognition. In: B. Wallace (Ed.) *The mind, the body and the world: Psychology after cognitivism?* (pp. 91–115). London: Imprint. Available at http://www.univie.ac.at/constructivism/riegler/44.

Riegler, A. & Bunnell, P. (eds.) (2011) The work of Humberto Maturana and its application across the sciences. Special Issue, *Constructivist Foundations, 6* (3). Available at http://www.univie.ac.at/constructivism/journal/6/3.

Schmidt, S. J. (1987). *Der Diskurs des Radikalen Konstruktivismus*. Frankfurt: Suhrkamp.

Schmidt, S. J. (1993). Zur Ideengeschichte des Radikalen Konstruktivismus. In E. Florey & O. Breidbach (Eds.), *Das Gehirn – Organ der Seele? Zur Ideengeschichte der Neurobiologie* (pp. 327–349). Berlin: Akademie-Verlag.

Scholl, A. (2010). Radical constructivism in communication science. *Constructivist Foundations, 6*(1), 51–57. Available at http://www.univie.ac.at/constructivism/journal/6/1/051.scholl.

Sillito, A. M., & Jones, H. E. (2002). Corticothalamic interactions in the transfer of visual information. *Philosophical Transactions of the Royal Society London B Biological Sciences, 357*, 1739–1752.

Simcock, G., & Hayne, H. (2002). Breaking the barrier? Children fail to translate their preverbal memories into language. *Psychological Science, 13*(3), 225–231.

Simon, H. A. (1969). *The sciences of the artificial*. Cambridge: MIT Press.

Stegmüller, W. (1971). *Hauptströmungen der Gegenwartsphilosophie, Band II*. Stuttgart: Kröner.

Tsou, J. Y. (2006). Genetic epistemology and Piaget's philosophy of science. Piaget vs. Kuhn on scientific progress. *Theory & Psychology, 16*, 203–224.

Varela, F. J. (1988). *Cognitive science: A cartography of current ideas*. Paris: CREA, Ecole Polytechnique. Republished in French as: Varela, F. J. (1989) Connaître. Les sciences cognitives. Tendances et perspectives. Paris: Seuil. In German: F. J. Varela (1990) Kognitionwissenschaft – Kognitionstechnik. Frankfurt: Suhrkamp.

Varela, F. J. (1996). Neurophenomenology: A methodological remedy for the hard problem. *Journal of Consciousness Studies, 3*(4), 330–349.

Watzlawick, P. (ed.) The invented reality: How do we know what we believe we know? Contributions to constructivism. New York: W. W. Norton.

Watzlawick, P., Weakland, J., & Fisch, R. (1974). *Change: Principles of problem formation and problem resolution*. New York: W. W. Norton.

Wittgenstein, L. (1922). *Tractatus Logico-philosophicus*. London: Routledge.

Wittgenstein, L. (1953). *Philosophical investigations*. Oxford: Basil Blackwell.

Chapter 14
Materialism

Luciano L'Abate

> I…view matter as a materialistic construct that has no application in the real world, though it may be useful for certain approaches to science. In the real world, matter, spirit, consciousness, awareness, nous, and mind are all one integrated whole. Matter never exists without some level of awareness, consciousness, or yearning. All matter yearns for greater levels of connectedness, freedom, awareness, consciousness, love, generosity, cooperation, and beauty, and what moves evolution is this yearning of all being to be fully actualized (Michael Lerner, *Tikkum*, November/December 2010, p. 34).

Materialism, according to *APA Dictionary of Psychology*, is the philosophical position that everything, including mental events, is composed of physical matter and is thus subject to the laws of physics. From this prospective, the mind is considered to exist solely as a set of cerebral processes. This philosophy can be traced back to ancient times but gained a new impetus from advances in the physical sciences beginning in the seventeenth century. Originally, the causes of behavior were to be found in the material of the body, particularly the nervous system. More recently, the focus has been the cerebral system: the brain.

Materialism was thought to be completely empirical from its early beginnings but in psychology has been researched as a personality characteristic. This term, as a philosophical paradigm, needs to be differentiated from materialism as the pursuit and acquisition of material goods and luxuries. This pursuit is typically perceived by the individual as a means and measure of personal worth and achievement, often at the expense of moral, psychological, and social considerations (Hartnett & Skowronski, 2008; Kasser, 2002; Van Hiel, Cornelis, & Roets, 2010).

By the same token, materialism needs to be differentiated also from Dialectical Materialism which is concerned with power and political issues rather than the underlying nature of behavior, human relationships, and objects (Berendzen, 2008; Fuller, 1946; Lektorskii, 2004; Tian, 2002).

L. L'Abate (✉)
Department of Psychology, Georgia State University, University Plaza, Atlanta, GA 30303, USA
e-mail: llabate3@bellsouth.net

L. L'Abate (ed.), *Paradigms in Theory Construction*, DOI 10.1007/978-1-4614-0914-4_14,

Psychological references about materialism *proper* could not be found. If that is the case, why consider materialism in this volume? Materialism as a philosophical paradigm, rather than psychological sources, will be pursued because it is the basis for quite a few chapters in this volume, for instance, more specifically Chaps. 18–20. Those chapters could be related to an overarching materialistic paradigm.

Historical Background

Originally, materialism (Fuller, 1946) was a philosophical:

> …doctrine that all aspects of the universe, including human life, can be reduced to explained in terms of matter in motion. The term is also applied to systems that, although they regard consciousness as irreducible to terms of physical energy, still consider it dependent upon matter for its existence and finds its processes and successive states explicable only when correlated with physiological processes and thus submitted to the laws governing physical motion and energy (p. 527).

Similar philosophical viewpoints related to Materialism could be Hard Determinism, Identity theory, Physicalism vs. Idealism and Immaterialism. These viewpoints will not be reviewed here because, perhaps with the exception of Identity Theory, they do not appear to have influenced, as far as I know, theory construction in psychology or related disciplines.

Two English philosophers are more relevant to materialism than others. John Stuart Mills, for one instance, discounted metaphysics as a philosophy because it could not answer important questions about reality and therefore it was impossible to validate. Knowledge cannot be extended beyond the empirical experience of "permanent possibilities of sensations" with the aid of memory and the association of ideas (Fuller, 1946, p. 393). Repetition of the same experience through our sensations and perceptions allow us to validate experience. This view is present as one criterion of the scientific method as *replication.* Without the possibility of replication, it is impossible to validate any experience. This criterion was present in Darwin's theory of biological evolution. Without replication of a phenomenon, if it happens only once, there is no proof of its existence.

Herbert Spencer followed Darwin's theory by focusing on biology and eventually on psychology and even sociology. One important aspect of his thinking was the relativity of all knowledge and, by implication, science as a continuous quest about tentativeness as a constant approximation to the truth of any reality. This kind of thinking influenced German naturalists as Trendelenburg and especially materialists such as J. Moleschott, L. Buchner, and E. Haeckel: "All three were agreed that consciousness could be reduced to terms of matter and physical energy" (Fuller, 1946, p. 412). This thinking was followed by Lange and Mach who linked philosophy to scientific concepts, such as the important of methods to evaluate concepts.

Thus, these philosophers in the eighteenth and nineteenth centuries predated and served as an historical backdrop to Vienna's positivistic school. At the beginning of the twentieth century that school supported operationalism in science, a position that has influenced greatly the development of psychology in the United States, especially the present phenomenal growth of neuropsychology as a discipline in its own rights, as shown especially in Chap. 19.

Materialism as a Philosophical Pursuit

Materialism has not been popular in philosophical circles as could be in psychology. It has received its share of criticisms within philosophical as well as religious and humanistic circles. Only a selected and admittedly scattered sample of these writings is included here to give readers a flavor about the nature of philosophical thinking about materialism. Very influential in this field has been the work of De Santillana and Zilsel (1941) who traced aspects of scientific rationalism to the eighteenth and nineteenth centuries. He emphasized the importance of what came to be known as "mechanistic materialism," tracing matter back to atomic particles and to physics. He was followed by Edgar Zilsel who traced modern classical empiricism to another English philosopher David Hume, who predicted the decline of mechanical conceptions of nature and the elimination of mechanical methods. I do not think this prediction came to pass, because in one way or another, I believe that psychological thinking is essentially materialistic even though it allows within its wide wings dialectical and nonempirical thinking (Rychlak, 1976).

In a controversial paper, Kerr-Lawson (2002) discussed Santillana's ontological categories of essence, spirit, and truth which constitute his realms of being. Kerr-Lawson contrasted between essence as being separate from existence, an issue that will be considered in greater detail in Chap. 20. Flamm (2009) examined specific aspects of the contemporary materialisms of philosophers Daniel Dennett and J.J.C. Smart that exhibited skepticism about the naturalism of the theory of knowledge that Kerr-Lawson believes in. Smart's beliefs in a materialist view of consciousness is affirmed by a realist restriction about mental claims. Dennett's attention to the mind–body problem of contemporary neuroscience is noted. Kerr-Lawson's contributions to the criticism of scientistic (sic) materialism and philosopher George Santillana's elaboration of naturalism are also discussed. Tiller (2006) focused on the concept of materialism of Santillana and his relationship with the pragmatist movement. Four formulations on matter of pragmatists against materialism are provided. Tiller listed the indispensable properties that result in the extensional identification of substance with matter in his book *Real of Matter*.

Moreno (2009) examined the disagreement in Kerr-Lawson's paper "Santillana on the Matter of Aristotle" and in John Lachs's paper "Substance and Matter: A Response to Angus Kerr-Lawson" over how to interpret philosopher George Santillana's materialism. The main issues of the disagreement on matter regarding matter as being unknown and the relation between substance and matter are considered. The distinction between language and world or between the terms and the reality for which these terms stand is also discussed.

Besides Santalliana's influence and Darwin's inevitably materialistic theory of evolution, Skinner (2006) challenged the neo-Darwinist physicalist position assumed by currently prevalent naturalizing accounts of consciousness. He suggests instead an evolutionary (Deweyan) understanding of cognitive emergence and an acceptance of mental capacity as a phenomenon in its own right, differing qualitatively from, although not independent of, the physical and material world. He argued that if we accept that consciousness is an adaptation enabling survival through immediate individual intuition of the world, we may accept this metaphysics as given. Methodological focus can then shift to investigating the, as yet untheorized, nature of consciousness itself as capacity/interconnectivity/potential. This article accepts Joseph Margolis's recent advocacy of a pragmatist approach that is "natural but not naturalizable" (Margolis, 2002, p. 7), that is, an anti-reductionist as opposed to an eliminativist position, but it seeks to develop this position further and to give it new direction.

Ivanov (2007) presented an analysis of problems connected with the understanding of the ontological status of mental phenomena in the universe by dealing with the opposite position given by Russian philosophers about dialectical materialism. He said that the reality of the mind is firmly described as an ideal reality that reflects material reality. Thus, the mind of man is the crucial object of contemporary scientific and philosophical research.

A traditional anti-metaphysical goal has been the materialist assimilation of human mental and personal qualities into the physical nature of the brain (Bussey, 2004). However, consciousness notably resists explanation in such a way. In an attempt to deal with this problem, a firmly realist view of the laws of nature is argued by Bussey. Aspects of consciousness are examined, showing that consciousness does not lie within the remit of physics. A survey of the thinking of Descartes, Aquinas and Duns Scotus is given to illustrate the nature of the mind–matter and mind–body problem, with further reference to Kant's position. By emphasizing that physical objects comprise form and matter, the medieval philosophers provide a starting point for a perspective that can incorporate modern developments in physics.

Bussey proposed that the concept of the "mental" be broadened to encompass the laws of nature, taken as mathematical ideas which determine the behavior and properties of physical things. The identification of what "matter" raises significant questions, but may involve quantum fields. Eddington postulated what can be called a "mental dimension," and this may provide a promising framework for uniting the laws of physics and our own mental nature. The main question is no longer the relationship between "mind" and "matter," for this is now at the heart of physics. It is more to do with human consciousness within this broader mental dimension.

Borodin (2004) focused on contemporary philosophical materialism and synergetics. Synergetics is the science of the self-organization of natural systems. The beginning of understanding of society as a self-organizing and self-reproducing system dates to the eighteenth century, to Adam Smith's theory of the reproduction of capital. Contemporary scholars draw two main conclusions from synergetics: the functioning of natural systems is reproductive in character and the reproduction of systems has a certain direction but is not subject to strict determination (see Chap. 10 for a recent expansion of this viewpoint). The theory of social production has also reached the same conclusions.

In a series of essays published in the 1990s, Lawrence Grossberg proposed a spatial-materialist cultural study, arguing that our key metatheoretical assumptions about reality, agency, ethics, and politics needed to be reconceptualized on a modern philosophical terrain – not within, or even against, the philosophical frameworks of modernism and postmodernism but outside them (Wiley, 2005). To develop this alternative terrain, Grossberg drew on the writings of Gilles Deleuze and Félix Guattari, whose work was based on a Spinozist ontology, a "monism of multiplicities." Grossberg argued that Deleuzean philosophy provides cultural studies with a way out of the epistemological problematic that has dominated critical theory. He relied on Deleuzean ontology to argue against modernist, postmodernist, and poststructuralist conceptualizations of identity and subjectivity. He proposed, instead, that cultural studies develop a mechanistic theory of agency. Grossberg also used Deleuzean philosophy to reconsider the ethics and politics of cultural studies, proposing a politics of "spatial becoming."

The foregoing essay seeks to highlight the productive conceptual moves that Grossberg made, to clarify some concepts that remain ambiguous in his approach, and to identify certain claims that merit further critical consideration. Grossberg's work repositioned cultural studies in relation to the discursive terrain of modern philosophy and theory, opening up new routes for thought and action. In doing so, it clarifies, rearticulates and, in many ways, radicalizes the critical interventions that reshaped cultural studies in the 1980s and 1990s. But Grossberg's most important contribution to cultural studies has remained implicit: by following Deleuze and Guattari, we are drawn into an affective state of theoretical affirmation and practical composition – a stance that is quite different from both modernist and postmodernist postures of critique.

Since the publication of Dynamic Basis of Geomorphology by Arthur geomorphology has become dominated by the process approach to inquiry. This approach treats geomorphological processes and process–form interactions as manifestations of mechanical stresses and strains acting on earth materials (Rhoads, 2006). Little attention has been given to the philosophical underpinnings of this approach, which correspond to the doctrine of *mechanistic materialism*, already mentioned earlier. Concerns about limitations of the process approach have been raised recently, including its inherent reductionism; difficulty in dealing with complex, large-scale geomorphic phenomena; and lack of a historical focus. This article presents an alternative perspective on the dynamic basis of geomorphology grounded in process philosophy. The central theme of process philosophy is that processes are ontologically and epistemologically more primary than substantive material objects.

A major implication of process philosophy is that mechanistic concepts, while epistemically valuable, represent abstractions that lack ontological depth. Evidence both from contemporary science and from human experience suggests that the notions of matter as unchanging substance and of dynamic change as matter in motion under mechanistic influences are flawed ontologically. Instead, this evidence indicates that the nature of reality, including geomorphological phenomena, is fundamentally processual. By casting aside the constraints of mechanistic materialism, a process perspective grounded in process philosophy embraces the entire spectrum of research in contemporary geomorphology and has the potential to open new avenues of thought and inquiry, perhaps in ways that may also lead to enhanced connections between human and physical geography.

van der Tuin and Dolphijn (2010) concentrated around three ways in which "new materialism" or "neomaterialism" – terms coined by DeLanda and Braidotti in the second half of the 1990s – can be called "transversal." New materialism is a cultural theory that does not privilege culture, but focuses on what Haraway would call "nature cultures." It explores a monist perspective of the human being, disposing of the dualisms that have dominated the humanities until today, by giving special attention to matter, as it has been so much neglected by dualist thought.

New materialism, a cultural theory inspired by the thoughts of Deleuze, that spurs a renewed interest in philosophers such as Spinoza and Leibniz, shows how cultured humans are always already in nature, and how nature is necessarily cultured, how the mind is always already material, and how matter is necessarily something of the mind. New materialism opposes the transcendental and humanist (dualist) traditions that are haunting a cultural theory, standing on the brink of both the modern and the post-postmodern era. The transcendental and humanist traditions, which are manifold yet consistently predicated on dualist structures, continue to stir debates that have a stifling effect on the field (think of the feminist polemic concerning the failed materialism in the work of Butler, and of the Saussurian/Lacanian linguistic heritage in media and cultural studies). New materialism allows for the conceptualization of the traveling of the fluxes of matter and mind, body and soul, nature and culture, and opens up active theory formation. The three "transversalities" concern disciplinarity, paradigms, and the spatiotemporality of theory.

Clarke (2009) discussed the relationship between neuroscience and the concept of the soul in response to a lecture delivered by professor Malcolm Jeeves in London, England on January 23, 2008. Clarke expressed approval of the view that there is an irreducible intrinsic interdependence between the occurrence in the cognitive and physical level. He cited that the different philosophies of mind–brain duality can be classified as either substance dualism or substance monism. He pointed out that Jeeves rejects the notion of eliminative materialism.

The thought of Wilfrid Sellars has figured prominently in recent discussions of the relationship between naturalism and normativity (O'Shea, 2010). On the one hand, some have appealed to Sellars' philosophy in defense of the thesis that what he called the normative "space of reasons" is in some sense sui generis and irreducible to the natural causal order described by the natural sciences. On the other hand, others have exploited equally central aspects of Sellars' philosophy in defense of the seemingly incompatible project of attempting to give an exhaustive scientifically naturalist account of mind and meaning, and perhaps of the nature of normativity itself. I contend that what Sellars described as "the Janus-faced character of languagings as belonging to both the causal order and the order of reasons" (Naturalism and Ontology) is the key to understanding his normative and pragmatist variety of naturalism. Sellars saw himself as having articulated a detailed philosophical perspective within which the normative aspects of meaning, knowledge, truth, and representation are themselves opened up, in principle, to naturalistic explanation.

In recent years, two-dimensional semantics has become one of the most serious alternatives to Millianism for the proper interpretation of modal discourse (Howell, 2008). It has origins in the works of a diverse group of philosophers, and it has proven popular as an interpretation of both language and thought. It has probably received most of its attention, however, because of its use by David Chalmers in his arguments against materialism. It is this more metaphysical application of two-dimensionalism that is the concern in this article. For though there is probably something salvageable from two-dimensionalism as a way to explain the content of thought, as a metaphysical tool it should be abandoned. In this paper, Howell aimed at establishing this point by reduction: if "metaphysical" two-dimensionalism is assumed, it can be shown to be false.

Tennant (2007) discussed the historical survey concerning the developments of philosophy of mind in relation of the tradition of "Naturphilosophie" and influence of "Ignorabimusstreit" on mathematics. It states that major themes of contemporary analytical and scientifically informed philosophy of mind have been completely expressed. Versions of materialism, supervenience and rudimentary functionalism, as well as mysterianism, were defended in the nineteenth century, however, they are no longer neglected. Furthermore, it is believed that the influence of physiologists and philosophers was widely felt ought to be long-lasting.

Tapper (2002) presented two ontological conceptions of sounds which include the unrepeated event (UE) and the repeatable object (RO). In the EU conception, sounds are considered events happening to and depending on material objects, and have temporal parts. The RO conception shares two principles with the EU conception, which state that sounds are particulars rather than universals and are not ordinary space-occupying objects. Lastly, it suggests that RO conception differs from the EU conception in terms of conceiving sounds to be kinds of objects rather than events.

Critiques of Materialism

As far as I can tell, in addition to criticisms about materialism within the discipline of philosophy, most critiques of materialism have come from religious and humanistic circles. For instance, Klima (2009) argued that Aquinas's conception of the human soul and intellect offers a consistent alternative to the dilemma of materialism and post-Cartesian dualism. He also argued that in their own theoretical context, Aquinas' arguments for the materiality of the human soul and immateriality of the intellect provide a strong justification of his position. However, that theoretical context is rather "alien" to ours in contemporary philosophy. The conclusion of the paper will point in the direction of what can be done to render Aquinas's position more palatable to contemporary philosophers.

The natures of mind and soul have been frequently discussed over the last decade in the journal *Science & Christian Belief* (Turl, 2010). The trend has been to move from a dualistic account toward some form of monism, while attempting to avoid the extreme of materialism with its perceived threat to rational and moral freedom. This article queries whether dualism really is dead and whether the new soul to which we are asked to subscribe is the soul of biblical teaching. Philosophical and metaphysical arguments are used to support the thesis that some form of dualism is still scientifically respectable, but the distinction of substance may be based in our ignorance of the nature of both matter and spirit.

Materialists argue that there is no place for God in the universe. Chance and contingency are all that structure our world. However, the materialists' dismissal of subjectivity manifests a flawed metaphysics that invalidates their arguments against God. In his essay, Peck (2003) explored the following: (1) How does personal metaphysics affect one's ability to do science? (2) Are the materialist arguments about contingency used to dismiss the importance of our place in the universe valid? (3) What are the implications of subjectivity on belief and science? To answer the first question, Peck examined the later years of Sir Alfred Russel Wallace, one of the cofounders of evolution through natural selection with Darwin. His belief in nineteenth-century spiritualism profoundly affected his standing in the scientific community. Peck also described the effect of spiritualism on Wallace's science. To answer the second question, he used his own work in mathematical modeling of evolutionary processes to show that randomness, and contingency at one level, can actually be nearly deterministic at another. Peck demonstrated how arguments about chance and contingency do not imply anything relevant about whether there is a designer behind the universe. To answer the third question I began by exploring a paradox of consciousness and show how the existence of subjective truths may provide a paradigm for sustaining a rational belief in God. These questions form the framework of a structured belief in a creator while yet embracing what science has to offer about the development of life on our planet.

According to Schneider (2005), psychology stands at a crossroads. Even more than in the time of Rollo May, psychology faces a dilemma of alarming moral and scientific proportions (see Chap. 15 in this regard). On the one hand, psychology is about to become a "biologism." This is a discipline – led by the high tech, consumerist model for living – that emphasizes measurement, materialism, and efficiency. On the other hand, psychology still has a chance – mainly through humanistic and depth orientations – to be a vibrant discipline. The question is, will we take this path to become the "queen of the sciences," as Nietzsche put it, or will we jeopardize 5,000 years of arts and humanities, centuries of depth analyses, and decades of awe-inspired practice. In this article, Schneider outlined the stakes in this dilemma, the players involved – from biopsychology to humanism – and an alternative, conciliatory vision that he called "awe-based" psychology.

According to him, the dilemma we face today is whether we will approach our subject matter with a sense of the magnificence and mystery of living (awe) or whether we will persist in making molehills of mountains and succumb to the glib, the well-packaged, and the instrumental; whether we will find the terms and resources to reflect life's profundity; or whether we will skirt its edges, resort to the commercial, and bow to the expedient. Schneider illustrated further how this dilemma need not be solved by an "us-them," "either-or" mentality but can be addressed dialectically with practical as well as visionary implications for the profession of psychology. Finally, the author summarized a few of these implications, both for our practices and contemporary lives.

Materialism Applied to Psychological Thinking

What can we conclude from this brief review of materialism as a philosophical and seemingly quite abstract paradigm? Can this paradigm be applied to psychological thinking in general and more concretely to the study of feelings in particular? Both questions can be answered if we make some assumptions about the role of materialism within a psychological framework in general and specific theory building in particular.

Assumptions About Materialism Applied to Psychology

There are at least three major assumptions that must be made if we attempt to apply materialism as a paradigm to psychological thinking. First, we must assume that psychology as a science requires different levels of analysis and interpretation. This assumption can and should be applied to materialistic thinking if we want to apply to psychological events. Second, we must assume that materialism as applied to psychological science requires different principles of causality. Third, we must assume that the movement from any abstract and extremely general paradigm such as materialism, if no others, is toward greater concreteness and specificity. Therefore, unless we can accept these three assumptions, our whole enterprise might remain empty of empirical evidence.

An example about how materialism can be applied to the study of experiencing feelings can be derived from work on emotionality and especially hurt feelings (L'Abate, 2011). There, it is assumed that hurt feelings are at the bottom of our existence and contribute to the development of psychopathology, constituting the so-called unconscious, when there is not sufficient joys to offset hurts (De Giacomo, L'Abate, Pennebaker, & Rumbaugh, 2010).

Levels of Analysis and Interpretation

Figure 14.1 shows how feelings can and should be studied at different levels of analysis and interpretation that in the past were the province of distinct scientific disciplines but that more recently are represented by specializations that span one or more biological and psychological disciplines, witness the growth of neuroscience

Complex Relational and Non-relational Feelings

Altruistic & Self-oriented Feelings

Basic Feelings

Gate Keepers

Experiencing Feelings

Cerebral Levels

Visceral Levels

Physiological Levels

Cellular Levels

Molecular Levels

Fig. 14.1 An Hourglass Model for experiencing feelings (adapted from L'Abate, 2011)

in the last generation. Since these feelings can be studied and located at the cerebral level (Xiao, Smith, Bechara, & L'Abate, 2012), this matter can be studied at different levels of analysis and interpretation.

When intimacy is defined behaviorally as the sharing of joys and hurts as well as the fear of being hurt, it becomes 1 of 16 models in Relational Competence Theory (RCT; Cusinato & L'Abate, 2012; L'Abate, Cusinato, Maino, Colesso, & Scilletta, 2010).

Principles of Causality

Levels of analysis and interpretation imply and involve directly three different principles of causality: (1) *equifinality* is present when in Fig. 14.1 we move up from the lowest molecular level to the centrality of experiencing feelings, each of these underlying and interrelated processes lead up to such an experience; (2) *equipotentiality*, on the other hand, is present when from experiencing feelings we move up at various levels of complexity reaching up to higher level of experiencing more delicate and nonrelational feelings; (3) both principles are reversed within a feedback process when we move down from the higher level of complexity to the central level of experiencing feelings through equifinality and from that central experience of feelings down to biological level through equipotentiality; and (4) *multipotentiality* when it is understood that it takes more than one cause to produce movements from down–up and from up–down. This process was discussed in detail by Capitelli, Guerra, L'Abate, Longo, and Rumbaugh (2009) and will be discussed in greater detail by Colesso in Chap. 20.

Abstraction and Generality as Guiding Principles in Theory Construction

Two dimensions have colored the original formulations of the hierarchical framework underlying the Table of Contents of this volume and the present classification of paradigms, as distinct from theories, models, and dimensions (L'Abate, 2009). According to two dimensions of abstraction and generality, both are present at the level of paradigms. Paradigms as worldviews are wide-ranging perspectives about what reality is like at the higher levels of abstraction and generality. As such, one paradigm could be related to various theories and from theories to models. As we move down from paradigms to theories, levels of abstraction and generality tend to become less abstract and less general. When we move from theories to models, abstraction and generality are present as being concrete and specific. Concreteness and specificity are found in the dimensions defining a model that can be evaluated and validated empirically.

For example, in this section and in foregoing paragraphs, we have moved from a clearly abstract and general philosophical paradigm as materialism down to a more concrete and specific paradigm defined by levels of analysis and interpretation, principles of causality, ending with a model of RCT defined by very concrete and specific models that can be evaluated empirically (Cusinato & L'Abate, 2012; L'Abate et al., 2010). Hurt feelings have been defined operationally and just opposed to joys, making the ratio of joys and hurt feelings that basic dimension to consider at conscious and unconscious levels.

Conclusion

From a seemingly abstract and general philosophical paradigm such as materialism, it is possible to describe it according to known levels of analysis and interpretation down to molecular levels. From these levels, it is possible to apply principles of causality and to describe how the experience of feelings in general and of hurt feelings in particular can be described according to feedback processes that include biological and psychological, conscious and unconscious levels.

References

Berendzen, J. (2008). Postmetaphysical thinking or refusal of thought? Max Horkheimer's materialism as philosophical stance. *International Journal of Philosophical Studies, 16*, 695–718.

Borodin, E. (2004). Contemporary philosophical materialism and synergetics. *Russian Studies in Philosophy, 43*, 28–46.

Bussey, P. (2004). Beyond materialism: From the medieval scholars to quantum physics. *Science & Christian Belief*, 16, 157–178.

Capitelli, M., Guerra, C., L'Abate, L., Longo, S., & Rumbaugh, D. M. (2009). Science or mind? Reductionism proper, optimal reductionism, emergent interaction, and synthetic integration. In L. L'Abate, P. De Giacomo, M. Capitelli, & S. Longo (Eds.), *Science, mind, and creativity: The Bari symposium* (pp. 1–22). New York: Nova Science.

Clarke, P. (2009). Neuroscience and the soul – A response to Malcolm Jeeves. *Science & Christian Belief, 21*, 61–64.

Cusinato, M., & L'Abate, L. (Eds.). (2012). *Advances in relational competence theory: With special attention to alexithymia*. New York: Nova Science.

De Giacomo, P., L'Abate, L., Pennebaker, J. W., & Rumbaugh, D. M. (2010). Amplifications and applications of Pennebaker's analogic to digital model in health promotion, prevention, and psychotherapy. *Clinical Psychology & Psychotherapy, 17*, 355–362.

De Santillana, G., & Zilsel, E. (1941). *The development of rationalism and empiricism*. Chicago: University of Chicago Press.

Flamm, M. (2009). Angus Kerr-Lawson and the ills and cures of scientistic materialism. *Transactions of the Charles S. Peirce Society, 45*(4), 467–483.

Fuller, B. A. G. (1946). *History of philosophy* (revised edition). New York: Henry Holt.

Hartnett, J., & Skowronski, J. (2008). Cash, money, woes: The match between a person's level of materialism and the materialistic (or non-materialistic) character of events alters affective forecasts. *North American Journal of Psychology, 10*, 647–664.

Howell, R. (2008). The two-dimensionalist reductio. *Pacific Philosophical Quarterly, 89*, 348–358.

Ivanov, A. (2007). On the problem of the ontological status of mental phenomena. *Russian Studies in Philosophy, 45*, 73–87.

Kasser, T. (2002). *The high price of materialism*. Cambridge: MIT Press.

Kerr-Lawson, A. (2002). The non-empiricist categories of Santayana's materialism. *Transactions of the Charles S. Peirce Society, 38*, 47–78.

Klima, G. (2009). Aquinas on the materiality of the human soul and the immateriality of the human intellect. *Philosophical Investigations, 32*, 163–182.

L'Abate, L. (2009). Paradigms, theories, and models: Two hierarchical models. In L. L'Abate, P. De Giacomo, M. Capitelli, & S. Longo (Eds.), *Science, mind, and creativity: The Bari symposium* (pp. 107–153). New York: Nova Science.

L'Abate, L. (2011). *Hurt feelings: Theory and research in intimate relationships*. New York: Cambridge University Press.

L'Abate, L., Cusinato, M., Maino, E., Colesso, W., & Scilletta, C. (2010). *Relational competence theory: Research and mental health applications*. New York: Springer-Science.

Lektorskii, V. (2004). Philosophy is the self-consciousness of a culture. *Russian Studies in Philosophy, 42*, 73–91.

Lerner, M. (2010, November/December). A spiritual approach to evolution. *Tikkun*.

Margolis, J. (2002). *Reinventing pragmatism: America philosophy at the end of the twentieth century*. Ithaca, NY: Cornell University Press.

Moreno, D. (2009). On the recent elucidations of Santayana's materialism by Angus Kerr-Lawson and John Lachs. *Transactions of the Charles S. Peirce Society, 45*, 491–505.

O'Shea, J. (2010). Normativity and scientific naturalism in Sellars' 'Janus-Faced' space of reasons. *International Journal of Philosophical Studies, 18*, 459–471.

Peck, S. (2003). Randomness, contingency, and faith: Is there a science of subjectivity? *Zygon: Journal of Religion & Science, 38*, 5–23.

Rhoads, B. (2006). The dynamic basis of geomorphology reenvisioned. *Annals of the Association of American Geographers, 96*(1), 14–30.

Rychlak, J. F. (Ed.). (1976). *Dialectic: Humanistic rationale for behavior and development*. New York: S. Karger.

Schneider, K. (2005). Biology and awe: Psychology's critical juncture. *Humanistic Psychologist, 33*, 167–173.

Skinner, J. (2006). Beyond materialism: Mental capacity and naturalism, a consideration of method. *Metaphilosophy, 37*, 74–91.

Tapper, A. (2002). Reid and Priestley on method and the mind. *The Philosophical Quarterly, 52*, 511–525.

Tennant, N. (2007). Mind, mathematics and the Ignorabimusstreit. *British Journal for the History of Philosophy, 15*(4), 745–773.

Tian, C. (2002). Tongbian in the Chinese reading of dialectical materialism. *Philosophy East & West, 52*, 126–145.

Tiller, G. (2006). The unknowable: The pragmatist critique of matter. *Transactions of the Charles S. Peirce Society, 42*(2), 206–228.

Turl, J. (2010). Substance dualism or body-soul duality? *Science & Christian Belief, 22*, 57–80.

van der Tuin, I., & Dolphijn, R. (2010). The transversality of new materialism. *Women, 21*, 153–171.

Van Hiel, A., Cornelis, I., & Roets, A. (2010). To have or to be? A comparison of materialism-based theories and self-determination theory as explanatory frameworks of prejudice. *Journal of Personality, 78*, 1037–1070.

Wiley, S. (2005). Spatial materialism. *Cultural Studies, 19*, 63–99.

Xiao, L., Smith, D., Bechara, A., & L'Abate, L. (2012). Biological foundations of hurt feelings: With special attention to neural mechanisms. In L. L'Abate (Ed.), *Hurt feelings: Theory and research in intimate relationships* (pp. 000–000). New York: Cambridge University Press.

Part IV
Particular-Specific Paradigms in Psychology

Chapter 15
Explicating and Exemplifying Empiricist and Cognitivist Paradigms in the Study of Human Learning

Sandra M. Loughlin and Patricia A. Alexander

It will come as no surprise to those who engage in research on human learning that the field suffers from definitional, conceptual, and methodological disorder (Hancock, 2007). All too often, researchers in this psychological domain use different terms to say essentially the same thing (or use the same term to mean entirely different things), investigate the relation between two or more constructs without a clear rationale or prediction, or use measures divorced from the nature or process of the phenomena under investigation (Marsh & Hau, 2007). Among the symptoms of this disorder are ambiguous terms (Dinsmore, Alexander & Loughlin, 2010), obfuscated constructs (Alexander, Schallert, & Hare, 1991), atheoretical research (Alexander, 2009), and inaccurate or invalid claims (Robinson, Levin, Thomas, Pituch, & Vaughn, 2007). Treatment of this pervasive and complex disorder is no small feat, requiring multidimensional and multiphasic intervention. One important step toward clarity and precision in the research on human learning, however, is being undertaken in this volume, namely untangling and clarifying the relation among paradigm, theory, and model.

In this chapter, we describe and differentiate between two important paradigms represented in the field of educational psychology, namely empiricism and cognitivism. Empiricism asserts that experience, particularly direct observation, is the only source of knowledge. Thus, in the study of human learning, research and theory emanating from this paradigm assume that activities and processes can be identified and meaningfully understood through direct measurements, most frequently through experimental or quasi-experimental designs. The theory of behaviorism (Skinner, 1953, Watson, 1924) is perhaps the clearest manifestation of empiricism within the psychological literature.

S.M. Loughlin (✉) • P.A. Alexander
Disciplined Reading and Learning Research Laboratory, Department of Human Development, College of Education, University of Maryland, College Park, MD 20742-1131, USA
e-mail: sandra.loughlin@gmail.com; palexand@umd.edu

L. L'Abate (ed.), *Paradigms in Theory Construction*, DOI 10.1007/978-1-4614-0914-4_15,

Yet, in modern psychology, empiricism in its "pure" form is relatively rare. Instead, the empiricist paradigm is often intertwined with some manner of cognitivism. Unlike behaviorists, who view learning as a response to environmental stimuli, cognitivists emphasize the inner mental activities and processes that guide learning, such as thinking, memory, knowing, and problem solving. Although changes in behavior or physical acts are important for cognitivists, they stand principally as the results or indicators of associated internal mechanisms of thought and processes of knowing.

However, these brief introductions to empiricism and cognitivism do not fully capture the complexity of orientations toward human learning. In fact, theories and models espousing both empiricist and cognitive roots vary widely along a number of dimensions, including beliefs about knowledge and the nature of knowing (Murphy, Alexander, & Muis, in press), differential emphases on the *who, what, where*, and *when* of learning (Alexander, Reynolds, & Schallert, 2009), and beliefs about the nature of evidence (Alexander and the Disciplined Reading and Learning Research Laboratory, 2010). As a result, rather than articulating these paradigms as static or unidimensional, we describe a family of empiricist and cognitive orientations toward human learning through historical, philosophical, and psychological lenses. Further, we attempt to disentangle the relation among paradigmatic iterations and associated theories and models of learning that populate the literature.

In addition to articulating and discussing the empiricist and cognitivist paradigms, we demonstrate their application in theory building. Specifically, we describe our ongoing development of the theory of trans-symbolic comprehension (TSC). TSC represents a substantial reframing of perspectives on comprehension that have dominated the empirical literature, requiring us to develop a new theoretical rationale and to construct and test models arising from it. We describe and reflect on that theory-building process as a way to explicate and exemplify the place of empiricism and cognitivism in the study of human learning, highlighting our movement from paradigm to theory to model. Finally, we discuss the complexities inherent in operationalizing paradigms in theory and research.

Untangling Paradigms, Theories, and Models

Before embarking on an explication of empiricism and cognitivism, however, we first turn to the central focus of this volume, namely untangling the relations among paradigms, theories, and models. Although this issue has been addressed extensively elsewhere (e.g., Hillix & L'Abate, this volume), we briefly reiterate to frame our discussion of these two paradigms and subsequent description of the developing theory of TSC.

As noted by James (1890), from infancy, our senses are assailed by a "great blooming, buzzing confusion" of information (p. 488). To make meaning of the chaos and complexity, we organize the world into categories, groups, and classes (Wertheimer, 1959). This drive to organize reality is reflected in the roots of philosophy, with both Plato and Aristotle addressing the inclination of humans to group objects and ideas according to their inherent properties (Ackrill, 1963; Ross, 1951). Dogs, the color red, holidays, and mathematics all represent groupings that allow us to make sense of the world. So do the paradigms, theories, and models.

In psychology, paradigms are broad sweeping conceptual constructions about the nature, process, or substance of psychological investigation (Hillix & L'Abate, this volume; Mertens, 2010). Alternatively called traditions of thought (Greeno, Collins, & Resnick, 1996), paradigms represent, at the highest, most abstract level, how we perceive human behavior, the world of the mind, and the study of their relation. For instance, psychologists with an empiricist orientation would argue that evidence of psychological processes can be observed, measured, and articulated. Likewise, psychologists with a cognitivist perspective emphasize the inner, mental activities of an individual.

Multiple paradigmatic orientations are represented in the field of psychology, as evidenced by the foregoing empiricist and cognitivist examples, as well as the chapters in this volume. Some paradigms represent with competing worldviews. For instance, behaviorism and cognitivism represent two ends of an external (environment)–internal (mind) continuum. Behaviorists view psychology as the study of human behavior, with the source of behavior to be found in the environment, not in the mind (Goetz, Alexander, & Ash, 1992). In contrast, cognitivists view psychology as the study of the mind, holding that any change in behavior is a consequence of a mental activity, such as thinking, memory, knowing, or problem-solving (Bransford et al., 2006). Not all paradigmatic orientations are mutually exclusive, however (Mertens, 2010). Thus, one can manifest both empiricist and cognitivist views, especially when non-radical or less extreme views of these orientations are instantiated. Specifically, one can believe that psychology should concern itself with the inner, mental activities of an individual and measure those activities with observable indicators, relying on physical evidence to document mental faculties or processes.

By mapping out the psychological landscape in this fashion, paradigms allow for the identification, organization, and legitimization of related theories and models testing those theories (Pemberton, 1993). In psychology, theories are speculative frameworks that attempt to explain or predict a number of interrelated psychological phenomenon (Pedhazur & Schmelkin, 1991; VardenBos, 2007). In other words, a theory is a set of propositions that provides a systematic account of a psychological occurrence. In contrast, a model is a graphic representation of a concept that allows for enhanced understanding and investigation (VardenBos, 2007). According to these definitions, theories and models serve the same purpose (i.e., to explain a phenomenon), and, following this rationale, the terms *theory* and *model* are often used interchangeably (cf., Cohen, Manion, & Morrison, 2007).

There is disagreement in the literature, however, about the appropriateness of using *theory* and *model* synonymously (Thomas, 1996; Tracy & Morrow; 2006). Ruddell, Ruddell, and Singer (1994), for example, make a clear distinction between theory and model, suggesting that a theory is dynamic and unobservable, while a model is static and more appropriate for examination. Hillix and L'Abate (this volume) expand on this distinction, arguing that theories and models are functionally different, operating at different levels of grain size. Drawing on the work of Feigel (1970), they hold that a theory is conceptually larger than a model and cannot be tested in its complete form. Yet, they argue, a theory possesses underlying, reducible dimensions that can be verified, both conceptually and empirically, by models. In other words, models test parts of a theory, but cannot test the whole theory at once. As such, a model represents a subset of a theory and must adhere to its scope, parts, and parameters.

According to Bredo (2006), one of the principal functions of psychology is to identify and correct conceptual confusion. Yet, in his examination of psychology, he notes the prevalence of "type III errors" (i.e., an error of conceptualization; Seidman, 1978), resulting in conceptual "lines" drawn in the wrong places, adopted too rigidly, or generalized beyond their intended scope. Paradigms, theories, and models are examples of these conceptual "lines" that help us organize and navigate the complex reality of psychology. Thus, it is necessary that those in the field are clear on the relation among them and use terms deliberately and consistently. Sadly, this is often not the case (L'Abate, 2005, 2009) and the resulting definitional, conceptual, and methodological chaos has created unnecessary ambiguity and difficulty interpreting the relation between lines of inquiry (Hancock, 2007).

This volume represents a necessary step toward greater clarity and precision by presenting a hierarchical approach to understanding the relation between paradigm, theory, and model and by articulating a number of paradigmatic orientations populating the psychological landscape. Two such paradigms, empiricism and cognitivism, are reflected in large proportion of the theory and research pertaining to human learning. Thus, as learning theorists and researchers, it is to these paradigms we now turn our attention.

Empiricism and Human Learning

At its core, empiricism signifies a family of perspectives on knowledge and knowing that rests on the premise that experience (not innate ideas) is the root of human understanding, especially when one learns to observe and seek evidence from that experience systematically and rigorously. From this position, empiricists contend that knowledge is acquired *a posteriori* through sensory experience. The physical world and human's connection to it become the means by which individuals acquire and refine knowledge. This sets empiricism most directly in contrast to rationalism that holds to *a priori* knowledge that demands no external justification, since such knowledge is innate or inborn (BonJour, 1998).

As educational researchers who engage in analytic inquiry, our work unquestionably pays homage to empiricism to some extent. In fact, the scholarly work in which we engage is commonly regarded as "experimental" in that it is assumed to derive from systematic analysis of physical evidence (i.e., data) carefully gathered, statistically analyzed, and objectively reported. Although we discuss more about the nature of empiricist methods in due course, we first want to consider the ancestry of empiricism and particular personages who have contributed to its prevalence within educational research and scientific inquiry.

Historical Overview

> All truths are easy to understand once they are discovered; the point is to discover them.
>
> Galileo Galilei

Aristotle and Galileo: Early Empiricists. Although John Locke, whom we will consider in more depth, is rightly credited with bringing "modern" empiricism to the fore in the late 17th century, especially in his efforts to counter the philosophies of rationalism and its mind–body dualism, the roots of empiricism trace far back in philosophical history. Empirical arguments were apparent in the philosophical debates that Aristotle had with his mentor and teacher, Plato, in ancient Greece – debates over the nature of knowledge and knowing (Murphy, Alexander, & Muis, in press). We see the hand of empiricism in the scientific studies conducted by renowned philosopher, astronomer, mathematician, and physicist, Galileo Galilei, as he sought to understand the universe and the earth's place in that universe. We briefly visit each of these historical figures to understand something of empiricism's path over the centuries.

As one of our philosophical progenitors, Plato held that there were ideal forms that existed beyond the physical realm, which was populated by flawed or imperfect iterations of these ideals. Thus, enlightenment for Plato, as captured in his *Allegory of the Cave* (*Republic*, 1993), required seeing beyond the imperfect patterns and images that were part and parcel of physical experience to reach a metaphysical state of understanding where ideal forms resided.

Unlike his teacher and mentor, Aristotle could not accept this untenable link between ideal or pure forms *beyond* the physical world and the weak or imperfect imaginations present *in* the physical world. Rather, for Aristotle, a theory of form required that one come to understand the "cause" or explanatory nature of a thing through analytic and reflective processes akin to a primitive scientific method (Gill, 2005). Aristotle also disputed the innateness of ideas, although these writings were less widely circulated. Nowhere is Aristotle's philosophical empiricism better exemplified than in his writing of *Categories*, a text that set out to explicate all the possible "kinds" that could be either the subject or predicate of a stated proposition (e.g., "the woman is brilliant") and was foundational to formal logic (Ackrill, 1963). Overall, Aristotle's argument for conceiving of experience and the analysis of that experience as the source of knowledge gave rise to empirical views that remained influential for centuries (Ross, 1951).

Applying the scientific method to the study of astronomical events, Galileo was a paragon of empiricism. As one of the most influential empiricists of his century or perhaps any other, Galileo set out to build tools such as the telescope and a military compass that would enhance his sensory capabilities and permit him to gather extensive and more detailed data about the physical universe. His meticulous calculations contributed to hypotheses about the motion of uniformly accelerated objects, and led him to his heliocentric theory of the solar system that conflicted with the geocentric views of the Catholic Church. Galileo was repeatedly called upon to repudiate his "scientific" claims and profess the ideas preached by the Catholic Church. Yet, no amount of social and religious pressure could convince Galileo to abandon his belief in physical evidence or the proof it afforded him. Through his empiricism, Galileo eventually replaced widely accepted Aristotelian categories of phenomena with a mechanical set that signaled the dawning of a new scientific era.

No man's knowledge here can go beyond his experience.

John Locke, *An Essay Concerning Human Understanding*, 1690, p. 98

Locke and "modern" Empiricism. As the discussions of Aristotle and Galileo illustrate, there has been a recurring tension between those who presume that thought and reasoning are best nurtured when individuals are able to rise above their humanness and their physicality and those who conversely argue that individuals' physical being and the experiences it affords is what empowers one's abilities to think and reason – to learn. This distinction is readily apparent in the contrast between the rationalists and the empiricists we referenced earlier. The intense philosophical debates between rationalists and empiricists continued for more than two centuries.

The rationalism of Descartes, Spinoza, Leibniz, and others represented a pursuit of knowledge and knowing through an intellect disencumbered by the distractions and confusions of the senses (Triadafillidis, 1998). By contrast, for empiricists like Locke, Berkeley, and Hume, it was precisely human experience, and individuals' capacity to discern pattern and evidence from that experience that gives rise to knowledge and knowing, as the opening quote by John Locke, the "father" of modern empiricism, suggests. In contrast to Descartes and Plato before him, Locke, like Aristotle, rejected the argument that ideas are innate and exposed through pure reason. Instead, Locke contended that the mind at birth resembles a *tabula rasa* (or blank tablet) written upon as individuals systematically sense, contemplate, and reason about information derived through sensation and by means of reflection (Wolterstorff, 1996).

The notion of experience that Locke conceived was both external (i.e., sensation) and internal (i.e., reflection) in nature. Through sensation, the individual is afforded understanding about both the primary qualities of physical objects (e.g., number, mass, or motion), which exist independent of the observer, and secondary qualities (e.g., colors or taste) that produce effects in the observer. Moreover, the ideas that form from these sensations and upon reflection can be either simple or complex.

Simple ideas are "uncompounded appearances," such as whiteness or softness. They are atomic and externally related. Complex ideas are of three classes:

- Ideas of substance, which are conceived of as stable support for the sensible qualities;
- Ideas of mode, which are conceived of as a form or property of things;
- Ideas of relationship, which are conceived of as the connecting of two or more ideas with one another.

One philosophical movement inspired by Locke's empiricism was *logical positivism* (cf., Murphy et al., in press). For logical positivists, knowledge consists of elements directly traceable *to* experience and, thus, must be verifiable *through* experience (Audi, 1999). Some logical positivists, who came to represent a more extreme form of empiricism, viewed thought as mentalistic expression and came to rely almost exclusively on observable human behavior as evidence of knowledge and thought (Murphy et al., in press). This near exclusive requirement for observability associated with *extreme empiricism* remained central to the tenets of *behaviorism*, a contemporary theory of learning that was built on the direct relation between environmental stimulus and physical response (Skinner, 1953; Watson, 1913, 1924).

Still, other less extreme forms of empiricism have dotted the philosophical and psychological landscape since Locke, including the *phenomenalism* of John Stuart Mill (Skorupski, 1989) where judgments are based on appearances or sensations, as well as the *radical empiricism* of William James (1912) and *instrumentalism* of John Dewey (1929). The variants of empiricism conceptualized by James and Dewey fall under the broader heading of pragmatism and are particularly noteworthy since they represent a melding of empiricism with rationalism—justification by experience with justification through mental reasoning.

Empiricism and Its Paradigmatic Implications

In their chapter on "The Role of Paradigms in Science and Theory Construction," Hillix and L'Abate (this volume) articulate the multiple interpretations of paradigms and paradigmatic thinking, as well as the argument for and against viewing the history of science or of psychology through the Kuhnian lens of paradigmatic change (e.g., Leahey, 1992). For us, it may not be as simple as distinguishing between theory and model change, even though deeply contested and seemingly antithetical theories and models of human learning and knowledge have long existed and will continue to exist in our field.

What can be learned from the brief historical overview of empiricism just offered – particularly when empiricism is juxtaposed to rationalism – is that, whether our humanness and the physical world are to be embraced or exiled in the pursuit of knowledge, the tensions between these emphases never go away. Empiricists and rationalists have likely walked among us ever since humans engaged in serious contemplation about knowledge and knowing. What has changed over the course of time has more to do with the rise and fall of empiricism's relative popularity as an explanatory system; whether empiricism or some theoretical alternative holds sway within a particular scholarly community. For that reason, we are inclined to see empiricism (or rationalism) as sitting at the intersection of an ontological orientation and an epistemic frame (Murphy, 2007). That is, this perspective not only orients individuals' understanding of the relations among objects (ontology) but also leads individuals to seek certain forms of evidence and justification over others (Murphy, Alexander, Greene, & Edwards, 2007).

Nonetheless, we are still able to examine empiricism and discern particular epistemic orientations toward knowledge and knowing, along with particular processes and procedures that mark one as "empiricist." For this articulation, we will assume a moderate stance within the family of epistemic perspectives. The processes and procedures include the following:

Observation. Given that empiricism is characterized by the claim that knowledge is based upon a relatively objective external reality, observation is one of the most fundamental tools in the empiricists' toolbox. Through observation, patterns are identified largely by means of induction, and questions arise that seemingly warrant systematic examination. Specifically, through induction, the particulars that emerge from being observant and attentive to the world and the experiences that world affords give rise to some manner of general pattern. In addition, evidence is sought through observations to verify or refute explanations drawn from physical experiences. Of course, what counts as accurate or valid observations is not arbitrary. Rather, those observations are governed by rules consistent with the form of empiricism espoused.

Hypothesis Generation. For many empiricists, their encounters, especially rich and repeated encounters with the physical world, become the data upon which they build tentative assumptions about relations among factors and forces embedded in those experiences. In effect, experiences permit hypothesis generation. From a more radical empiricist position, those hypotheses must be strictly induced or drawn out of individuals' encounters with the physical world, whereas from a more moderate stance there remains room for some abduction or perhaps even some deduction – reasoning processes applied in combination with induction. With abductive reasoning, or what Peirce (2010) called *guessing*, the spaces left by incomplete observations or induction are filled with logical inferences; hypothetical explanations about mechanisms or relations within the incomplete set of data.

Deduction, on the other hand, or the reasoning from general to particulars within the physical evidence, was more a part of hypothesis generation for the more liberal empiricists of the likes of James and Dewey who even parted ways on this issue with their fellow pragmatist, Peirce. For those who allowed for this interplay between induction, abduction, and deduction in hypothesis generation, an initial glimmer of a pattern could be sought in the observations through abduction, and the nature of that pattern which can then be mapped out through deduction. Induction may come into play through the mapping of hypotheses to the data in order to generalize from a limited set of observations. However, neither abduction nor deduction would be regarded as a valid process to those adhering to strict empiricist principles. Regardless of stance, however, the reason for hypothesis generation is to submit that tentative assumption to further testing through the accumulation of subsequent data.

Verification or Falsification. With a hypothesis generated principally by means of induction from experience, possibly with an sprinkling of abduction or deduction, the task of the empiricist is to systematically gather physical evidence that either verifies or falsifies the tentative assumptions formulated (i.e., hypothesis testing). Of course, it is essential that those data are not only systematically gathered during this phase, but also suitably chosen for the questions posed. The verification or falsification is undertaken in accordance with rules for the analysis of data, rules that serve to establish the rigor of the process. Thus, while induction and abduction stand as the primary mechanisms for generation of hypotheses, induction and deduction come more into play with hypothesis testing.

Theorizing. Finally, there is a place in empiricism for theory building. Fundamentally, theory in this situation is a claim about a pattern and its consistency that extends beyond the specific set of data gathered or the precise time or place in which those data were gathered. For empiricists, that theory can identify a mechanism or only go so far as to project a relation. Yet, as in all the processes we have discussed to this point, theory development must still conform to *a priori* rules for the generalizability allowed.

Cognitivism and Human Thought

Although empiricism and its counterpoint, rationalism, are philosophical views on the nature of knowledge and the justification of that knowledge, cognitivism is a psychological orientation that deals with the processes in which the mind engages. Those mental processes are present whether one is concerned with experiential interpretation or immersed in rational, logical thought. Those processes, which fill the spaces between sensory experience and the reflection on and response to those experiences, are part of the "black box" that has come to signify cognition.

We find express attention to cognition bases of human learning and development in the research and writing of the Gestalt school of psychology, the *Principles of Psychology* (Vols. I and II) authored by William James (1890), and subsequently in the developmental work of Jean Piaget (1930). However, cognitivism dramatically emerged as a movement in psychology during the 1970s and 1980s with the work of information-processing theorists like Herbert Simon (1989), cognitive developmentalists like John Flavell (1985), and cognitive scientists like John Anderson (1982). One of the primary catalysts for this growing concern for the "black box" of the human mind was the ever-expanding presence of computers and hypermedia technology (Goetz et al., 1992) and the concomitant interest in expertise (Alexander, Murphy, & Kulikowich, 2009). Collectively these forces sparked questions not only about what was evident in human behavior or performance, but also what unseen processes and mental activities contributed to such observable outcomes.

The Relation Between Cognitivism and Empiricism

We raise the topic of cognitivism because of the essential role it plays within the empirical research of many educational psychologists who investigate both *manifest* variables and *latent* variables. In keeping with the prior discussion of empiricism, manifest variables are those factors that can be directly measured through direct physical evidence (e.g., speed of reading or number of words written). We directly calculate how fast someone reads a segment of text or how many words they write in a given time frame, because speed or frequency are explicit measures. Thus, these are manifest variables. However, if we are interested in how well someone comprehends what they read or whether what they write communicates effectively, we must settle on indicators of these variables, since comprehension quality or communication effectiveness (i.e., latent variables) cannot be directly measured. Thus, aspects of empiricism (e.g., the gathering and systematic analysis of sensory and environmental data) became understandably linked to cognitivism, as physical indicators were sought as proxies for mental processes and as the methods of empiricism were applied to test hypotheses about latent variables pertaining to cognition, such as thinking, reasoning, and problem solving.

Historical Overview

> Thinking consists in envisaging, realizing structural features and structural requirements; proceeding in accordance with, and determined by, these requirements; thereby changing the situation in the direction of structural improvements.
>
> Max Wertheimer, *Productive Thinking*. 1959, p. 235

Gestalt Psychology and the Writings of James. During the latter half of the 19th century and into the early 20th century, there was an influential group of German theorists and researchers associated with the Berlin School. This group was committed to the hypothesis that the ability to discern sensory information from the environment was predicated on one's ability to *perceive*. This group, which included Kurt Koffka, Max Wertheimer, and Wolfgang Köhler, was the Gestalts, so named for the German word *estalt*, which loosely translates into whole, essence, or shape of an object or entity. Applying the scientific method and through repeated experimentation, these psychologists established a set of principles that demonstrated how the human mind imposed structure or order upon incomplete or ambiguous visual data from the environment.

What the Gestalts sought to demonstrate was that one merely did not pick up sensory data from the environment or experiences. Rather, according to this orientation, the mind helps to give shape or discern pattern from those environmental stimuli by formulating percepts consciously or most often automatically that lead to human response. The role of the unseen mind in what we perceive, therefore, is regarded as central to a cognitive orientation to human learning and development.

The mass of our thinking vanishes for ever, beyond hope of recovery, and psychology only gathers up a few of the crumbs that fall from the feast.

William James, *Principles of Psychology,* Vol. 2, 1890, p. 276.

William James was directly influenced by the work of the German school, as is evident in his writing of the *Principles of Psychology*. Delving directly into the "black box" of the human mind, James dedicated chapters of this tome to mental processes that he judged as foundational to the study of the psyche. Those foundational processes included attention; perception of time, space, and reality; reasoning; association; and memory. In arguing against the associationist schools of Herbart in Germany, and Hume, Mills and Bain in Britain, which he described as a "*psychology without a soul*" (p. 1), James wrote,

The experiences of the body are thus one of the conditions of the faculty of memory being what it is. And a very small amount of reflection on facts shows that one part of the body, namely, the brain, is the part whose experiences are directly concerned. If the nervous communication be cut off between the brain and other parts, the experiences of those other parts are non-existent for the mind. The eye is blind, the ear deaf, the hand insensible and motionless....The fact that the brain is the one immediate bodily condition of the mental operations is indeed so universally admitted nowadays that I need spend no more time in illustrating it, but will simply postulate it and pass on. The whole remainder of the book will be more or less of a proof that the postulate was correct. (1890, p. 4)

Although the research and writings of the Gestalts and William James appeared in the literature more than a century ago, their recognition of the place of "mind" in the space between experience and performance remains influential. Despite the waning investment in the internal workings of the mind with the expanding positivism of Thorndike and the behaviorism of Watson and Skinner, concerns for perception, attention, memory, and the like have never disappeared from the cognitive landscape.

Information-Processing Theory and Expertise. With the invention and popularization of the computer during the mid-20th century, there was again a rebirth of interest in the internal workings of the human mind, which had fallen into some disrepute with the rising influence of behavioral theories of human learning (Goetz et al., 1992). As with more extreme forms of empiricism previously discussed, behaviorists like Watson and Skinner, like Thorndike (1903) before them, viewed any investment in internal machinations of the mind as tantamount to "mentalism." But with the appearance of the computer, it became critical to understand what human minds do so that such processes could be programmed into the hardware of the machine.

Paradoxically, the more that psychologists and computer scientists focused their energies on trying to program computers to perform even simple human actions, the more that the computer and computer language became a metaphor for human cognition. This metaphorical influence is most evident in information-processing (IP) models. For instance, IP theorists talked about *input* from their environments and *outputs* from the mind, working memory, mental hardware, and storage capacity (Newell & Simon, 1972; Simon, 1989). They also examined what characteristics of individuals, such as their prior experiences, memory capacity, attentional patterns, or strategic capabilities, affect the way that information from the environment is inputted, processed, stored as knowledge, and subsequently outputted as behavior (Simon, 1978).

Because cognitivists and computer scientists needed to have idealized models of human processing to guide their analysis and programming, they became increasingly fascinated with experts and expert performance. What was it that experts in chess, medicine, or computer programming did mentally or what cognitive attributes did they possess that distinguished them from novices in the same field (Bransford, Brown, & Cocking, 1999)?

> The research on expert performance has, I believe, demonstrated convincingly that neither a magical bullet nor an innate talent can allow someone to attain an expert level of performance rapidly and effortlessly. Even for the most "talented" individuals, the road to excellence takes many years of daily deliberate practice to acquire the complex mechanisms and adaptations that mediate expert performance and its continued maintenance and improvement.
>
> K. Anders Ericsson (as interviewed by Schraw, 2005, p. 391)

We introduce the topic of expertise research here because of its systematic focus on the perceptions, reasoning, and problem-solving strategies of this exceptional group (Ericsson & Lehmann, 1996; Ericsson & Simon, 1993; Ericsson & Smith, 1991), which harkens back to the topics of the Gestalts and William James. This body of work has demonstrated that there is a marked difference between the way that experts and non-experts attend to, perceive, and reasoning about information. Specifically, experts can distinguish between relevant/irrelevant, important/unimportant information better; can capture the underlying structure of a problem that eludes non-experts; and process all this information at a much more rapid rate than non-experts (Alexander et al., 1992).

We do not want to leave our current discussion of cognitivism without noting that there are as many variants of cognitivism as there are of empiricism. We have devoted most of this overview to IP theory and expertise research. However, there are alternative forms of cognitivism that put more emphasis on either the mind of the individual or the social/cultural context in which that individual lives or works as the source of justification for what can be known. As a developmental psychologist, for example, Jean Piaget (1930) used his skills as a botanist to articulate a stage theory in which human capacity advanced from preoperational and sensory motor stages to concrete operational and then formal operational stages of cognition. Lev Vygotsky (1978; 1934/1986) took issue with the perspective on cognition advanced by Piaget (i.e., cognitive constructivism), which he estimated undermined the effects of social supports available within the context. Specifically, Vygotsky (1934/1986) argued that Piaget's stages reflected what individuals typically do on their own, whereas his social constructivism illustrated what individuals can potentially do when they are provided with guidance from a more significant other (i.e., zone of proximal development).

Although these two cognitivists differed on the role that others and the social context played, Piaget and Vygotsky shared a regard for the human mind and a concentration on the processes that unfold there. Their application of empirical methodologies resulted in somewhat varied interpretations about the nature of the "black box" and its development over the course of a lifetime.

Cognitivism and Its Paradigmatic Implications

As suggested, the methodologies of cognitivism have much in common with those previously outlined for empiricism—at least to the point where relevant data have been gathered and the systematic analysis of those data can be undertaken. Concerns for verification (i.e., reliability and validity) of data likewise remain, as do efforts to submit data in as unbiased manner to statistical tests to ascertain the pattern they reveal about human learning and performance. Yet, the paradigmatic implications for cognitivism are more complicated than for empiricism because of its interest in variables that are not directly manifest in nature. For that reason, we introduce several methodological principles that are in concert with most cognitive views.

Task Analysis. Because cognitivists are invested in understanding the underlying but unseen mechanisms that give rise to observe physical outcomes, they must engage in much more deduction and abduction than is evident in empiricism. Part of cognitivists' a priori analysis involves the identification of the internal mental processes that are likely to be required for some action or performance to occur. Such task analysis can be a very detailed accounting of the many mental processes entailed in a given action or performance (e.g., attending, remembering, or reasoning), as well as the various subroutines that a particular task might demand. Take the deceptively simply cognitive task of adding 2+2 and deriving the answer "4." For young children, such a task might require them to consider what the symbols "2" and "+" represent; calculating the quantity that results from this additive process; and, checking or verifying their product.

External Markers. In those instances when the focus of the cognitivists' work is on latent rather than manifest factors or forces, it is essential to identify external traces of the internal processes of import. Certainly, these external proxies for the specific factors or forces of true interest must be valid markers of the internal variables or constructs. For instance, let us say that we are interested in students' comprehension of text (i.e., their ability to derive meaning from linguistic information). As a marker of those internal comprehension processes, we might ascertain that students' ability to provide answers *about* the text would be indicative of the specific cognitive processes associated with comprehension.

Task Development. Deciding on the proxy data that would be valid external indicators of internal processes is but one step in the cognivitists' procedure. Tasks must be devised and administered that unearth such data in a manner that ensures the consistency and "truthfulness" of those data. Thus, cognitivists must spend time in the design of suitable instrumentation and in the statistical confirmation of the reliability and validity of the "evidence" that would be derived from that instrumentation. Let us return to the case of comprehension. Once it is determined that questions posed after reading would be appropriate indicators of comprehension processes, a set of questions must be constructed that effectively uncover the mental processes theoretically and empirically associated with comprehension (e.g., decoding, encoding, or inferring).

Model Construction. Because cognitivism often seeks to specify the relation between internal workings and external evidence, it can often entail the articulation of a process or relational model or hypotheses. Such process or relational models seek to specify the nature and interrelation of "black box" processes that are either associated with or predictive of the outcomes being measured. These models may specify not only the direction of hypothesized relations but also the theorized strength of those relations. For example, a cognitivist may hypothesize that students with more knowledge of the solar system will receive higher scores on questions that follow the reading of a text about the solar system, since comprehension is facilitated by the activation of relevant background knowledge. Then, cognitivist researchers would construct and administer a test about the solar system prior to students' reading about this topic and a measure of comprehension after the reading. That model would specify that higher prior knowledge should enhance comprehension and, thus, positively and significantly predict higher performance on the text questions.

Model Testing. As with empiricism, those adopting a cognitivist stance are held to the systematic testing of the models and relations they forward. Thus, data gathered by means of measures and instruments identified or constructed are submitted to statistical analysis. Moreover, that statistical analysis must be chosen to correspond to the nature of the questions posed, the data collected, and the relations or procedures specified. Again, as with empiricism, the outcomes will serve to either verify or falsify the claims made via the hypotheses or models forwarded. In addition, as we indicated with empiricism, there are well-articulated "rules" for hypothesis and model testing that must be followed if the outcomes are to be regarded as sound.

Theorizing and Modeling TSC

As the chapters in this volume illustrate, research and theories in psychology and related fields represent a wide variety of paradigmatic orientations, resulting in a rich and complex corpus of literature. When asked to contribute our perspective, we chose to describe both empiricism and cognitivism, not only because these paradigms are prevalent in our subfield, Educational Psychology, but also because they ground our own theoretical and methodological approach (Alexander, Schallert, & Reynolds, 2009).

Moreover, examining the relation between paradigm, theory, and model is particularly timely for us. At present, we are reconceptualizing the nature of comprehension, more particularly the processes that give rise to the understanding of symbolic representations, which is an elemental facet of human learning. Traditionally, comprehension has been viewed as the unearthing of meaning encrypted in linguistic symbols, most frequently classroom talk and written texts. However, we argue that the meaningful information encoded in alternative symbolic system (e.g., numeric, graphic, or pictorial) must also be unlocked through some of the same processes that are applied in the comprehension of oral or written language. This extended conceptualization has led us to develop a new theory of comprehension, the theory of TSC, and construct and test models arising from it.

This theory-building endeavor has required us to reflect upon our paradigmatic orientations, as well as interrogate and operationalize the fuzzy distinction between theory and model. In this section, we describe the process we undertook to develop and test the TSC in an effort to both illustrate the empiricist and cognitivist paradigms and to highlight the interrelation among paradigm, theory, and model.

Building the Theory

Our theory-building process began with the most basic of empiricist processes: observation. At the start of her professorial career, the first author was a reading teacher at an elementary school that took an arts-integration approach to instruction, wherein traditional subjects (e.g., reading or mathematics) were taught with and through the arts (i.e., visual art, music, dance, and drama). Thus, the first author often integrated paintings into her reading lessons. For instance, before having students read a poem by a Harlem Renaissance poet, she would have students look at and discuss a painting from the same period.

Over time, the first author began to observe a striking similarity between how students "read" paintings and how they read traditional text. In particular, students' discussions of the paintings evidenced their efforts at making connections, questioning, drawing inferences, and elaborating; some of the same comprehension processes necessary for understanding linguistic information. These observations raised questions about the nature of comprehension. Could it be that comprehension of linguistic text (i.e., print) and comprehension of a painting are dependent upon the same or similar processes? Moreover, might this set of general comprehension processes extend beyond linguistic and visual displays to other symbol systems, such mathematical or musical notation?

We set out to find the answer to those questions by first examining the literature for a theory that suggested the processes inherent in comprehending information across symbol systems. We found none. Rather, theories and models of comprehension were explicitly or tacitly nested in the context of linguistic processing (e.g., Graesser, Mills, & Zwaan, 1997; Kintsch, 1998), either excluding non-linguistic compositions or assuming the function in the same manner as text. Moreover, while we were able to identify a limited body of literature addressing nature and process of meaning-making in non-linguistic contexts (cf., Parsons, 1987), with few exceptions (e.g., van Kraayenoord & Paris, 2002), these investigations did not reference concepts or apply terminology common to traditional comprehension research. Thus, although it appeared that there was some shared desire to explore and describe meaning-making processes in different symbol systems, there was no unifying theoretical frame we could identify.

This lack of cohesion was also noted by Unsworth (2008). In his discussion of future directions for New Literacies research, Unsworth articulated the need or a "trans-disciplinary" framework that provides a unified resource for research in the comprehension of information presented in "meaning-making" systems, such as language, visual display, sound, and movement. He urged literacy educators, linguists, information, and media researchers, as well as psychologists to find common, compatible, or complementary theoretical frameworks that, together, inform the ways in which individuals process and understand information. In addition, he argued that this framework should allow researchers to pursue a focused study within a single meaning-making system, while building bridges between symbol systems.

Thus, it became clear that, while more than a century of research has illuminated the processes inherent in gaining meaning from text, the understanding of how individuals process information within and across symbol systems remained rudimentary. Further, it was evident that there was a need for a theoretical reconceptualization of comprehension that addressed the processes necessary for understanding both linguistic and non-linguistic information.

With this goal in mind, we returned to the literature with an empirical intent to examine conceptions of the nature of comprehension within and across symbol systems. Heeding the guidance of Unsworth (2008) and others (Boote & Beile, 2005; Maxwell, 2006), we cast a wide net to include literature in educational psychology as well as complementary literatures such as philosophy, semiotics, curriculum design, neuropsychology, museum education, and cognitive psychology.

The results of this extended review suggested the existence of both symbol-general and symbol-specific comprehension processes. For example, early conceptions about the nature and process of understanding (e.g., James, 1890/1950) suggest that there may be some significant similarity between the ways in which information is processed, regardless of the vehicle from conveyance. For example, attention is required for understanding a painting as well as a poem, just as associating new information with old is important in comprehending a statement in mathematics and in music. We hypothesized that comprehension, therefore, would include processes that transcend symbol systems, or *trans-symbolic processes*.

However, we also found several works implying that the manner in which information is presented (i.e., the symbology) influences processing. For instance, Sadoski and Paivio (1991) argued, based on Dual Coding theory, that at some level, words and pictures evoke different conceptual processes involving separate cognitive mechanisms. As suggested by this work and that of others (Mayer, 2001; Neisser, 1967), it seemed plausible that, in addition to trans-symbolic processes, comprehension would entail processes specific to the symbology in which information is presented (i.e., *symbol-specific processes*).

Building on the information presented in the literature and the observations of the first author, we developed the theory of TSC. Specifically, the TSC addresses meaning-making processes across and within symbol systems, suggesting that both trans-symbolic and symbol-specific processes are present in any act of comprehension. This expanded conceptualization of comprehension was not intended to negate or supplant the decades of research on text comprehension processes. Rather, our purpose in forwarding the TSC was to expand the frame of comprehension to reflect other forms of human communication, with the full expectation that, because of their trans-symbolic nature, many of the processes currently associated with traditional comprehension (i.e., print) would be evidenced in meaning-making endeavors in non-text symbologies.

Modeling and Testing the Theory

The TSC is conceptually large, in that it attempts to explain a complex phenonena (i.e., comprehension) across and within symbol systems. As such, it conforms to the definition of a theory previously discussed (Hillix & L'Abate, this volume), and investigating its robustness requires iteratively exploring and modeling particular relations. Thus, the next step in forwarding the TSC was to interrogate it by constructing and empirically testing a model arising from it. However, because the TSC was in its initial stages of verification, it was necessary to first determine whether *any* commonalities in comprehension processes exist between any two symbologies. Once this relation was clarified, subsequent relations could be addressed, and the scope TSC could be further articulated.

Model Building. We believed there was a logical first step in this verification process. In particular, we held that the robustness of the TSC could best be determined by investigating the relation between two symbol systems that are maximally distinct in their symbolic feature while being maximally similar in the processing they require and in the ideas they seem to convey. In other words, two symbol systems that seemed, on the surface, to be as different as possible, but still capable of communicating similar ideas. To make this determination, we drew on the work of Goodman (1976), who examined the nature and structure of symbol systems, what he called the "languages of art." Our analysis of his writings led us to construct a model of the TSC that brought us full circle – interrogating the relation between print and painting.

Print and paintings are, at first blush, quite different, most notably because they arise from different symbol systems. However, there are other important distinctions. For one, decoding (i.e., deciphering) print requires specific and necessary processes, such as sound/symbol associations, which must be executed in rather rigid, proscribed, and linear fashion (e.g., moving across the line of text and making required sweeps from line to line). When it comes to pictures, however, there can be more holistic, gestalt-like processing of the display, even though there may be some prosodic rhythm to the movement of the eye across the canvas.

Another important difference between print and paintings arises from the distinction between descriptive and depictive representations. According the Schnotz (2005), descriptions consist of symbols, which Peirce (1931/1958) defines as signs that have no similarity to their referent. Text and mathematical expressions are common kinds of descriptions. For instance, the word "ant" has no similarity to a real ant and the symbol 2 has no relation to two single units of an actual something.

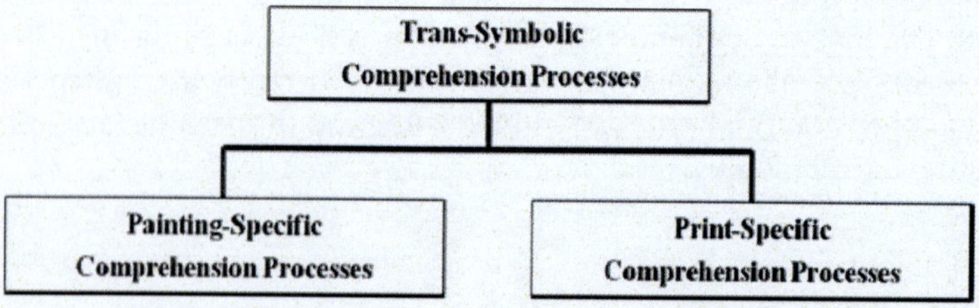

Fig 15.1 Model of the relation between trans-symbolic comprehension (TSC) processes and the comprehension processes particular to print and painting, specifically, as predicted by the theory of TSC

In contrast, depictions consist of icons, which are similar to or share a structural commonality with their referent. Pictures, paintings, and drawings are depictive representations, as are maps and bar graphs.

Schnotz (2005) also notes that descriptive and depictive representations are useful for different purposes. Descriptive representations such as text are more powerful than depictions in expressing some forms of abstract knowledge, such as the disjunctive *or* (e.g., the Marsh Harrier feeds on mammals *or* reptiles). On the other hand, descriptive representations are more informationally complete and are thus more useful for communicating a large amount of relatively concrete information efficiently. To return to the marsh harrier example, a picture of the bird stalking a mouse on the edge of a marsh communicates not only the fact that mice are preyed upon by marsh harriers, but also the size of the bird relative to the mouse, their orientation in space, the environment of the bird's common geographical area, and so forth. This example illustrates the common adage, "A picture is worth a thousand words" and highlights the fact that depictive representations can be more useful than descriptive representations for drawing inferences (Kosslyn, 1994).

Despite these differences, print and paintings share a significant similarity. They can communicate many of the same ideas. For instance, both authors and painters seek to convey descriptions, narratives, and abstractions such as emotion. Moreover, both print and paintings can be conceptually complex, utilize symbolism and metaphor, subject to multiple interpretations, and have layers of meaning.

Because of this unique relation between print and painting, it seemed apparent to us that, for the initial interrogation of the TSC, modeling their relation was an appropriate, rigorous first step. If we found significant overlap between the comprehension processes inherent in understanding print and paintings, so different in their symbolic display, then the existence of TSC processes would be affirmed. Further, these symbolic differences would likely give rise to processes specific to print or to painting, affirming the inclusion of symbol-specific processes in the theory. To this end, we constructed a visual representation of the hypothesized relation between print and painting (Figure 15.1). As evidenced from the model, it is a subset of the TSC and adheres to its scope, parts, and parameters.

Model Testing. Our next step was to systematically interrogate the model by seeking evidence of trans-symbolic and symbol-specific processes in individuals'

comprehension of print and paintings through four phases: evidence seeking, question formulation, task development, and data analysis.

Evidence Seeking. We drew on our both empiricist and cognitivist roots for guidance in seeking evidence of the model. Foremost, in keeping with our empiricist approach, we wanted to design a situation in which evidence of the comprehension processes would be observable and documentable. Further, because of our cognitivist orientation, we wanted to elicit these comprehension processes as they were unfolding. Thus we determined that the best approach for gaining observable, documentable, and real-time evidence of these comprehension processes was via a think-aloud protocol.

Think-aloud protocols are research methodologies in which participants perform a task while continuously reporting thoughts that occur during its implementation. Think-aloud protocols operate under the assumption that verbalizations are related to participants' concurrent thoughts that emanate from working memory (Ericsson & Simon, 1993). During a think aloud, participants in our research would thus be prompted to verbalize their thoughts while completing "reading" a painting and reading a text.

Question Formulation. With think aloud as our method for observing the desired processes, we turned to the development of questions to guide the investigation. Of primary concern to us was ascertaining the existence of the parts of the model (i.e., trans-symbolic and symbol-specific processes) in the context of print and paintings. Thus, we asked:

> What, if any, trans-symbolic and symbol-specific processes are evidenced when participants think aloud while comprehending print and paintings?

Further, as part of the verification process, we wanted to know whether individuals were consistent in their use of comprehension processes across symbol systems (i.e., Is a good comprehender of print also a good comprehender of a painting?). Thus our other research question was:

> What is the relation between individuals' manifest comprehension of print and paintings?

Task Development. The next step was developing a task to answer the research questions. We first determined that it was necessary to have participants think aloud with both print and painting. This approach would allow us to observe comprehension processes elicited in both symbol systems and to note patterns across and within them. In addressing the second research question, we made an effort to develop a task that, by its design, controlled for other, non-symbolic influences on comprehension. In other words, we wanted any differences between the manifest comprehension of print and paintings to be an artifact of their symbolic composition, not other factors. For instance, we were cognizant of the influence of prior topic knowledge on comprehension of print (Baldwin, Peleg-Bruckner, & McClintock, 1985; Graesser, Singer, & Trabasso, 1994) and so designed the task to control for it. Thus, we created topically themed think-aloud conditions, wherein participants engaged with a print and painting on the same general topic (e.g., a linguistic description and a painting depiction of a pastoral scene).

Data Analysis. The final step in the interrogation of a model is gathering and analyzing the data. As noted previously, the TSC is an ongoing project and, although we have completed a pilot study, we are currently in the data analytic phase of the full inquiry. For both the pilot and full studies, we are taking an empirical approach. First, after gathering think-aloud data from all participants, we will transcribe and code the utterances using an inductive, bottom-up approach. That is to say, because we cannot rely *a priori* on coding schemes derived from previous research investigating trans-symbolic and symbol-specific comprehension processes, we must examine the data for patterns that evidence these processes. Once these patterns emerge, we plan to quantify and subject them to statistical analysis. In keeping with both empiricist and cognitivist paradigms only those statistical procedures that fit the nature of the resulting data, while allowing us to address our overarching questions, will be applied. Further, only those interpretations warranted by the data will be forwarded. Although analysis of the full study is still underway, preliminary findings endorse the model, and thus the theory of TSC.

Discussion

As the foregoing account illustrates, our line of research is steeped in both the empiricism and cognitivism traditions. Our empiricist roots are evident in the observation of students "reading" paintings and the subsequent hypothesis that print and painting comprehension might be dependent on shared comprehension. It is also evidenced by our methodological approach to modeling and testing the TSC, specifically in the choice of think-aloud utterances as our source of data and our inductive, bottom-up approach to their analysis.

Our cognitivist approach is also illustrated. First, drawing on the TSC, we constructed a model examining the relation between print and painting comprehension. Specifically, our model suggested that comprehension of print and painting relies on shared processes, as well as processes specific to each symbol system. Then we tested the model by analyzing the task of comprehension, identifying external markers of it (i.e., think aloud utterances), developing an experimental task designed to elicit it from participants, and subjecting the resulting data to appropriate statistical analysis.

Although our approach and process reflect empiricist and cognitivist orientations, it is also clear that they are not pure representations of either. Because we are interested in latent, unobservable phenomenon, namely the internal processes of comprehension, we cannot call ourselves wholly empiricists, despite the fact that we believe in the power of observation and measurement. Unlike Galileo, who dealt with manifest variables, we cannot claim with certainty that we have accurately observed or measured the object of our investigation, because an individual's comprehension cannot be perceived by the senses or measured with precision. Instead, we are forced to rely on proximal observations (i.e., individual's utterances in the think-aloud protocol) and the assumption that these observations are true reflections of unobservable comprehension processes. In short, in order to investigate this important aspect of human learning, we must assume that what we can measure is reflective of what we really want to assess. Moreover, while we relied upon observations of utterances in our theory-building processes, we also relied on no small measure upon other, non-observable sources, that is, the writings of philosophers, psychologists, theorists, and researchers who, like us, are concerned with the nature and process of human learning.

Likewise, we cannot claim to be wholly cognitivists. As indicated previously, many of the cognitivist persuasion emphasize the social, motivational, and environmental context of an individual as important sources of justification for what can be known (e.g., Vygotsky, 1934/1986). In the explication of the TSC, however, we stayed within the mind of the one, using the individuals' manifest processes as the sole source of data. While we do not question the power of these environmental influences on the development and instantiation of comprehension, we make no attempt here to address them, positioning us at a distance from other cognitivists.

Thus, if our experience is representative, we suggest that the nature of paradigmatic orientation is not a black–white distinction, but rather a study in grays. Paradigms serve as flagposts in the psychological landscape, allowing researchers and theorists to position themselves and one another relative to the myriad of orientations represented in the field. We predict, however, that very few will find themselves adhering strictly to their paradigmatic roots. The complex nature of the human mind and behavior, the tension between latent and manifest variables, and the quantitative–qualitative question add significant complexities to the field of psychology and, in practice, blur the paradigmatic boundaries.

Conclusion

While we acknowledge and accept the limitations of paradigms in practice, we remain, nonetheless, convinced of their necessity in theory-building endeavors. As noted by Bredo (2006), one goal of psychology, and the related study of human learning, is to identify and correct conceptual confusion. Thus, it is essential that psychology acknowledges and treats its definitional, conceptual, and methodological disorder (Hancock, 2007). The articulation in this volume of important paradigms in the psychological literature, as well as the work of Hillix and L'Abate (Chapter 1 this volume) in untangling the relation among paradigm, theory, and model, constitutes an important step toward conceptual clarity and definitional precision.

Our process of conceptualizing and articulating the TSC are evidence of the elemental role of paradigms in theory building. We appreciate that, without the conceptual framework and operational roadmap provided by empiricism and cognitivism, our efforts to understand comprehension in its many symbolic iterations could not have transpired. By using these paradigmatic tools, we have found a path through the complex and treacherous terrain that describes human learning and performance and, as a result, are convinced that our efforts will contribute to the existing literature.

References

Ackrill, J. L. (1963). *Aristotle: Categories and De Interpretatione*. Oxford: Clarendon Press.

Alexander, P. A. (2009). *The need for theory in educational psychology: A case in point*. Wellington, New Zealand: Keynote at the Second Educational Psychology Forum.

Alexander, P. A., Reading, T. D., & Laboratory, L. R. (2010). *The challenges of developing competent literacy in the 21st century*. Washington, DC: The National Academy of Sciences.

Alexander, P. A., Murphy, P. K., & Kulikowich, J. M. (2009). Expertise and the adult learner: A historical, psychological, and methodological exploration. In M. C. Smith & N. DeFrates-Densch (Eds.), *The handbook of research on adult learning and development* (pp. 484–523). New York: Routledge.

Alexander, P. A., Schallert, D. L., & Hare, V. C. (1991). Coming to terms: How researchers in learning and literacy talk about knowledge. *Review of Educational Research, 61*, 315–343.

Alexander, P. A., Schallert, D. L., & Reynolds, R. E. (2009). What is learning anyway? A topographical perspective considered. *Educational Psychologist., 44*, 209–214.

Anderson, J. R. (1982). Acquisition of cognitive skill. *Psychological Review, 89*, 123–126.

Aristotle. (1979). *Metaphysics, Books I–IX* (H. Tredennick, Trans.). Cambridge, MA: Harvard University Press.

Audi, R. (1999). *The Cambridge dictionary of philosophy* (2nd ed.). New York: Cambridge University Press.

Baldwin, R. S., Peleg-Bruckner, Z., & McClintock, A. H. (1985). Effects of topic interest and prior knowledge on reading comprehension. *Reading Research Quarterly, 20*, 497–504.

BonJour, L. (1998). *In defense of pure reason: A rationalist account of a priori justification*. New York: Cambridge University Press.

Boote, D. N., & Beile, P. (2005). Scholars before researchers: On the centrality of the dissertation literature review in research preparation. *Educational Researcher, 34*, 3–15.

Bransford, J. D., Brown, A. L., & Cocking, R. R. (1999). *How people learn: Brain, mind, experience, and school*. Washington, DC: National Academy Press.

Bransford, J. D., Vye, N. J., Stevens, R., Kuhl, P., Schwartz, D., Bell, P., et al. (2006). Learning theories and education: Toward a decade of synergy. In P. Alexander & P. Winne (Eds.), *Handbook of educational psychology (Volume 2)*. Mahwah, NJ: Erlbaum.

Bredo, E. (2006). Conceptual confusion and educational psychology. In P. Alexander & P. Winne (Eds.), *Handbook of educational psychology (Volume 2)*. Mahwah, NJ: Erlbaum.

Dewey, J. (1929). *The quest for certainty: A study of the relation of knowledge and action*. New York: Milton, Balch & Company.

Dewey, J. (1913). *Interest and effort in education*. Boston: Riverdale.

Dinsmore, D. L., Alexander, P. A., & Loughlin, S. M. (2008). Focusing the conceptual lens on metacognition, self-regulation, and self-regulated learning. [Special Issue on Metacognition, self-regulation, and self-regulated learning: Historical roots and contemporary manifestations] Educational Psychology Review, 20, 391-409.

Ericsson, K. A., & Lehmann, A. C. (1996). Expert and exceptional performance: Evidence of maximal adaptation to task. *Annual Review of Psychology, 47*, 273–305.

Ericsson, K. A., & Simon, H. A. (1993). *Protocol analysis: Verbal reports as data*. Cambridge, MA: MIT Press.

Ericsson, K. A., & Smith, J. (1991). *Toward a general theory of expertise: Prospects and limits*. New York: Cambridge University Press.

Feigl, H. (1970). The "orthodox" view of theories. Remarks in defense as well as critique. In M. Radner & S. Winokur (Eds.), *Analyses of theories and methods of physics and psychology*. Minneapolis: University of Minnesota Press.

Flavell, J. H. (1985). *Cognitive development* (2nd ed.). Englewood Cliffs, NJ: Prentice-Hall.

Galilei, G. (1638). *Discourses and Mathematical Demonstrations Relating to Two New Sciences* (Discorsi e dimostrazioni matematiche, intorno à due nuove scienze. Leiden, Netherlands: Elsevier Publishers.

Gill, M. L. (2005). Aristotle's metaphysics reconsidered. *Journal of the History of Philosophy, 43*, 223–251.

Goetz, E. T., Alexander, P. A., & Ash, M. (1992). *Educational psychology: A classroom perspective*. Columbus, OH: Charles E. Merrill.

Goodman, N. (1976). *Languages of art: An approach to a theory of symbols*. Indianapolis/Cambridge: Hackett Publishing Company.

Graesser, A. C., Millis, K. K., & Zwaan, R. A. (1997). Discourse comprehension. *Annual Review of Psychology, 48*, 163–189.

Graesser, A. C., Singer, M., & Trabasso, T. (1994). Constructing inferences during narrative text comprehension. *Psychological Review, 101*, 371–395.

Greeno, J., Collins, A., & Resnick, L. (1996). Cognition and learning. In R. Calfee & D. Berliner (Eds.), *Handbook of educational psychology* (pp. 15–46). New York: MacMillan.

Hancock, G. R. (2007). Models for illuminating things otherwise unseen: Co-editor's introduction. *Contemporary Educational Psychology, 32*, 4–7.

James, W. (1890). *The principles of psychology (Vols. 1 and 2)*. New York: Dover Publications.

James, W. (1912). *Essays in radical empiricism*. New York: Longmans, Green, & Co.

James, W. (1955). *Pragmatism and four essays from the meaning of truth*. New York: Meridian.

James, W. (1997). Pragmatism's conception of truth. In L. Menand (Ed.), *Pragmatism: A reader* (pp. 112–131). New York: Vintage.

Kintsch, W. (1998). *Comprehension: A paradigm for cognition*. New York: Cambridge University Press.

Kosslyn, S. M. (1994). *Image and brain*. Cambridge, MA: MIT Press.

L'Abate, L. (2005). *Personality in intimate relationships: Socialization and psychopathology*. New York: Springer-Science.

L'Abate, L. (2009). Paradigms, theories, and models: Two hierarchical frameworks. In L. L'Abate, P. De Giacomo, M. Capitelli, & S. Longo (Eds.), *Science, mind and creativity: The Bari symposium* (pp. 107–122). New York: Nova Science Publishers.

Leahey, T. H. (1992). The mythical revolution of American psychology. *American Psychologist, 47*, 308–318.

Locke, J. (1979). *An essay concerning human understanding* (P. Nidditch, Ed.). New York: Oxford University Press.

Mill, J. S. (1869). *On liberty*. London: Longman, Roberts & Green.

Murphy, P. K. (2007). The eye of the beholder: The interplay of social and cognitive components in change. *Educational Psychologist, 42*(1), 41–53.

Murphy, P. K., Alexander, P. A., Greene, J. A., & Edwards, M. N. (2007). Epistemological threads in the fabric of conceptual change research. In S. Vosniadou, A. Baltas, & X. Vamvakoussi (Eds.), *Re-framing the conceptual change approach in learning and instruction* (pp. 105–122). Amsterdam: Elsevier.

Murphy, P. K., Alexander, P. A., & Muis, K. R. (in press). Knowledge and knowing: The journey from philosophy and psychology to human learning. In K. A. Harris & S. Graham (Eds.), *Handbook of educational psychology*. Washington, DC: American Psychological Association.

Parsons, M. J. (1987). *How we understand art: A cognitive development account of aesthetic experience*. Cambridge: Cambridge University Press.

Pedhazur, E. J., & Schmelkin, L. P. (1991). *Measurement, design, and analysis: An integrated approach*. Hillsdale, NJ: Erlbaum.

Peirce, C. S. (1958). The fixation of belief. In P. P. Wiener (Ed.), *Charles S. Peirce: Selected writings* (pp. 91–112). New York: Dover.

Peirce, C. S. (2010). *The writings of Charles S. Peirce, Vol. 9, 1890-1892*. Bloomington, IN: Indiana University Press.

Piaget, J. (1930). *The child's conception of the world*. New York: Harcourt, Brace & World. (Original published in 1926.)

Plato. (1993). *Republic/Plato* (R. Waterfield, Trans.). Oxford, England: Oxford University Press.

Manion, L., & Morrison, K. (2007). *Research methods in education* (6th ed.). London: RoutledgeFalmer.

Marsh, H., & Hau, K.-T. (2007). Applications of latent-variable models in educational psychology: The need for methodology and substantive synergies. *Contemporary Educational Psychology, 32*, 151–170.
Maxwell, J. A. (2006). Literature reviews of, and for, educational research: A commentary on Boote and Beile's Scholars before researchers. *Educational Researcher, 35*, 28–31.
Mayer, R. E. (2001). *Multimedia learning*. New York: Cambridge University Press.
Mertens, D. M. (2005). *Research and evaluation in education and psychology: Integrating diversity with quantitative, qualitative, and mixed methods* (3rd ed.) Thousand Oaks, CA: Sage Publications.
Neisser, U. (1967). *Cognitive psychology*. New York: Appleton, Century, Crofts.
Newell, A., & Simon, H. A. (1972). *Human problem solving*. Englewood Cliffs, NJ: Prentice-Hall.
Pemberton, M. A. (1993). Modeling theory and composing process models. *College Composition and Communication, 44*, 40–58.
Robinson, D. H., Levin, J. R., Thomas, G. D., Pituch, K. A., & Vaughn, S. (2007). The incidence of "causal" statements in teaching and learning research journals. *American Educational Research Journal, 44*, 400–413.
Ross, W. D. (Ed.). (1951). *Aristotle's prior and posterior analytics*. Oxford: Clarendon Press.
Ruddell, R. B., Ruddell, M. R., & Singer, H. (Eds.). (1994). *Theoretical models and processes of reading* (4th ed.). Newark, DE: International Reading Association.
Sadoski, M., & Paivio, A. (2001). *Imagery and text: A dual coding theory of reading and writing*. Mahwah, NJ: Lawrence Erlbaum Associates.
Schnotz, W. (2005). An integrated model of text and picture comprehension. In R. E. Mayer (Ed.), *The Cambridge Handbook of Multimedia Learning*. New York: Cambridge University Press.
Schraw, G. (2005). An interview with K. Anders Ericsson. *Educational Psychology Review, 17*, 389–412.
Seidman, E. (1978). Just, values and social science: Unexamined premises. In R. J. Simon (Ed.), *Research in law and sociology*. Greenwich, CN: JAI Press.
Simon, H. A. (1978). Information-processing theory of human problem solving. In W. K. Estes (Ed.), *Handbook of learning and cognitive processes* (pp. 271–295). Hillsdale, NJ: Lawrence Erlbaum Associates.
Simon, H. A. (1989). *Models of thought* (Vol. 2). New Haven, CT: Yale University Press.
Skinner, B. F. (1953). *Science and human behavior*. New York: Macmillan.
Skorupski, J. (1989). *John Stuart Mill*. London: Routledge.
Thomas, R. M. (1996). *Comparing theories of child development* (4th ed.). Pacific Grove, CA: Brooks/Cole.
Thorndike, E. L. (1903). *Educational psychology*. New York: Science Press.
Tracy, D. H., & Morrow, L. M. (2006). *Lenses on reading: An introduction to theories and models of reading*. New York: Guilford Press.
Triadafillidis, T. A. (1998). Dominant epistemologies in mathematics education. *For the Learning of Mathematics, 18*, 21–27.
Unsworth, L. (2008). Multiliteracies and metalanguage: Describing image/text relations as a resource for negotiating multimodal texts. In D. Leu, Corio, J., Knobel, M., Lankshear, C. (Ed.), *Handbook of research on new literacies* (377-405). Mahwah, NJ: Erlbaum.
van Kraayenoord, C. E., & Paris, S. G. (2002). Reading objects. In S. G. Paris (Ed.), *Perspectives on Object-Centered Learning in Museums*. Mahwah, New Jersey: Lawrence Erlbaum.
VandenBos, G. R. (Ed.). (2007). *APA dictionary of psychology*. Washington, DC: American Psychological Association.
Vygotsky, L. S. (1978). *Mind in society*. Cambridge, MA: Harvard University Press.
Vygotsky, L. S. (1986). *Thought and language*. (A. Kozulin, Trans.). Cambridge, MA: MIT Press. (Original work published in 1934)
Watson, J. B. (1924). *Behaviorism*. Chicago: University of Chicago Press.
Watson, J. B. (1913). Psychology as the behaviorist views it. *Psychological Review, 20*, 158–177.
Wertheimer, M. (1959). *Productive thinking*. New York: Harper.
Wolterstorff, N. P. (1996). *John Locke and ethics of belief*. Cambridge: Cambridge University Press.

Chapter 16
Humanism and Behaviorism

David Ryback

Imagine a hall with standing room only as B.F. Skinner – staunch proponent of radically deterministic version of behaviorism – and Abraham Maslow – one of the fathers of humanistic psychology – debate their respective perspectives. Such a debate never did take place but, if it did, each eminent scholar might proclaim a paradigm with which the world of psychology was quite familiar.

Maslow's model, were he to present one, would most likely demonstrate his Hierarchy of Needs:

Self-actualization
Self-esteem
Love, affection, and sense of belonging
Safety needs
Physiological needs

According to Maslow (1962), we all need to satisfy our physiological needs, such as oxygen, water, and food. Thus satisfied, we would then look for some safe social structure to allay our anxieties about safety. Our next search would be for love and affection as well as a sense of security about being part of a peer group. This would lead to a greater need, that of respect and esteem from such groups. Finally, having enough fresh air, water, food, security in our surroundings, love and affection, respect, and self-esteem, we would be ready for the highest state, that of giving to the world from an authentic self, with whatever talents or skills or unique gifts we have to offer that make us feel special and connected to the world at the highest level of altruistic sharing.

D. Ryback (✉)
EQ Associates International Suite 201, 1534 N. Decatur Rd., Atlanta, GA 30307–1022, USA
e-mail: David@EQassociates.com

L. L'Abate (ed.), *Paradigms in Theory Construction*, DOI 10.1007/978-1-4614-0914-4_16,

Skinner's structure – simple and succinct – put forth the model in which any behavior or response (R) followed by a rewarding stimulus (S^+), in a particular environmental stimulus setting (S^d) discriminated by the subject, would work to increase the probability of that particular behavior:

$$S^d \cdot R \rightarrow S^+$$

The response of interest could be as simple as depressing a bar, as complex as improved reading comprehension, or as abstract as altruistic leadership style. For complex behaviors, all that would necessary would be to analyze the behavior into its component responses, build the acquisition of these responses, and then synthesize the components into the orchestrated, complete desired behavior. This simple paradigm, according to Skinner, explained all our behaviors, how they were acquired and maintained over time.

Worthless Mind vs. Inner Truth: Two Opposing Paradigms: Hamlet and the Merchant of Venice

These two different models are taken from paradigms that are, in a sense, opposites of one another. Skinner's model is reductionistic, reducing all behaviors to simple response components, influenced by the rewards or punishments that result in the increased or decreased probabilities of any behavior under analysis. Maslow's model, though elegant in its simplicity, allows for more human variability and uniqueness and fosters a sense of human essence that transcends behavioral analysis. Skinner portrays human activity as a mere function of the surrounding stimuli. Maslow's view is to characterize human behavior as continually evolving into essential uniqueness. These two paradigms could not be further apart.

If Shakespeare were to explain, in his eloquent words, these two opposing paradigms, he might use the following quote to characterize the *behaviorist's* devaluation of thinking and mental process in general:

> His reasons are as two grains of wheat hid in two bushels of chaff—you shall seek all day ere you find them, and when you have them they are not worth the search.
>
> –Merchant of Venice

His quote to characterize the *humanistic* paradigm of authenticity might be:

> This above all: to thine own self be true,
> And it must follow, as the night the day,
> Thou canst not then be false to any man.
>
> –Hamlet

Paradigms of Value (or Not): Whole vs. Part

Though I use the terms "model" and "paradigm" freely, there is a difference. As I see it, the term "model" refers to an attempt at making a conceptual framework somewhat graphic, so we can wrap our minds around that concept quickly at some concrete level. The term "paradigm" is more abstract and inclusive, taking into consideration a broader conceptual approach that is more laden with the values of the community embracing any paradigm.

So Skinner's reductionist model is part of the behaviorist paradigm to which Occam's Razor is applied – no relevance is given to the human's conscious mind. People respond to surrounding temporal stimuli and thinking is a metaphenomenon which reveals nothing of importance in terms of the research on human behavior.

Maslow's Hierarchy of Needs is offered as a model within the paradigm of humanistic psychology which is diametrically opposed to that of the behaviorists. The humanistic paradigm looks to thinking and feeling as the essence of what is important in research on human behavior. The main postulates of the humanistic paradigm, according to Bugental (1964), maintain that humans cannot be reduced to their components, and emphasize conscious awareness, particularly of meaning, value, and creativity. Skinner, on the other hand, has no place, in the paradigm he offers, for value. The behaviorist paradigm looks to using the stimulus–response model as a set of tools for influencing behavior, both simple and complex. "When we can design small social interactions," he maintained, "and, possibly, whole cultures with the confidence we bring to physical technology, the question of value will not be raised" (1961, p. 545). Skinner went on to proclaim that his simple, yet powerful model of human behavior could revolutionize the science of psychology to reach undreamt rewards. "We have reached the stage … in which man can determine his future with an entirely new order of effectiveness" (1964, p. 485).

Historical Overview of Behaviorism: From Skinner Box to Pop-Up Ads

Before tackling some deeper question, let us travel back to earlier versions of their "uninterrupted calculus," to use Feigl's (1970) term. Both Skinner and Maslow had predecessors who laid the groundwork for their respective models and paradigms – termed the "objective" and "existential," respectively, by Carl Rogers (1960).

The Associationists, a direct outgrowth of the British empiricists of the 1800s, were not confined to Great Britain as there were also Associationists in the USA and in Russia about 1898. As early as 1749, David Hartley maintained that association derived from the contiguity of experience. In 1829, James Mill came forth with the concept of ideas linked by association. In Russia about 1890, Pavlov demonstrated the delayed conditioned reflex and experimental neuroses in animals, as he did research on physiological responses. A couple of decades later in America, John B. Watson founded the Behaviorist school. Beginning about 1912, Watson brought forth the paradigm that Skinner would adopt, discarding introspection in favor of complete reliance on objective data – stimuli and responses. This paradigm was also adopted by such scholars as John Dollard, Neal Miller, George Kelly, and many others (Woodworth, 1948).

If Watson is the father of the modern behavioral model, then Skinner is the father of the paradigm. He took the model – based on research on three particular species (rats, pigeons, and humans) – and built the paradigm of behavior manipulation through stimulus control. Skinner (1955–56) maintained that people were already manipulating their fellow humans. To him, denial of the importance of natural laws within his paradigm was an indication of immaturity in our culture. He maintained that one of the reasons we balked at the recognition of his reductionist paradigm in which we manipulate one another was because of the imperfection of existing techniques.

Sure enough, techniques of behavioral manipulation did improve over the decades, with dramatic results in improved reading skills (Ryback, 1971a, 1976a; Ryback & Staats, 1970) as well as in treating emotional disturbances (Andronico & Guerney, 1967), destructive behaviors (O'Leary, O'Leary, & Becker, 1967), and overcoming severe behavioral deficits in severely disturbed autistic children (Lovaas, Berberich, Perloff, & Schaeffer, 1966). The research from this radical paradigm did indeed return results.

Today, as advertisers purchase lists of consumers and target them in terms of their recorded purchasing habits, the shadow of Skinner can still be discerned on the ever-present Internet. From the point of view of the marketers, reinforcement comes in terms of greater sales, where S^d = customer's purchasing history, R = the placement of pop-up ads, and S^+ = greater sales figures. The marketers are rewarded for placing ads on the Internet by greater sales, particularly when they connect the ads to the Net user's history of past purchases and e-mailings (through "cookies" and such). From the point of view of the consumer, S^d = the pop-up ad, R = decision to purchase (or not), and S^+ = getting the service or product to consume. As computer technology grows, it appears that virtual reality becomes increasingly seductive, blurring the margin between that and flesh-and-blood reality. This makes it even more possible for those wishing to manipulate others' behaviors to do so, by rewarding them with visually, aurally, and even kinetically rewarding scenarios (Degener, 2005).

Not all such results, of course, are noble and desirable (Siegel, 2009). The research and applications deriving from the behavioral paradigm are like any other – neither inherently good nor bad – but individual proponents make them so. As I predicted some time ago, "it takes little imagination to create a scenario in which behavioral scientists become governing managers of our social environment" (Ryback, 1976b, p. 94). A far cry from the primitive Skinner box, in which an infant could have all its basic needs met automatically without human contact, the current virtual reality encompasses a geography limited only by Google's ability to photograph so much of our known world and recreate it on the screen of each of our home computers (Jones, 2007). We now have the technology to travel much of the world without human contact. Here, the S^d is the blank screen in front of us, the R is the mere push of some buttons, and the S^+ is the entire world in all its variety.

Historical Overview of Humanistic Psychology: Existentialism, Transparency, and the Third Force

The philosophy of existentialism (Camus, 1954; DeBeuavoir, 1953; Sartre, 1953) provided a framework for the beginnings of humanistic psychology. The "existential" paradigm in psychology start with the writings of Carl Rogers (1942, 1951, 1957) in which he revolutionized the "talking cure" of Sigmund Freud by interacting with the patient (soon to be called "client") on an emotionally authentic basis – anathema to the conservatively trained psychiatrists of the day. He was soon joined in his attack against Freudian psychoanalysis and the then-popular behavioral S–R treatment by such stalwarts as Rollo May (1950), Abraham Maslow (1962), and Sidney Jourard (1964), among others. This new approach, flying in the face of the then-widely accepted behavioral paradigm, advocated the supremacy of human awareness and open-hearted communication in all aspects of life, including psychotherapy (Reik, 1948), education (Rogers, 1977; Thomas, 2001) – including computer-based "e-learning" (Motschnig-Pitrik, 2005), business (Montuori & Purser, 2001; Ryback, 2000), and even international politics (Rogers & Ryback, 1984) and religion (Keawkungwal & Ryback, 1979).

Division 32 of the American Psychological Association represents its members dedicated to humanistic psychology. This division, born in 1971, has its history in Maslow's formulation of what he termed the "third force" during the late 1950s and early 1960s – a reaction to the methodological behaviorism and the Freudian bastion. At the invitation of Maslow and Clark Moustakas, two meetings were held in Detroit among those interested in such topics as self-actualization, creativity, and the nature of personal meaning. In 1961, the American Association for Humanistic Psychology was formed with the sponsorship of Brandeis University.

Maslow, May, and Rogers were important theoreticians guiding their new paradigm whose values aimed to replace what seemed to be missing in the contemporary psychology of the time. These three and others such as Gordon Allport, James Bugental, Charlotte Buhler, Floyd Matson, Gardner Murphy, and Henry Murray met together at the first invitational conference on humanistic psychology at Old Saybrook, Connecticut in 1964. Their goal was to adorn their fledgling paradigm with two main postulates:

1. A more complete image of the person than was held by reductionistic behaviorists and Freudians.
2. Research methods to study the whole person that were more experiential and personally meaningful than those currently in use.

Some Humanistic (and Very Human) Models: More Kindness

One of the models in this paradigm was offered by Maslow (1962, Chapter 6) who distinguished between two types of existence – Deficiency (D) and Being (B). According to Maslow, the controls advocated by the behavioral paradigm put an immediate brake to the expression of inner human nature, resulting in a D state (p. 185). The B state, on the other hand, allows for perceptions that tend to be "relatively ego-transcending, self-forgetful, egoless" (p. 74). It is, he added, "much more passive and receptive than active …" (p. 81), quite opposed to the manipulative, behavioral paradigm. Even the process of diagnosis or classification, part of the medical model, brings about resistance because it destroys one's individuality (Chap. 9). Maslow found support for this from a number of his colleagues, including Thomas Szasz, author of *The Myth of Mental Illness* (1960).

Another model within the humanistic paradigm, offered by Jim Bugental (2008), was quite similar to Maslow's. Bugental's model involved three postulates:

1. Full awareness in the moment
2. Choosing to invest in one's life with intentionality
3. Taking full responsibility for such choices with existential awareness of the consequences, including the potential tragedy that might be inevitable for any given choice

One more brief but powerful model might be appreciated if we can allow for a man's dying words. A few days before he died in 2008, Jim Bugental, affected by prior stroke, adamantly shared with some friends that "we need more kindness" in the world, with each other and with ourselves. According to those hearing the words (Bradford & Sterling, 2009), "He repeated this, speaking as much to himself as to us" (p. 318).

Paradigms' 15 Years of Fame: New Ways of Behaving

Many successful paradigms appear to have what I would call their "15 years of fame," just as Andy Warhol predicted everyone's claim to 15 min of fame (a strange prediction, yet proven true by YouTube). Freud's paradigm characterized by his id-ego-superego model reigned supreme in the mid-1900s and still survives, more so in parts of Europe than here in the United States. The behavioral paradigm saw its shining moments soon after, promising "an average duration of just sixteen sessions, instead of years of awkward silences" (Osnos, 2011, p. 56) and still survives, though under names no longer associated with the original S–R model. Then came the humanistic paradigm, shining brightest during the revolutionary 1960s. Referring to it as "the encounter culture," Ryback (2006) described it as

> a [then] new form of communication both among its individual members at large and within institutions academic, religious, and industrial. Openness about one's own needs, fears, and aspirations, and a sensitivity to respond to others' needs are the encounter ethic. Sensual awareness and verbal assertiveness are encouraged when appropriate. Experiential or emotional self-expression is the keystone. (p. 475.)

There was an inherently satisfying, even enjoyable aspect of this subculture. It appeared to fill a void in emotional interaction that characterized North American culture. What was this need? According to Rogers (1970):

> It is a hunger for relationships which are close and real; in which feelings and emotions can be spontaneously expressed without first being carefully censored or bottled up; where deep experiences—disappointments and joys—can be shared; where new ways of behaving can be risked and tried out (pp. 10–11).

An Earth-Saving Application of One Paradigm: Bringing Enemies to the Table

What practical uses could be derived from such a vulnerable paradigm? For one thing, it could be used to save the world from self-destruction, no small matter. Rogers & Ryback (1984), in a paper entitled, "One Alternative to Nuclear Planetary Suicide," described experiments using a humanistic approach to negotiation between hostile and warring factions. Such "experiments" involved health providers vs. consumers, Protestants vs. Catholics in Belfast, North Ireland, and Israelis vs. Egyptians in peace talks at Camp David. The postulates of this model for conflict resolution involved:

1. Casual human interaction at a neutral "retreat," including shared meals, casual contact during walks, fitness routines, evening participation in such games as chess.
2. Casual dress in a highly informal environment – no protocol, no assigned seating at meals.
3. Status is not a hindrance to personal communication at any level.
4. Privacy and confidentiality maintained within the retreat setting is assured, so as to encourage more honesty and candor, eliminating the possibility of press coverage. This allows for tentative statements as exploratory offerings.
5. There is no set agenda, though the general purpose of the meeting is profoundly well understood at the outset. Issues are considered as they naturally arise, instead of following a rigid format.
6. Emotionally laden statements are permitted and even encouraged if they lead to greater understanding and the possibility of resolution. The facilitator or host has the ultimate responsibility of managing such "outbursts" with gentle understanding and even humor, if that presents itself.
7. The facilitator or host must maintain an air of objectivity in terms of resolving differences. "I acted as a referee and put them back on track," explained President Carter (1982). "Strangely enough, every so often laughter broke out" (p. 353).

8. It is expected that the "pressure cooker" effect of being together day after day, attempting to resolve long-standing issues, might lead to intense feelings, sometimes clearly expressed anger. Ideally, this energy can be used to get to the "tipping point" where each side can more easily understand the frustration of the counterpart.
9. It is also expected that there may be moments – long ones – where the entire experience will be seen as an utter failure. Such "terrible moments" can often lead to a turning point toward success, the dynamics of which are difficult to explain. Sometimes it is the mere revelation of human vulnerability that shakes the advocates into a more receptive mode.
10. Such "terrible moments" are opportunities to transcend the impasse of human wills and reach out to the more "human" aspects beyond the issues at hand. This reveals a core component of the humanistic paradigm – the value of human meaning and understanding over points made in any conflict.
11. Most often, when the impasse is transcended through some emotional upheaval, the caution and hostility seem to become quickly transformed into trust and appreciation (even caring) between the formerly hostile counterparts.

I have taken the time to fully describe this process of conflict resolution in order to give credence to a paradigm that many consider trivial and self-serving. Though it often does seem to have such trappings, an astute use of this paradigm can have life-saving, and possibly Earth-saving, consequences.

Could Opposing Paradigms Coexist? A Question Worth Exploring

How ironic that this polarity between the "objective" and "existential" viewpoints has not yet been resolved, given that its proponents are psychologists, part of whose dedication is the resolution of conflict!

So is there a push to bring the two paradigms together, to resolve polar differences? In Chap. 1 of this book, Hillix and L'Aabate ask the question: Could two different paradigms unite and coexist with one another? This question is certainly worth exploring.

The two paradigms explored in this chapter – behaviorism and existentialism – have underlying theory the postulates of which, most would agree, are incompatible. What would it take to blend these underlying postulates? Can they be amalgamated to exist in a framework, or new paradigm, following the Hume-Mill tradition? Can the postulates of cause-and-effect, stimulus–response theory be considered to coexist with the postulates of spontaneous and unpredictable free will?

Two "opposing" paradigms can be compared to two rivers running in parallel toward the ocean, never meeting, with a considerable stretch of land between the two rivers. The land between the two rivers presents the challenge of combining the two paradigms (Ryback, 2010, Personal communication). The question here is whether we can successfully build a connection between the two rivers to connect them at some point, but this is always a challenge, similar to the challenges in constructing the Panama Canal. The challenge in joining behaviorism with humanism lies in their basic assumptions, i.e., determinism vs. free will. If System B (behavioral determinism, BD) exists, then System F (free will) cannot, and vice versa.

How Paradigms Can Handle the "Truth": Majority Consensus of Valued Peers

L'Abate (2009), referring to Kuhn's (1970) work, characterizes a paradigm as a set of metaphysical assumptions, to which many in the particular field adhere. As I see it, a paradigm is defined as a set of phenomena related to one another in a functional manner with a corresponding set of terms as labels defined along intrinsically functional lines, cohering to a unitary value structure, with an increasing number of adherents as it matures. Models can be offered to represent a succinct description of a paradigm, e.g., Einstein's $E=mc^2$, describing with a high degree of precision the relation between matter and energy in our physical universe.

If we could place paradigms in a three-dimensional framework, then the aforementioned incompatibility could be characterized as polarity on a dimension, this one being a determinism/free-will dimension. This helps us to comprehend that the incompatibility lies not so much in the paradigms themselves, but rather in the proponents of the paradigms. One cannot logically maintain both positions at once, but both positions are held by their respective proponents at one and the same time. In other words, it is quite clear that behaviorists and humanists hold positions incompatible with one another, but both argue their positions at one and the same time, revealing that the incompatibility is in the logic and minds of the respective proponents. While they while the time away arguing with one another, their respective domains of knowledge coexist with no chance of disappearing, no matter how eloquent their adversarial viewpoints. Confidence is equally shared among advocates of both paradigms. And "truth" is equally shared as well, as any well-stocked library will reveal.

The "truth" of any paradigm is comprised of the conceptual framework constructed by a majority consensus of valued peers within a scientific subculture. Subscribers to a particular paradigm use a set of rules to acquire majority consensus, primarily through the publication of results based on agreed-upon methods of scientific inquiry. It is only when some humanistic psychologists use qualitative (i.e., experiential) (Polkinghorne, 1989) rather than quantitative research data that the rules begin to change somewhat. But even qualitative data eventually must meet the criteria of conventional data analysis. Though slightly stretched, the consensus-acquisition rules are still compatible.

Determined to (Freely) Choose a New Paradigm: The Pure Data of Personal Experience

Even more basic to the "truth" of either paradigm are the pure data of personal experience. Whether you are a behaviorist or humanist, allow me to invite you to "analyze" the data of your own life experience along the free-will/determinism dimension. Allow me, as well, to use my own data as elucidation.

Between complete freedom of choice and passive acceptance of determinism are what I refer to as degrees of freedom. For any particular action or decision, determinism rules to the extent that I can identify the causal factors. What remains hidden from me allows for a sense of freedom of choice. That I am proposing this amalgamation of opposing paradigms is determined by my intellectual history and previous training. The exact form of expression of this advocacy is unknown to me and is revealed in a somewhat spontaneous and unpredictable fashion.

Yet I become aware of my philosophical bent and its determining factors only after the fact. Until I made this theoretical commitment, I experienced it as a matter of free choice. Having taken the plunge, I can search my past experience for determining factors and attempt to control, perhaps even predict, my future verbal output. To the extent that I can identify causal factors, I am clearly aware of the past environmental stimuli affecting my destiny. To the extent that part of my search remains hidden to me, I act in spontaneous and unpredictable fashion.

I must make clear here that what I am advocating is not merely the acceptance of the illusion of free will (i.e., how the behaviorist might explain this), but rather the transactional view that the conscious experience of free will is spontaneous and beyond deterministic factors on one dimensional level while my verbal behaviors – thinking and writing – can be described in deterministic terms on another. The difference in viewpoint is subtle but crucial.

Existential Behaviorism: A "Combined Paradigm" of Self-Determination

Freedom of choice is as valid an experience as that of feeling fatefully determined. Self-determination, a hallmark of most applied schools of psychology, involves the free choice of shaping one's own destiny (see, for example, Ellis (2009) and his use of the term, "Rational Emotive Behavior Therapy") using, perhaps, the knowledge of cause-and-effect principles of behavior, the *sine qua non* of all psychology. This approach, a combination of the two paradigms, creates a new paradigm to which the label "existential behaviorism" has been assigned (Ryback, 1971b, 1973).

In this "combined paradigm," the behavioral-deterministic and existential-free-will models are brought to converge within a single paradigm focusing on the potential for self-determination based on existential choice (Ryback, 1976c). Existential behaviorism integrates the technology of behaviorist methodology with the existential value of free choice. Here is what this new paradigm offers: Although all events can be understood to be determined in a universal or cosmic sense, one individual's experience of life – including yours and mine – may often either seem organized and sequentially ordered in a logical framework, or may be experienced as arbitrary and capricious, whether frustrating or joyous. It is from this individual perspective that existential behaviorism is best understood.

Quantum Mechanics and the Copenhagen Interpretation: "Connected with Emotions"

To appreciate the deterministic factors affecting our lives requires an acceptance of the postulates of scientific causality. To realize the existential nature of our life experience requires appreciation of an individual's inner world, including our own.

This "combined paradigm" has a history in the field of physics. The debate in physics between the proponents of classical and quantum mechanics has its parallels to the debate between the existential-free-will advocates and the behavioral determinists in psychology. In both cases, the existing paradigms are challenged by the "data" of their counterparts. In physics, one of the outcomes of the "measurement" problem has resulted in a "self-conscious" view of the nature of measurement. As Heisenberg (1958a, 1958b) put it, "The object of research is no longer nature itself, but man's investigation of nature." Psychologists would soon become keenly aware of this, as they examined the effects of their own influence on research outcome, influences such as outcome expectations (Rosenthal, 1961) and the unconscious influence of the experimenter on research subjects (Orne, 1961).

Within the field of physics, there were three main characters involved in the philosophical debate – Bohr, Heiseneberg, and Born. Framing a new paradigm largely influenced by the Danish philosophers Kierkegaard and Hoffding, Bohr (1934, 1958), working with Kramers and Slater in 1924, formed the "think-tank" of what was to become known as the Copenhagen Interpretation of quantum mechanics. Heisenberg's (1958a, 1958b) view of the Copenhagen Interpretation focused on the influence of the observing apparatus on the subsequent measurement – what we now know as the Heisenberg Principle of Uncertainty.

The third major founder giving birth to this new paradigm in physics was Born (1949), who wrote that "the concept of reality is too much connected with emotions to allow a generally acceptable definition." So we can see how the hard science of physics has in its philosophical underpinnings a strong existential substructure, and a concern with subjectivity and the influence of the observer.

According to the Copenhagen Interpretation model of quantum mechanics, at the interface between observer-apparatus and physical phenomenon being measured, there takes place an energy transaction between observer and measurement so that completely accurate measurement is impossible. "One of the fundamental tenets of quantum mechanics," as a recent article on quantum cryptography in *The Economist* (2010) put it, "is that measuring a physical system always disturbs it" (p. 12).

What is critical to appreciate here is that it is not the subjectivity of the observer that is at issue but rather the ***actual influence upon the measured process***. That in itself helps cure the impasse between determinism and free will; since completely accurate measurement is theoretically and conceptually impossible, then the door is opened to something other than strict determinism. This is a key point and allows for the transactional process by which determinism and free will are acting on the same stage at the same time.

This also helps us to understand how paradigms that may be on opposite poles of some dimension (such as determinism/free will), nonetheless may continue to influence one another. According to the Copenhagen Interpretation, the phenomenon to be measured can be isolated, except for contact with the observer-apparatus. Therefore, any measured phenomenon cannot be isolated from the external world. Even if the first observer-apparatus could be isolated, it would at least have to have interface with a second observer-apparatus if the original measurement were to have any public meaning. And since we cannot know each and every detail of the external world, we therefore cannot precisely measure the influence of the observation-cum-world on the phenomenon being measured.

Classical physics, just like Skinnerian determinism, purported to measure closed systems; quantum mechanics, like existential humanists, denies that precise and total human knowledge of closed systems can exist since, upon measurement, the system being measured ceased to be closed. "Knowledge of the 'actual' is thus, from the point of view of the quantum theory," wrote Heisenberg (1955), "by its nature always an incomplete knowledge … If we attempt to penetrate behind this reality into the details of atomic events, the contours of this 'objectively real' world dissolve—not in the mist of a new and yet unclear idea of reality, but in the transparent clarity of a mathematics whose laws govern the possible and not the actual" (pp. 27–28).

It is with this perspective that the Protagorean paradigm of existential behaviorism adopts the transactional view that both BD and existential free will share equal degrees of certainty in the "intuitive" amalgamation between conceptual formulation and sensory experience. Though the two paradigms are logically incompatible, the systems denoted by these paradigms are compatible in a complementary fashion. That is to say that, although the theoretical descriptions (or symbolic representations, as Popper (1963) called them) of the paradigms are logically incompatible because of the rules of logic of scientific language and syntax, the sets of experiential phenomena themselves, depicted by the labels "determinism" and "free will," take place and exist along independent and orthogonal dimensions.

In other words, although the theoretical formulations are largely incompatible because of scientific convention, both free will and scientific determinism are convincingly experienced through our senses and faculties and hence make up the information, in both cases, "through their connections with sense experiences … purely intuitive, not itself of a logical nature" and "… differentiates empty phantasy from scientific 'truth'" (Einstein, 1959, p. 13). To our human sensory abilities, both sets of phenomenal experience have equal justification in being interpreted as "reality."

A Vector Model for Existential Behaviorism: Splitting Perspectives

Grundy (1970) describes a doctrine by which different accounts of behavior, although from mutually exclusive frameworks, are nonetheless equally valid, i.e., "different ways of talking about the same events" (p. 130).

Along the lines of Grundy's Identity Hypothesis is a reworking of the Copenhagen Interpretation. Everett (1957), under the academic guidance of Wheeler (1957), offers a model which includes a noncollapsing state vector for the whole universe. This state vector may be split up into orthogonal vectors, each of which can be observed from mutually exclusive perspectives. This splitting into orthogonal vectors reflects a continual splitting of the universe into a multiplicity of mutually unobservable but equally real universes. This splitting is induced primarily by the act of human measurement, although the observer sees only the universe as he sees it and not the splitting process itself (DeWitt, 1970). For any particular moment in time, the observer can "see" only one of a number of worlds and not be aware of any of the others (Cooper & Van Vechter, 1969).

Everett's model can help to understand the syzygial nature of existential behaviorism as well as it transactional dynamic. Any personal experience – one's own or that of others – can be seen either from the behavioral-determinist or from the humanistic-existential perspectives. Although any particular set of behaviors can be characterized by a single state vector, this vector can be split into an existential vector and a deterministic vector, each of which can be understood from mutually exclusive perspectives, since the vectors are orthogonal to one another in this model. This splitting is a continual but not necessarily continuous process. That is, each conscious observation gives way to a split into one or the other of the two perspectives.

The Dynamic Perceptual Outcome Between Determinism and Free Will: Choosing a Compromise

The observer can then undertake to transact an intermediary state between the two polarities, since this intermediary vector (situated between the two orthogonal vectors of Humanistic Free Will [HFW] and BD, characteristic of the observation at a given time) is not inherently part of the original split. In more familiar terms, if I choose to integrate two otherwise opposing views, then I can do this if I so choose (see Smith, 1978 for similar approach). If I choose to allow a flexibility of mind and see how a personal and spontaneous decision can be made in the light of causal dynamics, then I can integrate that view. I am determined to write these words – that you are now reading – by the history of intellectual influence on me over the years, yet I also, at the same moment, enjoy – immensely – the sense of spontaneous creativity and free "flow" (Csikszentmihaly, 1990) that come with this creative intellectual enterprise.

In similar fashion, the fact that you take the time to read this is determined by your own history of intellectual stimulation. Yet, hopefully, there is a spontaneous intellectual process that you might also enjoy in what feels very "unprogrammed."

The complementarity of the dynamic between the two paradigms (BD and HFW), which coexist without contradiction despite the logical incompatibility between them, can be described as follows:

Consider a space of two coordinates of which the abscissa represents time and the ordinate represents the dynamic perceptual outcome (DPO), between the values 0 and 1, of causality. The limits of the ordinate are HFW at value 0 at the coordinate juncture and BD at value 1 in the positive direction. The DPO of any event at any given time will range between 0 and 1. This DPO is the perceived conceptual "compromise" between the two options – free will (0) and determinism (1) – determined solely by the observer. Although the HFW and BD paradigms exist simultaneously as potential DPOs, the observer can realize only one set of experiences at a given moment. The observer can focus on HFW or BD or on the transactional complementary dynamic between the two. In the last case, the observer realizes a value between 0 and 1 depicting the DPO as a focal realization between HFW and BD, which are characteristics of that particular observation, *not* of the phenomena themselves.

Twisting Our Fate: Authentic Causality

Faith is a matter of surrender, according to Garrison Kiellor (2010), our modern-day Mark Twain. Life's ups and downs can indeed seem determined by causal factors, or they may seem arbitrary. Yet the essence of psychological study and research is based on the assumption not only of the understanding of causal factors but also, in applied psychology, the use of such knowledge to improve all aspects of human life. Again, we have the paradox of using the knowledge of cause-and-effect determinism to improve life experience by making personal choices, sometimes against known environmental factors, often characterized as "challenges." We decide to "change" our fate, notwithstanding our acceptance, as social scientists, of the paradigm of scientific causality.

So what benefit does this new paradigm offer? It gives us, finally, a paradigm that most closely describes our true human experience of life – both the causal as well as the free-will aspects. Very few, if any, of the scientific community will deny the causal perspective. And very few, if any, will deny the personal experience of enjoying the experience of free will in day-to-day interactions, whether in creative endeavors or in social interactions. Finally, a paradigm exists that we can live with authentically.

Emotive Empathy vs. Cognitive Contingencies: A Continuum from 0 to 1

When offering friendship and support to colleagues or loved ones, our DPOs range from 0 at the free-will pole – existential concern and empathy with the other's emotional state – to the unitary value at the causality pole – sharing behavioral goals or modifications of behavior to overcome problems. At any given moment, our observations and conceptualizations range between 0 and 1. For example, I might begin at 0.9 as I ask the other to describe the problem, gathering rigorous, specific data, so I can better understand it, but soon find myself at 0.1 as I find myself drawn into the emotional affect of the individual's dilemma.

On the other hand, I might begin at 0.05 if an hysterical, sobbing youngster were to approach me with a grave concern, and remain at that level, or close to it as the crying continues, and then gradually work up to a value of 0.5 as I attempt to understand the facts leading up to the crisis. Clearly, it cannot be denied that the free-will/determinism dimension is not essentially different from an emotive-cognitive continuum. At value 0, emotive empathy reigns supreme while, at value 1, the strict behaviorist can be seen drawing up contingencies in the cool light of reason. Accordingly, most behaviorists would find themselves at values >0.5 while most humanists would find themselves at values <0.5.

The Unbearable Lightness of Being Flexible: Neurons, Genes, and the Savage Joy of Unpredictability

The enjoyment of life experiences would typically fall into the existential, free-will domain, yet neuroscientists such as Wolfram Schultz (2010) characterize such joy as emanating from the striatum region of the brain, which he refers to as the "pleasure center," sounding very more like a determinist. And free will, according to neuroscientist Daniel Salzman (2010), is "hijacked" by drug addiction, a finding backed by his fellow researcher, Nora Volkow (2010), who attributes a genetic causal factor of 50% to such addiction. When these scientists conclude that emotions are unconsciously derived, only after the pathways emanating from the midbrain reach the prefrontal cortex, it appears as if what is considered free will can be seen by these scientists as nothing more than an illusion of conscious awareness after the fact. In other words, free will is accepted as a conscious process, but only after the neural pathways of the brain have completed their predictable and predetermined circuits.

To put it in more human terms, romantic love – certainly considered as part of the existential-free-will domain – has been dissected into brain process as well.

Helen Fisher (2009), guru of love and sex research, dissects love into three categories: lust – a sexy feeling associated with activity in the caudate area of the brain, primarily the hypothalamus, stimulated by testosterone (in both men and women); attachment – leading to lifetime connection, associated with activity in the ventral palladium, driven primarily by vasopressin and oxytocin; and romance – that magic feeling of early love, associated with the ventral tegmental area, driven by dopamine, for most, but sometimes limited by the frontal cortex for more inhibited individuals. Yet this last component of love can be kick-started by novelty, adds Fisher.

On a highly personal note, Fisher shares how, despite her years of expertise on the dynamics of romance across cultures, she herself was not immune to the effects of romantic rejection. "I lost twenty pounds. My clothes looked ridiculous on me. I only had three hours of sleep at night" (p. 125). Even scientists, so adept at studying the deterministic cause-and-effect factors of human relationships, are not always immune to those very same dynamics, experienced with great existential angst. Fisher's DPO may seemingly have dropped from a purely scientific 1 of objectivity to an existential 0 of uncompromising, existential pain.

Moral decisions are typically seen as coming from the existential-free-will domain, yet neuroscientist Antonio Damasio (2010) sees "moral emotions," such as compassion and shame, as arising from the ventral prefrontal cortex. Many others agree, e.g., Iacoboni (2008), Cacioppo & Patrick (2008), that we are all cerebrally wired for social living. More and more, the new paradigm characterized by existential behaviorism is finding acceptance. As Nobel laureate neuroscientist, Eric Kandel, puts it: "The mere fact that most people are no longer [mind-brain] dualists is a major cultural advance" (2008, p. 17).

The Humanistic Free-Will/Behavioral Determinism dimension is not only a cognitive-emotive one. It is also control/noncontrol. Therefore, this complex dimension can be typified as free-will, emotionally based behavior on the impulsive or spontaneous side vs. the more rational, logically derived deterministic cognitive control. The control we are focusing on is not that which is opposite to passive, but rather that which is characterized by the difference between logical, objective control and spontaneous, emotional response.

The existential behaviorist's mainstay is the flexibility between the HFW and BD vectors, developing an overarching sense of comfort between the DPO resultant vector values of 0 and 1 over time or, put otherwise, along the free-will/determinism continuum. Decidedly, this is a more challenging task than that of restricting oneself to merely one of the two factors along the dimension.

Existential behaviorism does not place free will as subservient to an all-encompassing deterministic universe with humanistic existential responses as one of any number of human responses, all determined by the Grand Postulate of Causality. Rather, it gives equal import to the alternative that existential decisions (spontaneous and unpredictable) allow for complete freedom of movement and choice and, as well, to the self-determining nature of all humans.

This sentiment is eloquently expressed in F. Scott Fitzgerald's literary masterpiece, *The Great Gatsby* (1925):

> The test of a first-rate intelligence is the ability to hold two opposed ideas in the mind at the same time, and still retain the ability to function. One should, for example, be able to see that things are hopeless and yet be determined to make them otherwise.

Or, as writer Isaac Bashevis Singer (2011) is quoted to say, with a touch of irony:

> We have to believe in free will. We have no choice.

References

Andronico, M. P., & Guerney, B., Jr. (1967). The potential application of filial therapy to the school situation. *Journal of School Psychology, 6*(1), 2–7.

Bohr, N. (1934). *Atomic theory and the description on nature*. Cambridge: Cambridge University Press.

Bohr, N. (1958). *Atomic physics and human knowledge*. New York: Wiley.

Born, M. (1949). *Natural philosophy of cause and chance*. London: Oxford University Press.

Bradford, G. K., & Sterling, M. M. (2009). The journey is the goal. *Journal of Humanistic Psychology, 49*(3), 316–328.

Bugental, J. F. T. (1964). The third force in psychology. *Journal of Humanistic Psychology, 4*(1), 19–25.

Bugental, J. F. T. (2008). Keynote address to the Existential-Humanistic Institute. *Second Annual Conference*, San Francisco, CA, November 13, 2008.

Cacioppo, J. T., & Patrick, W. (2008). *Loneliness: Human nature and the need for social connection*. New York: W.W. Norton.

Camus, A. (1954). *The rebel*. New York: Knopf.

Carter, J. (1982). *Keeping faith*. New York: Bantam.

Cooper, L. N., & Van Vechter, D. (1969). On the interpretation of measurement within the quantum theory. *American Journal of Physics, 37*, 1212.

Csikszentmihaly, M. (1990). *Flow*. New York: Harper & Row.

Damasio, A. (2010, May 26). *Charlie Rose Brain Series*. Charlie Rose (Executive producer). New York: WNET.

DeBeuavoir, S. (1953). *The second sex*. New York: Knopf.

Degener, M. (2005). *Desert screen: War at the speed of light*. London: Continuum.

DeWitt, B. S. (1970). Quantum mechanics and reality. *Physics Today, 23*, 30–35.

Einstein, A. (1959). *Albert Einstein: Philosopher-scientist*. In P. A. Schlipp (Ed.). New York: Harper Torchbooks, Harper and Row.

Ellis, A. (2009). *All out*. Amherst: Prometheus Books.

Everett, H., III. (1957). "Relative state" formulation of quantum mechanics. *Modern Physics, 29*, 454.

Feigl, H. (1970). The "orthodox" view of theories. Remarks in defense as well as critique. In M. Radner & S. Winokur (Eds.), *Analyses of theories and methods of physics and psychology*. Minneapolis: University of Minnesota Press.

Fisher, H. (2009). Interviewed by A.J. Jacobs in his article, Do I love my wife? *Esquire, 151*(6), 120–125.

Fitzgerald, F. S. (1925). *The Great Gatsby*. New York: Charles Scribner.

Grundy, J. H. (1970). Is the brain a physical system? In R. Borger & F. Cioffi (Eds.), *Explanation in the behavioural sciences*. Cambridge: Cambridge University Press.

Heisenberg, W. (1955). The development of the interpretation of the quantum theory. In W. Pauli (Ed.), *Niels Bohr and the development of physics*. New York: McGraw-Hill.

Heisenberg, W. (1958a). *Physics and philosophy – The revolution in modern science*. New York: Harper and Row.
Heisenberg, W. (1958b). *The physicist's conception of nature*. London: Hutchinson.
Iacoboni, M. (2008). *Mirroring people: The new science of how we connect with others*. New York: Farrar, Straus and Giroux.
Jones, M. (2007). Google's geospatial organizing principle. *IEEE Computer Graphics and Applications, 27*(4), 8–13.
Jourard, S. M. (1964). *The transparent self*. Princeton: Van Nostrand.
Kandel, E. R. (2008). In an interview by S. Ayan in Speaking of memory. *Scientific American Mind, 19*(5), October/November 16–17.
Keawkungwal, S., & Ryback, D. (1979). Therapeutic approaches in Theravada Buddhism and existentialism. *The Bodhedrum English Section* (pp. 47–48).
Kiellor, G. (2010, August 28). From radio commentary on *A prairie home companion*.
Kuhn, T. S. (1970). *The structure of scientific revolutions*. Chicago: University of Chicago Press.
L'Abate, L. (2009). Paradigms, theories and models: Two hierarchical frameworks. In L. L'Abate, P. DeGiacomo, M. Capitelli, & S. Longo (Eds.), *Science, mind and creativity: The Bari Symposium* (pp. 107–122). New York: Nova Science Publishers.
Lovaas, O. I., Berberich, J. P., Perloff, B. F., & Schaeffer, B. (1966). Acquisition of imitative speech by schizophrenic children. *Science, 151*, 705–707.
Malraux, A. (1961). *The temptation of the West*. New York: Vintage Books.
Maslow, A. H. (1962). *Toward a psychology of being*. Princeton: Van Nostrand.
May, R. (1950). *The meaning of anxiety*. New York: Norton.
Montuori, A., & Purser, R. (2001). Humanistic psychology in the workplace. In K. J. Schneider, J. T. Bugental, & J. Fraser (Eds.), *The handbook of humanistic psychology* (pp. 635–645). Thousand Oaks: Sage.
Motschnig-Pitrik, R. (2005). Person-centered e-learning in action. *Journal of Humanistic Psychology, 45*(4), 503–530.
O'Leary, K. D., O'Leary, S., & Becker, W. C. (1967). Modifications of deviant sibling interaction pattern in the home. *Behavior, Research and Therapy, 5*, 113–120.
Orne, M. T. (1961). On the social psychology of the psychological experiment: With particular reference to demand characteristics and their implications. Paper presented at the Symposium on the Social Psychology of the Psychological Experiment. New York: American Psychological Association Convention.
Osnos, E. (2011). Meet Dr. Freud. *The New Yorker, 86*(43), 54–63.
Polkinghorne, D. E. (1989). Phenomenology and research methods. In R. S. Valle & S. Halling (Eds.), *Existential-phenomenological perspectives in psychology* (pp. 41–62). New York: Plenum Press.
Popper, K. R. (1963). *Conjectures and refutations: The growth of scientific knowledge*. New York: Harper Torchbooks, Harper and Row.
Reik, T. (1948). *Listening with the third ear*. New York: Farrar & Straus.
Roger, C. R. (1957). A process conception of psychotherapy. *American Psychologist, 13*, 142–149.
Rogers, C. R. (1942). *Counseling and psychotherapy*. Boston: Houghton Mifflin.
Rogers, C. R. (1951). *Client-centered therapy*. Boston: Houghton Mifflin.
Rogers, C. R. (1960). Two divergent trends. In R. May (Ed.), *Existential psychology*. New York: Random House.
Rogers, C. R. (1970). *Carl Rogers on encounter groups*. New York: Harper & Row.
Rogers, C. R. (1977). *Carl Rogers on personal power*. London: Constable.
Rogers, C. R., & Ryback, D. (1984). One alternative to nuclear planetary suicide. In R. F. Levant & J. M. Shlien (Eds.), *Client-centered therapy and the person-centered approach*. New York: Praeger.
Rosenthal, R. (1961). On the social psychology of the psychological experiment: The experimenter's hypothesis as unintended determinant of experimental results. Paper presented at the Symposium on the Social Psychology of the Psychological Experiment. New York: American Psychological Association Convention.

Ryback, D. (1971a). Cognitive behavior modification. *Canadian Journal of Behavioral Science, 3*(1), 77–87.
Ryback, D. (1971b). Existential behaviorism. *Canadian Psychologist, 12*(2), 243–247.
Ryback, D. (1973). A vector model for existential behaviorism. *Psychotherapy: Theory, Research and Practice, 10*(1), 5–9.
Ryback, D. (1976a). A "sweep" procedure for reading improvement. *Reading Improvement, 13*(2), Summer, 101–102.
Ryback, D. (1976b). Behavioral control pollution. *American Psychologist.*
Ryback, D. (1976c). Creative responsibility and the regulated society. *Journal of Pastoral Counseling, 11*(1), 54–57.
Ryback, D. (2000). Active listening: The quiet man who started it all. *AHP Perspective.*
Ryback, D. (2006). Self-determination and the neurology of mindfulness. *Journal of Humanistic Psychology, 46*(4), 474–494.
Ryback, D., & Staats, A. W. (1970). Parents as behavior therapy-technicians in treating reading deficits (dyslexia). *Journal of Behavior Therapy and Experimental Psychiatry, 1*, 109–119.
Salzman, D. (2010, April 21). *Charlie Rose Brain Series*. Charlie Rose (Executive producer). New York, WNET.
Sartre, J. P. (1953). *Existential psychoanalysis*. New York: Philosophical Library.
Schultz, W. (2010, April 21). *Charlie Rose Brain Series*. Charlie Rose (Executive producer). New York: WNET.
Siegel, L. (2009). *Against the machine*. New York: Spiegel & Grau.
Singer, I. B. (2011, February 18). Quoted in *The Week, 11*(502), 21.
Skinner, B. F. (1955–56). Freedom and the control of men. *The American Scholar, 25*, 47–65.
Skinner, B. F. (1961). The design of cultures. *Daedalus*, summer, 534–546.
Skinner, B. F. (1964). Man. *Proceedings of the American Philosophical Society, 108*, 482–485.
Smith, M. B. (1978). Humanism and behaviorism in psychology: Theory and practice. *Journal of Humanistic Psychology, 18*(1), 27–36.
Szasz, T. (1960). The myth of mental illness. *American Psychologist, 15*, 113–118.
The Economist. (2010). *Schrodinger's cat and mouse*, 396(8698):12.
Thomas, H. F. (2001). Keeping person-centered education alive in academic settings. In K. J. Schneider, J. T. Bugental, & J. Fraser (Eds.), *The handbook of humanistic psychology* (pp. 555–565). Thousand Oaks: Sage.
Volkow, N. (2010, April 21). *Charlie Rose Brain Series*. Charlie Rose (Executive producer). New York: WNET.
Wheeler, J. A. (1957). Assessment of Everett's "relative state" formulation of quantum theory. *Review of Modern Physics, 29*, 463.
Woodworth, R. S. (1948). *Contemporary schools of psychology*. New York: Random House.

Chapter 17
Existentialism

Mario Cusinato

The contents of this chapter discuss the relationship between paradigm, theories, and models in the existentialist perspective. For existentialism we mean a specific cultural climate in the twentieth-century old Europe as well as in the USA whose philosophical paradigm has its roots in some Greek philosophers, although the Danish thinker Søren Kierkegard and the French philosopher Jean-Paul Sartre are among its most important exponents. We will explore just three psychological approaches of this not homogeneous movement including many different schools of thought sometimes conflicting: the phenomenological perspective, the antipsychiatry movement, and existential psychotherapies. In this temporal and conceptual excursus, we will investigate whether the existentialist paradigm has generated theories and/or models that can be qualified as particular-specific or general-integrative.

The term "existentialism" was not initially used by any of the existentialist authors, and it does not appear in any of the fundamental texts of this tradition (neither in Sartre's *Being and Nothingness* (1943) nor in Gabriel Marcel describing Sartre and others); it was accepted by Sartre and de Beauvoir only in 1945. Merleau-Ponty never accepted the label wholeheartedly, whereas Heidegger vehemently rejected it. It is hence difficult to argue that existentialism represents a single, unified philosophical movement, although Sartre's famous comment that "existence precedes essence" is perhaps a good starting point, even if this dictum has also decidedly different purposes. Nevertheless, the mainstream sensibility can be clearly delineated by few words: human existence is notable for the fact that we are always ahead of ourselves, and "on the way," with various projects, intentions, and aspirations for the future. Rather than our identity being determined by our biological or social status, existentialism insists that our identity must be continually created, and

M. Cusinato (✉)
Centro della Famiglia, Via San Nicolo' 60, 31100, Treviso, Italy

Interdepartmental Center of Family Research, University of Padua,
Via Venezia 8, 35131, Padua, Italy
e-mail: mario.cusinato@unipd.it

L. L'Abate (ed.), *Paradigms in Theory Construction*, DOI 10.1007/978-1-4614-0914-4_17,

is a resultant emphasis on our freedom or, in the preferred philosophical vocabulary of the existentialists, our transcendence.

Existentialism, writes Friedman (1999), is before all a mood embracing a number of disparate philosophies united by a basic temper, that is a reaction against the static, the abstract, the purely rational, the merely irrational, in favor of the dynamic and the concrete, personal involvement and *engagement*, action, choice, and commitment, the distinction between *authentic* and *inauthentic* existence, and the actual situation of the existential subject as the starting point of thought. "Beyond this, the existentialists divide according to their views on such matters as phenomenological analysis, the existential subject, the intersubjective relation between selves, religion, and the implications of existentialism for psychotherapy" (p. 4).

One last important introductory remark, Friedman further points up: "Existentialism is a direction of movement toward particulars, but it is not and can never be an espousal of the particulars at the expense of all generality and abstraction" (p. 5). This statement expresses the central aspect of the chapter in which we will try to find an answer to the question: Is the existentialism a paradigm particular-specific or general-integrative? We will start right now by saying that a distinction still need to be made between philosophers, who are existentialists in the sense that they analyze existence according to general existential categories, and others who are existentialists in the very different perspective of pointing to the unique and concrete in existence that lies beyond all analyses in existence. In this bidirectional movement, some meaningful acquirements can be also found at a more psychological level.

Let us inquire in the first the cultural frame before we enter into more specific contents.

Existentialism as a Cultural Climate

The term *existentialism* principally indicates a specific cultural climate – *Weltanschaung* – in the old Europe, but also in the USA, of the twentieth century. Reynolds (2006) remembers some of its original elements:

> Existentialism, perhaps to an extent unprecedented in the history of philosophy, has managed to capture the attention of the general public. Estimates of the number of people at Jean-Paul Sartre's funeral in 1980 vary from 50,000 to 100,000, and this was well after his cultural and intellectual heyday. Simone de Beauvoir's famous treatise on the situation of women, *The Second Sex,* has been one of the most widely read non-fiction books of the twentieth century. Existential plays and novels – in particular Sartre's *Nausea* and Albert Camus's *The Outsider* – have been read voraciously and critically acclaimed. Sartre and his more academically inclined colleague Maurice Merleau-Ponty were the co-editors of the influential magazine *Les Temps Modernes,* which considered all things philosophical, political and aesthetic, providing an intellectual point of reference for much of France. Without quite the same mainstream accessibility, or the literary bent (notwithstanding his preoccupation with poetry), Martin Heidegger has been enormously influential on generations of philosophers, as well as people working in cognitive science and artificial intelligence, and his work has helped to spawn at least two very significant contemporary philosophical movements: hermeneutics and deconstruction (p. 1).

Novel literature, theater, and cinema worked as a big sound box. Existentialism, treated as a *lived philosophy*, is understood and explored through how one lives one's life rather than a "system" studied from books. Thereafter, much of the existentialist thought can be found in literary form (novels and plays), in films, and arts (Sanderson, 2004). The Existentialism entry (Crowell, 2009) at the Stanford Encyclopedia of Philosophy lists writers such as Parisians Jean Genet (1910–1986), André Gide (1869–1951), André Malraux (1901–1976), playwright Samuel Beckett (1906–1989), the Norwegian Knut Hamsun (1859–1952), and the Romanian Eugene Ionesco (1912–1994). Prominent artists such the Abstract Expressionists Jackson Pollock (1912–1956), Arshile Gorky (1904–1948), and Willem de Kooning (1904–1997) have been understood in existentialist terms, as have filmmakers such as Ingmar Bergman (1918–2007) and Jean-Luc Godard (1930–…).

"The list of those who might be considered forerunners of modern existentialism is inexhaustible, especially if one considers the whole range of literature, which has always had strong tendency toward the particular and the personal" (Friedman, 1999, p. 5). The Russian novelist Fyodor Dostoyevsky (1821–1881) in the middle of the nineteenth century protested against the common idea that the universe should be treated as a total, rational, comprehensible system of matter (Frank, 2009). Existentialism became popular during the World War II in France, and just after it. French playwrights have often used the stage to express their views, even during a Nazi occupation as Bernard Shaw (1856–1950). Existentialists felt that adopting a social or political cause was one way of giving purpose to a life. Jean-Paul Sartre is well known for the *Committed Theatre* (Contat & Rybalka, 1976) which is supposedly committed to social and/or political action. Other popular playwrights were Albert Camus (1913–1960) and Jean Anouilh (1910–1987). Right after the World War II, Paris became the theater capital of the west, and popularized a new form of surrealistic theater called *Theatre of the Absurd* (Esslin, 2004) probably under pressure of the gruesome revelations of gas chambers and war atrocities coming out of Germany after the war. Two of the most popular playwrights of this time include Samuel Beckett (1906–1989) and Eugene Ionesco (1912–1994).

There exists overlapping between a U.S. beat generation writers who slept in Paris, and felt it their spiritual front yard, and writers of road novels (Greasley, 2001). This overlapping is evident in the delayed French discovery of Western film noir (Hillier, 1985), during the decade of Nazi-Fascist censorship, when Truffaut and others in the Cahiers du Cinema indicated influenced novels and plays (Hillier) as well as in the 1950s, when, to a bit of extent too, a surrealist movement by André Breton (1896–1966) and others questioned a constituted reality and processed conceivable the isolation of novels protagonists by unmoral antiheroes (Brigstocke, 2001).

Motion-picture production (Pamerleau, 2009) has drawn plenty of the typical existential themes, like free will, personal identity, individualism, responsibility, mind vs. reality, and what really matters. It is enough to remember several directors as Tony Hancock with *The Rebel* (1961), Hiroshi Teshigahara with *Suna no onna* (1964), Mike Nichols with *The Graduate* (1967), Harold Ramis with *Groundhog Day* (1993), Hal Ashby with *Harold and Maude* (1971), Vincenzo Natali with *Cube*

(1997), John Patrick Shanley with *Joe Versus The Volcano* (1990), Luchino Visconti with *The Stranger* (1967), with Sam Mendez with *American Beauty* (1999), Richard Linklater with *Waking Life* (2001), David O. Russell with *I love Huckabees* (2004), Ron Underwood with *City Slickers* (1991). Still deserve to be remembered movies directed by Ingmar Bergman (1918–2007), Eric Rohmer (1920–2010), Alain Robbe-Grillet (1922–2008), Richard Linklater (1922–2008), David O. Russell (1958–…), Michelangelo Antonioni (1912–2007), and Woody Allen (1935–…).

Existentialist influences on visual arts are also important (See Encyclopedia of Irish & World Art, 2010). Artists existentialists like Antonin Artaud (1896–1948), Jean DuBuffet (1901–1985), Alberto Giacometti (1901–1966), Jean Fautrier (1898–1964), Wols, i.e., Alfred Otto Wolfgang Schulze (1913–1951), Bram van Velde (1895–1981), Jean Helion (1904–1987), Henri Michaux (1899–1984), Paul Cezanne (1839–1909), Germaine Richier (1904–1959) and Francis Gruber (1912–1948), Edvard Munch (1863–1944), Jean Hélion (1904–1987), Pablo Picasso (1881–1973), Henry van de Velde (1863–1957) have created masterpieces.

Existentialism as a Philosophical Paradigm

The existentialistic philosophy is particularly attentive to illustrate concrete human existence (Reynolds, 2006). While the forerunners of existentialism reach all the way back to Heraclitus, it is the nineteenth-century Danish Søren Kierkegaard (1813–1855) – as already mentioned – who is the recognized founder of the philosophy of existence through his emphasis on the existential subject – the *Single One*, whom Kierkegaard opposed to *the crowd*, Kierkegaard saw the Single One as set in a direct relation with a transcendent God, but before this relationship can come into being a person must have discovered his/her true inwardness. With all the passion of this inwardness, Kierkegaard clings to the *absurd* and attacks the *system*.

The already mentioned novelist Fyodor Dostoevsky (1821–1881) traces the wake of existential reflection through his awareness of *Angst,* the anguish and dread of human existence, his concern with the recovery of the human being's alienated freedom, his emphasis on the particular fact and the absurd, and his portrayal of the "man-god's" proclamation of self will. He anticipated with these issues the thoughts of Friedrich Nietzsche (1844–1900) and Jean-Paul Sartre (1905–1964). Nietzsche's teaching of "will to power," of "the death of God" – the loss of any absolute base for values, of the importance of the self, of man as a valuing animal, of the uniqueness of each person's way – all make him one of the foremost proto-existentialists (Young, 2010).

Embracing a number of disparate philosophical positions, existentialism can be described as a reaction to traditional philosophy with its emphasis on the static, the abstract, the objective, and the purely rational (Friedman, 1999). In its reaction, existentialism emphasizes the dynamic, the concrete, the subjective, and the personal. More specifically, existentialism opposes idealism, which gave priority to the idea over factuality and largely neglected the part of the philosopher himself in the construction of his philosophy. Existentialism stresses personal involvement and "engagement," action, choice, and commitment, and regards the actual situation of the existential subject as the starting point of thought. Revolting against the Cartesian view of the self as a thinking entity, existentialism is concerned with the existential subject in his wholeness and concreteness – the willing, feeling, thinking person, who decides and acts from the perspective of his particular life situation rather than from some universal vantage point provided by reason or history.

Despite great diversity, all existentialist thinkers of the twentieth century have a certain commonality that bind them together: "truth is found in-the-world" and, thereby, always begins with the concrete, in existence: "The root of existentialism is, of course, 'existence'" (Friedman, 1999, p. 6). Because grounded in existence, one's thought must necessarily be perspective and limited. The development of this core intuition leads to enucleate several common themes or postulates (Valle, King, & Halling, 1989): (a) the human being is a "being-in-the-world": the human kind of being is always already involved in meaningful projects with others and alongside things. As Heidegger would say, the human being is "there being" (*Dasein*), meaning that the human being exists as the projection of possibilities which open up as a world. In this sense, the human being is not "in the world" like a match is in a matchbox; rather, the human being is "in-the-world" in the sense that one is "in trouble" or "in a relationship"; (b) the human being is "thrown" into that "world" such that she finds herself in the midst of the "givens" of existence. One does not choose one's parents, the place of one's birth, or the fact that one will die, yet, despite these circumstances, the human being is faced with the freedom to respond to these "givens" of existence. In this sense, human beings can be said to be "responsable"; (c) the human being is always "with others." Even being alone can be said to be a mode of being-with-others, since one cannot be alone unless this is first understood secondarily as a being-away-from-others. Moreover, our being-with-others is always as a relationship of some sort, and, being so, we are both shaped by others and shape those others with whom we relate; (d) human beings are always "in-the-world" alongside things that exist as meaningful entities which, in one form or another, call to the human being as significant in terms of the human being's projection of possibilities. A thing is a thing when it matters to me in one form or another – when, as a thing, it enters into the clearing by which I am either helped or hindered on my way toward realizing my projects "in-the-world"; (e) human beings are not things. A thing does not exist as a "being-in-the-world" since, as a thing, it has no world. For a thing, nothing matters. Things can only matter for a human being, since it is only in the world of the human being that things can have meaning. Nevertheless, it is not uncommon to treat human beings as "things," such as with biology; (f) human beings are finite: as a "being-in-the-world," we recognize that death is a "not-to-be-outstripped" (inevitable) possibility. Death as such is the possibility of the end of all possibilities. Existence, therefore, is not limitless, but inevitably must face up to the mystery of the "nothingness," that which lies beyond what can be known as a "being-in-the-world." As a "being-towards-death," as Heidegger said, the human being becomes aware that she cannot have all the possibilities. Faced with the recognition of one's finitude, one also recognizes that one is always faced with choices, simultaneously eliminating thousands of other possible choices and never knowing with absolute certainty the rightness of the choice. This freedom carries the responsibility for own existence; (g) freedom, responsibility, and finitude face anxiety and guilt: choice may render a death to one's world and eliminates other choices.

All thinkers developed these postulates with an own original and specific road sometimes deeply diverging from each other:

> Many of these philosophers are existentialists in the very different senses of analyzing existence according to general existential categories and of pointing to the unique and the concrete in existence which lies beyond all analysis... Insofar as one can define existentialism, it is a movement from the abstract and general to the particular and the concrete. The understanding of a particular existentialist thinker in his particularity would be more important, therefore, than any amount of general definition. (Friedman, 1999, p. 4)

Existentialism as a Psychological Paradigm

From the very beginning, a number of existential thinkers has made significant contributions to psychology and psychotherapy: Kierkegaard with his concepts of *Angst* (anxiety or dread) and "sickness unto death" (the despairing will not to be oneself or the defiant will to be oneself) (Hong & Hong, 2000); Jaspers with his concept of limit situations and his critique of the objectification of freedom by psychoanalysis (Olson, 1994); Buber with his concepts of distance and relation, confirmation, "imagining the real," existential guilt, and healing through meeting (Friedman, 1999); Sartre with his concepts of bad faith, the liability of consciousness, freedom and choosing one's own project, existential analysis, and sex as the incarnation of subjectivity and the subjection of freedom to domination of the other (Douglas, 2003); Tillich with his concepts of ontological anxiety and the courage to be (Hammond, 1965).

Two key issues deserve to be highlighted for the development of psychological research: the existential subject and intersubjectivity. It is essential to become a real person, a *Single One*, an authentic human being, despite developmental constraints and age influences that objectify, depersonalize, alienate, or divide the human person. Martin Heidegger's treatment of the existential subject, though lacking the religious counter-pole of Kierkegaard's, is like Kierkegaard's in his contrast between the authentic self and "*das Man*" – the "One," or the "They" who is lost in the anonymity of what others think. Through resolute anticipation of one's death as one's nonrelational, not-to-be-outstripped, own most reality, one is called back to authentic existence. With Sartre, the existential subject becomes still more sharply defined: first, by cutting off from the self every link with Being or transcendence that would serve as a guide to value decisions; and second, through his dialectic between *pour soi* – the subjectivity of the self from within, and "for itself" – and *en-soi* – the self regarded objectively, whether by another or oneself, as "in itself," in the same sense that any definite object is. Martin Buber (1958) finds the authentic self in the sharing

Table 17.1 Intersubjectivity: convergence of existentialist thinkers

Kierkegaard S. (1813–1855)	Secondary and inessential
Heiddegher M. (1889–1976)	Dimension of the self
Sartre J. (1905–1964)	
Tillich P. (1886–1965)	
Buber M. (1878–1965)	Relation between man and man
Jaspers K. (1883–1969)	
Camus A. (1913–1960)	
Marcel G. (1889–1973)	Community makes possible the I-Thou relation
Royce J. (1855–1916)	

Elaborated from Friedman (1999), pp. 173–174

of the I–Thou relationship, in the Single One, who is responsible to its interpersonal Thou. One realizes one's unique direction ever anew in one's meeting with the concrete situation in which one finds oneself: "All real living is meeting."

The twentieth-century existentialists all recognize intersubjectivity in one form or another as essential to human existence (Friedman, 1999). Yet even here a distinction must be made between such thinkers as Martin Heidegger, Jean-Paul Sartre, and Paul Tillich, for whom the intersubjective tends to remain a dimension of the self, and such thinkers as Martin Buber, Gabriel Marcel, Karl Jaspers, and Albert Camus, who in one way or another see the relationship between person and person as central to human existence (Table 17.1).

Nevertheless, there are important differences even between Heidegger and Sartre. Heidegger's intersubjectivity remains the vague we of his "*Mitsein*" or "*Mitdasein*" – a being with others that expresses itself in solicitude but not in that sharp conflict between one particular person and another, which for Sartre enters into intersubjectivity. Martin Buber (1878–1965) criticizes Heidegger's "solicitude" as monological. On the other hand, Sartre's emphasis on conflict and particularity never take him beyond what Buber would call the "I–It" relation, in which I know the other only as a subject because I am aware of his looking at me as an object or because, in lovemaking, I am trying to possess his freedom and make it subject to my own. This also leaves Sartre short of the more rationalistic notion of *Kommunikation,* which is so central to the thinking of Jaspers, and of the existence "for others," which Maurice Merleau-Ponty (1908–1961) recognizes in his phenomenology. Like Sartre, Albert Camus (1913–1960) starts with Descartes's *cogito ergo sum,* yet he goes on to the dialogue between persons in which human beings really come to exist and sees both the interchange and the limitations of that dialogue as the key to real humanity. Gabriel Marcel arrives at a radical philosophy of I and Thou (*je et toi*), which he also understands as an approach to epistemology as well as to ethics and religion.

The psychological condition of human existence is particularly addressed in three lines of study and research efforts as part of the existential movement: the phenomenological approach, the antipsychiatric movement, and the existential psychotherapies.

The Phenomenological Approach

One of the most important influences on existentialism was phenomenology, which attempts to understand the world and man not through causal formulae and analysis, but through openness to the whole range of phenomena that are manifest, without asking whether they are "real" in some metaphysical sense. Yet, whereas Husserl's phenomenology investigated human consciousness and its intentionality, existentialists themselves were rather interested in existential situations as insecurity, anguish, depression, shame, tragedy, hope, solidarity, and love. Both Husserl's phenomenology and existentialism did not relate to the "reality in itself" according to Kant's thinking, but to the way reality appears to the subject that is open to it: they do not explain phenomena, they rather describe them (Valle & Halling, 1989).

Although the leading phenomenologists – Dilthey and Husserl – were not themselves existentialists, and many existentialists are not, or are only very secondarily phenomenologists, existentialism can not adequately understand without touching on phenomenology. Dilthey based his thought on the radical difference between the way of knowing practiced in the *Geisteswissenschaften* – the human studies such as philosophy, social sciences, and psychology – and that practiced in the *Naturwissenschaften* – the natural sciences. In the former, one cannot be merely a detached scientific observer but must also participate oneself, for it is through one's participation that one discovers both the typical and the unique in the aspects of human life that one is studying. At the same time, one must suspend foregone conclusions and the search for causality that marks the natural scientist in favor of an open attempt to discover what offers itself.

The man who raised phenomenology from an approach to philosophy to a systematic inquiring was Husserl (1859–1938). By the method of "parenthesizing," or phenomenological reduction, he replaced the detached subject and independent object of older philosophy with a field of knowing in which phenomena are accepted as pure phenomena without their independent existence being questioned. The existential status of the world arises "from me as the transcendental Ego," Husserl wrote (1950, p. 26), and the exploration of the field of transcendental experience becomes equivalent to the phenomenological knowledge of the world.

It was inevitable that Husserl's existentialist successors would either emphasize the direct experiential quality of his thought, as opposed to the idealist – like Merleau-Ponty – break with the transcendental ego altogether while retaining the method of phenomenology and the reality of intersubjectivity – like Sartre – or transform phenomenology from a method of knowledge into a "fundamental ontology," like Heidegger.

Both Sartre and Heidegger accepted Husserl's motto "To the things themselves" as an obstacle to any attempt to find Being behind phenomena; only an existential analysis of the existent, the *ontic*, would yield any knowledge of being. For Heidegger, this analysis was posited on his special use of *Dasein* – "the person's being there" in the world, thrown into a situation apart from which neither subject nor consciousness have any meaning. But while Sartre rejected Husserl's transcendental ego in favor of an impersonal consciousness which is nothing other than an emptiness or absence in the presence of the solid being of existents, Heidegger moved toward a concept of Being – the "ontological" – that was linked more with person than with consciousness – the person who becomes aware of herself as "there," "thrown," "in the world."

Jaspers, under the influence of Kant, Kierkegaard, and Dilthey, but independent of Husserl, developed their own phenomenology of "limit situations," which had an important influence on Heidegger. Later, Jaspers explicitly rejected "ontology" – the study of Being – in favor of "the Encompassing," or "the Comprehensive" – a way of knowing that transcends the subject–object relationship even more radically than phenomenology by recognizing that we can never grasp the whole reality since we are a part of it. Martin Buber took the phenomenology of Dilthey over into the kind of knowing that is central to the "I–Thou" relation – one which emphasizes the incomparable uniqueness of what is known in mutual relationship.

Existentialism has also had a significant impact on the postmodern emphasis on hermeneutics, i.e., the science of interpretation. Some thinkers have worked in this direction: Paul Ricoeur (1913–2005), Hans Georg Gadamer (1900–2002), Jurgen Habermas (1929–…), Emmanuel Levinas (1906–1995), Jacques Derridas (1930–2004), Mikhail Bakhtin (1995–1975), and others. *Hermeneutics* has been appearing in the psychological literature with increasing frequency in the last decades. Their hermeneutical perspectives are concerned with meaning and interpretation of messages and texts. At a basic level, hermeneutics can be defined as "the study of understanding, especially the task of understanding texts" (Palmer, 1969, p. 9). Nevertheless, understanding a text is not possible if one approaches it in a purely academic or intellectual fashion. Own personal, existential engagement with the text has got his/her critical role. Hermeneutics pertain to psychology if life is conceptualized as a text that human beings are constantly involved in reading and interpreting. In this perspective, hermeneutics and existential phenomenology have much in common: both are concerned with being faithful to texts and phenomena, respectively. The study of hermeneutics can be considered as a distinct contributor to psychology, in the context of research as well as therapy, particularly with two specific hermeneutic concepts, the hermeneutical circle and preunderstanding (Gadamer, 1975; Palmer 1969).

Existential-Phenomenological Thought in Psychology

When applied more specifically to human psychological phenomena, existential phenomenology became *existential-phenomenological psychology* (Spielberg, 1976; Valle & Halling, 1989) and, as such, that psychological discipline that seeks to explicate the *essence, structure,* or *form* of both human experience and human behavior as revealed through essentially *descriptive* techniques including disciplined reflection.

Examined in terms of the natural scientific viewpoint and its underlying assumptions, existential-phenomenological psychology raises several questions concerning: (a) capacity of purely descriptive techniques in revealing cause–effect relationships; (b) objective science grasping to reveal apparently subjective phenomena; (c) a distinction between a disciplined reflection and purely introspection. The basic problem is the same, i.e., the distinction between existential-phenomenological psychology and natural scientific psychology because they have upstream different assumptions.

The most critical issue left over from phenomenological psychologist is that people are not viewed as just objects in nature but in their total, indissoluble unity or interrelationship among individuals and their world. The existential man or woman is *more* than simply natural man or natural woman. In the truest sense, the person is viewed as having no existence apart from the world and the world as having no existence apart from persons. Each individual and his/her world are said to *co-constitute* one another. In traditional psychology, people and their environment are seen, in effect, as two separate and distinct things or poles. Existential-phenomenological psychologist rejects this conception in favor of the previously mentioned unity.

This total interdependency is based on the notion of *dialogue* (i.e., people and the world are always in a dialogue with each other): people are seen as partly active because they are always acting in their world in a purposeful way and partly passive because the world is always acting on them in a situational way. From this *dialogal relationship,* other insights derive: people must always make choices (condemned to choice). The only choice one cannot make is the choice of not choosing. From this, another assumption derives, i.e., the person with absolute free will does not exist as the totally objectified, completely determined individual. Each person is said to have *situated freedom,* i.e., the freedom (and obligation) of making choices within, and oftentimes limited by, a given situation that the world has presented to him/her.

Another critical issue involves how the *nature of consciousness* itself is regarded. In the everyday experimental work of the psychologist, the nature of consciousness as an issue is rarely raised. If it is brought up, consciousness is usually either equated with what makes up one's experience at a given point in time (i.e., consciousness is thought to be identical with what is presently being thought or felt), or it is thought of as a "container" that "fills up" with contents (new facts, sensations, feelings, etc.). In either case, consciousness is objectified. The notion *loses consciousness,* for example, implies that consciousness is *something* that can be lost (or regained at a later time, for that matter). In addition, consciousness is viewed as having a creative function. We speak of "digging things up from memory" or "synthesizing a new idea."

Existential-phenomenological psychologists reject this view of consciousness in favor of a perspective more in keeping with the already described view of human individual. For most of them, this different view of consciousness is a derivative of (if not identical to) the thinking of the phenomenologist Husserl. His insights and innovations are often difficult to understand, partly because they are so foreign to our natural scientific, causal view of the world. At the same time, these insights are more faithful to basic human experience than natural scientific interpretation. Nevertheless, his attention was not on the world as interpreted and thus created by scientific fact and theory, rather to the world of everyday experience as expressed in everyday language, i.e., with the world as given in *direct and immediate experience* ("Back to the things themselves" was his slogan). One could say that Husserl's domain was the domain of pure phenomena independent of and prior to any reflective interpretation, scientific or otherwise. This realm of naive experience is the *life-world*: the world lived by the common person. This life-world is the foundation upon which existential-phenomenological thought is built (no assumptions are made as to what might be behind or cause the life-world); in the truest sense, it is the beginning. It is not like the world of the natural scientist which is constructed or built up for explanatory purposes (e.g., the way in which simple ideas are first molded into hypotheses and then into complex theories through the direct employment of cause–effect thinking). The life-world is not a construction of consciousness: it is co-constituted or co-created in the dialogue of person and world. Hypotheses, theories, and the causal thinking on which they are based are, therefore, not the subject matter that constitutes the life-world per se as they are not given in direct and immediate experience. They are higher order, less basic, derived notions, i.e., they are, like all ideas, derivatives of the life-world.

Because the life-world is prior to and the foundation of reflective thought, existential-phenomenological psychologists describe the life-world as being of a *prereflective* nature. In this way, the *Lebenswelt* is both independent of knowledge derived from reflective thought processes, and yet, being prereflective (be/or -reflective), it is also the indispensable ground or starting point for all knowledge. Scientific knowledge could not exist without the prereflective life-world: here is the paradoxical nature of the *Lebenswelt*.

Existential-phenomenological psychologists assume, however, the peculiar position referred by Husserl (1962) as *transcendental attitude* (process called *bracketing*) in order to understand a given phenomenon. They attempt to explicitly suspend or put in abeyance their preconceptions and interact in a dynamic fashion. As discovered assumptions are bracketed, other assumptions can emerge, and so on. This process of bracketing and rebracketing is the manner in which one moves from the *natural attitude* toward the *transcendental attitude*, a process of *reduction* that renders the world as it is considered in the natural attitude to a world of pure phenomena. The existence of the natural world is not denied, rather one puts in abeyance one's belief that the world is independent of each individual person. The world, as a result of bracketing the natural attitude, becomes phenomenal in nature, a *world-for-consciousness*.

The notion of consciousness has a large part in this process. Consciousness is that forum in which phenomena show themselves or are *revealed*, always with reference to an object and characterized by *intentionality*. Whereas *intention* normally refers to a purpose or agenda that we have, *intentionality* addresses the ongoing dimension of our consciousness, that we are always in relation to that which is beyond us.

From what it has stated, it follows that existential-phenomenological psychology research is not subjective or introspectionistic in nature either. It is, in one way, a middle ground or, in a more active sense, a thrusting between the purely objective and purely subjective approaches. By treating perception as intentional in nature, the objective and subjective are seen as inseparable, one unable to exist without the other. It follows that the notion of causality in its linear or additive form is rejected. Only that which has its base in naive experience is real and, therefore, only that which is revealed or disclosed as pure phenomena is worthy of attention. The task of research is thereafter to reveal the structure of experience through descriptive techniques that permit to discover the meaning, i.e., the commonality running though the many diverse appearances of the phenomenon.

The aim of phenomenologically informed research is to produce "clear and accurate descriptions of particular aspects of human experience" (Reynolds, p. 2), as body experience (Moss, 1989), perception of alienation (Wertz, 1989), specific experience of memory (Georgi, 1989), being anxious (Fischer, 1989), experiencing valued relationships (von Eckartsberg, 1989). All researchers are expected to develop plans of study especially suited to understanding the particular experiential phenomenon, obviously holding to share a commitment to scientific values and the search for truth.

In particular, Fischer proposes (1989) a theory of personality assessment; the exposed way justifying his theory may offer the understanding links among levels of paradigm, theory, and data gathering. His starting point is the already mentioned, the comparison between the natural science paradigm and the human science paradigm: "The natural-science approach works well in that we clearly established that there is an orderliness, a comprehensibility of human affair, one that can be researched and then applied to constructive interventions into these affairs" (p. 160). The next development is consistent, although more or less shared. During these last decades, the American Psychological Association (APA), thanks to financial supports given by the federal government, has sustained and favored a big development in testing; in the meantime, more generations of psychologists have devoted themselves to construing, administering, and interpreting tests. Consequently, psychologists have identified themselves as scientists, proud of their methodological objectivity. This scientific expertise at test construction, research, and use has grounded both clinical psychology's claim to be a science and its unique contribution to the mental health team. Closeness to physicians and to the academicians also specialized in designing research strategy and statistical analysis of results has helped very much. These theoretical orientation and work application have permitted to explain the personality unity with three major kinds of theory: (a) motivational theories (Locke & Latham, 2002; Maslow, 1987; Reiss, 2004); (b) trait theories (Eysenck 1992a, 1992b; Goldberg, 1993; Matthews, Ian, Deary, & Whiteman, 2003; Mischel, 2003); and (c) learning theories (Mayer & Moreno, 1998).

His conclusion appears consistent to premises posed: "The natural-science paradigm has indeed established psychology as science, one that has demonstrated an orderliness to human affairs. But that demonstration has been a technological one. In large part, psychological data and theories have been artifacts of our technological methods" (p. 161) because: (a) in search for ultimate laws via statistical precision, psychologists ignore the holistic character of being human so that it is not reducible into parts; (b) although scores can be helpful in classifying people into general categories and in making related decisions about them, they do not assist in understanding the individual in his/her particular situation; (c) clinician interest is not so much in which preset categories the person best fits but rather in what light such artificially clear-cut organizational devices throw on the person's particular ways of being and on related dangers and options; (d) the formal training model for clinical psychologists (leading to PsyD degree) differs from academic research training (leading to PhD degree), both programs fully accredited by the APA.

The human-science paradigm choices a different way : (a) the unity of a person's life is preserved while still being an empirical, rigorous discipline; (b) perceived events are accounts for by describing their *whatness*, i.e., the differentiations that can be made within the whole; (c) "person-in-the-world" is the unit of study in order to conceptualize the particular ways people in general move through the worlds, transforming and being transformed by them; (d) research study starts with observations of, or reports from, individuals and then gradually abstracts commonalities; (e) narrative presentation of the findings preserves the structural whole, including the ways that different subjects exemplified the phenomena. In fact, this approach shows its preferences for qualitative research with case studies and grounded theory methodology (Glaser & Strauss, 1967) because "existentially phenomenologically grounded research and practices are consistently consonant with this theory" (Fischer, p. 164).

Antipsychiatric Movement

Antipsychiatry is a term used to refer to a configuration of groups and theoretical constructs that emerged in the 1960s and questioned the fundamental assumptions and practices of psychiatry, such as its claim that it achieves universal, scientific objectivity. Its igniting influences were Michel Foucault (1926–1984), R. D. Laing (1927–1989), Thomas Szatz (1920–…) and, in Italy, Franco Basaglia (1924–1980). The term was first used by the psychiatrist David Cooper (1967). Two central contentions of the antipsychiatry movement are that: (a) the specific definitions of, or criteria for, hundreds of current psychiatric diagnoses or disorders are vague and arbitrary, leaving too much room for opinions and interpretations to meet basic scientific standards; (b) prevailing psychiatric treatments are ultimately far more damaging than helpful to patients. What is interesting to highlight that the movement is closely related to the influence of existential paradigm and that has deeper roots (Galimberti, 1994).

At the beginning of the twentieth century, a new research area – named psychopathology – was born with the confluence of concepts and theories from psychiatry, psychology of medicine, neurology, and philosophy. Faced with these developments and found it impossible to discover "a germ of madness" or organic lesions to mental illness, it makes an attempt of emancipation from ruling by the mortgage and nosographic organismic, antipositivist with approaching the philosophy of Dilthey, Husserl, Heidegger, and particularly the thought of Jaspers and Binswanger. In fact, Jaspers has developed his *General Psycopathology* (1913) fully convinced of the sterility of all objectifying vision. He took up the distinction between explaining and understanding and concluded that the need for psychology to abandon any explanatory model, based on knowledge of the basic mechanisms that determine disease to pay attention to the understanding of mental illness. Madness is not quite *different* to normal, it is simply inadequate ways of relating to the world, or rather the result of the failure of a rational plan of life which could not tune with reality. Normality and pathology are the subjective mode that can only be understood in the empathic encounter between the therapist and the patient psychopathology, which cannot intervene therapeutically in order to modify the disease, but only in order to understand it.

By moving in the same conceptual horizon, Ludwig Binswanger (1881–1966) developed a new psychiatric paradigm: anthrop analysis or Daseinsanalyse. He studied in Zurich with Jung, went to Freud, and studied the philosophical proposals of Husserl (principle of intentionality) and Heidegger (Dasein or planning). *Daseinsanalyse* aims at understanding not to explanation and permits to see both sick and healthy of the same world with whom they relate in different ways. At epistemological level, there is a transformation of the concept of normality. This concept cannot identify with the norm, not the statistical average, nor a theoretical abstraction, nor can it be considered in its prescriptive dimension of normativity. The standard is constitutive, internal, individual, holding a particular project and determine the existence of events.

The Existential Psychotherapies

The phenomenological perspective, therefore, has revolutionized the understanding of psychopathology in general, opening epistemologically new horizons and feeding in the second half of the twentieth century humanistic psychotherapies. Moss (1989) condenses into a few lines these psychotherapeutic schools, accenting both their divergences and overlaps while other authors have reviewed it more broadly and reasoned (Bugental, 1987; Fischer & Fischer, 1983; Halling & Nill, 1995; May, Angel, & Ellenberger, 1958; May & Yalom, 1984; Yalom, 1980; Wyss, 1973).

Opening his review, Moss remembers the debt to Freud, Jung, Rank, and Adler, even if many of the insights that set phenomenological psychotherapy have been formulated in parallel fashion by creative dissenters within the psychoanalytic tradition.

Existential psychotherapy (Rollo May (1909–1994); James F.T. Bugental (1915–2008); Frankl (1905–1997)). Existential influences include the emphasis on the individual's freedom and responsibility for his/her own existence, the process of becoming an individual, the challenge to authenticity in existence, and the positive role of anxiety as a medium for change and growth. Psychotherapy process emphasizes the real relationship between the therapist and the patient, involving an authentic encounter in the present moment between two existing individuals. Psychiatric symptoms are seen as meaningful expressions of a disharmony in the patient's existence, which are to be understood and not merely suppressed. It advocates an openness to the tragic dimension of human life, and not merely a naive Pollyannalike optimism (Schneider, Bugental, & Pierson, 2000).

Daseinsanalysis. The existential movement in European psychotherapy took various names – existential analysis, anthropological psychiatry, *onto-analysis*, and the German name *Daseinsanalysis* (van Deurzen, 1997). Beside the best known Ludwig Binswanger (1881–1966), Victor von Gebsattel (1883–1976), Ervin Straus (1891–1975), Juerg Zutt (1929–…), and Eugène Minkowski (1885–1972) can be remembered.

After World War II, Boss (1963) forged his own viewpoint of Daseinsanalysis, and Wyss (1973) formulated the approach of "anthropological integrative" psychotherapy. Several European refugees from the Nazis brought a knowledge of existential analysis with them to the United States in the 1940s, but the real Americanization of existential analysis began with the 1958 publication by May, Angel, and Ellenberger of *Existence*, an anthology of works by European authors. Most Daseinsanalysts or existential analysts remain psychoanalytic in their basic clinical procedure, nevertheless abandoning the biological, instinctually oriented concepts of Freudian analysis and adopting a more existential interpretation of the patient's experience. Prominent American existential analysts include May (1977), Bugental (1987), and Yalom (1980).

Phenomenological therapy. The Dutch psychiatrist van den Berg (1980) defined phenomenology in psychiatry as the "science of divergent thinking" about people and their world. Its process is never content with a simple change in behavior; in fact, the patient is invited to see his/her world in a different light, to discover a novel perspective on life and relationships, and to recapture a sense of wonder in a fresh and vital way of perceiving. This transformed perception and interpretation of life events becomes the avenue for practical modifications in behavior and relationships.

Gestalt therapy. The emphasis is placed on the process of therapy and not on its content or issues. Role play, visualization, or an imagined reliving of an experience creates a greater emotional intensity in the therapeutic session. The patient is invited to a broadened *awareness* of his/her role in creating life and relationships, invited to a deeper *ownership* of that role, and challenged to *experiment* now with new modes of behaving and relating (Woldt & Toman, 2005).

Client-centered therapy. Carl Rogers (1902–1987) originated the first truly American school of psychotherapy abandoning the exclusive preoccupation with the patient's past and encouraging the therapist to respond nondirectively to the patient, to understand the patient's world of experience, and to reflect back in an affirming manner the patient's own feelings, thus enhancing the patient's self-awareness. The therapeutic procedure of client-centered therapy (Rogers, 1951) involves a continuous *reflection* or paraphrasing of the patient's experiences, with an emphasis on the feeling dimension. This nonjudgmental reflection serves to amplify the patient's awareness and acceptance of feelings. This concept of unconditional positive regard has broadened beyond an explicit approval of behaviors or statements to convey a deep respect for the person and for the *power of life* as it seeks its own way toward fullness and growth (Willis, 1985).

Experiential psychotherapy. It is a specific school of therapy and should not be confused with the broader movement of experientially oriented psychotherapies (Friedman, 1976). It draws on elements from phenomenological, Rogerian, and Gestalt psychotherapy but transforms them from a general approach into a specific method of psychotherapy with procedures, steps, and interventions. Gendlin (1968) began with research on the process of client-centered therapy and what he called the *experiential response*, a patient response that is felt deeply, bodily, and immediately at an emotional level. He refined this element in the therapeutic process further into a teachable procedure called *focusing*, which guides the individual to stay concentrated and attentive to the immediate flow of feelings about his or her total emotional situation (Gendlin, 1981). Gendlin (1973) and Mahrer (1983) have designed specific sequences of steps to enable the patient to enter an experiential state, to enhance the formation of a *felt sense* of one's situation, and to allow this felt sense to move toward deeper experiencing, self-encountering, and closure. Also important for the development of existential psychotherapy is the movement of dialogical psychotherapy, based on Buber's philosophical anthropology and his healing through meeting. This movement began with Hans Trüb (Friedman, 1999), who moved from Jung's dialectical psychology to Buber's dialogical anthropology. It has been carried forward by Farber (1965) with his *psychology of the will*, and after him by Maurice Friedman and Richard Hycner, co-directors of the Institute of Dialogical Psychotherapy in San Diego, and by Boszormenyi-Nagy and Krasner (1986) with their *Contextual Therapy* (1986).

Humanistic psychotherapy. Abraham Maslow (1908–1970) founded humanistic psychology as a "third force" within American psychology in an effort to throw the doors and windows of psychology wide open (Maslow, 1962). It became an eclectic meeting ground for divergent influences including existentialism, the psychology of consciousness, the encounter and sensitivity group movements, and transpersonal psychology. Maslow introduced many of the key themes of humanistic psychology and psychotherapy: the pursuit of self-actualization, personal growth, and full humanness; an emphasis on both being and becoming; and a recognition of the place of values within psychological science.

In his *The Search for Authenticity* (1965), Bugental summarizes the postulates of Humanistic Psychology, often quoted by other theorists: (1) Human beings cannot be reduced to components. (2) Human beings have in them a uniquely human context. (3) Human consciousness includes an awareness of oneself in the context of other people. (4) Human beings have choices and responsibilities. (5) Human beings are intentional and they seek meaning, value, and creativity.

On the American scene, these movements have converged to a great degree, endowing many humanistic or experientially oriented psychotherapists with a core of broadly shared beliefs about the psychotherapeutic process: (a) therapy must commence with an empathic sharing of the patient's experience of a personal world; (b) the therapist must encounter the patient in his/her full historicity – as a being with roots in the past, a home in the present, and aspirations for the future; (e) therapy proceeds most effectively through a lively, emotional, and immediate reexperiencing and mastery of key personal issues; (d) novel positive experiencing in the encounter with the therapist has a corrective therapeutic value in and of itself; and (e) the individual spontaneously moves toward health, growth, and fulfillment in a therapeutic atmosphere of safety, acceptance, and awareness.

Discussion

Considering the path shown, we seem to be facing with Janus for many aspects of the outset. The central paradox can be expressed by the question: Is existentialism a paradigm general-integrative or particular-specific? If we look at its influence on the cultural climate of the twentieth and this twenty-first century, we cannot underestimate the broad spectrum of its influence in every sector. The multiple streams can be traced back to the general framework of existentialism recognized sometimes explicitly, sometimes tacitly, and sometimes opposed and refused. We go into the merits of phenomenological research as the authors are seeking to present and motivate. The conflict between natural science and human science paradigm in many writings is continuous. For example, Romanyshyn and Whale (1989), taking the definition of psychology as a science given by Morgan, King, Weisz, and Chopler, (1986), have discussed the features polemically: rational, empirical, objective, quantitative, experimental, and systematic:

1. *Rational* means reasoned, *organized* knowledge, a criterion that is certainly desirable and necessary, "rational" has taken on a very special and very narrow meaning in psychology. Psychology is rational insofar as it organizes its knowledge as science, i.e., according to the attitude of science. In fact, the criterion of rationality is redundant as psychology is a science because it is rational, and it is rational *only* because it is a science. There are, however, other ways to be rational, other ways to organize knowledge; the discipline of history is only one example. Thus, psychology's exclusive identification of rationality with the attitude of science is itself an historical choice to organize its knowledge in a particular way.
2. The criterion of *empirical* also presents special problems because it, too, has taken on special meanings in psychology. On the one hand, the term *empirical* has become identified with "experiment," as opposed to its wider context of experience, and on the other hand, it has become identified with observation by the other. The first identity raises the problem of context because an experiment essentially means the isolation of phenomena. And the second identity prejudges what behavior and experience are. Psychology comes to identify the psychological with behavior that is understood in the narrowest way as a visible and exterior event in and of itself. Any understanding of behavior *and* experience as perspectives on human action is ignored.
3. The criterion of *objectivity* rests on the prejudice of the objective world. In adopting this criterion, psychology does not question the genesis or becoming of objectivity but simply takes it for granted. This position merely extends and builds on the unquestioned objectivity of everyday experience, leaving the origin of that experience unclarified. In this regard, psychology first proceeds to *ignore* commonsense experience and then to *correct* it through its styles of inference.
4. The *quantitative* aspect of psychology is a concrete way of carrying out its other criterion of being empirical. Numbers provide the neutral context of meaning that allows psychology to abstract a phenomenon from its given context of meaning.

Although this is not the only way that psychology achieves this abstraction, it is still true that measurement generally substitutes for meaning in psychology.

5. The criterion of being *experimental* is already discussed in the criterion of being empirical. One can add that the experimental procedures of psychology have been all-important in justifying its physiological style of inference, whereas the physiological investigations of psychology have found support in this criterion of being experimental.
6. The criterion of being *systematic* is an internal one because psychology believes itself to be systematic when it is being rational, empirical, objective, quantitative, and experimental in the ways we described. In the last resort, psychology is systematic in the way it *organizes* its knowledge, so that this criterion brings us full circle again to our original starting point: *the organization of knowledge is not neutral with respect to that knowledge.*

As owns conclusion, Romanyshyn and Whale (1989) present phenomenological perspective as an alternative option: "...psychology.... can understand itself as another kind of discipline, i.e., as a discipline that organizes its knowledge in another way. Whichever option it chooses, however, we think that, in the last analysis, the real issue is not the claim itself but rather the relevance of the plan of organization to its subject matter. And, in this regard, we believe that existential phenomenology makes two essential contributions. First, it insists that the demands of the subject matter shape the plan of organization of a discipline rather than the other way around. Second, it begins its investigation of the psychological with human action *as lived* rather than with an attitude of practiced ignorance toward it. In this regard, a systematic-critical investigation of the view of science that psychology adopts forces psychology to another *alternative* – existential-phenomenological psychology" (p. 38). Polkinghorne (1989) proceeds in the same line when states that: "The phenomenological research methods... are designed to yield clear and accurate descriptions of the structures of consciousness that constitute what appears in human experience. The research methods of other sciences are constructed to produce information about objects and human activities as they exist in themselves, outside of their appearance in human experience. Phenomenological methods are devised to investigate another realm of reality, the realm that comes into being at the intersection of consciousness and the world – human experience. Phenomenological methods have characteristics whose purpose it is to provide researchers clear access to this realm. These methods are also designed, however, to produce knowledge that meets the commitment to truth shared by all the sciences. The development of psychological research methods based in phenomenology is in a beginning phase. With more experience and effort, these methods will gain in sophistication and clarity" (pp. 58–59).

Despite this clear contrast between the two perspectives, the current literature review shows that this research perspective lends, as Reynolds (2006) explains, to investigate some specific issues such as: (a) freedom; (b) death, finitude, and mortality; (c) phenomenological experiences and "moods," such as anguish (or anxiety), nausea, and boredom; (d) an emphasis on authenticity and responsibility as well as the tacit denigration of their opposites (inauthenticity and bad faith); (e) a suggestion that human individuality tends to be obscured and denied by the common social mores of the crowd; (f) a rejection of any external determination of morality or value; (g) a rejection of the emphasis on rationality and progress that were foregrounded during the Enlightenment. For these research areas, the existential-phenomenological perspective appears as a particular-specific paradigm because the psychological content area is much wider.

When attention is moved to intervention or psychotherapeutic assessment, the theoretical framework remains substantially the same, despite the diverse school of thoughts previously mentioned. "Phenomenologically and experientially oriented psychotherapy decrees neither a specific technical attitude nor specific techniques. Rather, any therapeutic modalities or technique may be applied, as long as principles of mutuality, intersubjectivity, and kinship with the patient are respected, i.e., as long as the total human involvement of the therapist is not reduces to mere technique" (Moss, 1989, pp. 197–198). The attention is posed in the first case on therapist's values: they are decisive and will determine the difference between a mere mechanical imposition of technique and the therapeutic use of behavioral techniques as tools to facilitate a deeper process of genuine personal change. While the temporal process proposed by different schools has the same phases – opening, central, and conclusion – their specifications vary among schools. For example, *the imagery in movement method* (IMM, Valle & King, 1978) proposes four steps: *expression* (using any self-created art form such as sculpture, dance, poetry, or music), *mapping* (the exploration of each element as well as the overall organization of the created art product), *fantasy enactment* (the identification of most charged parts and report "what's happening"), and *verbal translation* (coming back to reality with the help of writing exercises). In conclusion, what is proposed concerns a somehow structured process, whose objective passages cannot be validated and reliable by all. Similarly, the dimensions of grounded theory methodology cannot assume a prescriptive quality (L'Abate, 1994).

Considering again the theory of personality assessment proposed by Fischer (1989), the author compares – like others already mentioned – the natural science with the human sciences and concludes his exposition with the attempt to overcome the limits of idiographic research through the enlargement offered by grounded theory. However, this road has been experiencing difficulties that cannot easily be overcome as its limitations have recently been highlighted. For example, Allan (2003) critiques the use of grounded theory as a research method and Kelle (2005) asserts that this research approach forces the empirical data. Basically, it does not ensure the passage from idiographic to nomothetic level.

Conclusion

Recalling the original question: "Is the existentialism a paradigm particular-specific or general-integrative?," it is oriented to the first alternative. In fact, it continually contrasts the two scientific modes of knowledge without trying to find any continuity. Second, existential scholars feel at ease in dealing with the contents mentioned by Reynolds and appear to be free of methodologies in finding ways to broaden their horizons.

The mentioned limits do not alter the attractiveness of this paradigm that in the last two centuries have repeatedly caught the attention of a large part of modern and postmodern communities with its commitment to so many tragic situations of human life. If fruitfulness is one of the criteria to appreciate any paradigm approach, then existentialism deserves to be taken into consideration. It may be a plant that can produce new original shoots as the paradigm of constructivism.

References

Allan, G. (2003). A critique of using grounded theory as a research method. *Electronic Journal of Business Research Methods, 2*(1), 1–10.

Boss, M. (1963). *Psychoanalysis and daseinsanalysis*. New York: Basic Books.

Boszormenyi-Nagy, I., & Krasner, B. R. (1986). *Between give and take: A clinical guide to contextual therapy*. London: Psychology Press.

Brigstocke, H. (2001). *The Oxford companion to Western art*. Oxford: Oxford University Press. ISBN 0198662033.

Buber, M. (1958). *I and thou*. Philadelphia: Temple University Press.

Bugental, J. F. T. (1965). *The search for authenticity*. New York: Irvington Publishers.

Bugental, J. F. T. (1987). *The art of psychotherapist*. New York: Norton.

Contat, M., & Rybalka, M. (Eds.). (1976). *Sartre on theatre*. New York: Pantheon.

Cooper, D. (1967). *Psychiatry and anti-psychiatry*. London: Paladin.

Crowell, S. (2009). *Existentialism* entry at the Stanford Encyclopedia of Philosophy (2009). Retrieved 2010 from http://plato.stanford.edu/entries/existentialism/

Douglas, C. (2003). *Sartre as biographer*. Harvard: Harvard University Press.

Encyclopedia of Irish & World Art (2010). Retrieved 2010 from http://www.visual-arts-cork.com/

Esslin, M. (2004). *The theatre of the absurd*. Garden City: Vintage.

Eysenck, H. J. (1992a). A reply to Costa and McCrae. P or A and C: The role of theory. *Personality and Individual Differences, 13*, 867–868.

Eysenck, H. J. (1992b). Four ways five factors are not basic. *Personality and Individual Differences, 13*, 667–673.

Farber, L. H. (1965). *Will and wilfulness*. New York: Basic Books.

Fischer, C. T. (1989). Psychotherapy and human experience. In R. S. Valle & S. Halling (Eds.), *Existential-phenomenological perspectives in psychology. Exploring the breadth of human experience* (pp. 157–178). New York: Plenum Press.

Fischer, C. T., & Fischer, W. F. (1983). Phenomenological-existential psychotherapy. In M. Hersen, A. E. Kadzin, & A. S. Bellack (Eds.), *The clinical psychology handbook* (pp. 489–505). New York: Pergamon.

Frank, J. (2009). *Dostoevsky: A writer in his time*. Princeton: Princeton University Press. ISBN 9780691128191.

Friedman, M. (1976). Martin Buber and Asia. *Philosophy East and West, 26*(4), 411–426.

Friedman, M. (1999). *The worlds of existentialism: A critical reader*. New York: Delmar Publishers.

Gadamer, H. G. (1975). *Truth and method*. London: Sheed and Ward.

Galimberti, U. (1994). *Psichiatria e fenomenologia [Psychiatry and phenomenology]*. Milano: Feltrinelli.

Gendlin, E. T. (1968). Client centered: The experiential response. In E. F. Hammer (Ed.), *Use of interpretation in treatment* (pp. 208–237). New York: Grüne & Stratton.

Gendlin, E. T. (1973). Experiential psychotherapy. In R. Corsini (Ed.), *Current psychotherapies* (pp. 317–352). Itasca: Peacock.

Gendlin, E. T. (1981). *Focusing*. New York: Bantam.

Georgi, A. (1989). Learning and memory from the perspective of phenomenological psychology. In R. S. Valle & S. Halling (Eds.), *Existential-phenomenological perspectives in psychology. Exploring the breadth of human experience* (pp. 99–114). New York: Plenum Press.

Glaser, B. G., & Strauss, A. L. (1967). *The discovery of the grounded theory: Strategies for qualitative research*. New York: Aldine de Gruyter.

Goldberg, L. (1993). The structure of phenotypic personality traits. *American Psychologist, 48*, 26–34.

Greasley, P. A. (Ed.). (2001). *Dictionary of Midwestern literature: The authors*. Bloomington: Indiana University Press.

Halling, S., & Nill, J. D. (1995). Demystifying psychopathology understanding disturbed persons. In R. S. Valle & S. Halling (Eds.), *Existential-phenomenological perspectives in psychology. Exploring the breadth of human experience* (pp. 179–192). New York: Plenum Press.

Hammond, G. B. (1965). *Estrangement: A comparison of the thought of Paul Tillich and Erich Fromm*. Nashville: Vanderbilt University Press.

Hillier, J. (1985). *Cahiers du cinema: The 1950's neo-realism*. Hollywood: New Wave.

Hong, H. V., & Hong, E. H. (Eds.). (2000). *Kierkegaard's writings, XXVI: Cumulative index to Kierkegaard's writings*. Princeton: Princeton University Press.

Husserl, E. (1950). *Cartesian meditations: An introduction to phenomenology*. Boston: Kluwer Academic.

Husserl, E. (1962 [1931]). *Ideas: General introduction to pure phenomenology*. London: Macmillans.

Jaspers, K. (1913). *Allgemeine Psychopathologie [General psychopathology]*. Berlin: Springer.

Kelle, U. (2005). "Emergence" vs. "forcing" of empirical data? A crucial problem of "grounded theory" reconsidered. *Forum: Qualitative Social Research, 6*(2), art. 27.

L'Abate, L. (1994). *A theory of personality development*. New York: Wiley.

Locke, E. A., & Latham, G. P. (2002). Building a practically useful theory of goal setting and task motivation: A 35-year odyssey. *American Psychologist, 57*, 705–717.

Mahrer, A. (1983). *Experiential psychotherapy: Basic practices*. New York: Brunner/Mazel.

Maslow, A. (1962). *Towards a psychology of being*. Princeton: D. van Nostrand.

Maslow, A. (1987). *Motivation and personality* (3rd ed.). New York: Harper and Row.

Matthews, G., Ian, J., Deary, I. J., & Whiteman, M. C. (2003). *Personality traits* (2nd ed.). New York: Cambridge University Press.

May, R. (1977). *The meaning of anxiety*. New York: Norton.

May, R., Angel, E., & Ellenberger, H. F. (Eds.). (1958). *Existence*. New York: Basic Books.

May, R., & Yalom, I. (1984). Existential psychotherapy. In R. Corsini (Ed.), *Current psychotherapies* (pp. 354–391). Itasca: F. E. Peacock Publisher.

Mayer, R. E., & Moreno, R. (1998). A split-attention effect in multimedia learning: Evidence for dual processing systems in working memory. *Journal of Educational Psychology, 90*, 312–320.

Mischel, W. (2003). *Introduction to personality* (7th ed.). Hoboken: Wiley. ISBN 0471272493.

Morgan, C. T., King, R. A., Weisz, J. R., & Chopler, J. (1986). *Introduction to psychology*. New York: McGraw-Hill.

Moss, D. (1989). Psychotherapy and human experience. In R. S. Valle & S. Halling (Eds.), *Existential-phenomenological perspectives in psychology. Exploring the breadth of human experience* (pp. 193–213). New York: Plenum Press.

Olson, A. M. (Ed.). (1994). *Heidegger and Jaspers*. Philadelphia: Temple University Press.

Palmer, R. E. (1969). *Hermeneutics: Interpretation theory in Schleiermacher, Dilthey, Heidegger, and Gadamer*. Evanston: Northwestern University Press.

Pamerleau, W. C. (2009). *Existentialist cinema*. Houndmills: Palgrave Macmillan.

Polkinghorne, D. E. (1989). Phenomenological research methods. In R. S. Valle & S. Halling (Eds.), *Existential-phenomenological perspectives in psychology. Exploring the breadth of human experience* (pp. 41–62). New York: Plenum Press.

Reiss, S. (2004). Multifaceted nature of intrinsic motivation: The theory of 16 basic desires. *Review of General Psychology, 8*(3), 179–193. doi:10.1037/1089-2680.8.3.179.
Reynolds, J. (2006). *Understanding existentialism*. Chesham: Acumen Publishing. ISBN 1-84465-043-X.
Rogers, C. (1951). *Client-centered therapy*. Boston: Houghton Mifflin.
Romanyskyn, R., & Whalen, B. J. (1989). Psychology and the attitude of science. In R. S. Valle & S. Halling (Eds.), *Existential-phenomenological perspectives in psychology. Exploring the breadth of human experience* (pp. 17–39). New York: Plenum Press.
Sanderson, G. (2004). Existentialism, globalization, and the cultural other. *International Education Journal, 4*(4), 1–20. http://iej.cjb.net
Sartre, J. P. (1943 [1992]). Being and nothingness. New York: Washington Square Press.
Schneider, K. J., Bugental, J. F. T., & Pierson, J. F. (2000). *The handbook of humanistic psychology. Leading edges in theory, research, and practice*. London: Sage.
Spielberg, H. (1976). *The phenomenological movement: An historical introduction* (2nd ed., Vol. 1 and 2). The Hugue: Martinus Hijhoff.
Valle, R. S., & King, M. (1978). *Existential-phenomenological alternatives for psychology*. New York: Oxford University Press.
Valle, R. S., & Halling, S. (Eds.). (1989). *Existential-phenomenological perspectives in psychology. Exploring the breadth of human experience*. New York: Plenum Press.
Valle, R. S., King, M., & Halling, S. (1989). An introduction to existential-phenomenological perspectives in psychology. In R. S. Valle & S. Halling (Eds.), *Existential-phenomenological perspectives in psychology. Exploring the breadth of human experience* (pp. 3–16). New York: Plenum Press.
van den Berg, J. H. (1980). Phenomenology and psychotherapy. *Journal of Phenomenological Psychotherapy, 11*(2), 21–49.
van Deurzen, E. (1997). *Everyday mysteries: Existential dimensions of psychotherapy*. London: Routledge.
von Eckartsberg, R. (1989). *Life world experience: Existential-phenomenological research approaches in psychology*. Washington: University Press of America.
Wertz, D. (1989). Approaches to perception in phenomenological psychology: The alienation and recovery of perception in modern culture. In R. S. Valle & S. Halling (Eds.), *Existential-phenomenological perspectives in psychology. Exploring the breadth of human experience* (pp. 83–98). New York: Plenum Press.
Willis, R. J. (1985). The 'life of therapy': An exploration of therapeutic method. *Psychotherapy in Private Practice, 3*(1), 63–70.
Woldt, A. L., & Toman, S. M. (2005). *Gestalt therapy: History, theory, and practice*. London: Sage. ISBN 9780761927914.
Wyss, D. (1973). *The psychoanalysis schools*. New York: Jason Aronson.
Yalom, Y. D. (1980). *Existential psychotherapy*. New York: Basic Books.
Young, J. (2010). *Friedrich Nietzsche: A philosophical biography*. Cambridge: Cambridge University Press.

Part V
Operational Paradigms in Psychology

Chapter 18
Information Processing

Piero De Giacomo, Luisa Mich, Carlos Santamaria, Laura G. Sweeney, and Andrea De Giacomo

Information processing and communication have been investigated in a large number of different disciplines: computer science, rhetoric, psychology, mathematics, sociology, anthropology, organizational economics, and many others. Information theory (Guiasu, 1977; Shannon & Weaver, 1949) and communication theory as specific disciplines have drawn an increasing interest since the inception of new communication channels and devices. Thus, we have a phenomenal variety of areas and contributions related to information processing and communication principles, theories and models, whose growth is now considered exponential (http://www.youtube.com/watch?v=cL9Wu2kWwSY; Berlo, 1960; Laswell, 1948; Ruben, 1984). Computer science has been investigating information processing as related to communication in the context of such disciplines as artificial intelligence, natural language processing, human computer interface, knowledge management, multimedia, and many others, often addressing also human-side aspects beside those considered in the first approaches, that included, for example, artificial languages to code algorithms in computer programs.

This chapter cannot include in detail all of above. Consequently, it will focus on highlights that show how pervasive the information processing paradigm is in all

P. De Giacomo (✉)
Department of Clinical Psychiatry, University of Bari, Via Fanelli, 239, Bari, Italy
e-mail: piero.degiacomo@tin.it

L. Mich
Department of Computer and Management Sciences, University of Trento, Via Inama 5, 38122, Trento, Italy

C. Santamaria
Department of Cognitive Psychology, University of La Laguna, Campus de Guajara s/n, 38071, La, Laguna, Tenerife, Spain

L.G. Sweeney
Avant Garde Books, 301 Briarvista Way N.E., Atlanta, GA 30329, USA

A. De Giacomo
Dipartimento di Neuropsichiatria Infantile, School of Medicine, University of Bari, Bari, Italy

L. L'Abate (ed.), *Paradigms in Theory Construction*, DOI 10.1007/978-1-4614-0914-4_18,

aspects of our lives and in particular how useful it is in interpreting human communication and behavior.

The Emergence of the Information Processing Paradigm

One of the most revolutionary ideas in science is the biological standpoint that asserts that any relevant operation observed in a living organism can be explained in physical terms and no immaterial force or principle should be taken into account. This view is often termed as *mechanism*. The demand of a general-purpose mechanism in evolutionary biology was widely satisfied by Darwin's natural-selection theory. The main algorithm of this theory made the plea for immaterial forces simply unnecessary to explain biological evolution.

The application of the mechanistic paradigm to psychology has been tried in several occasions. One of the most successful attempts was behaviorism (see Chap. 16). One of the main assumptions of behaviorism is that any sequence of actions executed by an organism can be described as a *reflex chain* where each of a series of movements may be linked through the stimulatory results of prior movements (each movement triggers the next movement in the sequence). The neurophysiologist Karl Lashley (Lashley, 1951) argued that this mechanism is unable to account even for most rudimentary human behavior which often occurs too quickly, without sensory feedback, and suggests the existence of internal plans for what will be done later. For example, the errors in language production suggest that speakers are planning and holding words in their mind while articulating previous words in the sentence (e.g., *Do you reel feally bad*?). A *plan* in this sense means that the current state of the organism can have some bearing on its future behavior.

This limitation of pure associationism revealed the necessity of an internal-states machinery to simulate simple sequential behavior. Fortunately, there already existed a clear description of a machine of this kind: the Turing machine. Alan Turing (1936, 1948) described the formal properties of a machine based on a finite number of conditions. As it is well known, the machine operates with a tape divided into squares. Each square is capable of bearing a symbol, and the tape runs forwards and backwards through the machine. The machine can read or write just on one square in a given moment. Turing called this square the *scanned square*, and the symbol on the scanned square, the *scanned symbol*. However, the scanned symbol affects further behavior of the machine on other symbols: it represents an *internal state* of the machine. Also, Turing described a universal machine so that any particular Turing machine can be implemented within it so that the universal machine includes the particular machine as part of its properties. A machine of this kind would be able to compute any computable sequence and is a clear precedent of the idea of a stored computer program.

Lashley presented his paper on the problem of sequential behavior in a conference where some of the most prominent pioneers of the science of computation were among the participants (the Hixon symposium; Jeffress, 1951; see also, Gardner, 1985). One of them was John von Neumann, credited precisely as the first person who described the idea of a machine with a stored program. The von Neumann architecture brings into play the idea of programs and data stored together in memory. Previous computing machines had to be physically designed independently for each task (a surviving example is a simple desk calculator). The von Neumann architecture acts as an instance of a universal Turing machine, and the concept of storing a program in the computer's memory had important consequences (in addition to programming without a tin solder or any other physical tool). One of them is that the program can modify itself as a result of data processing. This idea led von Neumann to think on artificial computers as devices that might be able to replicate themselves "self-reproducing machines" and act as living organisms.

Another brilliant applied mathematician shared this interest on the relation between artificial and natural computation with von Neumann. His name was Norbert Wiener and during World War II he developed automatic radar and missile guidance systems. This work raised his interest in the concept of *feedback*: particularly, how the output of a process can be compared with the goal, and the difference used to control further processing. Wiener founded the field of *Cybernetics* and defined it as the study of communication and control processes in living organisms and machines (Wiener, 1948). He defended the application of a simple structure based on the use of feedback as the main control mechanism in Biology and Sociology. This attempt was not particularly successful, but the main idea: the formal control systems might be applied to any data structure was a core assumption for early Cognitive Science and the Information Processing paradigm. The problem of planning sequential behavior could dissipate within this framework proposed by Turing, von Neumann and Wiener among others.

A related idea was presented by a young collaborator of these mathematicians: Claude Shannon. In his master's thesis, Shannon (1948) used Boolean logic in electric circuits and demonstrated that this kind of digital circuits can resolve any logical or numerical relationship. In 1948, after his interaction with Turing, von Neumann and Wiener, he published his famous mathematical theory of communication. The impact of this work for the information processing approach is striking. For the first time, information was treated as an entity independent of the substrate in which it is implemented. This perspective allowed scientists to develop a mechanistic approach focused on abstract concepts rather than matter (probably the first mechanistic but not materialistic approach in the history of science).

In mid twentieth century, experimental psychology was clearly governed by the behaviorist approach. However, by no means had it implied that every researcher was comfortable with behaviorist statements. Not only Lashley, but also Tolman, Hull and others have criticized some principles of behaviorism and were reluctant to reduce the mind to a "black box" but they still fear of the mentalist language. The impact of information theory and the computational theory of mind provided them with a immaterial but mechanistic language. As George Miller, one of the founders of cognitive psychology has expressed it: "[By 1956]...I had stopped pretending to be a behaviorist" (Miller, 2003, p. 141).

Information theory, cybernetics, and the early computational approaches allowed researchers to speak of things other than observable behavior. Information (an immaterial concept) can be measured, as in the seminal paper on the *Magic Number Seven* (Miller, 1956) where a limit for immediate recall was established for the first time. With the black box opened other concepts emerged easily from it. That was the case of Chomsky's (1957) Syntactic Structures which were also not behavior and proved more useful than Shannon statistical account (based on information theory) to explain natural language. Psychologists needed a catalyst to continue the mechanistic work initiated by behaviorism without its theoretical restrictions, and they found it in early computational theory. Probably the effect of this theory for the development of Cognitive Science in particular theories for specific fields is not so important, but at least in terms of speech freedom its contribution was invaluable.

Artificial Intelligence

The discipline of Artificial Intelligence was named after a desire more than a reality or a description of its immediate research field. A group of engineers and mathematicians viewed the newborn digital computer as the appropriate device to fulfill an old human desire or utopia: to create intelligent beings. They were optimistic and they transmitted their optimism to the institutions that should provide the funds and even to the general public (some of them were interviewed in Life magazine and similar generalist media, and Asimov started publishing his series of stories on robots as early as 1940s).

The mathematician John McCarthy coined the term in 1956 when he used it as the topic of a conference in Dartmouth College. The other organizers of the conference were Marvin Minsky, Claude Shannon, and Nathan Rochester. In spite of their relevance and later recognition as founding figures of the field, the conference more relevant paper was presented by Allen Newell, Herbert Simon, and Cliff Shaw (see, Newell, Shaw, & Simon, 1958). The paper presented their work on a "Logic Theorist" a program that simulates the skills of a human logic theorem prover: probably the first Artificial-Intelligence program in History.

Also the conference established the three basic assumptions of the theoretical core of Artificial Intelligence:

1. The recognition that thinking can occur outside the brain (i.e., machines).
2. The assumption that thinking can be understood in a formal and scientific manner.
3. The assumption that the best way to understand it is through digital computers.

The strictest enactment of the first assumption was the primordial work of McCullough and Pitts (1943) who presented a logical model of the neural activity. This work pioneered a branch of Artificial Intelligence that tried to emulate brain activity with simple units connected with each other such as neurons. Networks of this kind have special plasticity for learning. This branch had an initial expansion but was soon discontinued and resumed in early 1980s. Meanwhile, Artificial Intelligence devoted to the simulation of the product of the brain activity without attending to its structure.

In this line the IBM engineer Arthur Samuel (1959) created the first program that can learn to play a game at a human level, and in 1958 McCarthy designed the first programming language devoted to Artificial Intelligence: LISP.

Unfortunately, the optimism of the pioneers was not endorsed by the results. A machine was not able by the 1980s to do "any work a man can do," as Herbert Simon predicted in the 60s. Although computers are good in some formal tasks the progress has been minimal in most of the activities that look trivial for most human beings, such as recognizing a friend or enter an office and sit in the chair without bumping into the table. Perhaps we should first know how exactly we are able to do these tasks before teaching them to computers. One of the most relevant area of artificial intelligence is Natural Language Processing which goal is to develop tools and systems able to tackle a wide range of communication tasks: natural language understanding, translation, speech recognition, information retrieval, information extraction, natural language query, etc. Advances in all these areas highlighted pragmatics as critical to deal with the complexity of natural language in communication processes.

Pragmatics of Communication

A fundamental concept present in almost all the disciplines is pragmatics, which is relevant to understand the relational meaning of a message beyond its syntax (correct use of the rules of the languages) and semantic (intended meaning) (Chandler, 2002). Pragmatics is related to the information processing and communication efficacy, that is the capacity of communicating to accomplish a desired goal in human behavior and human relationships.

In the classical rhetoric of Aristotle a discourse can be given to "docere, delectare, flectere," that is to instruct the audience, gain their goodwill, and arouse their emotions (Herrick, 1990). Four objectives are identified in the most recent contributions of communication theory: control, motivation, information, and emotional expression (Scott & Mitchell, 1972). Some fundamental studies focused on the analysis of the pragmatics of communication identifying a limited number of relational patterns (Harper, Lymn Scoresby, & Boyee, 1977; Laing, 1970; Sluzki & Ranson, 1976; Watzlawick, Beavin, & Jackson, 1967; Watzlawick & Weakland, 1976); others authors investigated more specifically the effect of messages in terms of actions following an utterance or linguistic act (Austin, 1962; Searle, 1969, 1975; Winograd & Flores, 1987). A conceptual definition of information as "a difference that produces a difference" was introduced by Bateson (1979), who contributed to the success of the relational perspective in psychiatry and psychology.

Information Processing Technology (IPT)

Since when Norbert Wiener (1948) introduced the term "cybernetics," computer science and ICT (in this chapter also named Information Processing Technology, IPT) developed in an unforeseeable and unthinkable way, a phenomenon known also as information revolution (Porter & Read, 1998) and more recently also named "second information revolution," to underline the dramatic growth of online communication (Brock, 2003). Changes in the communication processing scenarios occurring in the last decade are frequently referred to as Web 2.0 (O'Reilly, 2005), a buzzword that indicates both technological and relational innovations. New tools, platforms, and a number of functionalities of increasing complexity have appeared on the scene. For an exhaustive review of such tools and services, the reader is referred to a variety of sources both online and offline. The goal of this section is to underline those issues that are relevant for researchers to investigate participant interactions and to apply models of the pragmatics of communication and in particular to apply the Elementary Pragmatic Model (EPM) described in the second part of this chapter.

This kind of studies can be useful in many and different areas; to name a few, online business models, customer relationships management (CRM), e-commerce, distant learning, teleworking, online psychiatry. Among the most recent, there are specific functionalities as online advertising, user profiling, user recommendation, online community management and control, and many more.

Firstly, a common advantage of interactions supported by ICT is given by the digitalization and recording of larger number of messages that can be retrieved, classified, tagged, and analyzed according to the chosen model. For example, for the EPM, e-mail messages can be tagged according to their pragmatic content as proposals/acceptation (1) or as refusal/denial (0) and then calculate relational patterns of the interacting participants. After that, applying the paradox table, we can foresee the outcome of a given relation (Colazzo, Mich, Malinverni, & Schäl, 1991) or suggest interventions to improve relationships, for example in an organization (Pereira, 1999). But to analyze online interactions, it is necessary to know which kind of tools and applications can be used for them and to highlight characteristics relevant to the analysis of the pragmatics of interactions. Before describing them, it is important to underline that not all the people are able to access online communication. Beside economics and educational barriers – often referred as digital divide (http://www.webopedia.-com/TERM/-D/digital_divide.html) – networks and hardware to use them are not uniformly available to all users (see, for example, data published on http://www.internetworldstats.com/ or on http://gandalf.it/data/data1.html).

As regard ICT-supported interactions, first of all, tools can allow participants to exchange only verbal – usually textual messages, like those more commonly used by email services – or also nonverbal messages – including multimedia content, e.g., video-call using a VoIP (Voice over Internet Protocol) service, as for example, Skype (http://www.skype.com) or Yahoo! Voice (http://voice.yahoo.jajah.com/), or Google talk (http://www.google.com/talk/). In particular, the Web supports a number of services to share not only texts, but also pictures, movies, and so on. In fact, there are specialized websites that have been designed to share a given type of "documents": YouTube (http://www.youtube.com) is based on a platform that allows to easily upload and share videos; Flickr (http://www.flikr.com) is specialized for publishing pictures and commenting them; FaceBook (http://www.facebook.com) was intended to retrieve contacts and share experiences, etc.

Then, interactions can be "one to one" (email), "one to many" (forum, newsgroup, personal blog), or "many to many" (talk, chat, or conference call). Large databases of messages exchanged on a given social network can be analyzed also to identify the multiplicity of existing relationships and have been studied by the new social network analysis discipline (Abraham, Hassanien, & Snásel, 2010).

Another parameter is related to the time: synchronous tools need the participants' presence at the same moment (e.g., chat, VoIP), while asynchronous ones allow the receiving participant to read and answer the messages also on a different time (e.g., email). Also participants involved in the relations can be different and a very important distinction is due to the possibility to create artificial participants, commonly named agents, as for example the avatar (alter ego) in video-games or artificial agent used in online auctions, as well as virtual personnel on services websites (e.g., answering to customers' questions on tourism website).

Finally, a user can interact using several devices. For example, to exchange email, besides Personal Computers (PC), one can use a mobile phone, a PAD (Personal Digital Assistant) or a smart phone (e.g., an I-Phone) that includes functionalities of both of them; to access websites and a television set can also be suitable, etc.

Two main trends emerge from the online communication scenario:

- Integration of the communication tools and functionalities on the same website platform.
- Convergence of the devices to access such platforms, usually social network websites.

The coexistence of a variety of information processing and communication tools is apparent in all the main social networks, so that, for example, on FaceBook a user can exchange comments and publish text messages, but can also publish photos, a functionality that was not native on that website. Moreover, to read email, publish a tweet on Twitter (http://www.twitter.com), watch a video on YouTube, send an SMS (Short Message Service), comment to a post on a blog, review an hotel, etc., a user does not have to necessarily use a PC, as he or she can do all that with a smart phone, that is with a mobile platform. Shopping online will become so widespread that face-to-face contacts between a buyer and a seller will become increasingly rare. Most objects (cars, clothes, shoes, personal items, etc.) will be stored in warehouses and sent to buyers through the mail.

Summarizing, all the characteristics described in this section have to be taken into account in any study of online communication in order to apply pragmatics models, to be able to define and identify:

- Participants involved in a given relation: number, role, and type (human or artificial).
- Granularity of the exchanged messages: email, instant messages, blog comments.
- Nature of the content: textual or multimedia content.
- Any constraints of the tools, as for example, posts (tweets) in Twitter are limited to 140 characters.
- How to access to the stored messages.

What is written above describes the general contexts for IPT, but given the relevant role of the World Wide Web and the Internet the rest of this section will focus more specifically on them, on the rise of educational and training programs, and on the incredible rise of instrumentation in this field, creating a completely new industry, new specialized disciplines, theoretical, empirical, and professional frameworks that will control and determine our lives for the rest of this century.

The World Wide Web and the Internet

This relatively new globally pervasive medium of communication and information where collaboration among people within and without national and geographic borders is fast becoming cheaper and easier (Tapscott & Williams, 2006; 2010). They coined the term "wikinomics" to explain that even amateurs can have access to world-class communication pools and world-wide markets, through this medium. It is possible for large groups of people who have never met to work together. Its applications to the "Delivery of low-cost interactive multimedia-enhanced behavioral interventions by computer and Internet can achieve unprecedented public health impact" (Pulier, Mount, McMenamin, & Maheu, 2007, p. 303). This conclusion is especially applicable when structured, distance writing interventions are composed of interactive practice exercises (L'Abate, 2011).

Mesch and Talmud (2010) demonstrated the positive effect of Internet connectivity on social involvement within the community through their longitudinal design examining the effects of the electronic bulletin board as utilized by two communities in Israel. They found that face-to-face meetings had less impact on civic participation and connection to neighbors than did the electronic forums. Research by Stern and Adams (2010) revealed that Internet usage positively affected the creation of social networks in rural communities. If such is the case, one may suppose that both long-distance students and counseling clients will be able to open up more readily over digital social networks than when dealing with others face to face. Furthermore, due to its convenience, it is likely that the iPad and similar Internet devices including the Smartphone will become the favored medium for education and behavioral interventions since these tablets permit users to communicate in a manner that will not impose upon others nearby. Discreteness is of the essence if practice exercises are to be readily available no matter where one is located across the globe and throughout the day. This rings true also in the realm of education for the reason that learning must take place on a constant basis in order to be connected to the continuous real world of information without frontiers.

Education

What is relevant to information processing and communication is the phenomenal development of completely new professions of Internet and computer technology experts with thousands of advanced degrees who can continue to specialize within their own disciplines. Not only are these Internet experts sources of growth for their online students, but they are also candidates for personal growth by adding additional specializations to their own repertoire. Access to online education via Internet, even Smartphones, provides educational leaders of tomorrow opportunities for personal and professional growth. Education via the wireless networks will undoubtedly open up new doors for the development of interdisciplinary studies such as integrated arts and writing and even online writing interventions immersed in avatar experiences.

National University in LaJolla, California, for instance, is a pioneer in wireless education without frontiers. The high school division has recently developed an online computer education class that permits students to access the online school through a Web address specifically created for the Smartphone, leaving students no excuses for not completing assignments on time or for not communicating on the discussion boards with their peers. Even PowerPoint presentations and films are readily available for asynchronous access. Directions for a final project are relayed to students through a virtual robotic animation stimulating interest. Naturally, e-mail is utilized for private conferences; however, the school could make further advances by utilizing instant messaging for real-time discourse. Although Internet schooling at National University and other schools has been thriving for the past 15 years, outstanding opportunities have recently emerged in new Smartphone formats.

For instance, Granville Stephens, a publishing company located in Australia, utilizes Smartphones to deliver English-to-speakers-of-other-languages courses on Smartphones in Asia. Given that miniature technology is very popular in Asia, the English courses have also sold well. Granville Stephens considered that people only recall 10% of what they hear, leaving the need to reinforce learning through visual slides, printed words, and accessible tests for instant grading and tracking (Rosen, 2009). If indeed people are interested in accessing learning through Smartphones and other wireless devices, there should be just as much fervor over the opportunity to seek counseling services in similar formats with concepts enhanced by visuals, compelling participants to answer all questions in a nonjudgmental environment.

According to Barley et al. (2008, p. 97), Computer-Assisted Instruction (CAI) has been proven effective in teaching at-risk students due to its nonjudgmental nature and how it gives frequent feedback needed to individualize instruction to meet students' needs. CAI takes the emphasis off the teacher and puts it on the student, where it rightfully should be. Just as technology removes teacher-domination of instruction, it could have the same impact upon counseling. Whereas CAI refers to the utilization of the computer, Smartphones and iPads offer the same applications to be opened up when participants are on the go. Therefore, the author proposes Smartphone/iPad-Assisted Counseling (SAC) to potentially take focus off the counselor and put it back on the client's needs no matter where or when needed. Counselors, exercises, and tests with feedback will be available 24/7 whenever and wherever needed.

Practitioners may take a look at ways in which computer therapy has already emerged. Computer therapy has been offered through McGill University's Sexpert which provided therapy to couples who were not hesitant to discuss intimate details of their relationships through a series of multiple choice questions. The developers believed that the writing process enabled participants to be relieved of their problems. Couples were much more open to answering questions presented in a program as opposed to answering them in a face-to-face therapy environment. Couples were much more willing to share private details with someone online whom they did not know than with people in face-to-face therapy (Wallace, 1999). Sexpert was also an educational process since participants grew to understand their own relationships better, and the effects were instrumentally measured.

Murphy (2011), in *Your Avatar, Your Guide*, proposes that avatars enable clients to live out experiences in a virtual environment that they often fear in their mundane worlds. Although one might be too fearful, she suggests, to go to a coffee shop, she may do so through computer simulation of virtual reality. Watching one's own look-alike Avatar accomplish tasks influences the subject for the better in what is known as the "doppelgänger effect." Murphy writes, "Just 3–5 min of watching your digital double can improve your social skills, calm your anxieties and help you make better lifestyle decisions" (Murphy, 60). Doppelgänger avatars assist in the modeling of behavior through the receipt of vicarious reinforcement from virtual peers and have been shown to be effective in both weight loss and motivating participants to exercise. Most of all, cyber-replicas assist people in attaining social cues to overcome fears and phobias (Murphy, 59–63).

Instrumentation

Instrumentation refers to the measurement of the impact of therapy conducted through online writing interventions, through avatar interventions, or through technology. Statistics programs quickly calculate research results leaving counselors and educators with more time than ever before to focus on the process itself rather than on the calculation of results. Aside from calculating differences, programs present results in highly visual formats available on iPods, iPads, and Blackberry Playbooks.

Scientists (Choi, 2011) are currently developing applications that will assist in research of ornithology, geology, molecules, time trees, species, and chemical bonds. Such applications will be available on both iPhones and "SciPhones," a word coined for Smartphones that now are valid research tools. The time tree shows when humans and chimps shared the last common ancestor, whereas a database of molecules stores a chemical bond that has been dragged by a finger across the screen. If such valuable data may be both stored and shared within Smartphones, then certainly psychological data may be easily stored and analyzed through similar technology.

Applications

Applications of traditional and mobile ICT include: (1) programs that perform major tasks for which a computer is used; (2) ways in which computer hardware or Smartphones are utilized in work and school settings. Currently, one finds many applications (software for accomplishing tasks) advertised on the Google apps Web site. However, little control is exerted to restrict publishers from producing apps that are sexually explicit to be published on Google. Conversely, more control and editing is exerted at MacIntosh's iPhone apps' Web site, http://www.apple.com/.

According to Microsoft, one can "download apps fast" and it only takes "a tap." In fact, today, there are approximately 350,000 apps available through the iPhone Web site. A Microsoft program is capable of recommending apps based on those the user has already downloaded, much like Amazon recommends books based on past purchases. On the bottom of the Microsoft Web page, one can click on icons representing apps for the cook, apps for the new year, apps for the great outdoors, and even apps for going out on the town. To date, no one has created an app for psychology other than iTunes University's Introduction to Psychology. Anything you can do on a computer, you can also do as you travel around with the iPhone. This includes video chat and HD video recording for recording lectures, and iTunes University sites for lectures (Apple.com. Apple Inc., 2011).

So what are the paradigms into which these technology applications are divided? Perhaps, apps fall into numerous paradigm-systems including: (1) the wireless paradigm; (2) the nonwireless paradigm; and (3) the ecological paradigm to prevent waste. Popular models include: (1) A behaviorist technology paradigm of multiple choice tests with learning applications; (2) A humanist paradigm consisting of apps utilizing one-to-one communication, permitting participants to accept one another despite differences. The uses of technology in work, school, and counseling settings include: (1) structured learning; (2) spontaneous receipt of news; (3) communicating with and receiving replies from others; and (4) constructivist creation or making meaning. Technology users engage in structured learning to deliberately attain knowledge that is practical for bettering each others lives. Those technology users who utilize technology to be informed do so in order to keep up with social trends, work trends, and general political movements. Most popularly, users desire to communicate with friends, family, and colleagues, about daily occurrences and needs. Lastly, users make new meanings through writing, programming, and other digital arts.

What amounts to a miraculous revolution in information processing is now under attack by governments that are increasingly attempting to assert their control and sovereignty over this medium that has no limits to its vast poker, such as China. Big companies are now building their own digital territories. An economic fragmentation and political balkanization is another attempt to decrease the immense and, therefore dangerous power of the Internet (*The Economist*, September 4th, 2010, pp. 11, 75–76, 61–62; Kelly, 2010; Wu, 2010). This epidemic growth has spawned new educational programs and a phenomenal rise in instrumentation (Apple.com. Apple Inc., 2011).

The Elementary Pragmatic Model

It is important to underline that IPT can be used also to facilitate the application of a pragmatic model to "prescribe" or suggest to participants to improve their attitudes. In the EPM attitudes correspond to a combination of 16 relational styles. To illustrate this kind of application is the goal now. This section gives a detailed description of the EPM; with an application of the EPM to generate new ideas, new mental constructs.

Within the context of an information processing perspective, to identify a systematic way to describe pragmatics in terms of relational patterns, De Giacomo and Silvestri (1982; Silvestri & De Giacomo, 1979) defined a model that introduces the elementary pragmatic as the unit of description of participants' interactions.[1] An elementary pragmatic interaction involves two participants that exchange bits of "pragmatic" information in terms of acceptance or refusal of a topic or proposal and can be formalized as triples of Boolean values. In fact, a given message contains relational information that can be coded according to the Boolean algebra as 0 (no) or 1 (yes): 0 if it corresponds to a refusal or negation to a given topic or proposal, 1 if it corresponds to an acceptation or agreement. So that we have, for example, 1 to code the first participant saying "I'd like to go to the cinema," 0 to code "I don't" for the second participant and 1 or 0 for the final position of the first participant (Silvestri et al., 1980).

The model also assumes that any participant behaves according to a "mechanism" that is (partially) independent of the topic of the interaction, that is the reaction to a given situation coded as binary couple is the same if the question is going to the cinema or deciding what to cook for dinner or where to go on the weekend, etc. This assumption derives from the observation of relational patterns that recur in real communications, so that even in common language a person is described as stubborn if he or she never changes his mind, or as "yes man" if he accept any proposal and ideas. According to the EPM, such patterns correspond to the 16 Boolean functions in two variables (Lefons et al., 1977). There are also some publications on the algebra of the model (Guerriero, 2009), as well as a link with Fuzzy Logic (Resconi & De Giacomo, 2004).

So far the descriptive levels of the model, but the EPM can also foresee how a given function will change interacting with another one: the results of the 16×16 (256) combinations are described by the paradox table. The table represents the prediction level of the model. Triples and functions constitute the descriptive level of the EPM. The paradox table allows foreseeing the impact of the elementary interactions on the relational patterns of the participants. Representing the evolutions of the participants' patterns, it is the main contribution of the EPM. Pragmatic actions are represented as sequences of 0 and 1; punctuating them we obtain the participants' coordinates; such coordinates allow to define a vector of probabilities of a participant to use one of the 16 functions; combining such functions we can foresee how a relational pattern will change. Using the paradox table in a prescriptive way, we can suggest different ways to think about a given topic: this is the application described in the last part of the paper.

Figure 18.1 (De Giacomo, 2011) shows a comprehensive representation of all the levels of the EPM: pragmatic actions are represented as sequences of 0 and 1, representing the participants' world (level 1). By delineating them, we obtain data allowing us to evaluate the participants' coordinates (level 2). These coordinates make it possible to define a vector of probabilities of a subject's using one of the 16

---

[1] The brief description of the EPM in this section is merely intended to be enough to allow an understanding of its application to generate new ideas. For the sake of brevity, many background details and explanations that are found in the cited papers have been left out.

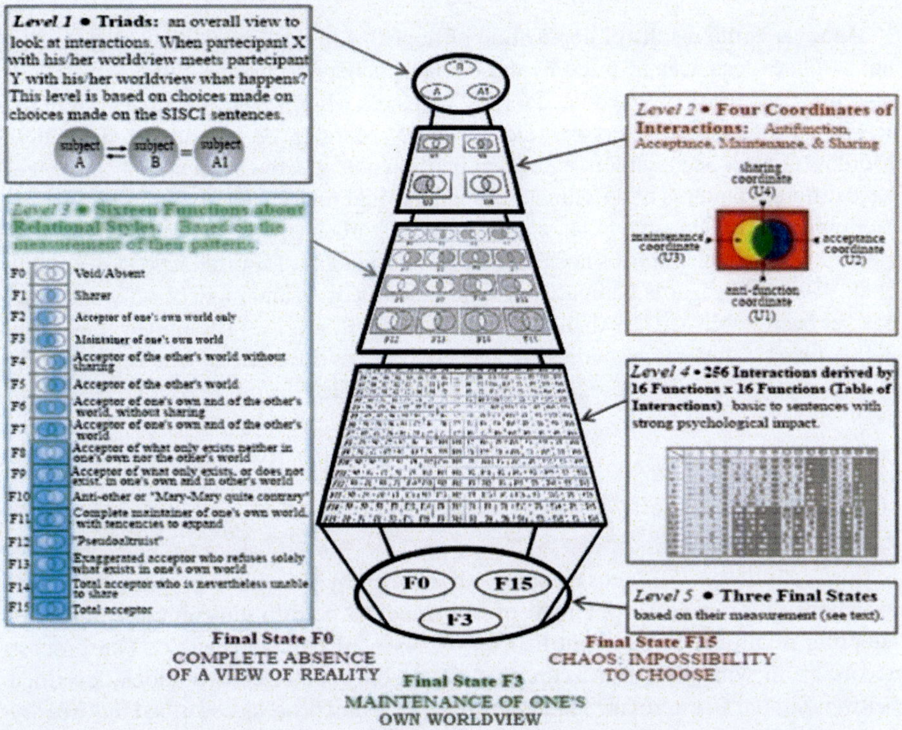

Fig. 18.1 The five levels of the elementary pragmatic model source: De Giacomo (2011)

functions (level 3). By combining these functions according to the paradox table, we can predict how a relational pattern will change (level 4). The table also predicts that long-term relationships will end in relational patterns consisting of only three final states (level 5). For example, the paradox table suggests that if you are doubtful you prefer to be with those who accept your world or to face situations with these characteristics (F6 || F5 = F3, where the symbol || stands for interaction).

So far, the EPM has been applied at all levels – to describe participants' patterns and to describe changes in such patterns (Silvestri, Mich, Pereira, & Ferreira, 1987) – and in different domains: psychiatry (an exhaustive review is given in L'Abate & De Giacomo, 2003; see also De Giacomo, De Giacomo, Margari, Masellis, & Santamato, 2008; De Giacomo et al., 2002b), economics (Pereira, 1999), problem solving (De Giacomo et al., 2002a), in different areas of computer science, including user modeling (Bison, Colazzo, Malinverni, & Mich, 1989), online communication (Colazzo, Mich, & Malinverni, 1992), and requirements elicitation (Mich, Anesi, & Berry, 2005). The application described in the last part of this chapter suggests using the paradox table in a prescriptive way to support the generation of new ideas. A similar approach was used to suggest phrases for relational therapies (De Giacomo et al., 2007, 2009). The application described in this paper to generate news is an innovative and original one.

As regards the results of application of the EPM in the clinical field, it should be noted that experiences inspired by these concepts have been gained in the treatment of schizophrenia (De Giacomo, Pierri, Lefons, & Mich, 1990; De Giacomo et al., 1997) and of anorexia nervosa (De Giacomo, Margari, & Santoni Rugiu, 1989). Moreover, it has been demonstrated that participants affected by mental disturbances have different patterns of coordinates, functions, and final states from those of normal participants (De Giacomo et al., 2007), while disturbed participants showing a clinical improvement also show a normalization of the pattern of coordinates (De Giacomo et al., 1986). As regards comparisons with other theories and models, only one work has yet been published (L'Abate, De Giacomo, McCarty, De Giacomo, & Verrastro, 2000). Other comparative studies are needed to assess differences, with respect to the theory of relational skills proposed by L'Abate and De Giacomo (2003).

A Program to Generate Ideas Based on the EPM

In this section, we will focus on the method whereby we built a program that aims to enhance the operative potential of our mind, to help to generate new ideas by adopting relational styles identified by means of the paradox table. The process according to which the EPM can give rise to the mental constructions described below is similar to the zigzag ladder between form and process described by Gregory Bateson (1979) (L'Abate & De Giacomo, 2003, pp. 6–10).

For example, a journalist who has to write an article after collecting information about the given topic (acquired via the Internet, books, magazines, dictionaries, etc., in short, after referring to all possible sources and to all the information already present in her/his own mind), can open the program we propose and see whether the article content and form is enriched as compared to what was present before the use of the program, and if new ideas other than those already jotted down appear or not.

This application owes its origin not only to its use in company courses, university lecturing and psychotherapy training courses but also to the first Author's experience of writing as a journalist, when he published a series of articles in a special column titled "Metti il cappello giusto" (Wear the right hat) on the daily newspaper Epolis Bari. The 74 articles, published weekly, were based on the operative methodology illustrated below.

The fourth level of the model is essential for the correct function of the program. This is the table that lists the result of all the possible combinations of the 16 functions (Table 18.1). In the program these functions, being aimed at nonmedical experts such as journalists, entrepreneurs, counselors and patients, are defined as "hats," using the term introduced by De Bono (1986), a world-famous expert on creative thinking, to indicate specific "mental frames."

To explain the use of our program, Table 18.1 includes a description of the behaviors corresponding to the 16 functions combined in the paradox table. For example, F1 is described as the sharer function, since it corresponds to a behavior in which only what is common to both interacting participants is accepted (in Fig. 18.1, F1 corresponds also to the logical "and" operation and to the Venn diagram where only the intersection is filled). There is also the doubter function, F6, so-called because in the Venn diagram only the right half-moon is filled, corresponding to the second subject's world, and the left half-moon, corresponding to the first subject's world, but not the intersection that represents what is common to both participants' worlds. In this case there is a situation of detachment, of separation between the two worlds, that can be interpreted as an oscillation between two positions, as typically occurs in cases of doubt or uncertainty (L'Abate & De Giacomo, 2003).

How to Use the Functions

Starting from the 16 functions, a first application of the EPM is the possibility of suggesting 16 "types" of articles. Once the meaning of the functions ("hats") has been understood, the user must associate the situation to be analyzed – and the features and characters involved – with one or more functions ("hats"). The first question to be asked is "What function does the situation we wish to analyze correspond to?" The link between writing texts and the 16 hats is the following article, published on Epolis Bari on 6 June 2009: *Journalists as painters with a blank canvas.*

> When adopting our model, the EPM, the following styles of journal articles emerge. First of all, like the painter in front of a blank canvas, let's put the article that has not yet been written (F0 for emptiness). Follows a highly realistic article (e.g. the description of a natural disaster) (Function F1 for pure sharing). This is followed by an article that reveals that there are some things the writer wishes to keep to her/himself (e.g. an aspect that is ignored, whereas the reader expected it to be dealt with) (Function F2 for withdrawal into one's own world). This is followed by an article that clearly delineates the journalist's world (political, civil, human) (Function F3 for maintenance of one's own world). Then follows an article of passive acceptance of the title by the journalist (e.g. an article written "to please the sovereign") (F4 for passive acceptance). This is followed by an article of complete identification with the world of the reader (F5 for acceptance of the other person's world). Then comes an article based on doubt (e.g. two opposite arguments are developed with an equal weight) (F6 for the doubter). This is followed by a mediation article (e.g. different arguments that are reconciled) (F7 for the mediator). Then there is an abstract, out-of-context article (e.g. the reader is led to think "but what has this got to do with what is happening?") (F8 for absolute abstraction). This is followed by an article in which it is clear that the journalist is pursuing a path with the aim of finding an original solution to the problem posed in the article (F9 for creatively pursuing a goal). After this is an article of opposition, of antithesis (e.g. an article that expresses an opinion that is exactly the opposite of an opinion held by another person or group) (F10 for Mary, Mary quite contrary). Then comes a dictatorial "fanatic" type of article, like those published under the rule of Mussolini (F11 for the dictator). After this comes a pseudoaltruistic article that holds an opinion that is apparently altruistic but actually serves the interests of the writer or the group he/she represents (F12 for the pseudoaltruist). There is also an article that reaches out too far toward the reader's world, that "fills the reader's spirit" (e.g. an unexpected win by the team of which the reader is a fan, that has brought the team into the top league) (Function F13 for reaching out too far into the other person's world). This is followed by a metaphorical article where the contents are expressed as a metaphor (e.g. that the article is a painting in which the 16 colors proposed should all be used at varying intensities) (F14 for the metaphorical). Finally, there is a chaotic article in which the various ways of writing here proposed are adopted all together unselectively, generating confusion in the reader (F15 for the total accepter).

Table 18.1 The table of functions

	A							
B	F0 void	F1 pure sharing	F2 withdrawal into his own world	F3 maintainer of his own world	F4 passive acceptor	F5 acceptor of other's world	F6 doubtfulness	F7 mediation
F0 void	F0 void	F0 void	F2 withdrawal into his own world	F3 maintainer of his own world	F0 void	F0 void	F6 doubtfulness	F7 mediation
F1 pure sharing	F0 void	F1 pure sharing	F2 withdrawal into his own world	F3 maintainer of his own world	F1 pure sharing	F1 pure sharing	F7 mediation	F7 mediation
F2 withdrawal into his own world	F0 void	F0 void	F0 void	F3 maintainer of his own world	F2 withdrawal into his own world	F2 withdrawal into his own world	F4 passive acceptor	F7 media-tionw
F3 maintainer of his own world	F0 void	F1 pure sharing	F0 void	F3 maintainer of his own world	F3 maintainer of his own world	F3 maintainer of his own world	F5 acceptor of other's world	F7 mediation
F4 passive acceptor	F0 void	F0 void	F2 withdrawal into his own world	F3 maintainer of his own world	F0 void	F4 passive acceptor	F2 withdrawal into his own world	F7 mediation
F5 acceptor of other's world	F0 void	F1 pure sharing	F2 withdrawal into his own world	F3 maintainer of his own world	F1 pure sharing	F5 acceptor of other's world	F3 maintainer of his own world	F7 mediation
F6 doubtfulness	F0 void	F0 void	F0 void	F3 maintainer of his own world	F2 withdrawal into his own world	F6 doubtfulness	F0 void	F7 mediation
F7 mediation	F0 void	F1 pure sharing	F0 void	F3 maintainer of his own world	F3 maintainer of his own world	F7 mediation	F1 pure sharing	F7 mediation
F8 abstraction	F0 void	F0 void	F2 withdrawal into his own world	F3 maintainer of his own world	F8 abstraction	F8 abstraction	F14 metaphoring	F15 total acceptor
F9 pursue a goal creatively	F0 void	F1 pure sharing	F2 withdrawal into his own world	F3 maintainer of his own world	F9 pursue a goal creatively	F9 pursue a goal creatively	F15 total acceptor	F15 total acceptor
F10 mary mary quite contrary	F0 void	F0 void	F0 void	F3 maintainer of his own world	F10 mary mary quite contrary	F10 mary mary quite contrary	F12 pseudoaltruism	F15 total acceptor
F11 dictatorship	F0 void	F1 pure sharing	F0 void	F3 maintainer of his own world	F11 dictatorship	F11 dictatorship	F13 go toward the other beyond measure	F15 total acceptor
F12 pseudoaltruism	F0 void	F0 void	F2 withdrawal into his own world	F3 maintainer of his own world	F8 abstraction	F12 pseudoaltruism	F10 mary mary quite contrary	F15 total acceptor
F13 go toward the other beyond measure	F0 void	F1 pure sharing	F2 withdrawal into his own world	F3 maintainer of his own world	F9 pursue a goal creatively	F13 go toward the other beyond measure	F11 dictatorship	F15 total acceptor
F14 metaphoring	F0 void	F0 void	F0 void	F3 maintainer of his own world	F10 mary mary quite contrary	F14 metaphoring	F8 abstraction	F15 total acceptor
F15 total acceptor (chaotic)	F0 void	F1 pure sharing	F0 void	F3 maintainer of his own world	F11 dictatorship	F15 total acceptor	F9 pursue a goal creatively	F15 total acceptor

F8 abstraction	F9 pursue a goal creatively	F10 mary mary quite contrary	F11 dictatorship	F12 pseudoaltruism	F13 go toward the other beyond measure	F14 metaphoring	F15 total acceptor (chaotic)
F7 mediation	F6 doubtfulness	F15 total acceptor	F15 total acceptor	F3 maintainer of his own world	F2 withdrawal into his own world	F15 total acceptor	F15 total acceptor
F6 doubtfulness	F7 mediation	F14 metaphoring	F15 total acceptor	F3 maintainer of his own world	F3 maintainer of his own world	F15 total | acceptor	F15 total acceptor
F5 acceptor of other's world	F4 passive acceptor	F13 go toward the other beyond measure	F15 total acceptor	F3 maintainer of his own world	F2 withdrawal into his own world	F13 go toward the other beyond measure	F15 total acceptor
F4 passive acceptor	F5 acceptor of other's world	F12 pseudoaltruism	F15 total acceptor	F3 maintainer of his own world	F3 maintainer of his own world	F13 go toward the other beyond measure	F15 total acceptor
F3 maintainer of his own world	F2 withdrawal into his own world	F11 dictatorship	F11 dictatorship	F3 maintainer of his own world	F6 doubtfulness	F11 dictatorship	F15 total acceptor
F2 withdrawal into his own world	F3 maintainer of his own world	F10 mary mary quite contrary	F11 dictatorship	F3 maintainer of his own world	F7 mediation	F11 dictatorship	F15 total acceptor
F1 pure sharing	F0 void	F9 pursue a goal creatively	F11 dictatorship	F3 maintainer of his own world	F6 doubtfulness	F9 pursue a goal creatively	F15 total acceptor
F0 void	F1 pure sharing	F8 abstraction	F11 dictatorship	F3 maintainer of his own world	F7 mediation	F9 pursue a goal creatively	F15 total acceptor
F7 mediation	F14 metaphoring	F7 mediation	F15 total acceptor	F3 maintainer of his own world	F10 mary mary quite contrary	F7 mediation	F15 total acceptor
F6 doubtfulness	F15 total acceptor	F6 doubtfulness	F15 total acceptor	F3 maintainer of his own world	F11 dictatorship	F7 mediation	F15 total acceptor
F5 acceptor of other's world	F12 pseudoaltruism	F5 acceptor of other's world	F15 total acceptor	F3 maintainer of his own world	F10 mary mary quite contrary	F5 acceptor of other's world	F15 total acceptor
F4 passive acceptor	F13 go toward the other beyond measure	F4 passive acceptor	F15 total acceptor	F3 maintainer of his own world	F11 dictatorship	F5 acceptor of other's world	F15 total acceptor
F3 maintainer of his own world	F10 mary mary quite contrary	F3 maintainer of his own world	F11 dictatorship	F3 maintainer of his own world	F14 metaphoring	F3 maintainer of his own world	F15 total acceptor
F2 withdrawal into his own world	F11 dictatorship	F2 withdrawal into his own world	F11 dictatorship	F3 maintainer of his own world	F15 total acceptor	F3 maintainer of his own world	F15 total acceptor
F1 pure sharing	F8 abstraction	F1 pure sharing	F11 dictatorship	F3 maintainer of his own world	F14 metaphoring	F1 pure sharing	F15 total acceptor
F0 void	F9 pursue a goal creatively	F0 void	F11 dictatorship	F3 maintainer of his own world	F15 total acceptor	F1 pure sharing	F15 total acceptor

How to Use the Upper Level of the Table of Functions

In the previous application of the EPM to generate new ideas, the link between the themes and the functions is relatively simple, whereas at this second level it appears more complex. The first step to be taken after exhausting all one's own resources and mental content, as well as all the information sources: dictionaries, treatises, search engines like Google, is not only to verify which function the situation to be analyzed corresponds to (see previous paragraph) but also to place it in a possible network of interactions, like those represented in the table of interactions (also named the paradox table or the table of functions) (see Fig. 18.1 and Table 18.1). This first attitude of creatively pursuing a goal, in other words relating a given situation to the table of functions ("hats"), is already in itself an interesting input and an example of possible creative operativity. In fact, it corresponds to a fundamental frame of our mental functions, and indeed in our view to "the heart of the human mind." As a first example, let's see how one can arrive, using the EPM, at this function of *creatively sharing a goal*. If we look at the table of interactions (Table 18.1) we can see which interactions, among the columns of vertical functions and the columns of horizontal functions, produce function F9 of creatively sharing a goal of interest, taking into account the composition of the table, where at the two extremes – in the appropriate spaces bordered by horizontal and vertical lines – there are the 16 functions as they become transformed, as they are modified by the interaction (in metaphorical terms "the patient"). Instead, in the vertical sense we have the functions that provoke a change in the other subject (metaphorically "the therapist").

To continue: we have said that the chosen theme is to creatively pursue a goal. The user wonders which function corresponds to creatively pursuing a goal. Answer: to function F9. At this stage the program will automatically reveal all the cases in which interaction yields F9, or the function of creatively pursuing a goal. In any case the reader, who does not have access to the program, can do this manually on the figure in the text, or better still on a photocopy of the figure. The result that emerges should be connected – as we shall see from the articles reported below – in a mental framework of adaptation and plausibility.

We have already stated that as the first example we have chosen the very suitable, immediate one of a user who is trying to find their way in the program with a precise goal, a result to be achieved associated with something abstract. It does not belong to their own world, it is strange but it is what will yield an original, creative imprint and so is necessary to achieve the goal. This mindset corresponds to function F9 for creatively pursuing a goal. It should be noted that this behavior is certainly a central aspect of the function of the human mind. The starting point of the example of a pathway through our program is, therefore, to identify the hat corresponding to what we are looking for; we have identified it as function F9 for creatively pursuing a goal. After identifying the function, we continue with the second part of the program based on the table of interactions.

Starting from these bases, when we look at the table we can see that function F9 for creatively pursuing a goal originates from passive acceptance (F4 || F9 = F9), from acceptance (F5 || F9 = F9), from doubt when it interacts with total acceptance (F6 || F15 = F9), when there is opposition to doubt (F10 || F6 = F9), when the metaphor ("the dream") interacts with doubtfulness (F10 || F6 = F9) or when the metaphor ("the dream") interacts with mediation (F14 || F7 = F9). If we replace the formulae with discourse we get the following text: "creatively pursuing a goal derives from acceptance of creatively pursuing a goal or from when doubtfulness interacts with total acceptance, or also from opposing doubtfulness, or from when the metaphor ("the dream") interacts with doubtfulness or with mediation." This text will be compared with the one the writer had in mind before to see if any idea emerges that was not present in their mind before, or else provides a different view of a mental pathway from the traditional one. To this pathway (pathway A in the program), that indicates the genesis, according to the formalism of the model of function F9, we could add another pathway (pathway B in the program) which observes how a certain function modifies the other functions. In this case we will use, as explained below, the horizontal strip that starts at the end of the table.

We report below two articles published in a journal using this program, showing in brackets the formulae of interaction among functions, and the relative result, that was not, of course, present in the published text.

All the roads that lead to Dictatorship, Epolis, Bari 18 March 2009

We intend to develop the theme of how one becomes a person centered exclusively and excessively on oneself. If a person has dictatorial tendencies this will be a natural end to their way of never enquiring about anything. If, instead, dictatorship is not natural to the subject, they will need to meet with active acceptance (F11 || F5 = F11) or passive acceptance to go down this road (F11 || F4 = F11). In fact, some people are satisfied with what they are and could never accept to become dictators. It would be possible for a doubter to become a dictator (F6 || F13 = F11), or for one who creatively pursues a goal if faced with events that reach out excessively toward their own world (F9 || F13 = F11): favorable events that carry them to the top of a particularly difficult ascent in a certain political, financial, scientific, or other field. Even Mary, Mary quite contrary, a systematic opposer, an anti-other person, can become a dictator when faced with an attitude of passive acceptance (F10 || F4 = 11). If their antithetical nature does not encounter any limit, any opposition, it can be transformed into dictatorship. Even those who are superacceptant of other people who are engaged in creatively pursuing a goal, when faced with great ambitions pursued by others in the scientific, humanitarian, or political field, can become themselves imbued with the idea, make it their own and carry it out with a high hand (F13 || F9 = F11). Finally, even dreamers and poets, when faced with others' passive acceptance or complete acceptance, could become dictators (F14 || F4 = F11, F14 || F5 = F11): their dreams are fulfilled (and often disappoint on awakening).

Doubt and its many options, Epolis Bari 10 June 2009

Doubt, uncertainty, with our mind oscillating among different possibilities, is a theme of great importance in our lives. Let's try first of all to describe it with a metaphor: navigating in fog, the gas tap not turned to 180°, not understanding if certain messages are positive or negative (F6). How does one enter into doubt according to our model of the mind (EPM)? First of all, when a person that accepts the other person's world meets a doubter (F5 || F6=F6). A second possibility is if an abstract person interacts with a very realistic one (F8 || F1=F6), or with a person creatively pursuing a goal in a determined manner (F8 || F9=F6). Another possibility of becoming a doubter is when a person creatively pursuing a goal interacts with an empty person (who provides 0 answers) (F9 || F0=F6). Even a Mary, Mary quite contrary, when they meet a person creatively pursuing a goal, can enter into doubt (F10 || F9=F6). Let's now see how to escape from doubt. A first possibility is when a doubter interacts with a very realistic person, in which case s/he will tend to mediate "on the horns of a dilemma" (F6 || F1=F7); instead, when interacting with a person who has withdrawn into their own world they will tend to become a passive accepter (F6 || F2=F4). When interfacing with a strong person they will tend to accept this other person's world (F6 || F3=F5). With a passive accepter they will tend to withdraw into their own world (F6 || F4=F2). When interacting with someone who accepts their world they will tend to maintain their own world, and become more themselves (F6 || F5=F3). If, on the other hand, a doubter meets another doubter the result is total annulment (F6 || F6=F0). When interacting with a mediator they will become realistic (F6 || F7=F1). When interacting with an abstract person they will become metaphorical (F6 || F8=F14). When interacting with a person creatively pursuing a goal they will fall into chaos (F6 || F9=F15). When interacting with Mary, Mary quite contrary they will become a pseudoaltruist (F6 || F10=F12). When interacting with a dictator they will become an excessive accepter of the dictator's world (F6 || F11=F13). When interacting with a pseudoaltruist they will become Mary, Mary quite contrary (F6 || F12=F10). When interacting with a dreamer they will tend toward abstraction (F6 || F14=F8). Finally, when interacting with a person in chaos they will become able to creatively pursue a goal (F6 || F15=F9).

Conclusions

A method for generating ideas and ultimately for developing the mind is presented, intended for use in various fields ranging from journalism to psychotherapy. It is based on the model of the mind in terms of relational patterns called the EPM, that can be developed on computer programs.[2] We are well aware that traditional psychologists and psychiatrists may see this as strange, abstract, far from historical views in our field but it is for this very reason that it could be useful to widen our mental horizons. As Nietzsche said "You must have chaos within you to give birth to a dancing star."

[2] A copy the computer program that can be used to follow the path can be obtained (in the Italian version only, at the moment) by writing to piero.degiacomo@tin.it.

References

Abraham, A., Hassanien, A. E., & Snásel, V. (Eds.). (2010). *Computational social network analysis: Trends, tools and research advances*. New York: Springer.

Apple.com. Apple Inc. (2011). Retrieved February 18, 2011, from http://www.apple.com.

Austin, J. L. (1962). *How to do things with words*. Cambridge: Harvard University Press.

Barley, Z., Lauer, P. A., Arens, S., Apthorp, H. S., Englert, K., Snow, D., et al. (2008). *Office of educational research and improvement, helping at-risk students meet standards: A synthesis of evidence-based classroom practices* (p. 97). Washington: U.S. Department of Education.

Bateson, G. (1979). *Mind and nature: A necessary unit*. New York: Dutton.

Berlo, D. (1960). *The process of communication: An introduction to theory and practice*. New York: Holt, Rinehart and Winston.

Bison, I., Colazzo, L., Malinverni, D., & Mich, L. (1989, September 13–15). The elementary-pragmatic model: A possible approach for setting up user model. *In proc. EC-CSCW'89, European conference on computer supported cooperative work* (pp. 389–401). London.

Brock, G. W. (2003). *The second information revolution*. Cambridge: Harvard University Press.

Chandler, D. (2002). *Semiotics: The basics*. London: Routledge.

Choi, C. Q. (2011). From iPhones to sciPhones. *Scientific American, 304*(3), 22.

Chomsky, N. (1957). *Syntactic structures*. The Hague: Mouton & Co.

Colazzo, L., Mich, L., & Malinverni, D. (1992). An application of the elementary pragmatic model to electronic communication. *Annals of Istituto Superiore Della Sanità., 28*(2), 245–252.

Colazzo, L., Mich, L., Malinverni, D., & Schäl, T. (1991, August 27–29). Interpretation of human relations in computer supported communication: A test with a pragmatic model. Elsevier science, North-Holland. *In proc. COSCIS'91 – IFIP-TC8 conference on collaborative work, social communications and information systems* (pp. 77–92). Finland: Helsinki-Hanasaari.

De Bono, E. (1986). *Six thinking hats*. New York: Little Brown and Company.

De Giacomo, P. (2011, October 14–16). Elementary pragmatic model. *Fifth national conference on system science: Methods, models, simulations and approaches toward a general theory of change*. Fermo (Italy). In Press.

De Giacomo, P., De Giacomo, A., Margari, F., Masellis, R., & Santamato, W. (2008). Methodological psychotherapy. In L. L'Abate (Ed.), *Toward a science of clinical psychology: Laboratory evaluations and interventions*. New York: Nova.

De Giacomo, P., L'Abate, L., Margari, F., Santamato, W., De Giacomo, A., Masellis, R., et al. (2009). A program to select sentences with strong psychological impact. *Journal of Contemporary Psychotherapy, 39*, 165–169.

De Giacomo, P., L'Abate, L., Santamato, W., Sgobio, A., Tarquinio, C., De Giacomo, A., et al. (2007). "Compass" sentences with strong psychological impact in family therapy: Preliminary investigations. *Journal of Family Psychotherapy, 18*, 45–69.

De Giacomo, P., Margari, F., & Santoni Rugiu, A. (1989). Successful one-session treatments of anorexia nervosa: A report on fifteen cases. *International Journal of Family Psychiatry, 1*(2), 123–132.

De Giacomo, P., Mich, L., Storelli, M., De Nigris, S., De Giacomo, A., Tarquinio, C., et al. (2002a, July 22–25). A method to increase students' cognitive potentialities using the elementary pragmatic model. *In Proc. Interactive learning through the world wide web – SCI2002* (pp. 1–12). Orlando.

De Giacomo, P., Mich, L., Storelli, M., De Nigris, S., De Giacomo, A., Tarquinio, C., et al. (2002b). A method of counseling on line using the elementary pragmatic model. In *Changing our minds* (pp. 1–6). New York: American Association for Technology in Psychiatry. AATP 2002, Philadelphia, Pennsylvania, May 18th 2002.

De Giacomo, P., Pierri, G., Lefons, E., & Mich, L. (1990). A technique to simulate human interactions: Relational styles leading to schizophrenic communication pattern and back to normal. *Acta Psychiatrica Scandinavica, 82*, 413–419.

De Giacomo, P., Pierri, G., Santoni Rugiu, A., Buonsante, M., Vadruccio, F., & Zavoianni, L. (1997). Schizophrenia: A study comparing a family therapy group following a paradoxical model plus psychodrugs and a group treated by the conventional clinical approach. *Acta Psychiatrica Scandinavica, 95*, 183–188.

De Giacomo, P., & Silvestri, A. (1982). *The elementary pragmatic approach to the study of interactional behaviour. New horizons in psychiatric research*. Kyoto: WPA.

De Giacomo, P., Silvestri, A., Pierri, G., Lefons, E., Corfiati, L., & Tangorra, F. (1986). Research on the effects of psychodrugs on human interaction. *Acta Psychiatrica Scandinavica, 74*, 417–424.

Gardner, H. (1985). *The mind's new science: A history of the cognitive revolution*. New York: Basic Books.

Guerriero, L. (2009). Appendix A. Mathematical description of the elementary pragmatic model. In M. Capitelli, P. De Giacomo, L. L'Abate, & S. Longo (Eds.), *Science, mind, and creativity: The Bari symposium*. New York: Nova.

Guiasu, S. (1977). *Information theory with application*. Great Britain: McGraw-Hill Inc.

Harper, J. M., Lymn Scoresby, A. L., & Boyee, W. D. (1977). The logical levels of complementary, symmetrical, and parallel interaction classes in family dyads. *Family Process, 16*(2), 199–213.

Herrick, J. A. (1990). *The history and theory of rhetoric, an introduction*. Boston: Allyn & Bacon.

Jeffress, L. A. (1951). *Cerebral mechanisms in behavior: The Hixon symposium*. New York: Wiley.

Kelly, K. (2010). *What technology wants*. New York: Viking Press.

L'Abate, L. (2011). *Sourcebook of interactive practice exercises in mental health*. New York: Springer-Science.

L'Abate, L., & De Giacomo, P. (2003). *Intimate relationships and how to improve them. Integrating theoretical models with preventive and psychotherapeutic applications*. Westport: Praeger.

L'Abate, L., De Giacomo, P., McCarty, F., De Giacomo, A., & Verrastro, G. (2000). Evaluating three models of intimate relationships. *Contemporary Family Therapy, 22*(1).

Laing, R. D. (1970). *Knots*. New York: Pantheon Books.

Lashley, K. S. (1951). The problem of serial order in behavior. In L. A. Jeffress (Ed.), *Cerebral mechanisms in* behavior. New York: Wiley. pp. 112–46.

Laswell, H. (1948). The structure and function of communication in society. In L. Bryson (Ed.), *The communication of ideas*. New York: Harper.

Lefons, E., Pazienza, M. T., Silvestri, A., Tangorra, F., Corfiati, L., & De Giacomo, P. (1977). An algebraic model for systems of psychically interacting participants. In O. Dubuisson (Ed.), *Proc IFAC workshop information & systems* (pp. 155–163). France: Compiegne.

McCullough, W. S., & Pitts, W. (1943). A logical calculus of the ideas immanent in nervous activity. *Bulletin of Mathematical Biophysics, 5*, 115–127.

Mesch, G., & Talmud, I. (2010). Wired Youth: The Social World of Adolescence in the Information Age. London: Routledge.

Mich, L., Anesi, C., & Berry, D. (2005). Applying a pragmatics-based creativity-fostering technique to requirements elicitation. *Requirements Engineering Journal, 10*(4), 262–274.

Miller, G. A. (1956). The magical number seven, plus or minus two: Some limits on our capacity for processing information. *Psychological Review., 63*, 81–89.

Miller, G. A. (2003). The cognitive revolution: A historical perspective. *Trends in Cognitive Sciences., 7*, 141–144.

Murphy, S. (2011). Your avatar, your guide. *Scientific American Mind., 22*(1), 58–63.

Newell, A., Shaw, J. C., & Simon, H. A. (1958). Elements of a theory of human problem-solving. *Psychological Review, 65*, 151–166.

O'Reilly, T. (2005). *What is web 2.0, design patterns and business models for the next generation of software*. Retrieved September 5th 2011, from http://oreilly.com/web2/archive/what-is-web-20.html.

Pereira, O. G. (1999). *Fundamentos de comportamento organizacional*. Lisbon: Fundação Calouste Gulbenkian.

Porter, A., & Read, W. H. (1998). *The information revolution: Current and future consequences*. Greenwich: Ablex Publishing Corporation.

Pulier, M. L., Mount, T. G., McMenamin, J. P., & Maheu, M. M. (2007). Computers and the internet. In L. L'Abate (Ed.), *Low-cost approaches to promote physical and mental health* (pp. 303–322). New York: Springer-Science.

Resconi, G., & De Giacomo, P. (2004). Logic communication and computation among agents. In R. Trappl (Ed.), A*ustrian society for cybernetic studies 1* (pp. 27–32). Vienna: University of Vienna.

Rosen, A. (2009). *E-learning 2.0: Proven practices and emerging techniques. Safari books online* (p. 141–142). Amacom Division of American Management Association. New York.

Ruben, B. D. (1984). *Communication and human behavior*. New York: Macmillan Publishing Co.

Samuel, A. L. (1959). Some studies in machine learning using the game of checkers. *IBM Journal of Research and Development, 3*, 210–229.

Scott, W. G., & Mitchell, T. R. (1972). *Organization theory: A structural and behavioural analysis*. Concord: Irwin.

Searle, J. R. (1969). *Speech acts*. Cambridge: Cambridge University Press.

Searle, J. R. (1975). A taxonomy of illocutionary acts. In K. Gunderson (Ed.), *Language, mind and knowledge* (pp. 334–369). Minneapolis: University of Minnesota.

Shannon, C. E. (1948). A mathematical theory of communication. *Bell System Technical Journal, 27*, 379–423.

Shannon, C. E., & Weaver, W. (1949). *The mathematical theory of communication*. Illinois: University of Illinois Press.

Silvestri, A., & De Giacomo, P. (1979). A computerized laboratory for a systematic approach to relational psychoteraphy. In B. Barber, F. Gremey, K. Uberla, & G. Wagner (Eds.), *Medical informatics*. Berlin: Springer.

Silvestri, A., De Giacomo, P., Pierri, G., Lefons, E., Pazienza, M. T., & Tangorra, F. (1980). A basic model of interacting participants. *Cybernetics and Systems: An International Journal, 11*, 115–129.

Silvestri, A., Mich, L., Pereira, O. G., & Ferreira, C. P. (1987). Simulation of the development of individual interactional patterns. *Cybernetics and Systems: An International Journal, 18*, 489–515.

Sluzki, E., & Ranson, D. C. (1976). *The double bind*. New York: Grune & Stratton.

Stern, M. J., & Adams, A. E. (2010). Do rural residents really use the internet to build social capital? An empirical investigation. *American Behavioral Scientist., 53*, 1344–1366.

Tapscott, D., & Williams, A. D. (2006). *Wikinomics: How mass collaboration changes everything*. New York: Portfolio.

Tapscott, D., & Williams, A. D. (2010). *Macrowikinomics: Rebooting business and the world*. New York: Portfolio.

Turing, A. M. (1936). On Computable Numbers, with an Application to the Entscheidungsproblem. In proc. *London Mathematical Society. 242*, 230–65.

Turing, A. M. (1948). Intelligent Machinery. reprinted in D.C. Ince (Ed.), (1992). Collected Works of A.M. Turing: Mechanical Intelligence. Amsterdam: North Holland (pp. 87–106).

Wallace, P. (1999). *The psychology of the internet* (pp. 203–205). Cambridge: Cambridge University Press.

Watzlawick, P., Beavin, J., & Jackson, D. D. (1967). *Pragmatics of human communication*. New York: Norton.

Watzlawick, P., & Weakland, J. H. (1976). *The interactional view studies at the mental research institute, PaloAlto*. New York: W.W. Norton & Company.

Wiener, N. (1948). *Cybernetics: Or control and communication in the animal and the machine*. Cambridge: MIT Press.

Winograd, T., & Flores, F. (1987). *Understanding computers and cognition: A new foundation for design*. Readimg: Addison-Wesley Professional.

Wu, T. (2010). *The Master Switch: The Rise and Fall of Information Empires*. New York: Knopf.

Chapter 19
Reductionism

Gary G. Berntson and John T. Cacioppo

The construct of reductionism has taken many forms, but a common feature is that higher level or otherwise complex systems can be understood in terms of (*reduced to*), the properties of the lower-level elements of which they are comprised. Reductionism as a philosophical doctrine could be considered in an abstract, metaphysical sense, apart from any pragmatic reference. In science, however, it is an important perspective on the relations between domains (physical vs. mental) or levels of organization (e.g., molecular, cellular, tissue, organ, organ system, individual organism, social systems). Consequently, it can impact broadly on issues such as how we approach science and what constitutes a scientific explanation. The issue becomes especially acute for multi-level interdisciplinary approaches in the neurosciences, where the goal is to integrate information across widely disparate levels of organization and function, and associated levels of observation and analysis.

A classical philosophical issue has come to be known as the *mind–body problem* – the relation between the mind and the brain, although it is more typically couched in terms of the relations between mental processes and physical or biological processes. This represents a particularly hard question for reductionism, because the mental and physical domains may appear so different. Philosophical positions have ranged from *immaterialism, idealism, or monistic idealism* (e.g., Goswami, 1995) which asserts that there is only, or that we can only know about, our subjective mental reality; to *reductionism*, which in some forms sees mental and physical processes as identities; to *materialism or eliminative materialism* (e.g., Churchland & Churchland, 1998;

G.G. Berntson (✉)
Department of Psychology, Ohio State University, 1835 Neil Avenue, Columbus, OH 43210, USA
e-mail: berntson.2@osu.edu

J.T. Cacioppo
Center for Cognitive and Social Neuroscience, The University of Chicago, 5848 S. University Avenue, Chicago, IL 60637, USA

L. L'Abate (ed.), *Paradigms in Theory Construction*, DOI 10.1007/978-1-4614-0914-4_19,

Feyerabend, 1963) which views the mind as a mere epiphenomenon or denies the existence of mental processes (see Chap. 15, Materialism). In some forms, immaterialism and materialism could both be considered essential denials of the mind–body problem. Other challenges to reductionism come from the doctrine of *vitalism* and *emergentism*. Vitalism or some forms of mentalism hold that the laws governing living organisms (or mental processes) are not reducible to the laws of physics. This is inherent in Cartesian *dualism*, which embraces both the physical and mental domains, but views them as very different sorts of things that exist in parallel. In contrast, emergentism maintains that combinations or interactions of lower level elements can give rise to novel (emergent) features or properties that are irreducible (e.g., Emmeche, Koppe, & Stjernfelt, 1997; see also Chaps. 21 and 22 on Emergence). In its strong form (*ontological emergence*), the emergent features are considered not to be inherent in the lower-level elements (e.g., O'Connor & Wong, 2005), whereas an *epistemological emergentist*, might only argue that those features could not be ascertained by studying the lower-level elements in isolation.

Although the mind–body problem might be a hard case for reductionism, issues of emergence apply more broadly to systems even in the absence of reference to the mental domain.

We will return to the mind–body problem later, but we turn first to an illustration of the pragmatic issues of reductionism and emergence in interdisciplinary science.

Emergence and Reductionism in Social Neuroscience

Phenomena can be studied at multiple levels of observation and analysis (e.g., the biological and the social) and knowledge derived from these multiple levels can mutually inform each other. That is the essential added-value in interdisciplinary research that cuts across levels of organization and analysis. Social and biological explanations of behavior, however, have sometimes been considered to offer distinct, or even incompatible, accounts. The Darwinian revolution and the ensuing focus on instincts generated mutual enthusiasm among social psychologists and psychobiologists for understanding the biological bases of social behavior. Unfortunately, instinct theory was not to survive as a viable explanatory system, and social psychology and psychobiology went rather separate ways (see Berntson & Cacioppo, 2000). Psychobiologists emphasized physiological processes and neural substrates, often eschewing functionalist theories or considering them subject to reductionism. In contrast, social psychologists emphasized multivariate systems, situational influences, and practical applications, while strongly rejecting biological reductionism. This lead to what some regarded as an unbridgeable rift between social and biological approaches (Scott, 1991).

The early American psychologist James (1950), however, was among the early voices asserting that neurophysiological processes underlie psychological phenomena and that developmental, environmental, and sociocultural factors influence the neurophysiological processes underlying psychological and social phenomena. Although these influences could be studied as neurophysiological transactions,

James recognized the diseconomies that would result if psychological phenomena were described only as neurophysiological events. In accord with James, the emergence of social neuroscience in the 1990s (the term was first introduced by Cacioppo & Berntson, 1992) demonstrated the mutual enrichment of both the social psychology and neuroscience by recognition and incorporation of the observations, theories, and understandings of distinct disciplines. For example, the hormone oxytocin is now recognized to both impact on, and be impacted by, social behavior – and the account of either the physiology or the social behavior would be incomplete without the other perspective. From a reductionist perspective, the biological, relative to the social, level might in some sense be considered more primary or fundamental. As noted by James, however, the lower-level analysis or representation may not always be optimal in capturing the essence of a phenomenon. Beethoven's ninth symphony could be characterized by the time- and space-varying patterns of movements of atomic and subatomic elements in the conducting media. That information, however, would fill a large book and may not fully capture the esthetic effect of the acoustic patterns on a human listener. There are different levels of observation, description, analysis, or explanation, each of which may offer an important perspective but some of which may be more relevant or efficient for a given purpose. In characterizing social pair-bonding in animals, for example, a behavioral descriptive analysis may represent a more relevant starting point than molecular studies of neurons. Although the latter may ultimately prove to be critical in fully understanding the mechanisms of social bonding, the significance or relevance of those mechanisms may not be apparent in the absence of a characterization of the social phenomena to be explained.

Oxytocin and Social Behavior: Mereological Perspectives

The observation that properties may emerge from aggregates, which are apparently not inherent to the constituent elements, has long been posed as an argument against reductionism, or at least some forms of reductionism. The neurohormone oxytocin, for example, has recently received much attention for its role in pair-bonding, trust, and social relations (Ross & Young, 2009). Oxytocin was first isolated and synthesized by Vincent du Vigneaud in 1953, for which he received the Nobel Prize in Chemistry (1955). Oxytocin is relatively simple, as biomolecules go, consisting of a sequence of nine amino acids (cysteine – tyrosine – isoleucine – glutamine – asparagine – cysteine – proline – leucine – glycine). Each of these amino acids is comprised of various combinations of precisely the same atomic elements (carbon, hydrogen, oxygen, and nitrogen), with the exception of cysteine, which also has a sulfur atom. What differentiates the rest of these amino acids, then, is not the constituent elements but the combinatorial organization of those elements. Indeed, two of them (leucine and isoleucine) have the exact same chemical composition ($C_6H_{13}NO_2$), differing only in how the elements are stereogeometrically arranged (Fig. 19.1). Yet, substituting isoleucine for leucine in oxytocin results in about a 95% reduction in potency. This story replays itself at a more macro level.

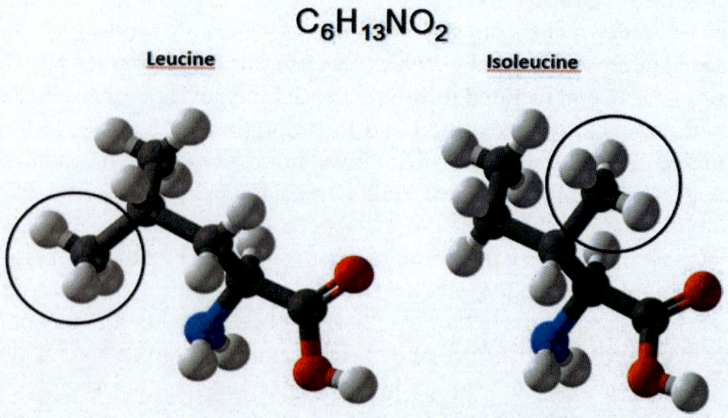

Fig. 19.1 The amino acids leucine and isoleucine are comprised of the same elements (carbon, hydrogen, nitrogen, and oxygen) in the same proportions, differing only in the location of the methyl group (*circles*) within the molecules (white = hydrogen, red = oxygen, black = carbon, blue = nitrogen)

Distinct combinations of carbon, hydrogen, oxygen, and nitrogen also comprise two additional amino acids, phenylalanine and arginine. If these two amino acids are substituted, respectively, for isoleucine and leucine in the oxytocin molecule, we have a different hormone, vasopressin, with totally distinct properties and actions (antidiuretic and vasopressor). Again, the functional impact of these hormones is defined not simply by the constituent atomic elements, but how those elements are combined into amino acids; and in turn how those amino acids are sequenced into molecules. One might argue that this is just another example of classical chemistry, but it goes beyond that. Even with the understanding of peptide biochemistry and the knowledge that oxytocin can have powerful effects on social processes (which are not predictable on the known chemical properties alone), it remains the fact that there is nothing intrinsically social about oxytocin. There is no inherent coding of social information or disposition in the molecule, no proper social characteristic. There is nothing in the chemistry of oxytocin or its constituent elements that could, in an obligatory fashion, predict its role in social processes.

Although peptide chemistry has been with us since amino acids first formed, the social role of oxytocin did not exist prior to the evolutionary sculpting of the brain. With a slightly different evolutionary trajectory, the functional contributions of oxytocin could be quite different. Indeed, in the absence of another molecule, oxytocin would be a socially vacuous sequence of amino acids. It becomes endowed with its social authority only by virtue of its interaction with another protein, the oxytocin receptor, and the neural sequelae triggered thereby. Recent evidence suggests that single nucleotide polymorphisms (SNPs) in the gene specifying the oxytocin receptor may alter these functions dramatically and dispose toward outcomes as diverse as diminished empathy and autism. Despite the obligatory receptor interaction, one might still argue that this is simply peptide chemistry – complex perhaps, but chemistry nonetheless. But there is another evolutionary contribution that goes well beyond peptide chemistry. That is the location of the oxytocin receptor within the brain and the specific neurons that bear this receptor. Research on voles illustrates the importance of this localization. The prairie vole is monogamous, displaying prolonged pair-bonding. This pair-bonding is attributable to oxytocin, acting on oxytocin receptors in a critical brain structure (the nucleus Accumbens) that underlies pleasure and reward (Gingrich, Liu, Cascio, Wang, & Insel, 2000). Evolution took a slightly different turn for the closely related montane vole, a polygamous animal that does not pair-bond. Although the montane vole also has oxytocin receptors in the brain, they are not expressed in the nucleus Accumbens (Insel, 1997) and hence do not have ready access to hedonic systems. The role of oxytocin in social processes is not dictated by peptide chemistry, and no amount of knowledge about peptide chemistry could predict this role, in the absence of knowledge about higher-level neural substrates and their evolutionary exigencies. The role of oxytocin in social processes is an emergent property of interactions in complex systems. While it might be argued that the social effects of oxytocin are indeed properties of the constituent elements of the relevant molecules (oxytocin and its receptor), the fact remains that those properties could only be known by studies at a higher level of organization and analysis and with a social neuroscientific perspective. That is the essence of epistemological emergentism.

Constituent elements may be crucial to the properties of aggregates, but the properties of aggregates may not always be determined from the proper features of the elements alone. Reductionism holds that higher-level organizations can be explicated in terms of (or reduced to) the properties, features, and combinatorial rules of the elements. However, if critical features or properties of the elements can be known only by studying the higher-level organization, a multi-level integrative approach may be optimal.

Multi-Level Perspectives and Calibrative Reductionism

New opportunities for scientific investigations of the working human brain are being created by developments in multimodal structural, hemodynamic, and electrophysiological brain imaging; sophisticated analyses of focal brain lesions; experimental manipulations of brain activity through transcranial magnetic stimulation and pharmacological agents; and emerging visualization and quantitative techniques that integrate anatomical and functional connectivity – in addition to information about neural processes at different scales of organization. Despite the increased sophistication and data yield from recent advances, however, an atheoretical exploration alone is not likely to yield many profound insights into the working mind. It is simply too complex to understand the neural basis of specific mental processes without well-designed tasks that isolate those processes. Moreover, because of the emergent properties of complex organizations, it is unlikely that an understanding of higher-level psychological processes can be satisfactorily derived from neurophysiology alone. Nor for that matter, by a restricted analysis of any single level of organization or function, whether it be psychological or neurobiological.

Some levels of analysis and description render relations more readily apparent and may be more optimal than others for characterizing specific relations. While an elegant dinner cuisine could be characterized by its chemical composition, even a chemist is more likely to use a food recipe than a chemical recipe in cooking. Because of the complexities of the phenomena and the problem of emergence, higher levels of analysis and description will always be important in behavioral, cognitive, and social neuroscience. At the same time, we have seen from the dead-end of behaviorism that psychological constructs must be tuned and evaluated by understandings of the organ that gives them rise.

Although the understandings of brain–behavior relations must be couched in the terms and constructs of biology, those terms and constructs will not likely be the extant biological constructs of today. Rather, biological understandings will need to be aligned and developed in the context of psychological findings (and vice versa), which will likely yield a rather different sounding biology from contemporary disciplines. Prior to the ready availability of computers in science, psychological studies were programmed by devices that implemented logic functions and other simple operations (AND/OR gates, inverters, switches, counters, timers, registers, etc.). Early versions of these devices were electromechanical, then transistor-based, and ultimately implemented in integrated circuits. Because those devices had to be interconnected to be useful, programming entailed the explicit mapping of devices or functions and their interconnections. That map, together with device rules and functions, constitutes a sufficient description and accounting of the circuit processes and operations.

Although brains are not like computers in many respects, a computer metaphor may be instructive. With the development of complex computer systems, the logic-gate mentality for characterizing its processes and operations is not efficient or particularly helpful – at least for a programmer. More useful will be considerations of information flow, computational processing, decision systems, memory and constructs of that sort. Such operations, of course, are all performed by semiconductor devices, but the low-level implementations are handled by hardware and the software programming language, with its subroutines and primitives that can be called or invoked as the need arises without reference to the underlying implementations. Similarly, an efficient and meaningful description, model or theory of a complex psychological process is not likely to be cast in the terms and constructs of neurophysiology and synaptology, although those levels of analysis certainly apply. More useful theories will be based on a higher level of functional description, constrained and calibrated by knowledge of the underlying physiology. Indeed, those theories will almost surely also shape the development and refinement of physiological constructs and the way physiology is characterized.

For both the computer and the brain, reductionism may be possible, although not necessarily useful in all cases for all purposes. In fact, it may be a clear hindrance to the extent to which eliminativist perspectives eschew studies of higher-level psychological processes. In contrast, what we term a calibrative reductionist approach may enhance knowledge and understanding of both higher-level and lower-level domains.

Calibrative reductionism. Eliminativism (or eliminative materialism) in its strongest form admits of no mental reality, and advocates the abandonment of such constructs. One problem with this perspective is that many interesting problems, with considerable personal and societal implications, revolve around mental or psychological processes. To "reduce" these to the language and concepts of neuroscience may not adequately capture the essence of the phenomena in question. A category error is likely, as there may not be a simple one-to-one mapping across domains. The identification of the brain mechanisms underlying specific psychological processes first requires the accurate specification of those processes. To the extent to which emergence is apparent, and as was the case for oxytocin and pair-bonding, adequate understanding may require observations at higher levels of organization to define the phenomena to be explained. An even stronger case has been argued for *downward causation*, or the role of emergent properties in the selection or shaping of lower-level elements (for example, through evolution; see Murphy, Ellis, & O'Connor, 2009).

Tens of thousands of years of evolution have deeply planted sociality, emotionality, and meaning-making in our genome, and this manifests notably in both our personal life and our societies. An early view was that our genotype (or the pattern of gene encoding in DNA) is the ultimate, bottom-up determinant of our phenotype – the characteristic expression of genetic information into physical, psychological, and behavioral manifestations (see Mayr, 1982). But it is now apparent that this bottom-up model is overly simplistic. Dennis Noble (2008) asserts:

> Relating genotypes to phenotypes is problematic not only owing to the extreme complexity of the interactions between genes, proteins and high-level physiological functions but also because the paradigms for genetic causality in biological systems are seriously confused.

It is now clear that the phenotype derives from an interaction between the genotype and the environment and that the environment and the phenotype, in a top-down fashion, can regulate the genotype (McGowan, Meaney, & Szyf, 2008; Noble, 2008). Epigenetic reprogramming of the genome and its expression in neurons (via methylation of DNA among other means) occurs during development in an environmentally dependent fashion (McGowan et al., 2008; Meaney, 2010). In addition to DNA methylation, the epigenome (the chromatin-based chemical environment of the DNA) can regulate gene expression by altering accessibility of the essential transcriptional machinery (see McGowan et al., 2008). This top-down regulation of the genome belies simple bottom-up models of causality and illustrates the complexity of biological systems, of particular note the human brain. The implication of these and related studies is that aspects of genetic expression, which were thought to be encapsulated within each living cell far from the reach of personal ties or social influences, are in fact subject to modulation by the social environment (Cacioppo, Berntson, Sheridan, & McClintock, 2000). This complexity poses challenges for how we are to understand neuropsychological processes and even what a fundamental understanding of these processes would look like and how it should be pursued. Given that any biological mechanism may play some role in a variety of behaviors and each behavior may derive from a complex set of mechanisms, explaining brain–behavior relationships presents an unprecedented level of scientific complexity. Psychological scientists are the experts in this specification and in creating protocols for isolating specific psychological processes in the normal waking brain, so psychological scientists have a central role to play in this endeavor.

As with behaviorism, however, an exclusive focus on a single level of analysis may not be optimal, as it may deprive the field of important constraints on theories or concepts that can come from alternative levels of analysis. The emerging approaches in computational neuroscience and network modeling hold promise in this regard. Understanding how networks might function, especially when constrained by knowledge of neural processes may yield the kind of multi-level perspective that truly informs mind–body relations. The terms, concepts, and theories of different disciplines, however, often develop in relative isolation. Consequently, integration of information across levels of analysis may be challenging in view of potential category errors and the misalignment of concepts. Consequently, the mere collection and description of data from different levels of analysis is not sufficient. What will be necessary is the alignment or calibration of data and concepts across disciplines and levels of organization, observation, and analysis – a task that is advanced by the study of the aggregate to determine the properties, features, and combinatorial rules of the elements that give rise to potentially emergent properties.

Optimally, this calibration would be a reciprocal, two-way process of multi-level alignments. Observations at the psychological or behavioral domain constitute much of the interesting grist for neuroscience. Of course, contemporary theories regarding psychological representations and processes are approximations that continue to evolve as new empirical data, methods, and models are developed, and major questions remain about these representations and processes. Neuroscience perspectives may help clarify and constrain these efforts in the psychological domain, and vice versa. This pursuit will contribute not only to a better understanding of the brain mechanisms underlying complex processes, but also to the evolution of sophisticated psychological theories. In contrast to eliminativism, calibrative reductionism seeks to mutually relate and align psychological concepts and understandings with lower-level analyses, thereby enriching both. Indeed, social neuroscience perspectives can likely contribute to the way scientists and philosophers think about the brain. Patricia Churchland (2008) notes:

> In the last two decades, neuroscience has profoundly transformed how we understand learning, decision making, self, and social attachment. Consequently, traditional philosophical questions about mind and morality have been steered in new directions.

Summary and Conclusion

The mind–body problem has been with us for some time, and will not be handily solved here. How do we relate mental to neural properties? A *property dualist* might assert that mental and physical properties are distinct, and while mental properties may arise from a physical entity (the brain) they remain irreducible (e.g., Jackson, 1982). Psychological processes arise from the brain, and for neuroscientists, understanding the underlying neurophysiology can inform or elucidate psychological processes. The question arises, however, as to whether they can ever be completely understood by reference to neurophysiology alone. If that is possible at all, it may be only to the extent that our concepts and theories of neurophysiology be shaped and informed (calibrated) by psychological or behavioral studies. This is the essence of calibrative reductionism. As in the case of epigenetics, where environmental and behavioral processes can impact genetic expression, there may be both bottom-up and top-down causation in complex systems such as the brain.

The concept of calibrative or reciprocal reductionism, as outlined earlier, is more of a metaphor or general strategy than a developed system for approaching multi-level integrative analysis. But it offers an important perspective on how one might think about explanations of psychological phenomena. Moreover, some general guideposts have been articulated for an implementation of this strategy to multi-level analyses (see Cacioppo & Berntson, 1992).

The principle of *Multiple Determinism* asserts that there are mutual back-and-forth influences among levels of organization and function in neurobehavioral systems. This principle stipulates that a target process or event at one level of observation (e.g., a neural response or an emotion) may have multiple antecedents within and across levels of organization. Social pair-bonding, for example, has many determinants extending from the cellular to the hormonal to the psychological or behavioral levels. Consequently, a comprehensive explanation of a psychological process may require attention to multiple levels of analysis. The principle of multiple determinism has an associated *corollary of proximity*. This corollary states that the complexity of the mappings between elements across levels of organization will increase with the number of intervening levels. Thus, a viable starting strategy might be to study relations among proximate levels, and then use that information to support further multi-level perspectives.

The principle of *Nonadditive Determinism* asserts that the properties of the collective whole are not always predictable from the properties of the parts, at least until the properties of the whole have been studied and documented across levels. This strategic principle further builds on the need for multiple levels of observation and analysis in understanding a complex phenomenon. This is a perspective that is highly consistent with calibrative reductionism, but not with models that are limited to upward causation. Finally, the principle of *Reciprocal Determinism* asserts that there are mutual back-and-forth influences among levels of organization and function in neurobehavioral systems. Consequently, the terms concepts and theories of different levels of organization and analysis may need to be reciprocally calibrated in order to yield a meaningful explanatory system.

References

Berntson, G. G., & Cacioppo, J. T. (2000). Psychobiology and social psychology: Past, present and future. *Personality and Social Psychology Review, 4*, 3–15.

Cacioppo, J. T., & Berntson, G. G. (1992). Social psychological contributions to the decade of the brain: Doctrine of multilevel analysis. *American Psychologist, 47*, 1019–1028.

Cacioppo, J. T., Berntson, G. G., Sheridan, J. F., & McClintock, M. K. (2000). Multi-level integrative analyses of human behavior: The complementing nature of social and biological approaches. *Psychological Bulletin, 126*, 829–843.

Churchland, P. M., & Churchland, P. S. (1998). On the Contrary: Critical Essays 1987–1997. Cambridge, Massachusetts: The MIT Press.

Churchland, P. S. (2008). The impact of neuroscience on philosophy. *Neuron, 60*, 409–411.

Emmeche, C., Koppe, S., & Stjernfelt, F. (1997). Explaining emergence: Towards an ontology of levels. *Journal for General Philosophy of Science, 28*, 83–119.

Feyerabend, P. (1963). Mental events and the brain. *Journal of Philosophy, 40*, 295–296.

Gingrich, B., Liu, Y., Cascio, C., Wang, Z., & Insel, T. R. (2000). Dopamine D2 receptors in the nucleus accumbens are important for social attachment in female prairie voles (Microtus ochrogaster). *Behavioral Neuroscience, 114*, 173–183.

Goswami, A. (1995). *The self-aware universe*. Los Angeles: Tarcher.

Insel, T. R. (1997). A neurobiological basis of social attachment. *American Journal of Psychiatry, 154*, 726–735.

Jackson, F. (1982). Epiphenomenal Qualia. *The Philosophical Quarterly, 32*, 127–136.

James, W. (1950). *Principles of psychology*. New York: Dover.

Mayr, E. (1982). *The growth of biological thought*. Cambridge: Harvard University Press.

McGowan, P. O., Meaney, M. J., & Szyf, M. (2008). Diet and the epigenetic (re)programming of phenotypic differences in behavior. *Brain Research, 1237*, 12–24.

Meaney, M. J. (2010). Epigenetics and the biological definition of gene x environment interactions. *Child Development, 81*, 41–79.

Murphy, N., Ellis, GFR., & O'Connor, T. (2009). Downward Causation and the Neurobiology of Free Will. New York: Springer.

Noble, D. (2008). Genes and causation. *Philosophical Transactions of the Royal Society A, 366*, 3001–3015.

O'Connor, T., & Wong, H. Y. (2005). The metaphysics of emergence. *Noûs, 39*, 658–678.

Ross, H. E., & Young, L. J. (2009). Oxytocin and the neural mechanisms regulating social cognition and affiliative behavior. *Frontiers in Neuroendocrinology, 30*, 534–547.

Scott, T. R. (1991). A personal view of the future of psychology departments. *American Psychologist, 46*, 975–976.

Chapter 20
Produced and Spontaneous Emergent Interactionism

Walter Colesso

Psychological interactionism is composed of a plethora of various studies among them it is difficult to navigate without any criteria of research or applications. In this chapter, a review of main psychological interactionistic studies is proposed in the light emergent interactionism and achieved through three steps:(1) the development of emergent interactionistic outlook in the psychological field as proposed by Sperry (1969) and recovered by Capitelli, Guerra, L'Abate, Longo, and Rumbaugh (2009) with a synthetic integration between produced and spontaneous emergent interactionism; (2) a new framework crossing spontaneous and produced emergent interactionism criteria and L'Abate's distinction among paradigm, theory, and model; (3) a review of any studies according to the new proposed framework.

We can find the epistemological background of the interactionistic approach in the philosophical thought of Dewey and Bentley (1949) when they addressed the topic of "linguistic disease" both in the general epistemology and in the specific field of logic. Their terminological investigation leads to establish broadly three levels of the organization and presentation of inquiry in the order of their historical appearance, labeled self-action, interaction, and transaction. They affect all human behavior in and with respect to the world representing the world itself as reported by people: (a) *self-action*: where things are viewed as acting under their own powers; (b) *inter-action*: where thing is balanced against thing in causal interconnection; (c) *trans-action*: where systems of description and naming are employed to deal with aspects and phases of actions, without any final attribution to "elements" or other presumptively detachable or independent "entities," "essences," or "realities" and without isolation of presumptively detachable "relations" from such detachable "elements."

W. Colesso (✉)
Psychidiagnostic Laboratory, Family Center Foundation,
Via San Nicolò, 60, 31100, Treviso, Italy
e-mail: wcolesso@libero.it

L. L'Abate (ed.), *Paradigms in Theory Construction*, DOI 10.1007/978-1-4614-0914-4_20,

The word "transaction" was introduced by Maxwell (1865) in describing physical events: "The most obvious mechanical phenomenon in electrical and magnetical experiments is the mutual action by which bodies in certain states set each other in motion while still at a sensible distance from each other" (p. 459).

Going back in time, we find other precursors of the interactionistic approach. For Koffka (1935), the relation between the geographical environment (or the stimulus pattern) and behavior involves a relation between the geographical and behavioral environment and the behavioral environment and the behavior. Also, both the geographical and behavioral environments are changed through actual behavior.

Murray (1938) focused on the organism–environment interaction and differentiated between physical and psychological environments. He argued that the stimulus situation is the total environment to which the "creature attends and reacts" (p. 40) usually responding as a patterned meaningful whole. In more detail, Murray explored any needs or motives such ambition and accomplishment, achievement and recognition, affiliation and affection, the interaction between needs and pressures in shaping behavior.

Lewin (1936, 1952) has considered the behavioral, psychological environment rather than the physical environment and noted that the physical environment can be identical for children and adults but the representation of any situation "real" for the individual is very different. The focus of study is on the *life space* of the individual "which consists of the person and the psychological environment as it exists for him" (p. xi). This information gives a richer understanding to the famous formulation $B=f(PE)=f(Lsp)$. Lewin uses the term "B" for "behavior" to describe behavior or any kind of mental event and includes "thinking, wishing, striving, valuing, achieving, etc." (p. xi). The whole situation or environment including the person is described as "(PE)" and this defines the psychological situation or "life space" (LSp). The person–situation relationship is mutually dependent with reciprocal interaction and should be considered "one constellation of interdependent factors" (p. 240).

Continuing the excursus, we find Gestalt concepts: there are "wholes," and the nature of "wholes" is not necessarily determined by the separate elements and the piecewise relations between elements. Gestalt theory examines the "whole" because it is more than its elements or parts, and it is the "whole" that is believed to shape and determine the part-processes not the other way around (e.g., Wertheimer, 1938).

Kantor clearly has applied interaction and transaction concepts to general psychology, as discussed at greater length in Mitch and Linda (2011): all psychological events consist of "interactions between organism and objects" (Kantor & Smith, 1975, p. 32). More specifically, "the psychologist is obliged to construct a descriptive unit simple and stable enough to enable him to understand what is essentially continuous and integrated. Such a descriptive tool he constructs in the form of a behavior segment" (p. 34). As Kantor went on to explain, "essentially the behavior segment is an abstraction designed to fixate a definite space-temporal event. This event can be analyzed into a series of factors operating in a specific framework which may be designated as a field or setting" (p. 34). In understanding the behavior segment, two of the just-mentioned factors are central, named by Kantor response function and stimulus function. The first is identified with an action of the organism and the second with an action of the stimulus object. Other factors participating in or contributing to a behavior segment include the contact media, the interactional history, setting factors, and specific factors of particular situations (Kantor, 1970).

Personality and social psychology are the two main development areas of interactionistic research. In personality psychology, interactionism refers to personality and situation dualism. In social psychology, interactionism is inherent in the field of the study defined as that branch of psychology concerned with persons and their relationships with others, groups, and with the society as a whole. Raynolds and colleagues recently explored the points of connection between personality and social psychology. Their aim is to propose an integrated interactionist approach, named *Dynamic Interactionism* (Raynolds et al., 2010), even if the followed open peer commentary (Brooks, Buhrmester, & Swann, 2010) highlights some difficulties to integrate and organize personality and social interactionist studies in a coherent framework. Capitelli et al., (2009) suggested also the emergent interactionism paradigm initiated by Sperry (1969) as a new point of view from which interactionistic studies may be usefully reviewed. In the same monograph, L'Abate (2009) proposes a hierarchical framework for general psychological studies defining and distinguishing these three theoretical concepts: paradigms, theories, and models for overcoming the current chaotic fragmentation and proliferation of research studies.

Based on this introduction, it seems useful to investigate the prospect of Sperry and the ensuing debate using the three-frame conceptual distinction reaffirmed by L'Abate. Hopefully, a new framework for organizing interactionistic studies in psychological area may be traced.

Proposal for a New Interactionist Framework

For a new framework of interactionistic studies, two criteria need to be defined: (1) a distinction among paradigm, theory, and model; (2) produced and spontaneous emergent interactionism.

First Criterion: Definition of Paradigm, Theory, Model, and Dimension

L'Abate (2009) defines these three theoretical concepts spread throughout the entire psychological landscape in theory and practice: paradigms, theories, and models in order to clarify among studies in psychology research.

Paradigms are viewpoints that emphasize an overall view of reality, behavior, and relationships. They are used to describe and explain physical and human behavior according to the all-encompassing view of including any possible variables considered to draw a reasonable portrait of human relationships (general paradigms), or to describe and emphasize specific aspects of behavior when one of these is hegemonic over the others (specific paradigms).

A *theory* is a speculative framework oriented to validation – both conceptual and empirical – through models.

Models are the irreducible components of a theory. They include the basic elements of a theory – its structure, processes, or contents – in order to limit and specify certain parts of the theory. The models are intended to represent the theory and they are valid insofar as the theory is valid. A model is characterized by *dimensions* quantified and quantifiable. Therefore, they are replicable lending themselves to direct any empirical validation.

Second Criterion: Produced and Spontaneous Emergent Interactionism

The label "Emergent Interactionism" was coined by Sperry (1969, 1988) who was engaged in neuropsychological investigations using the bisection of the corpus callosum of patients. He suggested subjective phenomena of conscious experiences, such as sensations, percepts, ideas, images, illusions, and feelings, exert causal influence on the sequence of events in the physical brain process. His surgical observations led to define a new paradigm: "The new model adds 'downward' to the traditional 'upward' micro determinism, and is claimed to give science a conceptual foundation that is more adequate, more valid, and more comprehensive. Further, this is achieved with no loss to previous benefits, making already powerful paradigm of science even strong." (…) "This entire case is seen to stand or fall on the validity of a changed concept of causal determinism. Instead of an exclusive bottom-up, micro determinism, we substitute a bi-directional model adding a reciprocal top-down, emergent or holistic form of downward control. To put it very simply, the old saying that 'the whole is greater than the sum of the parts' is just extended into concepts of inter-level determinism. Not only is the whole, as traditionally assumed, constantly control form below by its parts, but also, we claim, it works the other way around. The emergent whole, that is, constantly exerts downward control over its parts. The subsequent course and destiny of the parts, once joined together in a large whole, are determined thereafter by the new higher level, emergent properties of the whole." (Sperry, 1988, p. 1064–1065).

Sperry understands the emergent interactionism paradigm in neuropsychological area as referring to something new generated from a circular or bidirectional interaction of brain state (physical) and mental state (subjective). This two-way perspective can be extended to any interaction between the two poles: physical and mental, objective and subjective, observer and observable, individual and environmental, individual and collective.

In fact, Sperry's researches have stimulated a fruitful scientific debate. Fuente (1993) pointed the emergent interactionism role in mind–body problem field explaining three historical ways to justify the emergence of behavior in an organism: (a) materialism or monism proposes the behavior as strictly a function of physical events; (b) mentalism postulates the behavior as mediated only by mental or cognitive functions and emerging from strictly cognitive or mental experiences; (c) dualism suggests both material and mental events necessary for emerging behavior. He observed that Sperry's concept of emergent interactionism and downward causation not only further enhances the role of awareness, but also allows a more plausible interaction between objective and subjective states. Nevertheless, the emergent interactionism is such that a classical dualist position becomes untenable simply because the two variables are not mutually exclusive or separate. They are, by definition, inseparable because one gives rise and controls the other in a synergistic way.

Meltzer and Manis (1992) focused their considerations on the nature of emergent events and extend their analyses from personal to interpersonal processes. They do not have in mind the emergence of new properties at increasing levels of complexity of phenomena, as in biological evolution and in movement from the physical level to chemical level, to the biological level, to the psychological level, to the social level of phenomena. This form of emergence has as its antithesis the *reductionism*. They proposed the emergence of certain phenomena as "novel," not merely in the psychological sense of being unexpected, but in the theoretical sense of being unexplainable or unpredictable. Thus, emergent events are events that contain novel features not entirely derived from antecedents or previous experiences. This form of emergence is a major vehicle of indeterminacy, having as its antithesis the *determinism*. The behavior may be an unexpected, unpredictable, and emergent construct in the thought processes of the actor or actors. It is plausible that their conclusions are useful to distinguish two main study areas within emergent interactionism.

The studies proposing a predictable "new" emergent behavior (action, attitude, feeling, ability, etc.) from a circular interaction between person (personality trait or other individual variable) and environment (situations, stimuli, etc.) may be labeled as "Produced Emergent Interactionism." This perspective explains behavior as mechanical, deterministic, predictable with low voluntary response to environmental stimulus. The relation among dimensions can be expressed by the formula suggested by Kenrick and Funder (1991): Emergent behavior = person × perception of situation.

Spontaneous Emergent Interactionism" the research perspective which views the "new" behavior, generated between person and environment interaction, as unpredictable due to degree of individual choices or disturbance can be defined. It accounts for the behavior as influenced by external stimulus and lets to some degree of choice in the behavior, reducing the predictability of the resultant behavior:

$$\text{Emergent behavior} = \text{Person} \times \text{perceived environment} \times \text{degrees of variability.}$$

Crossing the produced-spontaneous interactionism with the distinction among paradigm, theory, and model, a grid organizing interactionist studies may be obtained. This new proposal overcomes the traditional framework, as discussed in greater details in L'Abate (2011).

A Review of Main Interactionist Literature According to the Proposed Framework

The purpose of this section is to integrate different viewpoints into one proposed framework.

Produced Emergent Interactionism

Following the publication of the book *Personality and assessment* by Mischel (1968), a major debate arose that led scholars to conceptualize three approaches concerning the individual and his/her interaction with the environment: personologism, situationism, and interactionism (Ekehammar, 1974). While at a paradigmatic level the three directions appear independents, at a theoretical and applicative level they are not so selective and each does not exclude totally the others, even partially overlapping. Each approach considers the new emergent interaction from person and situation with different weights assigned to the two components. In this way, the three paradigms may be considered as three different positions within the produced emergent interactionism.

Personologism Paradigm

This paradigm emphasizes stable intra-organismic constructs, such as "traits," "psychic structures," or "internal dispositions" as the main determinants of behavioral variation (e.g., Alker, 1972; Wachtel, 1973). Traits function as predispositional bases for response consistencies of behavior in a variety of different situations. This position may generally be expressed as $B = f(P)$, where B stands for behavior, P for person. It should be borne in mind that the situation is considered only by some personologist views in this paradigm. For example, Allport (1937) defined personality as partially "…adjustment to the environment" (p. 48) arguing the trait development as influenced by interpersonal feedback and environmental experiences (p. 291, 312). However, the person is emphasized and the relation between person and situation regarded as additive rather than interactive. Also, the trait view may be expressed as $B = f(P) + g(E)$, where B stands for behavior, P for person, and E for environment or its some part (e.g., situation) (Ekehammar, 1974).

The current personologist research is working to provide evidence that traits are powerful predictors of actual manifestation of traits in behavior (Fleeson & Gallagher, 2009), even if affected by situations. For example, Fleeson (2001, 2009) proposes the conception of *Trait as a Density Distribution of States*: behavior can be characterized as trait-relevant states, and over time, an individual's behavior forms a distribution. The central tendency is an adequate description of such distributions and the remaining variance behavior can be dismissed as an error. In this approach, a personality state is defined as a dimension with the same content and scale as a personality trait but assessing how the person is at the moment rather than he/she may be in general: The trait responsible for the behavior is caused by multiple sources, including the psychologically relevant features of situations, individual goals that are acting at the moment (Fleeson & Jolley, 2006).

Situationism Paradigm

This paradigm may be regarded as the antithesis of personologism and applies to those views emphasizing environmental (situational) factors as the main sources of behavioral variation (e.g., Bowers, 1973). External environmental factors are considered as the major determinants of personality and behavior. It is considered that most of the situationism proponents give an important role to individual differences. This position may generally be expressed as $B=f\,(E)$, where E stands for environment or its some parts (e.g., situation).

The theory "Self-Efficacy Mechanism" proposed by Bandura (1977, 1982) is associated to this paradigm: the human behavior is the product of an interaction between individual, environment, and behavior. Faced with a situation, individuals form an idea about both their abilities and skills (self-efficacy expectations) and the levels of performances resulted from the issuance of that behavior (outcome expectancies). These results will also influence expectations of outcomes and the self-efficacy. Outcome expectancies and self-efficacy will influence one another. The efficacy in dealing with one's environment is not a fixed act or simply a matter of knowing what have to do. Rather, it involves a generative capability in which component cognitive, social, and behavioral skills must be organized into integrated courses of action for serving innumerable purposes. Initiation and regulation of transactions with the environment are therefore partly governed by judgments of any operative capabilities. Perceived self-efficacy is concerned with judgments of how well one can execute courses of action required to deal with prospective situations (Bandura, 1982).

(Lawful) Interactionism Paradigm

This paradigm is presented as a summary of personologism and situationism and implies that neither the person per se nor the situation per se is emphasized, but the interaction of these two factors is the main source of behavioral variation. This position may generally can be expressed as $B=f(P,E)$ (Ekehammar, 1974).

Magnussen and Törestad (1993) note the fragmentation of interactional studies and propose an holistic paradigm to the functioning of the individual: an individual functions as a totality and each aspect of the structures and processes (perception, cognitions, plans, values, goals, motives, biological factors, conduct, and other aspects) takes on meaning from the role it plays in the total functioning of the individual. The interaction is the basic principle of operation of open systems at all levels of individual functioning: from the single cells and how they organize themselves into systems in the total organism to the functioning of an individual in relation to his/her environment. At level of psychological functioning, thoughts, feelings, actions, and reactions are aspects of a holistic process characterized by a "dynamic, [and] *lawful interaction*" (p. 438). This includes six basic principles: (a) *multidetermination*, as all activities of the system involve several underlying factors or conditions; (b) *interdependence* of factors according to each other without necessarily influencing each other reciprocally; (c) *reciprocity* of biological and individual factors operating together within the individual and of the way the individual relates to the environment; (d) *nonlinearity* of the interrelations among variables; (e) *temporality*, i.e., the continuous flow of interrelated, interdependent events in the system; (f) *integration* at all levels of dynamic, holistic processes and among operating parts coordinated to serve the goal of the membership system.

Within this paradigm, Mischel and Shoda (1995) have proposed a theory to reconcile those findings in the invariance of personality and the variability across the situation: a *cognitive-affective system theory*. The mains constructs of the theory are the features situation, the psychological mediating process, and the personality system. The situation or *features of situations* activates a set of internal reactions – cognitive and affective – based on the individual's prior experiences with those features. These are in turn encountered in the external environment but generated in thought, planning, fantasy, and imagination. The features encompass both social and interpersonal situations and intra-psychic situations in mood states and in the everyday stream of experience and feeling.

The situation is a perception of situation with the meaning built by the individual. The *psychological mediating process* underlines the individual differences in social behavior composed of relatively stable person variables, i.e., cognitive affective units. The *personality system* is given by those available cognitive and affective units organized in a distinctive network of interrelations. This personality theory justifies any variations in the individual's behavior across different situations: The behavioral variation, in connection with changing situations, constitutes a potentially predictable and meaningful reflection of the personality system, whose structure can remain stable across situations even if the personality state changes readily as the active situational features change or are encoded alternatively or transformed cognitively and emotionally (Mischel & Shoda, 1995). A model validating an aspect of the cognitive-affective theory is proposed by Mendoza-Denton, Ayduk, Mischel, Shoda, and Testa (2001) that shows evidences for the causal role of interactionist self-encoding – a specific affective cognitive unit – on the regulation of affect and social perception.

Relational Psychodynamic Paradigm

The more recent literature on dynamic psychology (Barbieri, 2009) introduces the relational psychodynamics as a paradigm in progress: at the basis of the mind activity there are not any drives but relationships and individual efforts to maintain ties with other people and to differentiate themselves from others. The nature of the mind is relational and interactive. The self cannot be conceptualized out of the matrix of relationships. The myth of the "isolated mind" is a mistake introduced by a drive perspective oriented to the individual and internal dynamics (Stolorow & Atwood, 1992). The precursors of this guideline can be considered Sullivan (1953), Fromm (1941), and Horney (1939) who moved the concepts of drive toward a relational perspective.

The hierarchical and relational theory developed from the two dynamic assumptions of space and time is proposed by L'Abate (2005; L'Abate, Cusinato, Maino, Colesso, & Scilletta, 2010). Space is characterized by the approach-avoidance tendencies occurring between and among individuals. Time is characterized by discharge-delay tendencies. These assumptions support the two individual relational competencies: the ability to love and the ability to negotiate. From these two postulates some models are derived focusing what and how effectively we deal with each other, with intimates and nonintimates, close/distant, committed/uncommitted, dependent/interdependent/independent, and short/prolonged relationships. The relational perspective shifts the focus from person–situation relation to relational competence–setting relation. Setting is not a generic environmental dimension, but those specific settings where relations take place: home, school/work, transit and transitory, or enjoyment and survival. Relationships are conceived as bidirectional rather unidirectional interactions with intimates and nonintimates in a continuous interdependent and reciprocal exchange of resources available to people. They are evaluated by how people feel, think, act, is aware, and evaluate proximal and distant subjective contexts (L'Abate et al., 2010).

According to this theory, it is impossible to disentangle the relational competence socialization from its relationships with intimates and nonintimates, because relational competence is circularly and contextually the product and producer at the hearth of those relationships. Socialization occurs through continuous interactions with significant and nonsignificant others (L'Abate, 2009). Because of the complexity of relationality, L'Abate has developed 16 models to conceptualize it. Among these models, the likeness continuum model refers to the self-identity differentiation. It is composed of a curvilinear developmental and dialectically related dimensional distribution with 3° of similarity and 3° of dissimilarity: (1) *Symbiosis* between related individuals (caretaker–child, partner, sibling). This degree is so strong that it may become symbiotic to the point that one individual cannot differentiate oneself from another self. (2) *Sameness* requiring blind or uncritical conformity to an authority figure, parent, or caretaker, without room for personal choice or deviations. (3) *Similarity* to some characteristics of the caretaker, such as attitudes, habits, identity, self, or physiognomy. This degree permits the freedom to choose and to have beliefs and customs different from those held by people in proximal relationships, that is, the freedom to be oneself without fear of criticism, recrimination, retribution, or punishment as long as one does not go against established normative laws, rules, and regulations. (4) *Differentness*, that is how an individual perceives self as being dissimilar in comparison with some attributes of parents, caretakers, or significant figures in one's life. Here is where the freedom to be diverse or unusual is permissible as long as it does not go against cultural and societal norms or restraints. Differentness can be conceived as and even equated also with diversity. (5) *Oppositeness* as a negative reaction to demand for conformity resulting in rebellion and rejection of rules and regulation from parents, caretakers, authority figures, or society. This degree flouts dictatorial and totalitarian rules and regulations, disregarding them as being as inconsistent, weak, and defective. (6) Finally, *Alienation* as being completely cut off from intimates and nonintimates. Here is where the individual is so different from cultural and political norms that no relationship exists between the individual and other people (L'Abate, 1976, 2005; L'Abate et al., 2010). The model has been validated by Cusinato and Colesso (2008).

The proposal of L'Abate is not alone. Several models were developed within interactional paradigm to predict behavior focusing interactions between individual and environment. For instance, Endler and coll. propose an interactional *model* to assess anxiety operationalized with the Endler Multidimensional Anxiety Scales, a self-report measure of trait and state anxiety (Endler, Edwards, Vitelli, & Parker, 1989). The model reflects the need to separate, methodologically and conceptually, the individual predisposition to a particular experience or behavior from the current experience of behavior in a specific situation. Spielberger (1983) emphasizes the considerable ambiguity of the conceptual status of anxiety because the term has been used to mean two very different types of variables, namely, state and trait. State anxiety has been defined as a transitory emotional response involving unpleasant feelings of tension and apprehensive thoughts (Spielberger, 1983). Trait anxiety is defined as a personality trait referring to individual differences in the likelihood that a person will experience state anxiety in stressful situations (Spielberger, 1983). At epistemological level, the "model" is a representation of the individual experience of anxiety defined as individual's *predisposition* and as individual's *actual response* in a specific situation: all this is referred to an undefined theory asserting behavior defined by traits and states. The model is a useful tool to measure anxiety, but does not explain how states interact states with traits.

Caspi and Roberts (2001) formulate an empirical *model* to explain the interactional developing of personality across the life span. There is a modest continuity of personality (characteristics unique to an individual) from childhood to adulthood, even if the changes are small in magnitude, and trait continuity and change in personality are not related simply ways to biosocial transitions (puberty, first job, marriage, parenthood, and menopause). They identify three mechanisms affecting continuity and change of personality across the life span: environment, genetics, and person–environment transactions expressed in social contexts, role experiences, and changing historical and cultural norms. The mechanism of development is often separated and can work in any time to engender continuity and change.

Spontaneous Emergent Interactionism

Various paradigms can be included in this section.

Mind–Society Paradigm

Raynolds et al. (2010) have recently reformulated the Mind and Society paradigm – labeled dynamic interactionism – starting from works of several scholars as Asch, Sherif, Lewin, Bruner, Tajfel, Turner, and the Gestalt tradition. Human systems are comprised of both individuals and groups: the person is both an individual and a group member. The *wholes* related to group life are emergent psychological processes overlooked or misattributed when the focus is the individual as an isolated part. Social norms, shared values, influence and persuasion, shared emotions, and shared goals are all collective products that arise through coordinated activities that shape the social system (i.e., power structures, laws, policies, conventions, culture, etc.) and structure in turn groups and individuals composing the system (valued or not valued, appropriate or inappropriate, right or wrong). Through this social interaction and shared activities, the mind (i.e., thoughts, emotions, memory, perception, and imagination) and the mental functioning are qualitatively transformed. It's the so-called mind–society interaction. Raynolds and colleagues consider as paradigm key these two theories:

In *Social Identity Theory*, Tajfel and Turner (1979) have used the concept of *personal identity* (the personal self: "I") to describe situations where individuals perceive themselves as distinct and different from others (available for comparison). On the other hand, the concept of *social identity* (the social self "we") is about an individual's knowledge that he/she belongs to certain groups together with some emotional and value significance to him/her of the group membership (Tajfel, 1972). Importantly, the term "social identity" does not necessarily refer to demographic, sociological, or role groups. The term can refer to psychological groups where an individual perceives him/her as a member because the group is self-relevant and self-defining (Turner, 1985).

More specifically about the *Self-Categorization Theory*, people can define or categorize themselves at different levels of abstraction; for example, at the *interpersonal* level (where self is defined as a unique individual relative to others available for comparison), at the *intergroup* level (where self is defined as being a group member in contrast to a relevant outgroup), and at the *superordinate* level (where self is defined as a human being in contrast to other life forms). The focus is placed in the explanation of the shift from "I" to "we" and of which identity (and its associated content) will become relevant and meaningful in a particular situation. The categorization process is a mechanism that provides stimuli (including the self) with meaning (Oakes, Haslam, & Turner, 1994).

When people categorize or identify self with a particular in-group, the norms, values, and beliefs that define the group are internalized and this process influences attitudes and behaviors of all group members. The greater is the identification with a particular group, he/she will behave more closely in line with the aspects defining that group. Under these conditions, other members can have an impact on one's own thoughts, attitudes, and behaviors, and an individual can have a significant impact on the actions of others. When individuals are considered to belong to the same class of stimuli, they are cognitively grouped as similar perceivers confronting the same stimulus situation. It is possible to define the process concerning how "others" come to affect one's own attitudes and behavior: (1) individuals define themselves as members of a distinct social category, (2) they learn or develop the appropriate, expected, desirable behaviors that are correlated with membership category differentiated from other categories (i.e., the stereotypical norm), (3) they assign the norms and attributes of the category to themselves (internalization), and (4) their behavior becomes normative as their category membership becomes more salient (Turner, 1987).

There is a broad convergence on the idea that a person can belong to many different groups and when a particular group becomes psychologically salient, the social norms related to that identity will shape the behavior in significant ways. Identities based on the group are recognized as variable and context dependent. This variability is often contrasted with the relatively stable and enduring features of personal identity or the personal self. The core point concerns the content generated to describe personal identity that depends on some comparative reference and can result in different (or the same) self-descriptors being generated depending on the context (also see Andersen & Chen, 2002).

When individuals compare themselves to others (interpersonal) rather than making assessments in isolation (intrapersonal), they are more likely to characterize themselves in a dispositional way. The interpersonal context accentuates similarities and differences between the person and the comparison other, leading to a strong sense of one's self-defining features. In this way, the comparative context has an impact on personal self-categorizations and such categorizations also can be variable depending on the comparison frame of reference.

Several *models* were created to test these theories. Works on stereotype or social identity threat clearly demonstrates, when people define themselves in relation to certain group memberships with negative meaning can have an impact on cognitive ability (i.e., intelligence) and performance on those dimensions defining the stereotype of the group (Steele, Spencer, & Aronson, 2002). Williams, Turkheimer, Magee, and Guterbock (2008) report also findings showing the contamination anxiety (an aspect of obsessive-compulsive disorder) not only affected by ethnic category membership (i.e., African American or European American) but also by that salient ethnic identity when relevant measures are completed. In a large-scale study of staff and students (7–10 years of age) of high schools realized by Bizumic, Reynolds, Turner, Bromhead, and Subaši (2009), social identity is significantly related to and mediates the relationship between organizational factors and individual psychological well-being (i.e., self-esteem, positive affect, and job involvement but also negative aspects such as depression, anxiety, loss of emotional control, aggressive and disruptive behavior).

Symbolic Interactionism Paradigm

One of the roots of social psychology is found in symbolic interactionism. Blumer (1969) has coined this term focusing on the ways people took control of their lives, as "acting people" in a society that is a "complex of ongoing activity." The two parts, symbol and interaction, produce meaningful interaction: the interaction involves giving symbolic value to the social objects that can be anything – physical objects, animals, history, language, ideas, emotions – as well as self and other people. According to Blumer, the individual in all of his/her everyday acts designs different objects to himself, giving them meaning, judging their suitability to his action, and making decisions on the basis of the judgment: People interpret and act on the basis of the symbols. *Symbols* are abstract meanings attached to things, people, and behavior so that they can have different meanings for different individuals. The important point is that individuals consciously and creatively evaluate, make decisions, and act. *Interaction* involves the self engaged in communication with self: selecting, suspending, regrouping, and transforming meanings in terms of the social context and the individual's intensions and interests (Blumer). But for Blumer, the most significant features of all human associations is that the participants "take each other into account" as a basis of conduct (1969). Society is a "complex of ongoing activity" involving collectively initiated "joint actions." Joint actions are constituted by the fitting together of the lines of behavior of separate participants as, for example, in a trading transaction, family dinner, wedding, games, or war. Joint actions have a history that is "orderly, fixed, and repetitious" because the participants have a common definition of the situation. Social interaction is an interaction between people, not between roles. Roles may affect, in some degree, the direction and content of action, but this is far cry from asserting action to be a product of roles.

In summary, the symbolic interaction paradigm can be resumed as follows: (1) human beings act towards things on the basis of the meaning things have for them; (2) the meaning of such things is derived from, or arises out, the social interaction one has with one's fellows, and (3) these meanings are handled in, and modified through, an interpretative process used by the individual in dealing with the things encountered. Individual and joint actions are framed by historical and cultural meanings, but there is always any room for creativity and improvisation. The social process in group life creates and upholds the rules, not the rules create and uphold life (Adams & Sydie, 2002).

Within this general framework, different positions about the emerging relationship between the individual and the society (environment) can be held. Mead (1934) has proposed the "theory" of self-consciousness and self-mediation defining consciousness as becoming other to oneself. By becoming other to self, self is able to mediate self's own reflexes. For Mead, self is already other from the perspective of other within social interaction and therefore, if self could take the attitude of the other (toward self), then we would have an explanation of consciousness (Gillespie, 2005). The aphorism that sums up the position of Mead said: "In the beginning there is social process," i.e., the social process shapes society, self, and social interaction, and each feeds back on the others (Stryker, 2008).

Burke (2004) proposed the *identity control theory* to link identities with the social structures. *Identities* are the sets of meanings people hold for themselves that define "what it means" to be who they are as persons, as role occupants, and as group members. These meanings constitute what is called an *identity standard* used as a reference with which persons compare their perceptions of self-relevant meanings in the interactive situation. When the perceptions match the meanings in a standard way, people are doing "just fine." Their identities are being confirmed or verified and they will continue to act as they are; no changes are required. When, however, it happens a *disturbance* changing the interactive situation and consequently the perceived situational meanings, standards do not match longer and people will act to remove the disturbance and restore the match in meanings between perceptions and standard. This way is named the *self-verification process*, heart of the identity control theory. No one can know in advance exactly what will be the behavior resulting from the relationship between perception and identity standard. Given that the disturbances cannot be expected, the behaviors counteracting them cannot be known in advance.

Yet another perspective. The *identity accumulation hypothesis* brought by Thoist (1983) is a *model* whereby many more identities an actor has, the less stress is expected to exhibit. The individual acquires a view of him/herself as an objective and meaningful social entity by taking the role of specific and then of generalized others. In this way, he/she perceives to be placed by others into recognized and meaningful social categories or social positions, i.e., those locations within systems of relationships (Merton, 1957). Sets of behavioral expectations – that is roles – are consequential to these positions. In this model, the self is conceptualized as a set of discrete identities – self-definitions in terms of occupied social positions and identities are claimed and sustained in reciprocal role relationships governed by behavioral expectations. The rights and duties of each interacting actor are normatively prescribed: when one knows who one is (in a social sense) then he/she knows how to behave. Role requirements give purpose, meaning, direction, and guidance to one's life. Greater is the number of held identities, stronger is his/her sense of meaningful and guided existence.

Structural Symbolic Interactionism Paradigm

Stryker (2008) offers a unique perspective to consider interactionism and qualifies it as "structural." Although society emerges from social processes, the organized society exists before the appearance of all new members. So the society shapes self and the social interaction. The framework of these processes can only be a sense of social structures as patterned interactions and relationships characterized by durability, resistance to change, and ability to reproduce. The social differentiation is inherent in the process as countering homogenization of interactional experience and the structures within the society. It is so justified the specification "structural": the society appears as organized systems of interactions and role relationships, complex mosaics of differentiated groups, communities, and institutions, cross-cut by a variety of demarcations based on class, age, gender, ethnicity, religion, etc. The different parts can appear interdependent and independent, isolated and insulated, cooperative and conflicting, highly resistant to change and inclined to change. The social life is not qualified by the salient features of the society as a whole, but by the relatively small networks of role relationships, perhaps most local. This image of society carries with it several implications. The greatest significance should be placed on the impact of social structures on social interaction and thinking structures as social boundaries makes it more likely people with differing backgrounds and resources to enable specific social relations. At last, the social structures at different levels and of various kinds operate as facilitators of and constraints on input and output from networks of interpersonal relationships.

Outlining his Identity Theory, Stryker (1968) specified his premise that the society impacts the self that influences the social behavior. Social behavior is specified by taking *role choice* – the opting by persons to meet expectations of one role rather than another – as that which the theory seeks to explain. The concept of identity salience is a specification of the self elaborated from a multifaceted view of the self. People are seen as having multiple identities [self is understood to include affective and conative as well as cognitive aspects (Stryker, 1968)] corresponding to organized systems of role relationships in which they participate. These identities require that persons are considered social objects with a position designated by others but internalized. These may be qualified self-cognitions tied to roles and, through roles, to positions in organized social relationships. In summary, they are cognitive schemas (Markus, 1977) affecting all cognitive and perceptual processes (Stryker & Serpe, 1994). They are not specific situations as people applied them in the several situations they experience affecting their conduct. In this way, identities are arranged in a hierarchy of salience that is formed gradually to the extent that a certain identity is invoked in many situations and the choice of role is assumed to be a consequence of identity salience.

The referent of commitment on a role is found in interaction networks and social relationships. Commitment, as ties to social networks, is a "small" unit of structure and a specification of that "society." One is committed to being that kind of person to the degree that his/her relationships to a set of others depend on being a particular kind of person and playing out particular roles. Commitment is hypothesized to be the immediate source of salience attached to identities (Stryker, 1968, 1980).

The results of many studies have provided strong support to models derived from this theoretical development. Moving from this perspective, Serpe (1987) has studied the relationship between commitment and identity salience with the passage of time and he established a mutual influence, even if the commitment appears to have greater influence on the salience than that of the salience on the commitment. Serpe and Strker (1987) traced the effect of primary identities of students moving from home to a new environment and noted that students tried to create in new settings any relationships with others allowing them to maintain those identities. More recently, Stryker, Serpe, and Hunt (2005) show racial/ethnic variations in commitment processes and evidence the role of large, essentially stratification features of the society in channeling persons through intermediate and/or direct levels of social structure to interpersonal networks. Finally, Lee (2005) has provided the evidence of gender similarities and differences in processes of identity change.

Radical Interactionism Paradigm

Athens (2009) reconsiders the theoretical position of Robert Park and proposes a new paradigm based on the *Social Domination Theory*. Park (1934) saw in the *dominance* the basic principle underlying the organization of human group life, from its first emergence on the earth to the present time. The dominance is the basic organizing principle beyond the fact if people and groups are the same or different race (Turner, 1967). Unlike both Blumer and Mead, Park does not place any temporal or racial restriction on the purpose of the operation of dominance as the base principle organizing the human group life. His interactionism is also at most radical founding the social existence.

In the developments proposed by Athens (2010), the proving grounds of human subordination are social acts because the subordination emerges and exists only in acts combining efforts of, at least, two people. When an individual or group participating in the social act steers the development direction according to their particular preferences, a situation of dominance results. If an individual or group sways the path of a social act's construction to their liking, then it could be said that they have dominated it. More specifically, they must steer not only the social object formed of the prospective social act and the attendant plan of action developed for the goal's accomplishment, but also the action plan of the subsequent execution. The domination thus appears in the design of a social act and in the subsequent supervision of its construction. In terms of social action, domination displays four defining characteristics: (1) certain individuals or groups exhibit subservient attitudes while others exhibit superior attitudes; (2) super-ordinate roles are always differentiated from subordinate ones; (3) individuals or groups seeking to perform the super-ordinate roles exhibit superior attitudes, whereas people resigning themselves to perform subordinate roles exhibit subservient attitudes; and (4) the individuals or groups performing subordinate and super-ordinate roles consciously assume, however accurately, each other's superior and subservient attitudes, and their conscious assumption of respective attitudes significantly affects the construction of the social act in which they are joint participants (Athens, 2002).

Domination is usually signaled *covertly* rather than *overtly* which can make difficult at least the immediate detection of its operation. Seldom, people make usually direct reference during the construction of a social act. On the contrary, they prefer to leave unspoken such matters. The domination operation becomes clearly visible when a conflict erupts on who should play the super-ordinate role in the construction of a social act.

Identity as a Life Story Paradigm

McAdams (1996) proposed a specific framework to take into account person and cultural context. Starting from Bruner (1986) – that argued that narrative is the natural mode through which human beings make sense of lives in the time – and Sarbin (1996) – narrative may be a new "root metaphor" for psychology as a whole – the author conceptualize the identity as a *life story*. Adult people living in our current societies construct their identities internalizing and evolving life narratives which include settings, scenes, characters, plots, and themes. Their lives may be viewed as narrated texts, known and 'read' as stories, framed through discourses, and told in culture a perspective that make a stable sense of identity. Adults argue to seek and maintain environments that fit their stories promoting behavioral consistency and reducing opportunities for change. The continuity is achieved by an individual selection and interpretation of situations even if the individual is free to choice which story to maintain or "write."

Summary and Conclusion

Interactionism is a plethora of studies and research that are generally discussed and classified according to criteria of personality and social psychology. This review has proposed a specific framework based on two criteria: (a) the emergent interactionism proposed by Sperry (1969) who does the distinction between produced and spontaneous; (b) L'Abate (2009) distinction among paradigm, theory, and model concepts.

After this part, some interactional perspectives are browsed: the personologism, the situationism, the lawful interactionism, and the relational psychodynamic approach have been included in the produced paradigms since the predictability of the "novel" behavior is emerging from individual–situation interaction. The mind–society approach, the symbolic interactionism, the structural symbolic interactionism, and the radical interactionism have been considered part of spontaneous paradigms: the emergent behavior from individual and environment is not theoretically predictable because of individual degree choices or disturbances. Table 20.1 shows the summary of the framework and the studies reviewed.

This framework is not meant to be an alternative to current proposals (e.g., Raynolds et al., 2010), but is presented as an aid to scholars and researchers to navigate between the multiple conceptual reflections and related research studies.

Table 20.1 Main interactionist research studies according to produced and spontaneous emergent interactionism

		L'Abate's distinction (2009) of		
		Paradigm	Theory	Model
Sperry's	Produced	Personologism	Trait as a density distribution of states conception (Fleeson, 2001, 2009)	
Emergent		Lawful interactionism	Cognitive-affective system theory (Mischel & Shoda, 1994)	Self-encoding and regulation of affect and social perception model (Mendoza-Denton et al., 2001)
Interactionism (1969)		Situationism	Self-efficacy mechanism (Bandura, 1977, 1982)	
		Relational psychodynamic	Relational competence theory (L'Abate, 1976, 2005, 2009)	Likeness continuum (L'Abate, 1976; Cusinato & Colesso, 2008) and others 15 models
				State-trait anxiety model; Endler et al. (1989)
				Empirical model of interactional developing of personality across the life span (Caspi & Roberts, 2001)
	Spontaneous	Mind–society approach	The social identity theory (Tajfel & Turner, 1979)	
		Symbolic interactionism	The identity control theory (Burke, 2004)	The identity accumulation hypothesis (Thoist, 1983)
		Structural symbolic interactionism	Identity theory (Stryker, 1968)	Psychological centrality and role choice (Stryker & Serpe, 1994)
		Radical interactionism	Social domination theory (Athens, 2010)	
		Identity as a life story (McAdams, 1996)		

References

Adams, B. N., & Sydie, R. A. (2002). *Classical sociological theory*. Thousand Oaks: Sage.

Alker, H. A. (1972). Is personality situationally specific or intrapsychically consistent? *Journal of Personality, 40*, 1–16.

Allport, G. W. (1937). *Personality: A psychological interpretation*. New York: Henry Holt & Co.

Andersen, S. M., & Chen, S. (2002). The relational self: An interpersonal social-cognitive theory. *Psychological Review, 109*, 619–645.

Athens, L. (2002). Domination: The Blind Spot in Mead's Analysis of the Social Act. *Journal of Classical Sociology, 2*, 25–42.

Athens, L. (2009). The roots of radical interactionism. *Journal for the Theory of Social Behavior, 39*, 387–414.

Athens, L. (2010). Human subordination from a radical interactionist's perspective. *Journal for the Theory of Social Behaviour, 40*(3), 439–468.

Bandura, A. (1977). Self-efficacy: Toward a unifying theory of behavioral change. *Psychological Review, 84*, 191–215.

Bandura, A. (1982). Self-efficacy mechanism in human agency. *American Psychologist, 37*(2), 122–147.

Barbieri, G. L. (2009). *Psicologia dinamica tra teoria e metodo [Dynamic psychology. Theory and method]*. Milano: Raffaello Cortina.

Bizumic, B., Reynolds, K. J., Turner, J. C., Bromhead, D., & Subaši , E. (2009). The role of the group in individual functioning: School identification and the psychological well-being of staff and students. *Applied Psychology: Annual International Review, 58*, 171–192.

Blumer, H. (1969). *Symbolic interactionism: Perspective and method*. Englewood Cliffs: Prentice-Hall.

Bowers, K. S. (1973). Situationism in psychology: An analysis and critique. *Psychological Review, 80*, 307–336.

Brooks, M. L., Buhrmester, M., & Swann, W. B. (2010). Discussing on "Interactionism in personality and social psychology: An integrated approach to understanding the mind and the behavior" by Reynolds, Turner, Branscombe, Mavor, Bizumic, and Subasic. *European Journal of Psychology, 24*, 483–500.

Bruner, J. S. (1986). *Actual mind, possible worlds*. Cambridge: Harvard University Press.

Burke, P. J. (2004). Identities and social structure: The 2003 Cooley-Mead Award Address. *Social Psychology Quarterly, 67*(1), 5–15.

Capitelli, M., Guerra, C., L'Abate, L., Longo, S., & Rumbaugh, D. M. (2009). Science or mind? Reductionism proper, optimal reductionism, emergent interactionism, and synthetic integration. In L. L'Abate, P. De Giacomo, M. Capitelli, & S. Longo (Eds.), *Science, mind and creativity. The Bari symposium* (pp. 1–22). New York: Nova Science Publishers.

Caspi, A., & Roberts, B. W. (2001). Personality development across the life course: The argument for change and continuity. *Psychological Inquiry, 12*(2), 49–66.

Cusinato, M., & Colesso, W. (2008). Validation of the continuum of likeness in intimate relationships. In L. L'Abate (Ed.), *Toward a science of clinical psychology: Laboratory evaluations and interventions* (pp. 337–352). New York: Nova Science Publisher.

Dewey, J., & Bentley, A. F. (1949). *Knowing and the known*. Boston: Beacon.

Ekehammar, B. (1974). Interactionism in personality from a historical perspective. *Psychological Bulletin, 81*, 1026–1048.

Endler, N. S., Edwards, J. M., Vitelli, R., & Parker, J. D. A. (1989). Assessment of state and trait anxiety: Endler multidimensional anxiety scales. *Anxiety Research: An International Journal, 2*, 1–14.

Fleeson, W. (2001). Toward a structure- and process- integrated view of personality: Traits as density distributions of states. *Journal of Personality and Social Psychology, 80*, 1011–1027.

Fleeson, W. (2009). On the difference between experience-sampling selfreports and other self-reports. *European Journal of Personality, 23*, 410–412.

Fleeson, W., & Gallagher, P. (2009). The implications of Big Five standing for the distribution of trait manifestation in behavior: Fifteen experience-sampling studies and a meta-analysis. *Journal of Personality and Social Psychology, 97*(6), 1097–1114.
Fleeson, W., & Jolley, S. (2006). A proposed theory of the adult development of intraindividual variability in trait-manifesting behavior. In D. K. Mroczek & T. D. Little (Eds.), *Handbook of personality development* (pp. 41–59). Mahwah: Erlbaum.
Fromm, E. (1941). *Escape from freedom*. New York: Avon.
Fuente, A. (1993). Perry's emergent interactionism as a new explanation for dualism. *Journal of General Psychology, 120*(1), 65–72.
Gillespie, A. (2005). G. H. Mead: Theorist of the social act. *Journal for the Theory of Social Behaviour, 35*, 19–39.
Horney, K. (1939). *New ways in psychoanalysis*. New York: Norton.
Kantor, J. R. (1970). An analysis of the experimental analysis behavior. *Journal of the Experimental Analysis of behavior, 13*, 101–108.
Kantor, J. R., & Smith, N. (1975). *The science of psychology: An interbehavioral survey*. Chicago: Prinicipia.
Kenrick, D. T., & Funder, D. C. (1991). The person-situation debate: Do personality traits really exist? In V. J. Derlega, B. A. Winstead, & W. H. Jones (Eds.), *Personality: Contemporary theory and research* (pp. 149–173). Chicago: Nelson-Hall.
Koffka, K. (1935). *Principles of Gestalt psychology*. London: Routledge.
L'Abate, L. (1976). *Understanding and helping the individual in the family*. New York: Grune & Stratton.
L'Abate, L. (2005). *Personality in intimate relationships: Socialization and psycho-pathology*. New York: Springer.
L'Abate, L. (2009). Paradigms, theories, and models: Two hierarchical frameworks. In L. L'Abate, P. De Giacomo, M. Capitelli, & S. Longo (Eds.), *Science, mind and creativity. The Bari symposium* (pp. 107–153). New York: Nova Science publishers.
L'Abate, L. (2011). Essentialism. In L. L'Abate (Ed.), *Paradigms in theory construction*. New York: Springer.
L'Abate, L., Cusinato, M., Maino, E., Colesso, W., & Scilletta, C. (2010). *Relational competence theory: Research and mental health applications*. New York: Springer Science.
Lee, J. D. (2005). Do girls change more than boys? Gender differences and similarities in the impact of new relationships on identities and behavior. *Self Identity, 4*, 131–147.
Lewin, K. (1936). *A dynamic theory of personality*. New York: McGraw-Hill.
Lewin, K. (1952). *Field theory in social science: Selected theoretical papers*. New York: Harper & Row.
Magnussen, D., & Törestad, B. (1993). A holistic view of personality: A model revised. *Annual Review of Psychology, 44*, 427–452.
Markus, H. R. (1977). Self-schemata and processing information about self. *Journal of Personality and Social Psychology, 35*, 63–78.
Maxwell, J. C. (1865). A dynamical theory of the electromagnetic field. *Philosophical Transactions of the Royal Society of London, 155*, 459–512.
McAdams, D. P. (1996). Narrating the self in adulthood. In J. Birren, J. E. Ruth, J. J. F. Schroots, & T. Svenson (Eds.), *Aging in biography: Exploration in adulthood development* (pp. 131–148). New York: Springer.
Mead, G. H. (1934). *Mind self and society from the standpoint of a social behaviorist*. Chicago: University of Chicago.
Meltzer, B. N., & Manis, J. G. (1992). Emergence and human conduct. *The Journal of Psychology, 126*(4), 333–342.
Mendoza-Denton, R., Ayduk, O., Mischel, W., Shoda, Y., & Testa, A. (2001). Person x situation interactionism in self-encoding (I am ...when...): Implications for affect regulation and social information processing. *Journal of Personality and Social Psychology, 80*(4), 533–544.
Merton, R. K. (1957). *Social theory and social structure. Revised Edition*. New York: The Free Press.
Mischel, W. (1968). *Personality and assessment*. New York: Wiley.

Mischel, W,, & Shoda, Y. (1994). Personality psychology has two goals: Must it be two fields? *Psychology Inquires, 5*, 156–158.
Mischel, W., & Shoda, Y. (1995). A cognitive-affective system theory of personality: Reconceptualization situations, dispositions, dynamics, and invariance in personality structure. *Psychological Review, 102*(2), 246–268.
Mitch, F., & Linda, J. H. (2011). Interbehaviorism. In L. L'Abate (Ed.), *Paradigms in theory construction*. New York: Springer.
Murray, H. (1938). *Explorations in personality*. New York: Oxford University Press.
Oakes, P. J., Haslam, S. A., & Turner, J. C. (1994). *Stereotyping and social reality*. Oxford: Blackwell.
Park, R. (1934). Dominance. In E. Hughes (Ed.), *Human communities* (pp. 159–164). New York: The Free Press.
Raynolds, K. J., Turner, J. C., Branscombe, N. R., Mavor, K. I., Bizumic, B., & Subasic, E. (2010). Interactionism in personality and social psychology: An integrated approach to understanding the mind and the behavior. *European Journal of Personality, 24*, 458–482.
Sarbin, T. (Ed.). (1996). *Narrative psychology. The storied nature of human conduct*. New York: Praeger.
Serpe, R. T. (1987). Stability and change in self: A symbolic interactionist explanation. *Social Psychology Quarterly, 50*, 44–55.
Serpe, R. T., & Strker, S. (1987). The construction of self and the reconstruction of social relationships. In E. J. Lawler & B. Markovsky (Eds.), *Advances in group processes: Theory and research* (Vol. 4, pp. 41–82). Greenwich: JAI Press.
Sperry, R. W. (1969). A modified concept of consciousness. *Psychological Review, 76*(6), 532–536.
Sperry, R. W. (1988). A paradigm made strong. *Neuropsychologia, 36*(10), 1063–1068.
Spielberger, C. D. (1983). *Manual for the state-trait anxiety inventory (Form V)*. Palo Alto: Consulting Psychologists Press.
Steele, C. M., Spencer, S., & Aronson, J. (2002). Contending with group image: The psychology of stereotype and social identity threat. In M. Zanna (Ed.), *Advances in experimental social psychology* (Vol. 34, pp. 379–440). San Diego: Academic.
Stolorow, R. D., & Atwood, G. E. (1992). *Context of being: The intersubjective foundations of psychological life*. Hillsdale: Analytic Press.
Stryker, S. (1980). *Symbolic Interactionism: A Social Structural Version*. Menlo Park, CA: Benjamin/Cummings.
Stryker, S. (1968). Identity salience and role performance: The relevance of symbolic interaction theory for family research. *Journal of Marriage and Family, 30*, 558–564.
Stryker, S. (2008). From Mead to a structural symbolic interactionism and beyond. *The Annual Review of Sociology, 34*, 15–31.
Stryker, S., & Serpe, R. T. (1994). Identity salience and psychological centrality: Equivalent, overlapping or complementary concepts? *Social Psychology Quarterly, 57*, 16–34.
Stryker, S., Serpe, R. T., & Hunt, M. O. (2005). Making good on a promise: The impact of larger social structures on commitments. In S. I. Thye & E. J. Lawler (Eds.), *Social identification in groups* (pp. 93–124). Oxford: Elsevier.
Sullivan, H. S. (1953). *The interpersonal theory of psychiatry*. New York: Norton.
Tajfel, H. (1972). Experiments in a vacuum. In J. Israel & H. Tajfel (Eds.), *The context of social psychology* (pp. 69–119). London: Academic.
Tajfel, H., & Turner, J. C. (1979). An integrative theory of intergroup conflict. In W. G. Austin & S. Worchel (Eds.), *The social psychology of intergroup relations*. Monterey: Brooks-Cole.
Thoist, P. A. (1983). Multiple identities and psychological well-being: A reformulation and test of the social isolation hypothesis. *American Sociological Review, 48*(2), 174–187.
Turner, R. (Ed.). (1967). *Robert E. Park on social control and collective behavior: Selected papers*. Chicago: University of Chicago Press.
Turner, J. C. (1985). Social categorization and the self-concept: A social cognitive theory of group behaviour. In E. J. Lawler (Ed.), *Advances in group processes* (Vol. 2, pp. 77–121). Greenwich: JAI Press.

Turner, J. C. (1987). Rediscovering the social group. In J. C. Turner, M. A. Hogg, P. J. Oakes, S. D. Reicher, & M. S. Wetherell (Eds.), *Rediscovering the social group: A self-categorization theory* (pp. 19–41). Oxford: Blackwell.

Wachtel, P. L. (1973). Psychodynamics, behavior therapy, and the implacable experimenter: An inquiry into the consistency of personality. *Journal of Abnormal Psychology, 82*(2), 324–334.

Wertheimer, M. (1938). The syllogism and productive thinking. In W. D. Ellis (Ed.), *A source book of Gestalt psychology* (pp. 274–282). London: Paul Kegan.

Williams, M. T., Turkheimer, E., Magee, E., & Guterbock, T. (2008). The effects of race and racial priming on self-report of contamination anxiety. *Personality and Individual Differences, 44*, 746–757.

Chapter 21
The Emergence of Reason, Intelligence, and Language in Humans and Animals

William A. Hillix, Duane M. Rumbaugh, and E. Sue Savage-Rumbaugh

Emergence as an overarching paradigm requires an introductory analysis about the importance and types of causality. As L'Abate (2005) commented, there are at least three different types of causality or important principles necessary for theory construction: equifinality, when many courses lead to one single outcome, equipotentiality, one single cause can lead to many different outcomes, and multicausality, when many causes lead to many different outcomes.

The Importance of Causality in Theory Construction

Questions about causality are so striking that they can make us overlook the almost magical ability of humans and animals to detect and manipulate causal relationships in their everyday adaptation to the environment. Philosophers are just as mystified about causality as the Federal Aviation Authority (FAA) and forensic specialists, although the nature of their puzzlement is different. David Hume's analysis of the meaning of causality set the stage for all subsequent discussions of causal relationships. He freed the concept from an absolutist perspective by examining the operations that led to a belief that one event was causally related to another. The effect had to follow the cause invariably, and the effect had to be contiguous with the cause.

W.A. Hillix (✉)
Emeritus Professor, San Diego State University, 11028 Explorer Road, La Mesa, CA 91941, USA
e-mail: ahillix@cox.net

D.M. Rumbaugh
Emeritus Professor, Georgia State University, 402 Lincoln Avenue, Highland Park, NJ 08904, USA

E.S. Savage-Rumbaugh
Great Ape Trust of Iowa, 3131 Park Fleur, Unit 804, Des Moines, IA 50321, USA

L. L'Abate (ed.), *Paradigms in Theory Construction*, DOI 10.1007/978-1-4614-0914-4_21,

John Stuart Mill (as cited in White, 1995) contributed to the analysis of causality by suggesting three methods that could be used to infer causal relationships. First is the method of agreement. If all instances of a phenomenon share only a single circumstance, that circumstance must be the cause (or effect) of the phenomenon. Second is the method of difference; if the occurrence and nonoccurrence of an event differ in only a single circumstance, that circumstance must be the cause, or effect, or part thereof, of the event. The third method is simply a combination of the first two. Mill's suggestions have been imported into experimental psychology from philosophy, and still provide a logical basis for psychological experimentation.

Although these philosophical analyses were a beginning, they left many questions unanswered. One might begin by asking just how contiguous the cause and effect had to be. Second, can the requirement that the effect invariably follow the cause be imposed in practice? There might be some unobserved difference in the causal circumstances, or a failure to observe the effect, that would lead to an apparent failure of the cause to produce the effect.

These issues lead immediately to other questions about the dimensions of causal relationships. We already have two: the temporal relationship between the cause and the effect, and the probability that the effect will follow the cause. These are the properties that lead to the formation of the idea of cause, which, for Hume, was a mental construct rather than something that existed in the external world. A third property is the complexity of the causal situation, i.e., how does one define the circumstances that stand at the center of Mill's analysis? At one extreme, the flash of a light in a classical conditioning experiment, followed by the delivery of food, should, according to Hume's criteria, lead to the idea of cause in the mind of the subject. In this case, the cause is a simple stimulus. In the case of human courtship, a multiplicity of favorable situations might lead eventually to a successful engagement, while a change in one circumstance could lead to a breakup. Such cause–effect relationships are extremely complex, with many different temporal delays between parts of the situation and the eventual effect.

A favorite simple example of a causal relationship is that between billiard balls: the cause is one ball striking another, and the effect is the resulting change in speed and direction of both balls. Michotte's (1963) classic work on the perception of causality is a modification of this simple instance of causality. Michotte studied the "launching effect" that subjects perceived under various circumstances when two objects interacted. The first object was seen to cause the movement of the second when the second moved off immediately after being struck by the first, as one would expect. But Michotte studied many values of closeness of approach, and many delays of the movement of the second object after the first approached, thus making a parametric study of the circumstances under which Hume's mental construct of causality was formed and of the circumstances under which a causal relationship was or was not perceived.

Important as Michotte's work is, it elucidates only a relatively simple type of causal perception, and it provides little information about the development of causality.

A central figure in this respect, as in other areas of development, was Jean Piaget (1930, 1974). Piaget agreed with many previous and subsequent students of causality that the conception of causality arose from voluntary action. That is, the child is initially incapable of conceiving of causal relationships between events. Gradually the child discovers that he or she is capable of voluntary movement; the power of movement is discovered and eventually generalized from the internal to the external world. Thus, the concept of action on the world (initially, on one's own body) is inextricably intertwined with the concept of causality. Piaget and others believed that they had discovered the origins of Hume's mental construct in the experience of the child. Without that experience of self-movement, the belief in causal powers in the external world would not, according to these thinkers, arise. Although these suggestions are attractive, they probably overlook some nativistic influences. Recent experimentation with infants indicates that they are surprised when unsupported objects don't fall, or supported objects do. Thus, human infants appear to be preprogrammed to expect at least some causal relationships between events.

Piaget thought that children's understanding of causality passed through various stages (cf. White, 1995), just as other aspects of their development progressed through stages. In stages 1 and 2, the child learns that he or she has efficacy, that certain results (like nursing) can be brought about, but the child has no understanding of how this is achieved. In stage 3, the child learns to distinguish cause and effect, for example by visual observation of the position of the hands after experiencing the attempt to move them. Later the child comes to believe that she has lasting powers of her own, and finally to infer causes after observing effects. It would seem that this last stage of causal understanding is very helpful in linguistic interactions, when it is necessary to connect actions of listeners with the language generated by speakers. The behavior of bonobos and parrots in language-rich environments (see below) indicates that these animals have probably reached Piaget's final stage, stage 6, in which causes can be inferred from the observation of effects.

In philosophical discussions, there is a tendency to ignore the role of the pool player in the billiard ball example, but in reality all causal events are just selected parts of a continuous chain of events in the world, some of which involve actions by organisms. There is a very important, indeed critical, distinction between causes that are dependent upon the behavior of organisms and causes that are independent of them. We can use a cue stick to move a billiard ball, but volcanoes erupt and earthquakes cause tidal waves quite independently of the behavior of organisms. Psychological interest tends to be concentrated on the former, since we cannot prevent the latter type of effect. We can, however, move out of the way of eruptions and tidal waves if we are forewarned; these are just dramatic instances of the general problem of avoiding untoward consequences. Everyday examples are predator avoidance, avoiding extreme heat and cold, and keeping out of the way of cars. The adaptation of organisms is completely dependent upon actions that are consistent with the causal structure of the environment. In another view of causal relationships, the behavior of an organism may be an effect, rather than a cause. In the case of classical conditioning, salivation is the effect of stimulation, initially only to the unconditioned stimulus, but later to the conditioned stimulus as well. It is as if the organism is saying "I know that the light causes the delivery of food."

Nature and human artifice both present many cases of complex causal relationships, from airplanes and automobiles through computers to ecological systems and zebra metabolism. The ability to detect and take advantage of causal relationships in the environment may be the most significant single index of adaptive intelligence across animal species. However, humans have a huge quantitative advantage in their ability to intervene in and create new causal relationships.

The philosopher, Robert van Gulick, said the following about the function of causal laws (1993, p. 246): What then is the function of causal laws in our cognitive economy? We are biological organisms with needs, goals, and wants. Our ability to succeed in their satisfaction is enhanced by possessing an accurate representation of the environment with which we must interact to achieve those results. In addition to possessing means by which we can pick up information about our present situation and store information picked up in the past, we must be able to make reliable predictions about the future and form plans of action that will enable us to determine or at least influence how the future will develop.

Animals share with humans the need to make causal inferences in an uncertain universe.

Egon Brunswik (1943, 1952) emphasized the uncertain, probabilistic nature of perceived causality in his formulation of "probabilistic functionalism." Brunswik, in contrast to Hume, recognized that there are no "sure things" in the real world as we usually observe it, and organisms are reduced to making best guesses about the validity of cues. These best guesses are used to select behaviors that are likely to be adaptive. The best-adapted organisms are those that are most successful in selecting the responses with the greatest likelihood of success in their ecological situation.

B. F. Skinner (1948) brought a related perspective to the same problem when he called many behaviors "superstitious." To call a behavior superstitious implies that the organism has attributed causal power to an action that had only a fortuitous relationship to an effect. Thus, the idea of causality may arise in such circumstances, so all attributions of causality are to some extent superstitious. The placebo effect and belief in psychic predictions are both based on "superstitious" attributions.

It is often adaptive to persist in responding in the face of initial failure, because circumstances may change from one attempt at a causal action to the next. For example, the chimpanzee who finds a tree void of ripe fruit on one occasion may return to find fruit there later, so the effects of the same approach behavior are quite different on the two occasions. The rat on a fixed-interval schedule of reinforcement will find, after the interval has timed out, that a response will now cause a pellet to be delivered, when earlier it did not. The general point is that the environment is not fixed; new possibilities emerge with changes in values of variables and their interactions. Organisms that are flexible enough to respond to the changes that occur over either long or short time intervals can behave more adaptively (see the discussion of transfer effects below). In almost all cases, there are hidden connections that connect "causes" to "effects." In the case of the rat the whole operant apparatus, with its internal clock that completes a circuit when the interval "times out," connects the rat's bar press to the delivery of the pellet. Because organisms generally are unaware of what is happening in these chains, it may pay to obey the aphorism – "If at first you don't succeed, try, try again." The organism generally must engage in a chain of behaviors if it is to bring about an effect. Operant psychologists have sometimes established long chains of behaviors that are only rewarded after the last step; for example, an animal might be taught to run through a tube, climb a ladder, jump across a chasm, slide down an incline, and finally press a bar in order to obtain a pellet.

Suppes' (1984) definition of causality explicitly recognizes the contingent nature of causality by defining it probabilistically. An event C is defined as a cause of an event E provided that C occurs earlier than E and that *P*(E/C) is greater than *P*(E). According to Suppes' definition, flipping a light switch is a cause of a light going on, even though sometimes the electricity is off, or the switch has malfunctioned, or someone has cut the wires.

Types of Causal Relationships

It might be useful to divide causal relationships into four classes. First are the purely naturalistic relationships that occur in the physical (including biological) environment. Rocks fall and crush plants and animals, the moon and sun pull earth's water and cause tides, and so on. Second are relationships between the behavior of organisms and their physical consequences: the ape eats fruit and feels fuller, the carpenter misses the nail and smashes a finger, the mechanic fixes the car. Third are relationships between events in the environment and the behavior of organisms: seeing the red fruit causes the ape to climb the tree, the appearance of the light causes Pavlov's conditioned dogs to salivate, the mechanic sees the results of a compression test and says that the car needs a motor overhaul. Fourth are relationships between the behaviors of two organisms; the proper interaction between two bonobos leads to copulation. Note that this relationship could be subsumed, from the point of view of the individual organism, under the second or third category, but it is useful to separate causal situations involving multiorganism interactions from those involving only a single organism. The fourth class is of particular interest because it is the class that is involved in communicative relationships, of which research on the acquisition of the rudiments of human language by animals is a particular example. The later parts of this discussion, as the reader may have guessed, concentrate on this fourth class of causal relationships.

Ideal and Real Causality Detectors

Let us consider an ideal causality detector and its attributes. First, it should be able to sort out from its morass of experiences those events that are reliably concurrent and those that are reliably sequential. It should be able to detect patterns in the occurrences of events. Second, the cause–effect association should be able to span long periods of time between cause and effect, in either time or number of intervening steps. Going beyond detection, an ideal organism should be able to intervene in causal chains and through its behavior bring about desired goals. The organism should be able to sort out coincidental associations from those that are causal.

Animals as Causal Analysts: Tolerating Delays in Response

Real causality detectors, or course, are far from ideal. One indication of an organism's ability to detect causal relationships is the ability to discern and accurately remember events that occur either simultaneously (e.g., what goes with what) or sequentially (e.g., what precedes another event). Delayed response to a stimulus (category 3 of causal relationships) is one indicant of this kind of memory. The longer an animal can wait after observing the placement of food, and still go to the location of the food, the better its memory for the association between some stimulus and the presence of food. Hume might be pleased to learn that most animals in formal test conditions cannot remember this association for 30 s, although animals with larger brains generally do better than animals with smaller brains.

Dramatic exceptions to this memory limitation in animals do exist, however. One is the now-famous ability of animals to remember an association between the smell and taste of a new food and a subsequent illness (Garcia, Ervin, & Koelling, 1966; Revusky & Garcia, 1970), an example of how biological preparedness can facilitate the formation of certain kinds of associations that are vital to survival. There is now no question that animals can form some associations over far longer periods than was thought previous to Garcia's work. And, consistent with how critical an event is to their survival in nature, many animals are prepared to form some kinds of associations more quickly and over longer time delays than they are other types of associations (Garcia, McGowan, Ervin, & Koelling, 1968).

Expressed in causal terms, animals can form some kinds of causal connections much more easily than other kinds; it is easy for them to determine that food with a particular taste causes nausea, but difficult for them to determine that a bright light or a sound is associated with nausea. They are also likely to learn about relationships between places and between conspecifics, as well as about signals that announce the imminent appearance of a predator. Evolution has made them "ready to perceive and to learn" about things that provide a real threat to their survival and reproductive fitness.

Great apes can remember some one-time associations for days and, indeed, several weeks. The language-competent chimpanzee, Panzee (Menzel, 1999; also, Hillix & Rumbaugh, 2004; Rumbaugh & Washburn, 2003) has a remarkable ability to remember and communicate the location of a hidden food to a "blind" observer after extended periods of time. After she watches one experimenter hide a food outside her cage, she later will solicit the attention of another person, a friend or caregiver, who does not know the location of the food. Through use of her word-lexigrams and gestures, Panzee can, and does, communicate the name of the food that is hidden outdoors to that person and show him or her where to go to get it. Her highly reliable pointing and gaze to the food site make it very easy for the "blind" experimenter to find the food that was hidden earlier. She developed this ability spontaneously by combining skills that she had learned earlier in other programs. She can even use televised portrayals of the food being hidden in the forest as the basis for directing an individual to get the food for her.

Simple and Emergent Abilities

Psychology has made several mistakes during its historical development. One has been to insist that behavior must be made to appear simple! Pavlov and Skinner shared the belief that the unconditional stimulus (for Pavlov) and the reinforcer (for Skinner) established specific responses to either specific stimuli or specific contexts. The immediate and pragmatic effects of reinforcement are so impressive and powerful that it has led to the mistaken belief that learning always involves only behavioral change. To the contrary, we believe that learning is a state in the central nervous system that can enable organisms to generate novel and creative ways to take advantage of causal relationships. Although the classical/respondent conditioning of Pavlov and the operant conditioning of Skinner are good science, they are but the beginnings, not the ends, of what is needed to understand what is learned and how it influences behavior – in brief, how and why behavior changes and how and why creative means of solving problems emerge (Rumbaugh, Washburn, & Hillix, 1996).

There is a clear sequence of abilities to detect and affect causal relationships, as in the classical conditioning experiment. First, the subject can remember the connection between a bell and food by making a new response to the conditioned stimulus. But even in classical conditioning there are several strong indications of cognitive operations. Rescorla (1988), for example, demonstrated that, once a conditional stimulus becomes effective in eliciting the conditioned response, the addition/coupling of another stimulus to the established conditional stimulus rarely accrues any effectiveness as a conditional stimulus. It is suggested that the subject perceives the addition of the second stimulus as irrelevant – hence something that can be ignored, even though it precedes the unconditional stimulus.

Second, many animals affect causal chains by responding to stimuli/signals (i.e., a discriminative stimulus) indicating that a particular chain is available. Third, animals recognize relationships between causal situations by generalizing learned responses to similar situations, thereby greatly increasing the ability to adapt. Fourth, animals transfer responses to situations that differ on more than a single stimulus dimension. Responses may be used in new contexts because of one or more principles that relate them. Their transferring of responses and/or learning can produce adaptation and problem-solving effectiveness that is disproportionately greater than was the learning that was available for transfer! Later we will discuss evidence that larger-brained primates have transfer-of-learning skills that are superior to those of smaller-brained primates.

A fifth ability is the ability to learn or construct through trial and error or insight a chain of responses that lead to a goal. This ability is often combined with an ability to respond to relationships between stimuli – for example, to choose the largest available stimulus, regardless of its absolute size.

Observational Learning and Animal Language Skills

Until recent decades, psychologists were generally willing to accept the hypothesis that humans were able to learn by observation – after all, they had been doing it for thousands of years that we know of, but operant psychology was so dominant in the early days of animal language research that observational learning was almost completely neglected until Bandura (1971) resurrected it. Two related programs of research on animal language drastically changed beliefs about the abilities of animals and the best ways to teach them the rudiments of human language. First, Sue Savage-Rumbaugh and her colleagues (Savage-Rumbaugh, 1986; Savage-Rumbaugh & Lewin, 1994) tried to teach Matata, a female bonobo, to use lexigrams by using traditional operant techniques. They had almost no success. But Matata's adopted son, Kanzi, observed the training situation in the course of his early development. To the experimenters' surprise, Kanzi thereby not only learned all that comprised Matata's tutoring sessions, but also knew how to use lexigrams effectively to make requests and to propose games that were not part of Matata's sessions. Remarkably, he also learned the specific meanings of several dozen spoken words! As he did so, however, his learning and inductions were "silent" in that there was nothing in his overt behaviors that indicated to anyone that he was learning very complex relationships and skills. Only when Matata was separated from him for breeding elsewhere, when he was slightly over 2 years of age, did that fact become clear – and it did so on the very first day of work with him.

At nearly the same time, Pepperberg (1990a, 1990b) was training the African gray parrot, Alex, using what she called the "model-rival" approach (Pepperberg, 1985). That involved having Alex observe linguistic interactions between two people, who served alternatively as "models" and "rivals" for Alex, as they talked about and exchanged a set of objects that they hoped would interest Alex. Alex could observe and enter the conversation at will, and he eventually attained the ability to ask for, choose, and describe the color, shape, and materials of the objects with which he was familiar.

Alex appears to have adapted his ability to ask for objects to attain other goals. The senior author visited the Pepperberg laboratory and observed Alex repeatedly asking for food objects, for example: "Want corn. Want grapes, Want anana." Animal language researchers generally obey the commands of their subjects, lest the latter cease their linguistic behaviors. The Hayeses learned the consequences of refusing requests when their chimpanzee, Viki, stopped using photographs for requests after they failed to take her for car rides every time she presented them with a photograph of a car.

Thus, Pepperberg's assistants ran around the lab delivering the foods Alex requested, each of which he picked up briefly and threw away. It seemed clear that

Alex was using his language to produce activity in his minions. He was using his ability to make requests to produce entertainment, as well as to produce food.

Panbanisha's pointing and Alex's requests are both examples of what Skinner called "mands" (short for "commands"), and both fit his general definition of verbal behavior, that is "behavior that depends for its reinforcement on the intervention of another organism" (Skinner, 1957). An ability to direct other organisms to intervene in causal sequences to the benefit of the "speaker" was a quantum leap in the evolutionary history of organisms, an ability that humans possess to a remarkably greater degree than any other animals.

Nevertheless, the unprecedented successes of the chimpanzee, bonobo, and parrot animal language projects and their replications have significantly affected our view of animals' abilities to detect and affect causal chains. It now appears that animals have an *emergent* ability to learn through observation, and that what they learn in such situations further affects their ability to learn about causal chains and predict consequences.

Language-competent chimpanzees appear to have an enhanced ability to retain information about earlier stimuli, and thus to bring about desired effects by setting up their own causal sequences through the use of symbols. We do not know why language training brings about this emergent ability to bridge temporal delays (e.g., episodic memory as with Panzee, the chimpanzee), but it is certainly an emergent ability and probably functions in ways very similar to those used by humans.

Another type of emergent ability is the ability to respond to relationships between stimuli, for example, to select the larger of two circles, rather than just a specific size of circle. There was a long and sometimes rather acrimonious controversy about whether animals "really" responded to the relationship, or whether their choices were determined by overlapping generalization gradients, but the exact mechanism of response is irrelevant for the present purpose. Animals do respond to relationships and to complex qualities of stimuli; this opens up new categories of comprehensible causal relationships, beyond responses to absolute values of stimuli.

Yet another category of response is rule based, elegantly exemplified by Harry F. Harlow's (1949) research on "learning to learn." Harlow showed that animals could learn complex tasks that were related conceptually, rather than in terms of either specific or relational stimulus characteristics. His work, primarily with monkeys, showed that some animals progressed from an initial stage in which they required hundreds of trials to solve a problem (e.g., choosing an object that concealed a reward when two objects were displayed as a pair) to a final stage in which they solved the problem in a single trial. In this final stage, if they chose correctly on the first trial with a new pair of objects, they continued to choose the object on subsequent trials; if they chose incorrectly, they immediately switched their choice on all subsequent trials with that pair. Thus, there was a progression from gradual, trial-and-error learning, reminiscent of Thorndike's description of learning, in the early stages, to insightful learning, reminiscent of Köhler's description of insight learning, in the final stages. The animals learned that they could cause food to be available by moving the correct complex stimulus aside. We mentioned earlier that evidence indicates that chimpanzees and parrots, at least, have the ability to understand causal relationships between language-using models, as well as the ability to take action for themselves in these language-using contexts. Thus animals, like humans, can make use of all of the four classes of causal relationships that we suggested earlier.

Bonobos (*Pan paniscus*) like Kanzi and Panbanisha, chimpanzees (*P. troglodytes*) like Panzee, and even the parrot Alex learned to use symbols to manipulate situations after being reared from shortly after birth in language-structured environments (i.e., in environments where speech and other symbols are used to explain what has happened and is happening and to announce imminent events).

The bonobos and chimpanzees learned to use lexigrams on a symbol board to request foods, to be taken to locations, and also merely to comment on what they saw or heard. They also announce what it is that they would like to do, such as to take food to a specific ape friend, or to scare another with a mask, or that they would like to visit another group of apes in another building. In the case of all of these animals, it appears that it is better to regard them as causality-detecting animals than to regard them as animals who associate stimuli and responses if they are rewarded for the correct responses.

Also instructive on this point is the report that Savage-Rumbaugh failed to teach the chimpanzees Sherman and Austin when she reinforced them with the same food for correct "naming" of different objects (Savage-Rumbaugh, 1986). However, when they were first given a little of the food to be named, so that there was a consistent causal relationship between the symbol chosen and the first food received, the chimpanzees quickly "caught on" and learned to name 21 different foods, despite then receiving the same food for each correct naming response.

Kanzi – Observational Learning of Language and Spontaneous Speech Comprehension

Formal training of the kind used successfully with the chimpanzees Sherman and Austin, and intended for Kanzi, was canceled when, as a 2½ year old infant, Kanzi gave evidence of having learned by observation that which Matata, his adoptive mother, had failed to learn through the course of months and months of formal drill and training. It was a consequence of that observation that the formal training intended for Kanzi was canceled before it began. Instead, Kanzi was introduced to a life in which throughout his waking hours he was part of a social and language-structured life style. People talked in English to Kanzi and touched the appropriate lexigrams on his keyboard whenever they coincided with words of spoken utterances. Kanzi was encouraged both to observe others' use of the keyboard and even to use it himself; however, he was never denied objects or participation in activities if he failed to do so. Caregivers used the keyboard to comment on events (present, future, and past), to communicate with each other concerning their intentions or needs, and, in particular, to talk about anything that appeared to be of interest to Kanzi.

Kanzi's lexigram vocabulary increased to 149 words by the time he was 5½ years old. Even before that time, unlike Lana, Sherman, Austin, and Matata, Kanzi appeared to comprehend human speech – not just single words, but sentences as well.

The story of Kanzi (Savage-Rumbaugh, Shanker, & Taylor, 1998) is remarkable in other ways as well. As a result of his observation of Matata's training, with no formal training directed to him, Kanzi learned not only to use lexigrams, but also to understand several hundred words of English (Savage-Rumbaugh & Lewin, 1994). No ape specifically trained in language via discrete trial presentations is known to comprehend such a corpus (Hillix & Rumbaugh, 2004). Kanzi showed that he comprehended sentences that he had never heard before. Both Kanzi and Alia, a precocious 2½-year-old child, were given 660 novel oral commands (Savage-Rumbaugh et al., 1993). For example, they were asked to take a specific object to a stated location or person ("Take the gorilla (doll) to the bedroom"; "Give Rose a carrot"), to do something to an object ("Hammer the snake!"), to go somewhere and retrieve a specific object ("Get the telephone that's outdoors"), and so on.

Across trials many different objects were present, and the subjects were asked to act upon them in a variety of ways. This was done to ensure that "compliance with a request" was *not* simply a result of subjects doing whatever was obvious with the materials before them. Thus, when asked to "Get the lettuce that's in the microwave (oven)," Kanzi found not only lettuce, but also a variety of other things in the oven.

At the completion of the study, when Alia was 2½ years old and Kanzi about 9 years old, Kanzi's comprehension of novel spoken sentences was remarkably similar to Alia's. On the very first presentation of novel requests, both subjects were about 70% correct. Thus, Kanzi's comprehension of spoken English was commensurate with that of a precocious 2½-year-old child.

Kanzi's remarkable ability to comprehend English sentences demonstrates that he can analyze very complex acoustic stimuli, probably by detecting subtle changes in the structure of the formants that define phonemes. Savage-Rumbaugh's analyses also make a strong case for the conclusion that Kanzi decoded the structure of requests and did far more than infer what should be done on the basis of understanding just a word or two of each sentence of request. Thus, when asked to pour the water in the jelly vs. pour the jelly in the water, he behaved appropriately. He responded far better than chance on requests of several different types. Clearly some animals (at least Kanzi) are capable of carrying out causal sequences, as well as of requesting that they be carried out (at least Alex, Panbanisha, and Kanzi), and have some ability to understand the syntax of requests.

Language Increases Other Skills

The evidence indicates that the opportunity to observe linguistic interactions between language-using models enhances the ability to understand other types of causal relationships. After his exposure to a language-rich environment and consequent ability to understand several hundred unique English utterances, Kanzi learned after a very short period of observation that a joystick could be used to control the position of an icon in the game of Pacman (a video game). It appeared that Kanzi had understood via his experiences with symbols that very "unnatural" relationships between causes and effects were possible, so that it was not difficult for him to accept the relationship between pressure on a joystick and the position of the Pacman symbol. Other language-competent apes also learned how to use the joystick by observation; however, those without substantial language skills learned to manipulate a joystick only after special and protracted special training to do so. Operant conditioning training was necessary when an animal did not have at least entry-level language skills in comprehension and production.

Among their several accomplishments, three chimpanzees (Lana, Sherman, and Mercury) have acquired impressive counting skills. They accurately count out up to eight dots as appropriate to the value of the Arabic target number on each trial. Each trial is unique in the random selection of the quantity and the array of dots availed to the subject for its counting. Dots can be selected in any order. Once selected, they disappear. Accordingly, the subject has no pattern or any other cue that it can produce to speciously "count" to the value declared by the target number – which also is randomly selected from the values 1–8 across trials.

Syntax and Its Comprehension

Although there is room for debate about the degree to which Kanzi and Panbanisha understand complex syntax, it is generally recognized that they possess clear semantic capacity at a high level, impressive understanding of spoken English, the ability to read and decode written lexical symbols, the ability to answer questions, the ability to express themselves through, at minimum, something like "proto-syntax," and that they acquired these skills through the basic process of enculturation into a *Pan/Homo* society, without specific training. It is also recognized that they can generalize to novel exemplars, and that tests of English word comprehension given over headphones rule out all potential inadvertent cues. Additionally, they can respond even to words said with a non-English accent or intonation by speakers whom they have never seen before.

The influence of syntax for the bonobos was first revealed when a sentence was used to explain something to Kanzi that was incongruent with what the speaker (the third author in this case) intended to convey. Prior to that incident, the author believed that Kanzi's comprehension of syntax was limited, as critics suggested.

On this occasion, the author was seated by the keyboard and chopping up some food for Kanzi when she happened to cut her thumb with the knife. At the moment the accident occurred, Kanzi was looking away, engaged in play with a toy. When he turned and realized that the author's thumb was bleeding, he assumed an expression of surprise and concern as though he were wondering how this could have suddenly happened. Since the word "cut" was not on the keyboard, the author said "Knife hurt Sue" at the keyboard. In response, Kanzi looked amazed, produced a loud alarm sound and knocked the knife across the room.

That was not Kanzi's normal response to minor injuries. The author realized that she had syntactically conveyed something quite different from what had actually happened, not what she had intended to convey. By saying "Knife hurt Sue" she implied that the knife had acted of its own accord. Had she said instead, "I cut myself with the knife," Kanzi's response likely would have been to groom the injury, as he had done many times with similar previous injuries. However, Kanzi responded, not to the intended meaning, but to the syntax of the actual utterance, and he treated the knife as if it were an agent. Had Kanzi been sensitive only to the words "knife" and "hurt," and not to syntax, he could not have attributed agency to the knife.

The fact that he did so, in spite of the author's intent that he do otherwise, implied that syntax as a force in Kanzi's consciousness had been greatly underestimated. When she realized what had occurred, she quickly explained to Kanzi, in English, that the knife really did not hurt her, but that she had cut herself accidentally. Kanzi looked suspicious for a few moments, but then acted as though the knife were no longer a threat, and asked to groom the wound.

This incident led to the extensive series of tests in which Kanzi was given over 600 novel sentences designed to test his grammatical sensitivity. All contextual cues were eliminated from these tests, and they were interspersed with play across a series of months to determine the extent to which Kanzi was responding to syntactical structure, as well as to semantic content. These tests showed that he responded appropriately to a number of strictly syntactical relationships.

For example, Kanzi was sensitive to word order, responding appropriately and differentially, to sentences such as "Pour X in Y" vs. "Pour Y in X." The sentences given were the precise inverse of one another (e.g., "Pour the juice in the egg" vs. "Pour the egg in the juice"), in order to eliminate the possibility that he was only performing the most likely action. Substantial time elapsed between presentations of these kinds of sentences.

In addition, Kanzi and Panbanisha were sensitive to sentences requiring recursion such as "Get the X that is in Y" – where numerous exemplars of X were located in alternative locations. Only if the bonobos recognized that a particular X was distinguished by its syntactically defined relationship to Y could sentences containing recursions be comprehended. For example, in the sentence "Get the tomato that's in the microwave" – getting the tomato that is in front of one and getting something else from the microwave, or simply indicating the microwave, will not suffice. The importance of a syntactical structure such as recursion lies in the fact that the relationship between the content words (X and Y) is specified by a word (in this case the word "that") whose sole function is syntactical. There exists no external semantic referent for "that" which is defined apart from the structure of the sentence.

Supporting the view that it was the syntax and not the individual words that accounted for their behavior was the fact that when recursion was not utilized (as in the ambiguous sentence, "Go to X and get Y" vs. "Go get Y that's at X") incomplete or incorrect responses were significantly more probable. "Go to X and get Y" frequently resulted in Kanzi doing two things without coordination: he would go X but he would get a Y from some other place or not get it at all. Thus, the bonobos did not just understand the syntactical device of recursion; without it, they treated ambiguous sentences as precisely that – ambiguous.

During these tests many sentences were employed that could not be properly processed without a grasp of the syntactical indication of agency. For example, Kanzi was asked, "Can you feed your ball some tomato?" In this case the "ball" was a round sponge ball with a pumpkin face which was on the floor along with other objects, including other balls. Kanzi turned the ball around till he found the face and then pretended to place the tomato on the ball's mouth, as though the ball could eat the tomato. Had he been shown how to do this, it would have been less impressive, but no one had shown Kanzi the face that was hidden in the ball, nor treated the ball as a face.

In this instance, Kanzi was able to pretend that the ball was as an animate agent capable of eating, and he was able to do so solely because of the role of agency assigned to the ball in a single sentence.

Had Kanzi, as critics argued, only responded to the words ball and tomato, how could such a complex behavior have ensued? He had to have understood the verb "feed," and the syntactical implication that the ball was to be the object of a pretend action with a tomato. Understanding only ball and tomato would have led to handing the two objects to the experimenter.

Stages of Causal Understanding and Parameters of Animal Intelligence

Pavlovian classical conditioning and Skinnerian operant conditioning are so ubiquitous – observed in organisms from humans to insects – that "readiness to be conditioned" is not useful as a measure of an animal's ability to understand causal relationships. Those mechanisms are basic to behavior throughout the animal kingdom. On the other hand, more complex emergent processes and behaviors are more promising indices of different levels of intelligence in animals, as they are in people.

Emergent behaviors are distinct from, yet probably ultimately founded on, the basic conditioning phenomena of Pavlov and Skinner (Rumbaugh, Washburn, & Hillix, 1986). Emergent behaviors and processes are not specifically reinforced during discrete trial training procedures and are not conditioned to specific conditioned/discriminative stimuli. Rather, they emerge (e.g., become evident) quite unexpectedly, and quite typically are initially discovered as organisms face new problems/contexts. Emergent behaviors involve new patterns of behavior that appear to be based on principles or rules. This is not to say that emergent behaviors must lack well-defined antecedents, as we saw in Harlow's learning-to-learn paradigm. Emergents are believed to have their foundations in the collective experiences relevant to the new problem at hand that is to be solved. Murray Sidman (1994), for example, showed how novel relations emerge (i.e., transitivity and equivalence), not only between stimuli but also between responses, between rewards, and all combinations of these, from rather extensive training on a much smaller set of relations. These relational combinations are characteristic of languages, which when fully developed offer the possibility of a nearly infinite number of utterances comprised of a large number of words that can occur in a large number of orders to construct a huge number of sentences that can, in turn, be combined with other sentences to express an indefinite number of causal relationships, among other things that also can be expressed.

Animal Intelligence and Brain Complexity

Significantly, emergents appear more readily when species' brain complexity is relatively high and early environmental conditions provide the opportunity for behavioral interaction. Brain complexity is related to the brain's mean absolute size and to its cytoarchitectonics. Size is more easily measured at present and has been shown to correlate reliably (Spearman $r = 0.79$) with 12 primate species' abilities to transfer small increments of learning to altered test situations (Rumbaugh, 2002). Specifically, when animals were trained to a criterion of 87% correct vs. 67% correct on a discrimination problem, and the correct and incorrect stimulus were then reversed, the large-brained animals benefited from the increased experience on the original problem; they reversed their choices more quickly than did the smaller brained animals, who tended to stay with their original choices much longer. With the increased amount of learning (e.g., 84% correct rather than only 67% correct), the smaller brained primates were sorely compromised in switching their choice of objects on transfer-test trials; however, the same increase in level of learning profoundly enhanced the transfer of learning for the large monkeys and great apes! Frequently, their transfer of learning performances was substantially better than the amount that could be accounted for directly on the basis of what had been learned prior to transfer.

There is a similar probabilistic shift from basic stimulus–response associative learning to relational learning (i.e., principle or rule learning) as one goes from prosimians to monkeys to apes. Whether this relationship holds true within other orders of mammals is not known, though much literature attests to the advanced intelligence of complex-brained mammals such as whales, dolphins, sea lions, and elephants, for instance.

Early Rearing Contexts Are Crucial

A general principle that applies to primates is that their general social and cognitive competence, and notably their proficiency in discerning and affecting reliable cause–effect relationships as adults, is very dependent upon early environment. In particular, their capacity to learn complex tasks, to be good breeders and caretakers of young, and to be competent members of social groups is determined in large measure by how they are reared (Davenport, Rogers, & Rumbaugh, 1973; Gluck, 1979). More generally, primate taxa develop best if individuals are raised in social groups with others of their species who have had good and, insofar as possible, natural rearing experiences.

It appears that primates are very vulnerable to inadequate rearing and impoverished environments during infancy in two ways. The first is that what are basically preprogrammed response tendencies, specific to each species, need to be elicited from the primate at times during its development when it is biopsychologically ready to respond to events of a kind that it would quite certainly encounter in nature. For chimpanzees, these responses are most efficiently and completely elicited in interactions with other chimpanzees.

The second way in which primates are vulnerable is that, to the degree that a species has a brain that is extraordinarily large, given the size of its body, it needs rich and varied experiences during infancy and youth for optimal development. If primates are reared in a socially impoverished rearing environment, their play behaviors likely will be reduced, if not absent. As a consequence, their trustworthiness probably will be low and their tendency toward aggressive behaviors will be increased as they mature.

The whole complex of environmental and social experiences, and especially "play" behaviors, can be viewed through the causal lens as an opportunity to learn about causal connections and how much the individual's actions can affect them.

Henry Nissen initiated a study in which chimpanzees born and reared under very impoverished conditions for the first 2 years of life were compared in a variety of tasks with others that were group reared (see Davenport et al., 1973). At the age of 14 years, when assessed through use of Transfer Index tests (Rumbaugh & Pate, 1984), it was clear that the chimpanzees reared in impoverished environs for the first 2 years were extraordinarily deficient in transfer skills. On the other hand, they did not differ from chimpanzees who were reared in social groups, during the first 2 years of life, in their *simple discrimination learning* performances – that is, when they were required to learn, for example, that the choice of a red square results in a prized piece of food (i.e., banana) whereas the choice of a blue triangle does not. What they differed in was their ability to "learn how to learn" and the very important capacity to transfer to an advantage what was learned in one situation to another. Thus, experience during infancy is critical to the long-term development of the chimpanzees' complex learning abilities. Data available on rhesus (monkeys) (*Macaca mulatta*) suggests that their complex learning skills are less compromised than are the chimpanzee's when the monkeys are reared in severely restricted environments. This, in turn, suggests that, the more complex the species brain, the more important it is that rearing occur in appropriately stimulating environments. This principle has clear implications as to how adverse early rearing environments can compromise the learning abilities of human children, who are likely to be even more affected than chimpanzees by their rearing environments.

Television. The ability to comprehend cause–effect relationships is clearly revealed in chimpanzees' abilities to extract specific information from televised materials. In order for chimpanzees to extract meaning from what they see on television, they generally must learn that what they see on the screen is relevant to the world they have known, or in which they currently live (Savage-Rumbaugh, 1986). This is a process of discovery, of perceiving cause–effect relationships between events. If chimpanzees are allowed to see and hear real-world events on television, while they also view these events directly, they are likely to "discover" that the television also can portray parts of the world that they otherwise will never see (Savage-Rumbaugh, 1986). For example, chimpanzees can garner important information from television about the locations of foods/incentives and about activities in areas beyond their view. On one occasion, their fear from the sounds of chain saws, emanating from outside their quarters and out of direct view, was abated when they were given a televised view of the work crew cutting a tree.

Seeing the source of the threatening sound allowed curiosity to supplant fear, which could have happened only if the causal connection was understood. Also, chimpanzees learn to prefer watching specific videotaped recordings – those of other chimpanzees, especially ones they know and that consequently have meaning to them. They develop preferences not only for specific videotapes, but also for scenarios. They are capable of communicating "what comes next," in a favorite scenario, by selecting pictures appropriate to the scene that is about to be portrayed.

Video-Formatted Tasks and Competence with Joysticks

Research at our laboratory (Washburn & Rumbaugh, 1992) with two 5-year-old rhesus monkeys (*M. mulatta*) supports the observation that under certain conditions they would rather work for food than obtain it "for free." These rhesus monkeys were given a "video menu" from which they could select one of a set of familiar tasks on which to work. Once selected, the task remained available to them for five trials. The menu then appeared once again for the subject to choose either the same or some other task for the next five trials, and so on. For this research, no food privation schedules were used for either training or testing. If the task they requested was presented on the screen, their performance (both in terms of accuracy and rate) significantly increased. By contrast, it dropped significantly if there was no relationship between tasks requested and what was presented to them. It is suggested that the monkeys were making a real-time assessment of known cause–effect relationships and lost willingness to work when they could no longer rely on those relationships.

As noted, one of the options on the video menu was for "free" food. This choice resulted in five pellets of free food, one every 5 s. Under this condition, the rhesus gradually came to prefer this choice, but when choice of free food resulted in 30 min of pellets being delivered every 5 s and in their computer being disabled for work on other tasks of their choice, free food as an option became *non*preferred. Thus, one could argue that the monkeys needed to exercise their causal powers more often than every 30 min! This observation also suggests that working on video tasks is intrinsically rewarding. . In another situation where the rhesus had more food immediately available than could be consumed by even two or three monkeys, they also still kept working on their video tasks!

Significantly, both apes and rhesus monkeys respond predictively on their computer-monitored interactive tasks, which include pursuit of erratically moving targets.

In other words, as they chase a moving target, they don't head for "where it is," but rather to "where it will be," making appropriate allowance for the speed with which they can move their cursors, as well as for the trajectory and speed of the target.

Intelligence from a Comparative Perspective

Intelligence, whether it be in human or nonhuman, surely provides for enhanced understanding of cause–effect relationships between things and events of the world and for the capacity to assess whether relationships are fortuitous or causal.

The construct of intelligence is used to explore the general topic of "individual differences" among humans, which includes personality, aptitudes, and so on. Intelligence here will be viewed as the ability to achieve, the ability to learn complex problems, the ability to transfer learning to a significant advantage, and the ability to respond cleverly in novel situations. Because of evolutionary continuity and our close relationship to the great apes (Andrews & Martin, 1987), in particular, evidence of "psychological continuities" between humans and apes might be obtained, just as evidence of anatomical and biological continuities has been recognized for more than 100 years. Most psychologists of today hold that intelligence is the product of *interactions* between heredity and experience (i.e., environment). Intelligence is not "fixed"; it varies in its scope, profile, and efficiency across the life span of an individual.

Complex learning skills and novel modes of solving problems provide data possibly relevant to an understanding of primate intelligence. Also, primates' *relative potential for achievement* in very complex areas of cognitive operations, such as those entailed in transfer tests of learning, language, and the mastery of numbers can be helpful.

Correlates of Intelligence: Processes of Learning, Transfer of Learning, and Early Environment

Köhler's (1925) classic studies clarified the role of perceptual reorganization in the chimpanzee's efforts to solve problems. For example, Köhler suspended a banana so that the chimpanzees could see but not reach it. Would the chimpanzees' perceptual reorganization of the components of the problem result in stacking boxes so as to obtain the incentive? Would they use a stick to get it? What kinds of ineffective or inappropriate tactics would be attempted? These questions were of interest to the Gestalt psychologists, who were interested in "insightful" perceptual learning, rather than in the gradual elimination of errors as in animal maze learning.

The contrasting perspectives of the Gestaltists, represented by Köhler, and the behaviorists in the early part of this century can be amalgamated in the light of more recent research (Rumbaugh & Pate, 1984). The learning of specific responses to specific stimuli, a la behaviorists, is here termed *stimulus–response associative learning*. By contrast, the problem-solving achievements of Köhler's apes will be abbreviated as *relational learning*, in which the subject is presumed to learn about the relationships between stimuli, objects, and the potential consequences of responding to them in various ways.

Experience is necessary for both forms of learning. For stimulus–response associative learning to occur, there must be repeated opportunities (i.e., trials) to learn, where learning is indicated by a gradual elimination of errors. For Köhler's relational learning, the chimpanzee had to have prior experience with boxes and sticks to learn about their attributes so that, subsequently, these items might be perceived as tools for solving problems. They needed to know about the attributes of sticks in order to use them innovatively in new contexts.

Of monkeys tested to date, only rhesus (*M. mulatta*) gives evidence of being able to transfer small amounts of learning to a marked advantage. Their primary learning mode is not as likely to be of a relational type as it is for the great apes, but, given protracted and varied learning opportunities in automated testing apparatus with complex tests, there is recent evidence that they can shift from a predominant stimulus–response associative learning mode to a relational mode (Washburn & Rumbaugh, 1991). In a separate study, it was found that no other learning measurement correlated more highly with rhesus (*M. mulatta*) monkeys' achievement levels on a battery of 18 psychological tasks than does their transfer-of-training skills, as measured by the Transfer Index. Consequently, it is suggested that there might, indeed, be a global or general factor in nonhuman primate intelligence, as has been argued for human intelligence. One way of describing this "g factor" would be to say that it involves the ability to detect and affect a broad range of causal relationships.

The great apes and the larger monkeys are more frequently noted for being particularly "clever" than are the smaller monkeys and prosimians. It is probable that their fundamental abilities to transfer even small amounts of learning to new problems are the basis for their general cleverness. It is proposed that cleverness and creativity are generally a positive function of brain complexity in primates, provided that early rearing is in a stimulating and logically structured environment.

Causal Chaining and Communication

One aspect of intelligence is the ability to construct for oneself, rather than to be taught explicitly, how to structure a sequence of causal actions in order to bring about an effect. Even animals that would usually be considered rather unintelligent, like wasps, are capable of executing long sequences of responses. They prepare nests and then attack, kill, transport, and arrange several spiders in position in a nest, after which they lay eggs and close the nest. Because wasps are not able to adjust to any change in this pattern, even a slight relocation of the spiders by an observer will cause the wasp to start anew, back at the beginning of the sequence! Thus, there is reason to doubt that any kind of logical process or consciousness of goals guides their actions. Their behavior is perhaps best classified as "biologically smart" rather than "psychologically intelligent" (Rumbaugh & Washburn, 2003).

The situation is surely different with great apes. Köhler's best chimpanzee student, Sultan (Köhler, 1925), was able to construct the following sequence of goal-directed actions for himself without specific training: (1) pick up the very short stick made available by Köhler, (2) use it to drag within reach a second somewhat longer stick, (3) join the two sticks by inserting the end of one into the hollowed end of the other, (4) use this longer stick to drag within reach a third stick, (5) join it to lengthen further the one constructed from the first two sticks, and (6) use the three joined sticks to drag in food that was otherwise beyond Sultan's reach! Conscious intent is strongly suggested as the basis for this highly creative behavior.

Savage-Rumbaugh (1986) reported that the chimpanzees, Sherman and Austin, engaged in novel and creative behaviors that corrected deficiencies in one another. Sherman, for example, forcibly took Austin's hand and used one of Austin's fingers to select a word-lexigram appropriate to the food that, through prior pointing, Sherman clearly wanted Austin to request so that he, Sherman, could "serve" shared portions to Austin and himself. Sherman also would correct Austin when Austin, in error, gave him a tool other than the one requested and needed by Sherman to obtain food from one of several puzzle boxes. Also, when Sherman mistakenly asked Austin for an end-wrench rather than a stick to poke food out of a long plastic tube, he compensated for his mistake by giving a manual blow to the handle of the wrench, thus making it fly as a missile through the tube and knock the food out the other end, where he had positioned his other hand to catch the "food missile" as it literally flew out the far end of the 1-m tube. And when Lana was being "toilet trained" and had no more urine to squirt into a specific pan, she creatively spit into it – then looked expectantly for her chocolate reward.

Humans are, of course, the preeminent creators of causal chains. Neither the planning nor assembly of the great pyramids of Egypt of history nor the giant passenger planes of today could be achieved without the coordinated efforts of throngs of people who work and plan together. Language expresses multiple causal possibilities for their joint evaluation. But do animals ever exhibit even a rudimentary ability to create superordinate causal chains that require true communication?

The beginnings of this emergent ability can be seen in cooperative hunting, engaged in by several species that include lions, chimpanzees, and wolves. However, cooperative hunting is coordinated through each individual's direct observation of the behavior of other hunters in the group, not (at least not demonstrably) through the exchange of symbolic information.

Perhaps, for our purpose here, the best experimental instances of coordinated social communication by chimpanzees are reported in a paper entitled, "Can a chimpanzee make a statement?" (Savage-Rumbaugh, Pate, Lawson, Smith, & Rosenbaum, 1983). Sherman and Austin, in turn, made unaccompanied visits to the end of a corridor where they had visual access to an array of foods and drinks that changed unpredictably across trials. Then Sherman or Austin had his only opportunity to survey the prized foods and drinks and covertly to select which one he wanted.

Next, they returned to an adjoining room in which the experimenter and the other chimpanzee were with a large keyboard that had well over 150 word-lexigrams, some of which were of the foods and drinks just visually surveyed. The returning chimpanzee then (1) through use of the keyboard had to declare which food or drink it had covertly selected, (2) return unaccompanied to the corridor alone, (3) select a food/drink, (4) return to the adjoining room, and (5) there show the experimenter the food/drink that it chose to bring back with it. Only if the food/drink brought to the experimenter by the chimpanzee was one and the same as that declared in step "1" was the item shared and social "congratulations and thanks" expressed – both by the experimenter and by the second chimpanzee. Clearly, not only had a complex causal chain been mastered, but also one that contained a crucial step that seemingly would be impossible to condition or train. That step is the covert, yet demonstrable, selection of a specific food or drink and its symbolic encoding so that it then might be declared at the keyboard (step 1). Though one might conceive of how to condition/train this behavioral sequence, the fact is that it was not trained! The subject brought that capacity as an emergent skill to the experiment by virtue of its prior and extensive history of working with symbols in situations that required social communication and the coordination of social behaviors for goals to be achieved (Savage-Rumbaugh, 1986).

In juxtaposition, the causal chains of the wasp and statement behaviors of our chimpanzees define the extremes afforded by "biological smartness" and by "psychological intelligence and symbolic processes," respectively. Surely it has been the emergence of social causal chains of this kind that has enabled the elaborations of highly creative, foresightful plans that, in turn, have defined the bases of human culture and technology. In keeping with Darwin's postulate of psychological as well as biological continuity between animals and humans, we conclude that the differences between animal and human psychology are reflections of emergent processes generated by quantitative evolution. The emergent processes certainly can and do put in place not only behavioral changes but also neuroanatomical changes. To the degree that the neuroanatomical changes come under the control of genetics, they can serve to ensure qualitative differences between even closely related species, as between chimpanzee, bonobo, and human. To the degree that emergent processes afford changes only at the behavioral level, they, nonetheless, can provide the foundations for marked differences between individuals and their ability to perform complex tasks, such as the comprehension of human speech and the fabrication of stone tools – some of Kanzi's several accomplishments!

Just how learning and behavior come about so as to serve adaptation to changing environments and reproductive fitness quite apart from the perspectives of decades past that have emphasized S–R habit or bonding and the fixedness imposed by reinforcement has been addressed anew by Rumbaugh, King, Beran, Washburn, and Gould (2008). They posit the formation, not of bonds or habits between stimuli and responses, but rather the formation of amalgams, where amalgams are comprised of temporally contiguously occurring stimuli/events.

To the degree that a constituent member of a given amalgam is salient, strong, and with well-defined response-eliciting properties, it plays a commanding role interactively with other constituents of a given amalgam. Thus, to the degree that constituent members of amalgams have those attributes, and quite regardless of whether a given member has been intended to serve as a reinforcer, a reward, a resource, a secondary reinforcer, or a conditioned reinforcer, it commands dominance in the determination of the amalgams' attributes. Reinforcers are thusly redefined in relative standing with all other contiguously occurring stimulus events. The amalgams thus are the basic units of learning in salience theory. In turn, they define what has been learned and as a unit of life experiences they enter relationally with other amalgams into the formation of a knowledge base. From that base come emergents of limited scope in nonhuman animals or ones of commanding scope as seen in the highest creative forms of behavior of which we as humans are capable. Salience theory, thus, incorporates the specific effects of instincts and conditioned behaviors into relational systems deemed necessary for emergent creative behaviors and capabilities (Hillix & Rumbaugh, 2004; Rumbaugh & Washburn, 2003).

Summary

Our analyses lead to the conclusion that causality is a strikingly complex concept. It certainly is more complex than one might think after reading David Hume's account. Causes may be relational and probabilistic, they may be complex and delayed, they must occur in total contexts, and a sufficient cause in one context may be insufficient in another. It also is true that there are several types of causal origins – as when an environmental event brings about an action vs. when an action brings about an environmental change. Organisms develop an increasing ability to understand causes with increasing experience, perhaps starting with the recognition of their own ability to bring about consequences through their own actions and, if extant, their symbolic capacities; they understand that they have "causal powers," and appear to have a need to exercise these powers even when it is not necessary to do so. The ability to understand and affect causal chains passes through developmental stages, of which six have been suggested. The great apes, if not other species as well, appear to reach, with us, the highest stage, at least in elementary form.

Adaptive intelligence involves the ability to understand and manipulate causal relationships – the basic mechanism of intelligence. Psychological smartness and intelligence have been made possible by an extraordinary enlargement of the brain, elaboration of the cortex, and the organizing effects that patterns of early experiences have upon the brain. And language capacity, both in humans and chimpanzees, dramatically increases the capacity to respond flexibly and persistently to environmental events, and makes possible social cooperation in producing causal chains that are necessary in order to achieve goals.

The capacities of animals other than human are simply too complex, too creative, and too efficient in their novel application to be attributed simply to the connection of stimuli to responses by reinforcement.

As the primate brain's complexity increases, relational learning becomes more probable than stimulus–response associative learning; and learning and behavior have become increasingly dependent upon observation and the organizing properties of the brain, a function for which it has evolved, rather than upon action alone. The brain maximizes the organization that can be brought to bear, even upon experiences that are related by principles and not by specific details.

Biology, notably in the great apes and in humans, has provided for the potential to "learn how to learn" and also to transfer learning proficiently. And as this trend progresses, at some point there emerges sentience, the ability to determine whether it is more likely that events are related fortuitously or that they are causally related. These important progressions have provided the foundations of language and thus of human culture and technology, and are present in their earlier stages in various nonhuman life forms.

Acknowledgment The preparation of this manuscript and much of the research herein reported were supported by grants from the National Institute of Child Health and Human Development (HD-06016 and HD-060563). Additional support from the College of Arts & Sciences, Georgia State University, also is gratefully acknowledged.

References

Andrews, P., & Martin, L. (1987). Cladistic relationships of extant and fossil hominoids. *Journal of Human Evolution, 16*, 101–108.

Bandura, A. (1971). *Social learning theory*. New York: General Learning Press.

Brunswik, E. (1943). Organismic achievement and environmental probability. *Psychological Review, 50*, 255–272.

Brunswik, E. (1952). The conceptual framework of psychology. *International Encyclopedia of Unified Science, 1*, 1–102.

Davenport, R. K., Rogers, C. W., & Rumbaugh, D. M. (1973). Long-term cognitive deficits in chimpanzees associated with early impoverished rearing. *Developmental Psychology, 9*, 343–347.

Ferster, C. B., & Skinner, B. F. (1957). *Schedules of reinforcement*. New York: Appleton-Century-Crofts.

Garcia, J., Ervin, F. R., & Koelling, R. A. (1966). Learning with prolonged delay of reinforcement. *Psychonomic Science, 5*, 121–122.

Garcia, J., McGowan, B. K., Ervin, F. R., & Koelling, R. (1968). Cues – Their relative effectiveness as a function of the reinforcer. *Science, 160*, 794–795.

Gluck, J. P. (1979). The intellectual consequences of early social restriction in rhesus monkeys (*Macaca mulatta*). In G. C. Ruppenthal (Ed.), *Nursery care of nonhuman primates*. New York: Plenum.

Harlow, H. F. (1949). The formation of learning sets. *Psychological Review, 56*, 51–65.

Hillix, W. A., & Rumbaugh, D. M. (2004). *Animal bodies, human minds – Ape, dolphin, and parrot language skills*. New York: Kluwer Academic/Plenum.

Köhler, W. (1925). *The mentality of apes*. London: Routledge & Kegan Paul.

L'Abate, L. (2005). *Personality in intimate relationships: Socialization and psychopathology*. New York: Springer-Science.

Menzel, C. R. (1999). Unprompted recall and reporting of hidden objects by a chimpanzee (*Pan troglodytes*) after extended delays. *Journal of Comparative Psychology, 113*, 426–434.

Michotte, A. (1963). *The perception of causality* (T. R. Miles, Trans.). New York: Basic Books.

Pepperberg, I. M. (1985). Social modeling theory: A possible framework for understanding avian vocal learning. *The Auk, 102*, 854–864.
Pepperberg, I. M. (1990a). Cognition in an African Gray parrot (*Psittacus erithacus*): Further evidence for comprehension of categories and labels. *Journal of Comparative Psychology, 104*, 41–52.
Pepperberg, I. M. (1990b). Conceptual abilities of some nonprimate species, with an emphasis on the African Grey parrot. In S. T. Parker & K. R. Gibson (Eds.), *"Language" and intelligence in monkeys and apes: Comparative developmental perspectives* (pp. 469–507). Cambridge: Cambridge University Press.
Piaget, J. (1930). *The child's conception of physical causality*. London: Routledge & Kegan Paul.
Piaget, J. (1974). *Understanding causality*. New York: Norton.
Rescorla, R. A. (1988). Pavlovian conditioning: It's not what you think it is. *American Psychologist, 43*, 151–160.
Revusky, S., & Garcia, J. (1970). Learned associations over long delays. In G. H. Bower (Ed.), *The psychology of learning and motivation: Advances in theory and research* (Vol. 4, pp. 1–83). New York: Academic Press.
Rumbaugh, D. M., King, J. E., Beran, M. J., Washburn, D. A., & Gould, K. L. (1988). *A salience theory of learning and behavior: International Journal of Primatology, 28*, 973–996.
Rumbaugh, D. M., & Pate, J. L. (1984). *The evolution of cognition in primates: A comparative perspective. In Roitblat, Bever, & Terrace* (pp. 435–439). New York: Columbia University Press.
Rumbaugh, D. M. (2002). Emergence and rational behaviorism. *Eye on Psi Chi, 6*, 8–14.
Rumbaugh, D. M., & Washburn, D. A. (2003). *The intelligence of apes and other rational beings*. New Haven: Yale University Press.
Rumbaugh, D. M., Washburn, D. A., & Hillix, W. A. (1986). Respondents, operants, and emergents: Toward an integrated perspective on behavior. In K. Pribram & J. King (Eds.), *Learning as a self-organizing process*. Hillsdale: Erlbaum.
Savage-Rumbaugh, E. S. (1986). *Ape language: From conditioned response to symbol*. New York: Columbia University Press.
Savage-Rumbaugh, E. S., & Lewin, R. (1994). *Kanzi: At the brink of the human mind*. New York: Wiley.
Savage-Rumbaugh, E. S., Murphy, J., Sevcik, R. A., Brakke, K. E., Williams, S., & Rumbaugh, D. M. (1993). Language comprehension in ape and child. *Monographs of the Society for Research in Child Development, 58*(3–4), 1–242.
Savage-Rumbaugh, E. S., Pate, J. L., Lawson, J., Smith, S. T., & Rosenbaum, S. (1983). Can a chimpanzee make a statement? *Journal of Experimental Psychology: General, 112*, 457–492.
Savage-Rumbaugh, E. S., Shanker, S., & Taylor, T. J. (1998). *Apes, language, and the human mind*. New York: Oxford University Press.
Sidman, M. (1994). Equivalence relations and behavior: A research story. Boston: Authors Cooperative.
Skinner, B. F. (1948). "Superstition" in the pigeon. *Journal of Experimental Psychology, 38*, 168–172.
Suppes, P. (1984). *Probabilistic metaphysics*. Oxford: Blackwell.
Van Gulick, R. (1993). Who's in charge here? In J. Heil & A. Mele (Eds.), *Mental causation*. Oxford: Clarendon Press.
Washburn, D. A., & Rumbaugh, D. M. (1991). Rhesus monkey (*Macaca mulatta*) complex learning skills reassessed. *International Journal of Primatology, 12*(4), 377–388.
Washburn, D. A., & Rumbaugh, D. M. (1992). Testing primates with joystick-based automated apparatus: Lessons from the Language Research Center's Computerized Test System. *Behavior Research Methods, Instruments, and Computers, 24*, 157–164.
White, P. A. (1995). *The understanding of causation and the production of action. From infancy to adulthood*. East Sussex: Lawrence Erlbaum Associates.

Chapter 22
Essentialism

Luciano L'Abate

While essentialism originally was uniquely rhetorical in philosophy, based on arguments rather than evidence, in the last generation essentialism has been the subject of a great deal of psychological research. This research raises serious questions about essentialism actually being considered as a theory rather than as a paradigm. Therefore, essentialism may have started as a paradigm, at least in philosophy, but has emerged as an influential psychological theory worthy of inclusion in this volume.

The essence of behavior, human relationships, and "things" is to understand them according to their fundamental nature and not just on their appearances. Please note that neither this paradigm nor Chap. 14 on Materialism were included in the original classification presented by Hillix and L'Abate (2011). These paradigms, however, belong to two different sections of this volume for reasons that will be made explicit in each chapter. Since the psychological literature on essentialism is surprisingly vast, for reasons of necessarily limited space allotted, this chapter will be devoted to only a representative review of selected and more recent studies about this paradigm/theory.

According to the American Psychological Association Dictionary of Psychology (2007), essentialism is defined as a philosophical position that things (or some things) have essences, that is, they have certain necessary properties without which they could not be the things they are or seem. Essentialism is the position rejected by other philosophical schools, like Marxism, Postmodernism, and Post-structuralism that human beings have an essential nature that transcends such factors as social class, gender, and ethnicity. An essential property is a characteristic of an idea or entity that is basic or fundamental to its nature or existence. For example, being female is an essential property of mothers, whereas being tired is an accidental property.

L. L'Abate (✉)
Department of Psychology, Georgia State University, University Plaza, Atlanta, GA 30303, USA
e-mail: llabate3@bellsouth.net

L. L'Abate (ed.), *Paradigms in Theory Construction*, DOI 10.1007/978-1-4614-0914-4_22,

Historical Background

Briefly, essence was used in metaphysics to designate that which makes a thing what it is and nothing else, in contrast to the qualities, or accidents, which attach superficially and for the time being to the thing and may be detached from it. Contrasted also with the existence, or factual being or "thereness" of a thing in and for itself, or its presence as an experimental fact. By some, the essence of a thing is lodged in that which it shares in common with of things of the same sort (e.g., Plato); by others in that which makes it individual and concrete (e.g., Aristotle) (Fuller, 1946, p. 526).

Ripken's (1980) philosophical lectures mentioned essentialism several times, though the main topic was naming and identity. Essentialism appears as part of the discussion of essential properties (of both individual things and classes or kinds of things; Putnam, 1973) is notable in philosophy for introducing the "Twin Earth" thought experiment. The article does not mention the term essentialism. Instead, it focuses on a book which Putnam edited, the second of a two-volume series (Putnam, 1975). In this book, Putman expanded on the Twin Earth thought experiment first introduced in the article cited above. Putnam sets up the experiment by asking the reader to consider an imaginary planet identical to Earth in every aspect except for one (he chose chemical composition of water as the difference). He called this "Twin Earth." He goes on to argue that the word for "water" in Earth English as well as in Twin Earth English would not mean the same thing, because water on the two different planets is not the same chemical compound. This idea supports a branch of philosophy called semantic externalism, which argues that the meaning of a term depends, wholly or partly, on elements external to the person speaking or understanding the term, as suggested by Schick (2003) and Wittgenstein (Forster, 2004).

The Role of Essentialism in Psychological Specialties

Surprisingly, essentialism has exerted a much wider influence on psychological thinking and especially cognitive, cultural, developmental, and social psychological thinking than most psychologists, including this writer, are aware of. Indeed, a few psychologists questioned about this paradigm confessed complete ignorance about its existence and influence. Nonetheless, as shown in this chapter, this theory is widely known within a selected number of psychologists.

Cognitive Psychology

Medina and Orton (1989), for instance, considered the implications of the distinction between the more accessible, surface, aspects of representations and the less accessible, deeper, aspects for the nature of similarity and its role in cognition. They argued that, although it can be a powerful heuristic for various cognitive tasks, there are limitations to using similarity with respect only to surface properties and that there are problems with ignoring the relation between surface similarity and deeper properties.

Braisby, Franks, and Hampton (1996) argued that the essentialist approach to word meaning has been used to undermine the fundamental assumptions of the cognitive psychology of concepts. Essentialism assumes that a word refers to a natural kind category in virtue of category members possessing essential properties. In support of this thesis, Kripke and Putnam deploy various intuitions concerning word use under circumstances in which discoveries about natural kinds are made. Although some studies employing counterfactual discoveries and related transformations appear to vindicate essentialism, Braisby et al. added that the intuitions have not been investigated exhaustively. In particular, discoveries concerning the essential properties of whole categories (rather than simply of particular category members) are critical to the essentialist intuitions. The studies reported these investigators examine such discovery contexts, and demonstrate that words and concepts are not used in accordance with essentialism. The results are, however, consistent with "representational change" views of concepts, which are broadly Fregean in their motivation. They concluded that since essentialism is not vindicated by ordinary word use, it fails to undermine the cognitive psychology of concepts.

Braisby et al. (1996) presented two externalistic adaptations of psychological placeholder essentialism, a strict externalist and a hybrid externalist view, which are experimentally testable. These authors examined claims that undermine externalism, and argued that the study fails in its aims. They conducted two experiments, the results of which undermine internalism and the hybrid theory, and support strict externalism. Their conclusion is that lay speakers' natural kind concepts involve a belief in an external category essence, which determines reference. Some experimental studies have recently claimed to undermine semantic externalism about natural kind terms (Jylkkä, Railo, & Haukioja, 2009). However, it is unclear how philosophical accounts of reference can be experimentally tested.

Mahalingam (2007) synthesized two perspectives on essentialism: cognitive and social. The cognitive essentialist perspective argues that our bias to appeal to the psychological belief that categories have innate essences enables us to make inferences about social categories such as race, caste, and gender. The social essentialist perspective argues that essentialist thinking serves the needs of those in power to justify existing social and economic hierarchies. Examining the relationship between essentialism and power, that article expands the framework for folk sociology, incorporating cognitive and social essentialist accounts to examine folk theories of social differences between caste, race, gender, and class, with a particular emphasis on social marginality and cultural narratives. Further, that article discusses the relevance of the folk sociology perspective to the study of children's emerging understanding of their social world.

Psychological essentialism is an ordinary mode of category representation that has powerful social–psychological consequences. Prentice and Miller (2007), for instance, reviewed those consequences, with a focus on the distinctive ways people perceive, evaluate, and interact with members of human categories they essentialize. Why and when people engage in this mode of thinking remain open questions.

Variability in essentialism across cultures, categories, and contexts suggests that this mode of representing human categories is rooted in a naturalistic theory of category origins, combined with a need to explain differences that cross category boundaries.

People who believe that human attributes are immutable (entity theorists) display preferences for stereotype-consistent over stereotype-inconsistent information (Plaks, Stroessner, Dweck, & Sherman, 2001). How do people respond to information that counters a stereotype? Do they approach it or avoid it? Four experiments showed that attention to stereotype-consistent vs. -inconsistent information depends on people's implicit theories about human traits. Those holding an entity theory (the belief that traits are fixed) consistently displayed greater attention to (Experiments 1 and 4) and recognition of (Experiments 2 and 3) consistent information. whereas those holding an incremental (dynamic) theory tended to display greater attention to (Experiment 1) and recognition of (Experiment 3) inconsistent information. This was true whether implicit theories were measured as chronic structures (Experiments 1, 2, and 4) or were experimentally manipulated (Experiment 3). Thus, different a priori assumptions about human traits and behavior lead to processing that supports vs. limits stereotype maintenance.

Bastian and Haslam (2007) argued that entity theories are components of broader sets of beliefs that represent differences between people in terms of underlying essences. Consistent with expectations, the authors found that measures of essentialist beliefs (i.e., in the biological basis, discreteness, informativeness, and inherence of human attributes) covaried with the presence of entity theories. Essentialist beliefs as a set are related to stronger preferences for stereotype-consistent information, and this effect was not reducible to the effect of entity theories. The authors proposed that research on implicit person theories can be placed within an encompassing framework of psychological essentialism.

Implicit person theory research can be conceptualized within the framework of psychological essentialism (Bastian & Haslam, 2008). Essentialist beliefs are associated with entity theories and both predict phenomena such as stereotyping. The present research extended previous work on the links between implicit theories and social identity processes, examining how essentialist beliefs are associated with social identification and processes related to prejudice and intergroup perception. After developing a new measure of essentialist beliefs in Study 1, Study 2 showed that these beliefs were associated with negative bias towards immigrants, particularly when participants were primed with an exclusive social identity. In Study 3, essentialist beliefs among immigrants moderated their adoption of Australian identity as a self-guide during acculturation. Essentialist beliefs therefore play a significant role in the psychology of social identity.

Essentialism is an ontological belief that there exists an underlying essence to a category. Kashima, Kashima, and Bain (2010) advanced and tested in three studies the hypothesis that communication about a social category, and expected or actual mutual validation, promotes essentialism about a social category. In Study 1, people who wrote communications about a social category to their ingroup audiences essentialized it more strongly than those who simply memorized about it. In Study 2, communicators whose messages about a novel social category were more elaborately discussed with a confederate showed a stronger tendency to essentialize it. In Study 3, communicators who elaborately talked about a social category with a naive conversant also essentialized the social category. A meta-analysis of the results supported the hypothesis that communication promotes essentialism. Although essentialism has been discussed primarily in perceptual and cognitive domains, the role of social processes as its antecedent deserves greater attention.

Developmental Psychology

Gellman (2003) argued that children have an essentialist cognitive bias (i.e., that they believe that things and people have an underlying "true" nature that gives them their identities), and that this bias affects the way that children learn and make sense of the world around them. Bloom (2004) argued that children are naturally born dualists. Babies find it perfectly natural for a person to begin moving without coming into physical contact with anything but are surprised if an object moves under the same conditions. Out of the dichotomy between things and people grows our conviction that body and mind are distinct entities – that physical things are driven by principles such as solidity and gravity and immaterial minds are driven by emotions and goals. Bloom illustrated how we manipulate our dualism. We are able to see the same object as part of either the physical world or the mental world. We think differently about a painting by Vermeer and one that was painted to look like a Vermeer. We judge not only the physical product but also the creative act that led to the product, an act that is intimately tied to the goals of the creator.

We can thus see a thing as more than a thing. But we can also see a person as less than a person. Disgust is a reaction to the physical, and when we use the term to describe our reaction to people, we are, in effect, turning those people into physical objects – we can then scorn them, ignore them, or even kill them. In skillful prose that weaves together clinical research, literature, philosophy, neuroscience, and captivating examples from children (some of the best are from his own family), Bloom makes the case that responding differently to physical things and to immaterial minds is adaptive, an unsurprising product of evolutionary pressures. Moreover, the capacity to respond to the minds of others, which has developed during evolutionary time, has led to unexpected by-products during historical time.

These by-products are some of our most interesting and distinctive traits – the ability to construct religions, value art, and hold moral beliefs, to name just a few featured in the book. Bloom also makes a thought-provoking case for historical progress, not only in our dealings in the physical world but also in the moral world – we are "nicer to one another than we used to be." In addition to outlining change over evolutionary and historical timescales, Bloom tackled change over ontogenetic time. His examples from the world of developmental psychology are some of his best, since this is his own discipline. Bloom is careful not to claim that babies are full-blown dualists. The bias to see objects as distinct from people provides the foundation for a dualist stance, but that orientation needs to be fleshed out by children as they interact with members of their culture.

According to this review by Goldin-Meadow (2004), she argued that Bloom (2010) did not consider the possibility of a culture that eschews dualism – a culture in which it does not make sense to ask whether the mind affects the body (or vice versa) because the two are one and the same. Are there no dualist cultures? Could there be? What type of historical trajectory might lead to no dualism? What type of developmental path would children born into no dualist cultures follow? Although it does not raise these questions, Bloom's far-reaching and provocative book brings novel speculations of this sort into bold relief and thus maps out the terrain for a new generation of thinkers.

In two studies, Poling and Evans (2002) had 6–12-year-old children (Study I: $N=58$; Study 2: $N=38$) and adults (Study 2: $N=22$) rank order intentional, teleological and essentialist explanations for different behaviors of living-kind groups representing a range of biological kinds from plants to humans. Overall, humans elicited more intentional explanations, insects and plants elicited more essentialist explanations, and intermediate taxa, such as ungulates, elicited more teleological explanations. Children made fewer fine-grained taxonomic distinctions than adults, and the youngest children tended to reject essentialism. The 6–7-year-old children preferred to reason about living-kind behaviors from an intentional and teleological perspective; only towards the end of the elementary school years did children seem to incorporate a biological essentialism. Neither adults nor children were exclusively bound to a particular mode of explanation, but exercised "causal flexibility" across different behavioral contexts.

Gelman, Ware, and Kleinberg (2010) hypothesized that generic noun phrases ("Bears climb trees") would provide important input to children's developing concepts. In three experiments, 4-year-olds and adults learned a series of facts about a novel animal category, in one of three wording conditions: generic (e.g., "Zarpies hate ice cream"), specific-label (e.g., "This zarpie hates ice cream"), or no-label (e.g., "This hates ice cream"). Participants completed a battery of tasks assessing the extent to which they linked the category to the properties expressed, and the extent to which they treated the category as constituting an essentialized kind. As predicted, for adults, generics training resulted in tighter category-property links and more category *essentialism* than both the specific-label and no-label training. Children also showed effects of generic wording, though the effects were weaker and required more extensive input. Gelman et al. discussed also the implications for language-thought relations for the acquisition of essentialized categories.

Social Psychology

Haslam (1998) examined the extendibility of natural-kind concepts and essentialist thinking into the social domain from two angles. First, to what degree is the extension of natural-kind concepts and essentialism warranted epistemologically and second why is it that laypeople's intuitive understandings of human kinds commonly favor such accounts. Both will be tackled in a psychological point of view. Haslam argued that the extension of natural-kind concepts into the social domain may be less problematic than many theorists believe, because natural kinds, at least those in the biological realm, are not what theorists imagine them to be. Biological kinds must be understood in nonessentialist ways, and natural-kind concepts can be relieved of the implication of essentialism. Moreover, essentialist claims have several distinct and loosely bound conceptual components, and accounts of human kinds can profitably retain some components without thereby yielding to essentialism.

Social psychologists studying intergroup perceptions have shown an increasing interest in essentialist thinking (Verkuyten, 2003). Essentialist beliefs about social groups are examined as cognitive processes and these beliefs would serve to rationalize and justify the existing social system. Discourse analyses on racism have emphasized that problems of racism are to a large extent problems of essentialism. Antiessentialism has emerged as an emancipatory discourse in the challenge of hegemonic representations and oppressive relations. The present study examines how, in group discussions, ethnic Dutch and ethnic minority people define and use essentialist notions about social groups. Both Dutch and ethnic minority participants engaged in an essentialist discourse in which an intrinsic link between culture and ethnicity was made. However, there were also examples where this discourse was criticized and rejected. This variable use of (de-) essentialism is examined in terms of the conversation's context and issues at hand, such as questions of assimilation, group provisions, cultural rights, and agency. The main conclusion of this article is that essentialism is not by definition oppressive and that de-essentialism is not by definition progressive. The discursive power of (de-)essentialist group beliefs depends on the way they are used and the context in which they appear.

Cultural Psychology

According to the psychological essentialism perspective, people tend *to* explain differences between groups by attributing them different essences (Leyens et al., 2001). Given a pervasive ethnocentrism, this tendency implies that the human essence will be restricted to the ingroup whereas outgroups will receive a lesser degree *of* humanity. Therefore, it is argued that people attribute more uniquely human characteristics to the ingroup than to the outgroup. These researchers focused on secondary emotions that constitute such characteristics. Study 1 showed that members *of* high- and low-status groups attribute more positive secondary emotions to the ingroup than to the outgroup. Study 2 verified that the differential attribution extended also *to* negative secondary emotions. No exemplars *of* emotions were provided in Study 3. Instead, participants had *to* estimate *the* means *of* two distributions *of* numbers that supposedly represented characteristics of the ingroup and of the outgroup. *The* results *of* this third experiment illustrated the reluctance to attribute secondary emotions to the outgroup. These findings can be discussed from the perspective of psychological essentialism.

The research by Verkuyten and Brug (2004), conducted among ethnic minority ($N=109$) and majority ($N=649$) adolescents in the Netherlands, examined the endorsement of multiculturalism in relation to ethnic ingroup identification, perceived group essentialism, and protestant ethic ideology. Results revealed that ethnic minorities were more in favor of multiculturalism than Dutch participants. Furthermore, ethnic identification, group essentialism, and protestant ethic were independently related to multiculturalism. However, these relations were all moderated by ethnic group status. High ethnic identification, essentialist beliefs about minority groups, and protestant ethic were related to lower endorsement of multiculturalism among the majority group. In contrast, high ethnic identification and essentialist beliefs about one's own ethnic minority group were related to stronger endorsement of multiculturalism among the minority groups, but not protestant ethic. Perceived essentialism of the majority group showed no effects. It is concluded that social psychology is able to make a valuable contribution to issues related to the management of cultural diversity.

In three studies, Zagefka, Pehrson, Mole, and Chan (2010) tested the effects of essentialist beliefs regarding the national ingroup in situations where a perpetrator group has inflicted harm on a victim group. For members of the perpetrator group, it was hypothesized that "essentialism" has a direct positive association with "collective guilt" felt as a result of misdeeds conducted by other ingroup members in the past. Simultaneously, it was hypothesized to have an indirect negative association with collective guilt, mediated by perceived threat to the ingroup. Considering these indirect and direct effects jointly, it was hypothesized that the negative indirect effect suppresses the direct positive effect, and that the latter would only emerge if perceived "ingroup threat" was controlled for.

This hypothesis was tested in a survey conducted in Latvia among Russians ($N=70$) and their feelings toward how Russians had treated ethnic Latvians during the Soviet occupation; and in a survey in Germany among Germans ($N=84$), focusing on their feelings toward the Holocaust. For members of the victim group, it was hypothesized that essentialism would be associated with more anger and reluctance to forgive past events inflicted on other ingroup members. It was proposed that this effect would be mediated by feeling connected to the ingroup victims. This hypothesis was tested in a survey conducted among Hong Kong Chinese and their feelings toward the Japanese and the Nanjing massacre ($N=56$). Results from all three studies supported the hypotheses.

Past research has demonstrated a broad association between prejudice and essentialism (Morton, Hornsey, & Postmes, 2009). However, research has also shown that essentialism and prejudice are not always linked in the same way – sometimes essentialist thinking is associated with prejudice, but sometimes it is not. The aim of the present research was to explore experimentally how prejudice might relate to essentialist beliefs about race differently depending on how race is being used (inclusively or exclusively) and who is the implied target of such treatment (ethnic minorities or the white majority). Study 1 ($N=178$) demonstrated that, although prejudice among white Australians is typically related to essentialist beliefs about

Aboriginal identity, this relationship disappeared when racial criteria were used to exclude someone for "being white." Under these conditions, prejudiced participants expressed opposition to such treatment and de-essentialized race. Study 2 ($N=198$) broadly replicated this pattern in a British context and indicated that prejudiced participants' de-essentialism of race was due to a stronger emphasis on values of equality under the same conditions. These results demonstrate that prejudiced people endorse essentialism when it can be used to exclude others (who they want to exclude), but reject essentialism when it is used to exclude them.

Holtz and Wagner (2009) investigated a total of 4,997 postings on an extreme right-wing Internet discussion board with regard to the groups and themes mentioned. The most frequently mentioned target groups were Africans, Jews, Muslims, Poles, and Turks; the most prominent themes and contexts were conspiracy, criminality, exploitation, threats to German identity, infiltration, mind control and harassment, procreation, rape, and sex. Holtz and Wagner (2009) analyzed in detail postings about Africans/Blacks and Jews, that is: target groups that were the most clearly connected to particular themes. The analysis reveals that extreme right-wing discourse essentializes the target groups of Jews and Africans/Blacks and ascribes them immutable group-specific attributes that effectively make them "natural kinds." The group of Jews appears as a kind of their own with super-human powers and influence. Africans and Blacks are despised, firstly because their essential characteristics prohibit them to be categorically mixed with Germans (i.e., to become German by nationality) due to their incompatible essence, and secondly when they procreate with Whites. Such procreation produces "bastards" that are met with disgust. Holtz and Wagner argued that essentialist thinking about social and ethnic groups explains a good part of their rejection by right-wing followers.

Are human individuals universally seen to be more real entities than social groups? Although the individual may be seen to be more entitative than social groups in the West, it is unclear whether this is the case in other cultures, especially in East Asia (Kashima et al., 2005). Two aspects of perceived entitativity are distinguished: psychological essentialism (belief in the presence of essence-like unchangeable properties) and agency (perception that a social entity is an agent), and examined for four social targets (individual, family, friendship group, and society) in three English-speaking cultures (Australia, UK, and USA), three East Asian cultures (Hong Kong, Japan, and Korea), and two continental European cultures (Belgium and Germany). In all cultures, the individual person was seen to possess essence-like unchangeable characteristics more than social groups (i.e., essentialized). As for agency, the individual person was seen to be more agentic than groups in Western cultures, but both individuals and groups were conferred an equal level of agency in East Asia. Individuals may be universally more essentialized than friendship groups and societies, but not always seen to be more agentic than social groups. Implications of the results for conceptions of individualism and collectivism are discussed.

Many hold that ordinary race-thinking in the USA is committed to the "one-drop rule," that race is ordinarily represented in terms of essences, and that race is ordinarily represented as a biological (phenotype- and/or ancestry-based, nonsocial) kind. Glasgow, Shulman, and Covarrubias (2009) investigated the extent to which ordinary race-thinking subscribes to these commitments. It also investigated the relationship between different conceptions of race and racial attitudes. Participants included 449 USA adults who completed an Internet survey. Unlike previous research, conceptions of race were assessed using concrete vignettes. Results indicated widespread rejection of the one-drop rule, as well as the use of a complex combination of ancestral, phenotypic, and social (and, therefore, nonessentialist) criteria for racial classification. No relationship was found between racial attitudes and essentialism, the one-drop rule, or social race-thinking; however, ancestry-based and phenotype-based classification criteria were associated with racial attitudes. These results suggest a complicated relationship between conceptions of race and racial attitudes.

In line with as subjective essentialist view of stereotypes, Yzerbyt, Rocher, and Schadron (1997), proposed that two classes of factors that mutually reinforce each other influence the perception of groups. The perception of a strong level of similarity and organization among group members (i.e., group entitativity) suggests the existence of a deep essence that would account for the detected regularities. Conversely, the existence of naive theories regarding the presence of an underlying core encourages the search for resemblances and connections within the group. After a short presentation of the recent literature dealing with the concepts of entitativity and essentialism, Yzerbyt, Corneille, and Estrada (2001) reviewed a series of studies from their laboratory showing the impact of entitativity on essentialism as well as the influence of essentialism on entitativity. These investigators also provided empirical evidence for this bidirectional process from both the outsider and the insider perspective. Finally, they examined the potential role of cultural differences both in the ascription of a fundamental nature to an entitative assembly of people and in the use of a priori naive theories to create surface similarity among group members. As a set, these efforts point to the importance of taking into account the constant dialogue between perceivers' theory-based explanations and group members' perceptual characteristics if one wishes to understand group stereotypes.

Ghandnoosh (2010) examined how participants interpret a cultural practice commonly associated with a race other than their own. Determining if, how and why participants experience such "cross-cultural" forms in racialized terms can clarify whether these practices promote tolerance or *essentialism* in everyday life, and whether they enable appropriation in the field of cultural production. Through interviews and participant observation with non-African-American women learning and teaching hip-hop dance, the author captured a spectrum of participant views. Most dancers see hip-hop as African American in its origins. But while novices often speak of an inherent or learned authenticity among blacks, experts rarely express racialized views of the dance's contemporary practice. Experts' views are shaped by personal ability, exposure to dancers whose ability is not racially patterned, and exposure to others who accept their skill. How dancers act on these interpretations challenges common associations of racialized views with tolerance, and nonracialized views with appropriation.

Bernstein, Sacco, Young, Hugenberg, and Cook (2010) argued that social belonging is an essential human need. Belonging to social groups serves an important role in shaping our social identities. Nonetheless, research indicates that exclusion by ingroup and outgroup members seems equally aversive. Bernstein et al. tested the hypothesis that unlike more trivial groups (e.g., smoking or computer preferences), highly essentialized groups may lead to differential effects of ingroup vs. outgroup rejection. Consistent with this, exclusion and inclusion by racial ingroup members (relative to racial outgroup members) exacerbated the sting of rejection and the glow of inclusion (Study 1). In a second study, direct manipulations of essentialist beliefs about ingroups and outgroups (i.e., political affiliations) led to the same results. These results offer a novel demonstration that essentialized ingroup–outgroup distinctions enhance the sting of social exclusion and the positivity of social inclusion.

Application of an Essentialist Perspective to a Model of Relational Consequence Theory

The whole nature of essentialism, as noted in Materialism (see Riegler, 2011) is to look at the intrinsic characteristics of cognitions, development, and human relationships. Its closeness to actual data and the absence of an overarching paradigm, perhaps as far as similar goals with materialism, makes one wonder what would happen if such a perspective were applied to one Model[2] of Relational Competence Theory (RCT). That model views relationships according to a vertical perspective that aims at viewing relationships in their complete essence (Cusinato & L'Abate, 2011; L'Abate, Cusinato, Maino, Colesso, & Scilletta, 2010).

The essentialist perspective views human relationships according to two major levels of analysis and interpretation (see Riegler, 2011), each level with two sublevels: Description and Explanation. Description is composed of two sublevels. The first sublevel is the Public facade, our superficial and temporary presentational impression formation and management. The second phenotypical sublevel is how we behave in the privacy of our homes, under the prolonged stress of personal and interpersonal intimate interactions.

Explanation is composed of two sublevels. The first genotypical sublevel is supposed to "explain" why we behave the way we do, including consistencies and inconsistencies between the two descriptive sublevels and those two sublevels with one's genotype. Historically, this explanatory sublevel was composed by terms such as ego, self-esteem, or similar inferred constructs. In RCT the genotype can be represented by self-identity defined by a continuum of Likeness or Resemblance validated by Cusinato and Colesso (2008). The genotype is produced by a second sublevel that contains an explanation according to our intergenerational, generational, and developmental backgrounds.

Conclusion

As discussed in greater detail in Hillix, Rumbaugh, and Savage-Rumbaugh (2011), Essentialism at best fits very closely to the original definition of Operational Paradigms included in this Part IV of this volume. However, essentialism is really emblematic of an operational theory. This theory is without an overarching paradigm because it meets the criterion of being defined by concrete and specific empirical operations. Furthermore, the theory is characterized by various models with underlying dimensions. These operations evaluate the theory itself, and essentialistic theory might be independent of a paradigm, except perhaps for Materialism.

References

Bastian, B., & Haslam, N. (2007). Psychological essentialism and attention allocation: Preferences for stereotype-consistent versus stereotype-inconsistent information. *Journal of Social Psychology, 147*, 531–541.

Bastian, B., & Haslam, N. (2008). Immigration from the perspective of hosts and immigrants: Roles of psychological essentialism and social identity. *Asian Journal of Social Psychology, 11*, 127–140.

Bernstein, M., Sacco, D., Young, S., Hugenberg, K., & Cook, E. (2010). Being "in" with the in-crowd: The effects of social exclusion and inclusion are enhanced by the perceived essentialism of ingroups and outgroups. *Personality and Social Psychology Bulletin, 36*, 999.

Bloom, P. (2004). *Descartes' baby: How the science of child development explains what makes us human*. New York: Basic Books.

Bloom, P. (2010). *How pleasure works: The new science of why we like what we like*. New York: W. W. Norton.

Braisby, N., Franks, B., & Hampton, J. (1996). Essentialism, word use, and concepts. *Cognition, 59*, 247–274.

Cusinato, M., & Colesso, W. (2008). Validation of the continuum of likeness in intimate relationships. In L. L'Abate (Ed.), *Toward a science of clinical psychology: Laboratory evaluations and interventions* (pp. 337–352). New York: Nova.

Cusinato, M., & L'Abate, L. (2011). *Relational competence theory: With special attention to alexithymia*. New York: Nova Science Publishing.

Forster, M. N. (2004). *Wittgenstein on the arbitrariness of grammar*. Princeton: Princeton University Press.

Fuller, B. A. G. (1946). *The history of philosophy: Revised edition*. New York: Hency Holt and Company.

Gellman, S. A. (2003). *The essential child: Origins of essentialism in everyday thought*. New York: Oxford University Press.

Gelman, S., Ware, E., & Kleinberg, F. (2010). Effects of generic language on category content and structure. *Cognitive Psychology, 61*, 273.

Ghandnoosh, N. (2010). 'Cross-cultural' practices: Interpreting non-African-American participation in hip-hop dance. *Ethnic and Racial Studies, 33*, 1580.

Glasgow, J., Shulman, J., & Covarrubias, E. (2009). The ordinary conception of race in the United States and its relation to racial attitudes: A new approach. *Journal of Cognition & Culture, 9*, 15–38.

Goldin-Meadow, S. (2004). Review of the book Descartes' baby: How the science of child development explains what makes us human by Paul Bloom. *The New England Journal of Medicine, 351*, 1365.

Haslam, N. (1998). Natural kinds, human kinds, and essentialism. *Social Research, 65*, 291–314.
Hillix, W. A., & L'Abate, L. (2011). Paradigms in science and theory construction. In L. L'Abate (Ed.), *Paradigms in theory construction*. New York: Springer.
Hillix, W. A., Rumbaugh, D. M., & Savage-Rumbaugh, S. (2011). The emergence of reason, intelligence, and language in humans and animals. In L. L'Abate (Ed.), *Paradigms in theory construction*. New York: Springer.
Holtz, P., & Wagner, W. (2009). Essentialism and attribution of monstrosity in racist discourse: Right-wing internet postings about Africans and Jews. *Journal of Community & Applied Social Psychology, 19*, 411–425.
Jylkkä, J., Railo, H., & Haukioja, J. (2009). Psychological essentialism and semantic externalism: Evidence for externalism in lay speakers' language use. *Philosophical Psychology, 22*, 37–60.
Kashima, Y., Kashima, E., & Bain, P. (2010). Communication and essentialism: Grounding the shared reality of a social category. *Social Cognition, 28*, 306–328.
Kashima, Y., Kashima, E., Chiu, C., Farsides, T., Gelfand, M., Hong, Y., et al. (2005). Culture, essentialism, and agency: Are individuals universally believed to be more real entities than groups? *European Journal of Social Psychology, 35*, 147–169.
L'Abate, L., Cusinato, M., Maino, E., Colesso, W., & Scilletta, C. (2010). *Relational competence theory; research and mental health applications*. New York: Springer-Science.
Leyens, J., Rodriguez-Perez, A., Rodriguez-Torres, R., Gaunt, R., Paladino, M., Vaes, J., et al. (2001). Psychological essentialism and the differential attribution of uniquely human emotions to ingroups and outgroups. *European Journal of Social Psychology, 31*, 395–411.
Mahalingam, R. (2007). Essentialism, power, and the representation of social categories: A folk sociology perspective. *Human Development, 50*, 300–319.
Medina, D. L., & Orton, A. (1989). Psychological essentialism. In S. Vosniadou & A. Ortony (Eds.), *Similarity and analogical reasoning*. New York: Cambridge University Press.
Morton, T., Hornsey, M., & Postmes, T. (2009). Shifting ground: The variable use of essentialism in contexts of inclusion and exclusion. *British Journal of Social Psychology, 48*, 35–59.
Plaks, J. E., Stroessner, S. J., Dweck, C. S., & Sherman, J. W. (2001). Person theories and information-seeking: Preferences for stereotypic vs. counterstereotypic information. *Journal of Personality and Social Psychology, 80*, 876–893.
Poling, D., & Evans, E. (2002). Why do birds of a feather flock together? Develop-mental change in the use of multiple explanations: Intention, teleology and essentialism. *British Journal of Developmental Psychology, 20*, 89.
Prentice, D., & Miller, D. (2007). Psychological essentialism of human categories. *Current Directions in Psychological Science, 16*, 202–206.
Putnam, H. (1973). Meaning and reference. *Journal of Philosophy, 70*, 699–711.
Putnam, H. (Ed.). (1975). *Mind, language and reference: Philosophical papers* (Vol. 2). New York: Cambridge University Press.
Riegler, A. (2011). Constructivism. In L. L'Abate (Ed.), *Paradigms in theory construction*. New York: Springer.
Ripken, S. (1980). *Naming and necessity*. Boston: Harvard University Press.
Schick, F. (2003). *Ambiguity and logic*. New York: Cambridge University Press.
Verkuyten, M. (2003). Discourses about ethnic group (de-)essentialism: Oppressive and progressive aspects. *British Journal of Social Psychology, 42*, 371–391.
Verkuyten, M., & Brug, P. (2004). Multiculturalism and group status: The role of ethnic identification, group essentialism and protestant ethic. *European Journal of Social Psychology, 34*, 647–661.
Yzerbyt, V., Corneille, O., & Estrada, C. (2001). The interplay of subjective essentialism and entitativity in the formation of stereotypes. *Personality & Social Psychology Review, 5*, 141–155.
Yzerbyt, V. Y., Rocher, S. J., & Schadron, G. (1997). Stereotypes as explanations: A subjective essentialistic view of group perception. In R. Spears, P. Oakes, N. Ellemers, & A. Haslam (Eds.), *The social psychology of stereotyping and group life* (pp. 20–50). Oxford: Basil Blackwell.
Zagefka, H., Pehrson, S., Mole, R., & Chan, E. (2010). The effect of essentialism in settings of historic intergroup atrocities. *European Journal of Social Psychology, 40*, 718–732.

Part VI
Conclusion

Chapter 23
Criteria and Processes in Paradigm Selection

Luciano L'Abate and Laura G. Sweeney

The purpose of this chapter is to review critically the usefulness of differentiations made in an earlier publication (L'Abate, 2009) that was basic to the Table of Contents for this volume. How does that classification hold up in front of the many paradigms introduced then and now in this volume? To select a particular paradigm over others and make sense of such a bewildering array of paradigms, each claiming an important position in theory-construction, four different criteria will be necessary. Two criteria of abstraction and generality are descriptive. Two processes of equipotentiality and equifinality are explanatory because they integrate various levels of observation and interpretation.

The basic issue, as discussed from the outset of this volume (Chap. 1), is how to differentiate paradigms from theories, and theories from models, as well as models from paradigms. If we cannot differentiate and define clearly each one from the other two, confusion will continue to occur and overlap among paradigms will continue to occur as well. Can one specific paradigm lead to or connect with a specific theory? How is this connection or link to be made: Rationally, empirically, or rhetorically? Consequently, we must go back to the original definitions (L'Abate, 2009) to see whether we can improve on them on the basis of the contributions contained in this volume. Another question related to the overlap among many paradigms is: Can we avoid such an overlap once we define paradigms according to a hierarchical classification reached earlier (L'Abate, 2009) and contained in this volume?

A specific paradigm needs to show a direct one-to-one relationship with a specific theory in order to function as a *bona fide* paradigm. If there is no direct relationship between the paradigm and a theory, what happens to "orphan" paradigms? For

L. L'Abate (✉)
Department of Psychology, Georgia State University, University Plaza, Atlanta, GA 30303, USA
e-mail: llabate3@bellsouth.net

L.G. Sweeney
Avant Garde Books, 301 Briarvista Way N.E.,
Atlanta, GA 30329, USA

L. L'Abate (ed.), *Paradigms in Theory Construction*, DOI 10.1007/978-1-4614-0914-4_23,

instance, a direct relationship was shown between materialism and levels of analysis and interpretation in dealing with hurt feelings. The same relationship occurred between essentialism (Chap. 22 this volume) and levels of description and explanation for Model[2] of Relational Competence Theory (RCT) in L'Abate et al. (2010).

Abstraction and Generality in Theory Construction

The same two criteria that were used to distinguish paradigms from theories, and theories from models (L'Abate, 2009), levels of abstraction and generality, are also working within the classification of paradigms. General integrative paradigms included in Part II are much more abstract and general than particular, specific paradigms contained in Part III, while paradigms in that section are somewhat more abstract and general than operational paradigms in Part IV. One could argue whether Materialism (Chap. 14 this volume) should not be included in the last Part IV because it is relatively specific and relatively concrete. The answer to that possible argument is that there may not be specific operations that can be performed in Materialism, while concrete and specific operations characterize both Essentialism (L'Abate & Sweeney, 2011), Reductionism (Colesso, 2011), and Emergence (L'Abate, 2011a).

Abstraction is one dimension that permeates the whole framework presented here and elsewhere (L'Abate, 2009; Hillix & L'Abate, 2011), going from abstraction to concreteness proper. Generality is a second dimension necessary to understand the original framework, going from generality to specificity proper. According to these two criteria, starting with general-integrating and moving down the level of abstraction to a more concrete level to Part II and eventually down to Part III, concreteness is seen in operational paradigms where it is almost impossible to separate the overarching paradigm from theory, dimensions, or applications.

Specificity in science and psychology is achieved when paradigms, theories, models, and dimensions are defined concretely and differentiated clearly one from the other (L'Abate, 2009, 2012). As long as there is confusion about what a paradigm, a theory, or a model is, it will be impossible to achieve such as criterion (Ambrose, 2006). Over time, scientists have developed an effective investigative process that includes the acceptance of particular basic presuppositions, methods, content, and theories.

These presuppositions are the philosophical foundations of scientific thought and do much to define the field's worldview. These fundamental assumptions can be esoteric for many and can become a source of conflict when they are not commonly shared with other points of view. Such presuppositions affect the observations, the conclusions drawn, and the positions taken. Furthermore, in some cases presuppositions in science have undergone important shifts in meaning, causing an increasing dissonance. We argue that disputes in religion and science often are due to these very basic differences in philosophy that are held by members in different communities. To better understand the nature of science and its differences with religious views, presuppositions rather than conclusions should be articulated and examined for validity and scope of application (Ambrose, 2006).

Hierarchy in Paradigms-Theories-Models-Dimensions

Again, as already discussed in Hillix and L'Abate (2011), Overton (2011), and L'Abate (2011b), and by L'Abate (2009), another exemplary definition of a paradigm contains the same error of equating a paradigm with a model. For instance:

> "I define a *paradigm* as a *model* shared by most but not all members of a scientific community, designed to describe and interpret observed or inferred phenomena, past or present, and aimed at building a testable body of knowledge open to rejection or confirmation. In other words, a paradigm captures the scientific thinking of the majority but most of the time it coexists with competing paradigms – as is necessary if new paradigms are to displace old paradigms" (Shermer, 1997, p. 39).

Ruse (1989) discriminated among four types of paradigms: (1) *sociological*, as discussed in Sweeney (2011); (2) *psychological*, as contained in most other chapters of this volume, especially Part II; (3) *epistemological*, where there is a direct connection between the paradigm and its hypotheses determining directly the operations, problems, and techniques used theoretically and empirically in models, as in the case of De Giacomo, Mich, Santamaria, Sweeney, and De Giacomo (2011), Berntson and Cacioppo (2011), Colesso (2011), and Hillix, Rumbaugh, and Savage-Rumbaugh (2011); and (4) *ontological*, perhaps as seen in Chaps. 15, 16, and 17 contained in Part II. In this type of paradigm "… any paradigm is as good as any other paradigm because there is no outside source of corroboration" (Shermer, 1997, p. 40).

To evaluate and select a paradigm, one must ask the same questions asked of theories, models, and dimensions according to how and how much each of these components fulfill the basic criteria of any scientific enterprise: (1) *verifiability* in terms of being susceptible to empirical validation; (2) *applicability* in terms of what populations, realms, settings, and venues does a particular paradigm, theory, or model encompass and relate to; (3) *redundancy* in the sense of how and how much does one particular paradigm, theory, or model link, relate, and correlate with other paradigms, theories, or models; (4) *fruitfulness* in the sense of how and how much research and applications have been generated by a particular paradigm, theory, or model; and (5) *longevity* in the sense of how long a particular paradigm, theory, or model has lasted within a specific scientific and professional community.

Given these criteria, here are the questions to ask in regard to paradigms, theories, models, and even possibly dimensions:

Verifiability

How does this criterion apply to a paradigm?
How does this criterion apply to a theory?
How does this criterion apply to a model?

As a whole, a paradigm as a world view is usually not directly verifiable. If that is the case, how can it be linked to a theory except by proclamation and personal preference? A theory in and of itself is evaluated through its models. However, one has to demonstrate that models derive directly from the theory and that the dimensions used to evaluate a model derived directly from that model and no other model. By the same token, a theory needs to demonstrate its link(s) to a specific paradigm by inclusion, and not to other competing paradigms by exclusion.

Applicability

How does this criterion apply to a paradigm?
How does this criterion apply to a theory?
How does this criterion apply to a model?

What population, setting, venue, culture, or language does a paradigm apply to? For instance, while Materialism (Chap. 14 this volume) applies to any organic and inorganic substance, Essentialism (L'Abate & Sweeney, 2011) applies directly to cognitive, developmental, and cultural identities, demonstrating its being reconsidered as a theory rather than a paradigm.

Redundancy

How does this criterion apply to a paradigm?
How does this criterion apply to a theory?
How does this criterion apply to a model?

Since this requirement is relatively new to psychological thinking, it may need further explication. If one were to think of the framework presented here in terms of an organizational chart of any commercial, industrial, military, and religious institution, then this framework will highlight that in any such organization all its components or members are all interrelated. What effects one component in the system effects all the other components. That said, all components overlap with each other to the point that one component can take over, if and when necessary, for another component. All components have to show that they are accountable (verifiable and applicable) for the particular function they are supposed to fulfill. Goals may be the same. Functions may be different but pursuing the same goals links together all component members. Such an overlap is visible among many paradigms contained in this volume. However, the paradigm that is specifically linked to a theory, models, and dimensions, or applications would assume a much more important position in the present hierarchy of paradigms.

An example of redundancy applies to the whole conception of love as a multidimensional construct to the point that in RCT (L'Abate, Cusinato, Maino, Colesso, & Scilletta, 2010) love can be considered from a variety of models in terms of: (1) Model[4] about an assumption of Space within a dimension of distance and approach-avoidance; (2) Model[7], Being Present emotionally and instrumentally to Self and intimate Others; (3) Model[9] about a Conductive-Creative style in intimate relationships; (4) Model[11] about attributing and bestowing a sense of importance to Self and intimate Others; and (5) Model[15] about Intimacy as sharing joys and hurts as well a fears of being hurt.

This multidimensionality and redundancy among different ways to express and share love is supported by the work of Ortigue, Bianchi-Demicheli, Patel, Frum, and Lewis (2010). They found that different types of love derive from different patterns of neural activation. While all types of love activate subcortical reward systems, including both dopamine and oxytocin responses, different kinds of love trigger different brain and nervous system networks. Unconditional love between a parent and a child, for instance, triggers activation in the midbrain; passionate love activates about a dozen different brain areas including the thalamus, hippocampus, and occipital-temporal cortex.

Fruitfulness

How does this criterion apply to a paradigm?
How does this criterion apply to a theory?
How does this criterion apply to a model?

Whether research emanates from a theory or a model, does it derive also from a specific paradigm or any paradigm will do? Of course, more proponents of any paradigm point to research that would support such a paradigm. However, there are exceptions. For instance, Kantor's inter-behavioral paradigm (Fryling & Hayes, 2011; Chap. 20 this volume), as interesting and exciting as it may be, seems to have failed to produce specific research that would validate its existence. Is this a reason why inter-behaviorism has lost favor in the psychological community? Could one conclude that that paradigm was rich in rhetoric but poor on evidence? By the same token, how about other paradigms? Which paradigm leads to direct empirical validation and which did not? We leave this answer to the judgment of readers themselves.

Longevity

How does this criterion apply to a paradigm?
How does this criterion apply to a theory?
How does this criterion apply to a model?

Evolution has clearly outlasted the onslaught of criticisms and rebuttals over the last century, not to include religious substitutions based on superstition and questionable religious and moral grounds. However, its direct derivation, evolutionary psychology is still very much alive and active and empirical in its approach. By the same token, the psychoanalytic paradigm, which is still alive in some disciplines, such as anthropology (see Overton, 2011), and clearly in some specialties in the mental health field, is rich on rhetoric but poor in evidence, for some but not all of its parts (De Giacomo, L'Abate, Pennebaker, & Rumbaugh, 2010).

There is no way we can even begin to answer each question given above. However, they remain in the foreground in trying to select which paradigm, theory, or model is more adequate than the others.

Equifinality and Equipotentiality in Levels of Observation and Interpretation

In the face of such an embarrassment of riches in paradigms, theories, models, and dimensions, one cannot help wondering whether Kuhn was incorrect in arguing for an overarching paradigm. Whether correct and incorrect, what are the principles, rules, or suggestions to pare down such an embarrassment of riches found here and elsewhere? Have contributions to this volume helped answer or clarify this issue or render it more complex, complicated, and apparently unsolvable. To answer this question we need to go back to the introduction in Chap. 21 this volume about causality. Can a paradigm *cause* a given theory or even more specifically, can a paradigm be so intimately connected to operations to the point that it is impossible to discriminate where a paradigm begins, except historically, where it ends and where it is extended into operations completely overlooking a theory. Is there a one-to-one relationship between a given paradigm with a given theory, and even more concretely and specifically a given model with given dimensions?

A positive reframing of an embarrassment of riches tantamounts to a richness of possibilities, making selective choices more difficult to achieve. What criteria are one to use to evaluate paradigms, theories, and models? Answers have been given earlier. However, what are processes that allow us to go from the bottom to the top and from the top to the bottom? This process includes two types of causality (L'Abate, 2011a), *equipotentiality*, one cause may produce different effects, and *equifinality*, many effects end up into one cause. For instance, as illustrated in Fig. 23.1, we can proceed from the top down to the bottom, or we can proceed from the bottom to the top. There are no rules we know of that allow us to choose one direction more than the other. Both directions are acceptable in science (Capitelli, Guerra, L'Abate, Longo, & Rumbaugh, 2009), as discussed also in Chap. 21.

Even more specifically, both directions are visible in levels of observations and interpretation already included in Chap. 8, Fig. 8.1, and in Fig. 23.2 here. The Biopsychosocial (L'Abate, 2011c) and Reductionism (Colesso, 2011) are examples of how we can go from the bottom up to higher levels of functioning. Many paradigms included in this volume, and especially the Inter-behavioral (Chap. 10), Riegler (2011), L'Abate (2011d), Cusinato (2011), and De Giacomo et al., (2011), represent ways to go from the top to the bottom.

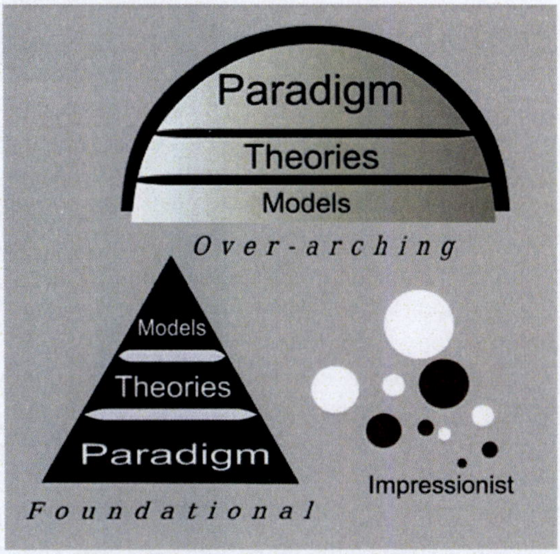

Fig. 23.1 Levels of inquiry in theory construction. Drawing by Laura G. Sweeney

Fig. 23.2 Levels of observation and interpretation in theory construction. Adapted from L'Abate (2011a–d)

Complex Relational and Non-relational Feelings

Altruistic & Self-oriented Feelings

Basic Feelings

Gate Keepers

Experiencing Feelings

Cerebral Levels

Visceral Levels

Physiological Levels

Cellular Levels

Molecular Levels

From the Bottom Up: Equipotentiality

Just to make sure that we are grounding both processes on evidence rather than rhetoric, we shall use some examples from the bottom up and from the top down. Mirror neurons, for instance, from the bottom up equipotenally, at the cellular level are becoming topical and their relevance for clinical hypnosis (relational level) is immense (Antonelli & Luchetti, 2010). The gap between empirical certainty and biological evidence is narrowing and new perspectives are opening up for the understanding of mankind and thought. Mirror neurons, empathy, and hypnosis, accordingly propose a novel paradigm for hypnosis, starting from the analysis and commentary of the relevant literature from databases as well as from monographs and expert opinion. The mirror system appears to unite a wide array of phenomena within the same neuronal system, ranging from elementary behavior such as facilitating responses to higher cognitive functions, imitative learning and action, and language understanding.

Mirror neurons can help us to understand the neuronal basis of empathy and bridge the gap between the sciences of biology, psychology, philosophy, sociology, educational theory, and anthropology. The discovery of mirror neurons provides strong evidence in favor of long-recognized concepts of modern clinical hypnosis regarding the signifcance of the therapeutic relationship. The mirror neuron system paradigm gives us the opportunity to fully reappraise a more anthropologically correct form of medicine based on human agents before technical agencies. These studies represent an opportunity for practitioners of hypnosis to reflect on a novel paradigm, which may be more unifying than previous ones, on the basis of the natural functions of the human being.

From the Top Down: Equifinality

Two examples on how one can go from the top down to lower levels of observation and interpretation may illustrate the process of equifinality. For instance, chronically lonely people may be at higher risk for certain types of inflammatory disease because their feelings of social isolation trigger the activity of pro-inflammatory immune cells (Cole et al., 2011). In their analysis of 93 older adults, researchers screened for gene function among different types of immune cells and found that genes originating from two particular cell types – plasmacytoid dentritic cells and monocytes – were over-expressed in chronically lonely individuals, compared with the remainder of the sample. These cell types produce an inflammatory response to tissue damage, and are part of the immune system's first line of defense, which produces an immediate inflammatory response to tissue damage. It's the same inflammatory response that, over the long term, can promote cardiovascular disease, cancer, and neurodegeneration.

This report provides further evidence of how lifestyle and social environments can impact human health. In addition, the researchers of this study suggested that evolutionarily ancient immune system cells may have developed a molecular sensitivity to our social environment in order to help defend us against socially transmitted pathogens.

In a second instance, previous research indicates that long-term meditation practices are associated with altered resting electroencephalogram patterns, suggestive of long-lasting changes in brain activity (Lazar et al., 2005). These investigators hypothesized that meditation practice might be associated with changes in the brain's physical structure. Magnetic resonance imaging was used to assess cortical thickness in 20 participants with extensive insight meditation experience, which involves focused attention to internal experiences. Brain regions associated with attention, interoception, and sensory processing were thicker in meditation participants than matched controls, including the prefrontal cortex and right anterior insula. Between-group differences in prefrontal cortical thickness were more pronounced in older participants, suggesting that meditation might offset age-related cortical thinning. Finally, the thickness of the two regions correlated with meditation experience. These results provide the first structural evidence that experience-dependent cortical plasticity is associated with meditation practice.

Conclusion

As much as De Giacomo has insisted in calling his Elementary Pragmatic Model (EPM) (Chap. 18 this volume), according to the definitions and classifications derived from them, EPM fits into the criteria of a theory according to its clearly hierarchical structure, different and specific levels of analysis, and dimensions or functions. The fundamental level constituted by three functions (Fig. 18.1) allows one to go from the top to the bottom according to the synthetic integration originally proposed by Capitelli et al. (2009) and in Chap. 21. Starting with seemingly abstract assumptions (Fig. 19.1, Level 1), one goes down to rather concrete and specific functions according to processes of equifinality and equipotentiality, discussed above.

Given all of the information gathered in this volume, what can we conclude based on the criteria of abstraction and generality and processes of equifinality and equipotentiality? The answer to this question is unequivocal. Information processing (Chap. 18 this volume) links a vast and practically immense paradigm into unimaginable applications that pervade our lives to degrees never imagined a generation ago. Cloud computing is the latest example of how networks can use "remote computers for all types of information technology, from simple file storage to complex enterprise systems…" thus reducing costs of ownership of expensive information technology infrastructures. Examples of this type illustrate how information technology may become a commodity just like electricity (Roka, 2011).

When this paradigm is linked to a theory (EPM) the only conclusion available to us is to declare information processing the paradigm for this century. Table 23.1 summarizes how this paradigm starts with various realms of activity down to specific and minute applications in every field of endeavor. The pervasiveness of information processing in much that transpires necessitates that lifelong learners pursue ongoing education and self-understanding, to some extent, by digital means.

Table 23.1 Hierarchy in information processing

Paradigm	Information processing		
Media: talk	Hybrid (Skype)	Writing	Nonverbal
Applications			
Education		Internet	Play/sports
Psychotherapy		Mil(itary)	Manual labor
Religion		Org(anizational)	Sex
		Com(mercial)	
		Individual (net)	
Dimensions: millions		Millions	Millions

References

Ambrose, L. M. (2006). Large-scale contextual influences on creativity: Evolving academic disciplines and global value systems. *Creativity Research Journal, 18*, 75–85.

Antonelli, C., & Luchetti, M. (2010). Mirror neurons and empathy: Proposal of a novel paradigm for hypnosis. *Contemporary Hypnosis, 27*, 19–26.

Authors. (2011). Loneliness and immune system functions. *Proceedings of the National Academy of Sciences*, Feb. 7, 13–17.

Cole, S. W., Hawkley, L. C., Arevalo, J. M. G. & Cacioppo, J. T. (2011). Transcript origin analysis identifies antigen-presenting cells as primary targets of socially regulated gene expression in leukocytes. *Proceedings of the National Academy of Sciences, 108*, 3080–3085.

Capitelli, M., Guerra, C., L'Abate, L., Longo, S., & Rumbaugh, D. M. (2009). Science or mind? Reductionism proper, optimal reductionism, emergent interactionism, and synthetic integration. In L. L'Abate, P. De Giacomo, M. Capitelli, & S. Longo (Eds.), *Science, mind and creativity. The Bari symposium* (pp. 1–22). New York: Nova Science Publisher.

Colesso, W. (2011). Produced and spontaneous emergent interactionism. In L. L'Abate (Ed.), *Paradigms in theory construction*. New York: Springer.

Cusinato, M. (2011). Existentialism. In L. L'Abate (Ed.), *Paradigms in theory construction*. New York: Springer.

De Giacomo, P., L'Abate, L., Pennebaker, J. M., & Rumbaugh, D. M. (2010). From A to D: Amplifications and applications of Pennebaker's analogic to digital model in health promotion, prevention, and psychotherapy. *Clinical Psychology & Psycho- therapy, 17*, 355–362.

De Giacomo, P., Mich, L., Santamaria, C., Sweeney, L. G., & De Giacomo, A. (2011). Information processing. In L. L'Abate (Ed.), *Paradigms in theory construction*. New York: Springer.

De Santillana, G., & Zilsel, E. (1941). *The development of rationalism and empiricism*. Chicago: University of Chicago Press.

Forster, M. N. (2004). *Wittgenstein on the arbitrariness of grammar*. Princeton: Princeton University Press.

Fryling, M., & Hayes, L. J. (2011). Interbehaviorism. In L. L'Abate (Ed.), *Paradigms in theory construction*. New York: Springer.

Fuller, B. A. G. (1946). *A history of philosophy: Revised edition*. New York: Henry Hold and Company.

Hillix, W. A., & L'Abate, L. (2011). Paradigms in science and theory construction. In L. L'Abate (Ed.), *Paradigms in theory construction*. New York: Springer.

Hillix, W. A., Rumbaugh, D. M., & Savage-Rumbaugh, S. (2011). The emergence of reason, intelligence, and language in humans and animals. In L. L'Abate (Ed.), *Paradigms in theory construction*. New York: Springer.

L'Abate, L. (2009). Paradigms, theories, and models: Two hierarchical frameworks. In L. L'Abate, P. De Giacomo, M. Capitelli, & S. Longo (Eds.), *Science, mind, and creativity: The Bari symposium* (pp. 107–153). New York: Nova.

L'Abate, L. (2011a). Essentialism. In L. L'Abate (Ed.), *Paradigms in theory construction.* New York: Springer.

L'Abate, L. (2011b). Anthropology. In L. L'Abate (Ed.), *Paradigms in theory construction.* New York: Springer.

L'Abate, A. (2011c). Sociology. In L. L'Abate (Ed.), *Paradigms in theory construction.* New York: Springer.

L'Abate, L. (2011d). Materialism. In L. L'Abate (Ed.), *Paradigms in theory construction.* New York: Springer.

L'Abate, L. (2012). *Clinical psychology and psychotherapy as science.* New York: Springer-Science.

L'Abate, L., Cusinato, M., Maino, E., Colesso, W., & Scilletta, C. (2010). *Relational competence theory: Research and mental health applications.* New York: Springer-Science.

L'Abate, L., & Sweeney, L. G. (2011). Criteria and processes in paradigm selection. In L. L'Abate (Ed.), *Paradigms in theory construction.* New York: Springer.

Lazar, S. W., Kerr, C. E., Wasserman, R. H., Gray, J. R., Greve, D. N., Treadway, M. T., et al. (2005). Meditation experience is associated with increased cortical thickness. *NeuroReport, 16,* 1893–1897.

Ortigue, S., Bianchi-Demicheli, F., Patel, N., Frum, C., & Lewis, J. W. (2010). Neuroimaging of love: fMRI meta-analysis evidence toward new perspectives in sexual medicine. *The Journal of Sexual Medicine, 7.* doi: 10.1111/j.1743-6109.2010.0199.x

Overton, W. F. (2011). Evolving paradigms: Retrospective and prospection. In L. L'Abate (Ed.), *Paradigms in theory construction.* New York: Springer.

Riegler, A. (2011). Constructivism. In L. L'Abate (Ed.), *Paradigms in theory construction.* New York: Springer.

Roka, P. (2011). A cloud is growing over information technology. *Atlanta Journal-Constitution,* D2, 4.

Ruse, M. (1989). *The Darwinian paradigm.* London: Hutchinson.

Rychlak, J. F. (1968). *A philosophy of science for personality theory.* Boston: Houghton MIfflin.

Rychlak, J. F. (Ed.). (1976). *Dialectic: Humanistic rationale for behavior and development.* New York: S. Karger.

Schwartz, F. (Ed.). (1974). *Scientific thought and social reality: Essays by Michael Polanyi.* New York: International Universities Press.

Shermer, M. (1997). *Why people believe weird things.* New York: MJF Books.

Sweeney, L. G. (2011). Education. In L. L'Abate (Ed.), *Paradigms in theory construction.* New York: Springer.

Index

L. L'Abate (ed.), *Paradigms in Theory Construction*, DOI 10.1007/978-1-4614-0914-4,

R

T

U

V